International Scientific and Research Conference on Knowledge-based Technologies in Development and Utilization of Mineral Resources (KTDUMR 2018)

IOP Conference Series: Earth and Environmental Science
Volume 206

Novokuznetsk, Russia
5 - 8 June 2018

Editor:

Viktor Fryanov

ISBN: 978-1-5108-7604-0
ISSN: 1755-1307

Printed from e-media with permission by:

Curran Associates, Inc.
57 Morehouse Lane
Red Hook, NY 12571

proceedings.com

Some format issues inherent in the e-media version may also appear in this print version.

No changes have been made to the content of these proceedings. There may be changes to pagination and minor adjustments for aesthetics.

Printed by Curran Associates, Inc. (2019)

For permission requests, please contact the Institute of Physics at the address below.

Institute of Physics
Dirac House, Temple Back
Bristol BS1 6BE UK

Phone: 44 1 17 929 7481
Fax: 44 1 17 920 0979

techtracking@iop.org

Additional copies of this publication are available from:

Curran Associates, Inc.
57 Morehouse Lane
Red Hook, NY 12571 USA
Phone: 845-758-0400
Fax: 845-758-2633
Email: curran@proceedings.com
Web: www.proceedings.com

Table of contents

Volume 206

International Scientific Conference "Knowledge-based technologies in development and utilization of mineral resources"

5–8 June 2018, Novokuznetsk, Rusian Federation

Accepted papers received: 8 November 2018
Published online: 27 November 2018

Preface

Papers

IOP Conf. Series: Earth and Environmental Science **206** (2018) 011001 doi:10.1088/1755-1315/206/1/011001

Preface

The conference proceedings contain the selected papers presented at the International Scientific Conference "Knowledge-based technologies in development and utilization of mineral resources" held on June 5-8, 2018 within the framework of the International Specialized Exhibition of Mining Technologies *Russian Coal and Mining* in Novokuznetsk, Russia. The conference was dedicated to the 400th anniversary of the city of Novokuznetsk.

The conference was organized by Siberian State Industrial University (Novokuznetsk, Russia), the Federal Research Center for Coal and Coal Chemistry of the Siberian Branch of the Russian Academy of Sciences (Kemerovo, Russia), Scientific Research Institute for Industrial Research and Environmental Safety in the Mining Industry (Kemerovo, Russia), the University of Science and Technology Liaoning (Anshan, China), Expo center "Kuzbass Fair" (Novokuznetsk, Russia). The event was actively supported by the Administration of Kemerovo Region and Novokuznetsk, the Russian Foundation for Basic Research.

The conference is a recognized scientific and practical forum for professional discussion of problems related to the state and prospects for the development of new technologies, machinery and equipment that ensure the rational use of mineral resources, industrial and environmental safety of mining enterprises. The event was attended by scientists and specialists from different institutions, companies and coal mining enterprises from Russia, Germany, India, Italy, the Donetsk People's Republic, India, Italy, Tajikistan. Active participation of PhD students and young scientists is particularly noteworthy.

More than 100 papers were submitted to the Conference Organizing Committee. The present issue of *IOP Conference Series: Earth and Environmental Science* contains 56 specially selected research papers covering the following topics studied by leading domestic and foreign R&D organizations, representatives of the mining and processing industry, universities: creation of digital technologies for performance of production processes dangerous for personnel with the use of robotic manipulators; improvement of traditional methods and technical means of mining and processing of mineral resources based on the results of mathematical modeling of the main and auxiliary processes parameters, including coal seam gas drainage, prevention of geodynamic phenomena.

The Organizing Committee expresses deep gratitude to all organizers and participants of the conference for their active participation and discussion of the theoretical foundations and results of the practical application of innovative ways and means of integrated use of mineral resources in the safe working conditions, which is a significant contribution to the development of mining science and practice. Also the Organizing Committee would like to thank the Russian Foundation for Basic Research for its financial support (project No. 18-05-20026).

Vice-chair of the Organizing Committee
and Editor-in-chief
Prof. Dr. Viktor Fryanov

IOP Conf. Series: Earth and Environmental Science **206** (2018) 011002 doi:10.1088/1755-1315/206/1/011002

Peer review statement

All papers published in this volume of *IOP Conference Series: Earth and Environmental Science* have been peer reviewed through processes administered by the proceedings Editors. Reviews were conducted by expert referees to the professional and scientific standards expected of a proceedings journal published by IOP Publishing.

IOP Conf. Series: Earth and Environmental Science **206** (2018) 012001 doi:10.1088/1755-1315/206/1/012001

Triple helix: from theory to realities

S M Nikitenko[1], E A Pakhomova[2], A I Kermite[3] and L I Voronina[1]

[1]Kemerovo Institute (branch) of Plekhanov Russian University of Economics, 39 Kuznetskiy ave, Kemerovo, 650992, Russia
[2]Dubna State University, 19 Universitetskaya street, Dubna, 141980, Russia
[3]Teknik Mesin Otomotif, Institut Teknologi Indonesia, Jl. Raya Puspiptek Serpong. Kota Tangerang Selatan, 15320, Indonesia

E-mail: nsm.nis@mail.ru

Abstract. The article is devoted to the forms of interaction between business, scientific organizations and authorities on the development of partnership in the conditions of the regional economy. The role of intellectual property objects (IPO) in the innovative development of territories is analyzed. The practical experience of projects implementing, based on a preliminary study of the problems and potential of municipalities, as well as the transfer of IPO, is presented.

1. Introduction

For leading Russian and foreign scientists and economists, the task of finding tools for interaction between business, government, science and society is still relevant. The solution is aimed at ensuring the efficiency of the national economy. "The social triangle of actions" for big business is considered as a possible way to solve this problem. Public-private partnerships (PPP) deal with investment projects, the "triple innovation helix" deals with innovative projects and corporate social responsibility deals with social projects carried out simultaneously with core activities [1]. At the same time, the main thesis of the "triple helix" model implies that science can occupy a parity position with the government and business in the system of innovative development [2]. Henry Itskowitz, the author of the concept of the "triple helix", believes that new types of scientific institutions should become the locomotives of modernization, developing the country with the direct assistance of business and authorities.

This thesis for the conditions of Russia is supported by the methodology for determining the innovative potential of socio-economic objects of different hierarchical levels, developed on the basis of descriptive examples of the empirical basis of the mechanisms for the emergence and development of innovations collected by H. Itskowitz, and approved at the macro level between 2010 and 2015. The methodology demonstrates that over the given period the knowledge-intensive factor made the greatest contribution to the innovation potential [8]. The methodology is based on the concept of space institutions, which allows considering it an important starting point for modeling. "However, to work ... with real objects, it is needed to move from the model plane to the level of institutional interaction with the identification of the institutional gaps, since when trying to shift the representation in concepts and models on real objects in domestic conditions it turns out that the gaps make it difficult to develop working mechanisms of organizational interaction" [9]. Thus, they should be classified for convenience of the subsequent analysis with recommendations for overcoming them.

IOP Conf. Series: Earth and Environmental Science **206** (2018) 012001 doi:10.1088/1755-1315/206/1/012001

2. Data and methods

It should be noted that there is an opinion that in Russia the resource dependence, the narrow cultural horizon of leaders, the absence of leaders, the passivity of the authorities and, often, of science itself, slow down the process of development of interaction between science, government and business. The examples of such gaps are identified by the classification as "gaps, the nature of which lies in the interaction of different hierarchical levels; gaps which occur at the entrepreneurship level at forming a relationship between the development of science and the introduction of developments in production; gaps which are caused by the contradictions and differences in the levels of complexity of the institution and the institutional environment", etc. [10]. It is assumed that the main task of the government is to improve the legislative base of innovation activity and create conditions for public-private partnership, that is, to attract business to cooperation with universities in the field of creating and introducing new domestic developments and technologies, when the government is ready to share risks and R & D and technology introduction costs with business [3]. Here there are "the gaps caused by the absence and / or halfness of the normative and methodological base" [10], as well as in the institute of Intellectual Property.

Intellectual property as a potential point of growth and the basis for interaction in a certain territory, practically in all countries (including Russia) is generated by the three clusters: state research centers, laboratories, public academies (in Russia it is the Academy of Sciences with its fundamental orientation and research planning horizon for decades); corporation research centers (in Russia they present branch applied sciences tightly connected with production); universities [4].

The results of intellectual activity in the form of intellectual property objects have both scientific and technical novelty and market potential, ensuring the production of a knowledge-intensive product in demand on the market. Patent protection ensures product competitiveness for a certain time, but the processes of IPO commercialization depend on the stage of knowledge readiness for use by the consumer, not only in the sphere of material production, but also in making managerial decisions in the areas of finance, marketing, human resources management, etc., as well as on the production potential, the complexity of patent protection (the presence of an umbrella patent and the so-called "cluster of inventions") and other factors.

In Russia, the mechanisms for securing rights to intellectual property created when performing work under a state or municipal contract are regulated by the Civil Code of the Russian Federation (Article 1373). The right to obtain a patent and the exclusive right to an invention, a utility model or an industrial design created in the performance of works for municipal needs belong to the organization that performs the municipal contract (the executor) if the municipal contract does not provide that this right belongs to the municipality or both to the executor and the municipal formation.

Thus, the Russian legislation allows the subjects of the federation (SF) and municipal formations (MF) to form a portfolio of intellectual property (patent portfolio), while using different options for ownership, which should be defined in each individual contract, based on the conditions of expediency.

In accordance with the current Russian legislation, SF and MF may assign rights to:

- results of intellectual activity related to the interests of the regional (municipal) economy;
- intellectual property objects, concerning which SF and MF have the intention to independently develop a prototype into the industrial application and the finished products sale;
- intellectual property objects, which SF and MF can profitably sell or license.

The assessment of the innovation potential of SF (MF) can be made on the basis of two factors: the level of production potential (availability of basic and circulating funds, labor and natural resources, scientific and technical information, etc.) and the level of their scientific and technical potential (patent activity, volume of the available scientific and technical reserve, the structure and level of staff qualifications, the number of scientific schools and the corresponding system of knowledge transfer). Schematically the results of the evaluation are shown in figure 1.

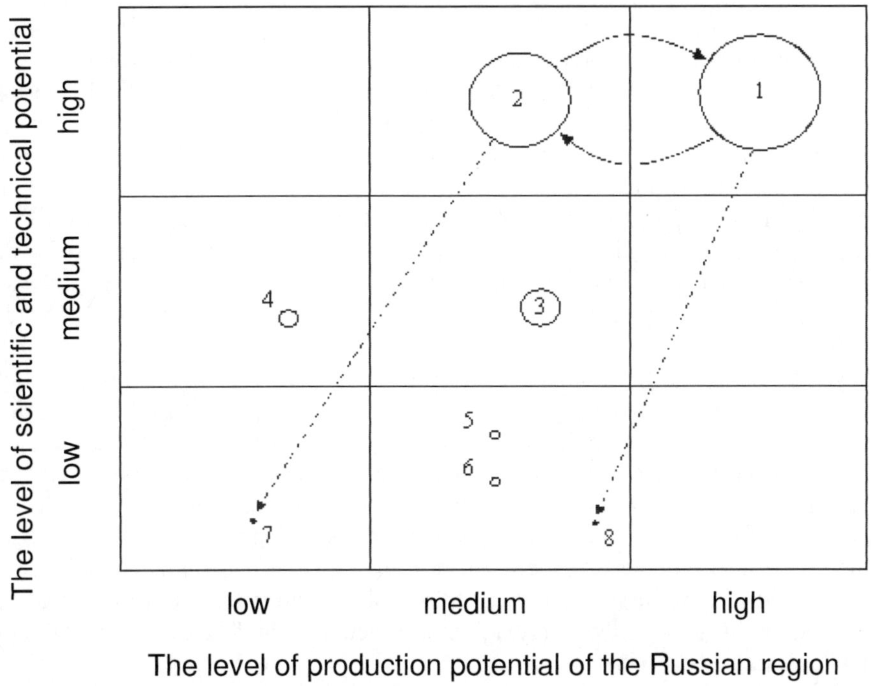

Figure 1. The chart of intellectual activity of the region.

During the formation of projects based on IPO, the following factors were taken into account: availability of necessary natural resources, compliance (excess) of new technologies to the world level, economic, social, environmental efficiency of projects, potential synergies through complementarity of projects, which is especially important for depressed areas.

Harmonization of IPO portfolio management processes provides for the transfer of knowledge from publicly-legal entities with high scientific potential to depressed areas with a certain kind of natural resources. Diffusion of the IPO promotes the leveling of the socio-economic development levels of individual MF in a region and the transfer of their economy to an innovative path.

Realization of rights to IPO can be carried out by the SF (MF) by adding to the authorized capital the value of rights to patented technologies, which can serve as the beginning of the formation of new high-tech companies. Without monetary payment, it is possible for the founder to contribute the market value of the IPO right as a contribution to the authorized capital and acquire blocks of shares (controlling or simple ones). This approach allows to IPO commercializing, for example, through the formation of joint ventures: strategic alliances, enterprises based on franchising (a complex of exclusive rights is transferred to objects of different nature), etc. Expansion of intangible assets and the use of IP rights as collateral is not the only list of issues that require their consistent implementation.

The main tasks of the government (municipal) policy to involve IPO, created at the expense of state (municipal) funds, in the economic turnover, are the orientation of scientific developments, R & D work and industry on the creation of high-tech, competitive products commercially profitable for the economy and life activity of SFs (MFs) and the spread of market relations in the sphere of the intellectual property use.

The government (municipal) policy on the involvement of IPO in the economic circulation is designed to strengthen and expand commercial, industrial and scientific-technical cooperation between scientists and entrepreneurs, scientific and industrial organizations, regardless of the form of their ownership, and protect both their interests and the interests of the territories. This policy is aimed at developing the principles of public-private partnership based on scientific knowledge.

The existing foreign and Russian experience confirms that the private sector is more sensitive to the needs of the market. At the same time, "a private entrepreneur" realizes that the commercialization of scientific developments is a long-term process, requiring considerable effort and cost to obtain possible profits. Therefore, at the initial stages of this process, it is very important to involve support structures funded or co-financed by the public sector. It is also important to ensure the reciprocal movement of requests for innovation from the territories and enterprises seeking to solve the problems and tasks of their development, and innovations from innovative and active research institutes and design bureaus, universities and small innovative enterprises focused on obtaining commercial results from their activities.

Practically in all Russian regions some natural minerals are mined, and in all municipalities there is a question of interaction with extractive companies on the principles of PPP and on the basis of a project approach. Municipalities are already actively looking for structures that would help them develop and support territories development programs on the principles of public-private partnership, based on the project approach and integrated natural resource development. They are prompted to this by the Federal Law of June 28, 2014 No. 172-FZ "On Strategic Planning in the Russian Federation".

The authors have some practical experience in the formation and implementation of projects in municipalities, based on a preliminary study of the problems and potential of municipalities [5, 6, 7]. The key direction, in this case, is the introduction of effective technologies and modern approaches at the municipal level in the following areas:

- resource and energy saving (including the reduction of energy consumption at the municipal level);
- use of secondary energy resources, including forest and mining waste;
- optimization of housing and communal services infrastructure;
- sustainable and environmentally oriented agriculture;
- development of small business;
- development of social potential and capital of the local community.

The following main principles of the project approach are used:

1. A set of projects (a group of interrelated projects that are carried out within one or several compactly located municipal entities) is formed from a group of applicant organizations that should include the municipal entity or city administration representatives and representatives of various sectors: business, non-governmental and non-profit organizations (NPOs), as well as representatives of the social sphere, namely education and health.
2. A set of projects unites from two to five different projects. Projects alone cannot be financed.
3. Projects should be implemented at the municipal level, facilitate the involvement of citizens in local government and be carried out with the participation of different sectors of society (NPOs and administrations of municipal organizations).
4. Each project should include practical measures aimed at sustainable development of the territory.
5. The set of projects should be aimed at concrete results that contribute to the sustainable development of the territory, i.e. they would significantly improve the natural and energy resources management and reduce the threat to human health from environmental pollution; the results of the projects should be presented in specific measurable indicators.
6. The work on projects should be conducted publicly and openly; the results of a project set should be made available to the general public.
7. Project sets should facilitate the involvement of the public, especially young people, in specific activities.

8. Project sets mean reinvestment to the future; all communities should demonstrate their readiness and ability to effectively reinvest their savings, resulting from the implementation of energy-saving (or other) projects, in the development of their communities.

This approach is aimed at developing different models of mutually beneficial cooperation of all sectors of the local community in the interests of its sustainable development, creating favorable conditions for the participation of small and medium-sized businesses in programs for sustainable social and economic development of the territories.

At the same time, both innovative and already tested models of the municipalities' development are getting expanded on the territory of the Russian Federation.

For example, according to the available experience, the use of energy and resource-saving technologies to create officially recognized mechanisms for investing savings into other priority projects of the community with active public participation is an effective way of creating new tools and mechanisms for the development of municipalities.

As practice has shown, it is important to maintain the following priorities in the development of territories development programs based on the project approach:

1. Development of municipalities; municipalities should propose Municipal Project Sets aimed at improving the quality of life of community members and the local economic situation in general, and promoting transparency and sustainability of positive structural changes taking place in the community. Consortia should offer projects based on an assessment of the needs of their community. It is encouraged to attract other resources of the administration, local businesses, NPOs and citizens to facilitate the more successful implementation of the proposed initiatives.

2. Implementation of practical measures and achievement of concrete results for increasing the sustainability of the process of local and regional development.

3. Strengthening of partner alliances and cooperation of various stakeholders; that means the involvement of local municipal authorities and representatives of two other sectors, NPOs and business, as well as representatives of organizations providing services to the community, such as educational and medical institutions, in the process of community development; involvement of the population in the implementation of the projects through social partnership is a strategic direction.

4. The implementation of projects sets in the same territory contributes to the effectiveness of project implementation and the achievement of synergies from activities in the fields of resource management, increasing community activity and building partnerships.

5. Participation of citizens in the community life; that implies involving community members in the decision-making process, protecting their own interests and public examination of new legislative initiatives to increase activity and transparency of community processes, and motivating young people to participate in community life.

6. Interaction between the public and local government bodies; that includes the creation of new and the promotion of proven sustainable mechanisms for financing socially significant projects by reinvesting the savings (obtained from the introduction of energy and resource-saving technologies at municipal facilities) on the basis of local and regional legislative mechanisms.

7. Use of community resources; that implies the work with communities that have expressed a desire to introduce new experience using their own resources.

The main tasks to be solved by the project approach are:

1. Implementing practical projects aimed at improving the quality of life of the population in local communities according to formulated priority topics based on the involvement of local residents in the process of solving community problems (introducing effective technologies for the development of municipal energy, housing, agriculture and small business);

2. Launching mechanisms for collegial decision-making by all stakeholders on balanced development of the local community, including the use of savings for sustainable development;

3. Approbating management mechanisms for funds saved and attracted during the projects implementation;

4. Encouraging further dissemination of project experience in the region.

3. Results and discussion

The analysis of the companies' selected patents showed that the main direction of their research and technological activity is the development of technologies and equipment related to the enrichment of coal.

At the same time, these patents were registered over the past three years, and the lifetime of the technology is 8 years, therefore, the companies are constantly improving their own developments. Moreover, the patents are valid, i.e. justified in force, which proves that they are used in coal production.

At the third stage of our study, the existing patents were evaluated in terms of their "breakthrough character". The studied patents cannot be attributed to "transforming innovations". According to experts, the patents refer to the obsolete technology, and match the level of the 1970s – so, they cannot be attributed to "transforming innovations" nor can they shape the value chain.

To create and implement "transforming innovations" in the coal industry, firstly, it is necessary to cooperate with coal companies from related industries, which will increase the added value of the final product. The closest to the coal industry in Kemerovo Region are the enterprises of mechanical engineering, small chemical production, energy industry, and developers of automated control algorithms. Secondly, new forms of cooperation are needed, for example, innovative enterprises in related industries to form value chains with the region's largest coal companies which will develop new technologies taking into account the identified strategies for conducting research activities. Thirdly, measures are needed to encourage companies to form and use "transforming innovations".

4. Conclusion

As a conclusion, the following results, obtained during the project sets implementation, could be listed:

- local administrations, NPOs, business structures and the public are developing and approving mechanisms for long-term joint public and private financing of their communities development on the principles of funds consolidation and transparency in making decisions about investing resources;
- there is an active public involvement, especially of young people, in the decision-making process related to the planning of their communities development and the resources allocation, which contributes to the institutionalization of civic initiatives;
- the stability of communities in the long-term perspective is improving, the problems of energy and resource saving, environmental management, public health, environmental education, responsible approach to business development, creation of jobs, etc. are being addressed.

Acknowledgements

This article was prepared with the financial support of Russian Foundation for Basic Research (No. 16-06-00054, the project "Instrumental and methodical approach to the adaptation of the triple helix model for Russia given the historical retrospective").

References

[1] Shestak V P 2016 *Finance: Theory and Practice* **20(4)** 6–16 DOI:10.26794/2587-5671-2016-20-4-6-16
[2] Dezhina I *"Triple helix" in the Innovation System of Russia* http://institutiones.com/innovations/265-q-q-.html
[3] http://pppcenter.ru/29/novosti/events/rossiya-buksuet-po-spirali-rbk-daily.html
[4] Itskowitz H 2010 *Triple helix. Universities – Enterprises – the State. Innovation in Action* (Tomsk: Tomsk State University of Management Systems and Radio Electronics) p 238
[5] Nikitenko S M 2008 *Fundamental Research* vol 12 33–36
[6] Nikitenko S M et al 2009 *Formation of Effective Mechanisms of Innovative Development of the*

Region's Economy. Series Innovation. Entrepreneurship. Investments (Kemerovo)
[7] Nikitenko S M et al 2009 *Regional Economy: Theory and Practice* **24** 17–26
[8] Istomina S V et al 2018 *National Interests: Priorities and Security* vol 14 1(358) 97–120
[9] Istomina S V et al 2018 P*roblems of the Configuration of the Global Economy of the 21st Century: The Idea of Socio-economic Progress and Possible Interpretations. Collection of Scientific Articles* vol 1 (Krasnodar) pp 102–109
[10] Pakhomova E A et al 2017 *Philosophy of Economy. Almanac of Center for Social Sciences and Economics Faculty of Lomonosov Moscow State University. Special issue* December 2017 pp 357–368

Modeling influence of fault zone on stress distribution in rock mass with regard to the accumulated elastic energy

S V Lavrikov, O A Mikenina and A F Revuzhenko

Chinakal Institute of Mining, Siberian Branch, Russian Academy of Sciences, 54 Krasny ave, Novosibirsk, 630091, Russia

E-mail: lvk64@mail.ru

Abstract. Based on the mathematical model of structurally inhomogeneous rock mass, the problem of stress distribution nearby a disjunctive break ahead of the stoping face is considered. The model describes the property of rocks to accumulate elastic energy in the form of internal self-balancing stresses. The problem is solved numerically in the quasi-static formulation using the finite element method. It is shown that the presence of the fault zone can provoke release of the accumulated elastic energy and, thereby, change stress distribution in surrounding rock mass.

1. Introduction

It is common knowledge that stress state of rocks and stability of underground openings are governed by the internal structure of rock mass. One of the essential properties of a structurally non-uniform rock mass is the ability to accumulate external action energy in the form of internal self-balancing stresses. A real rock mass is composed of hierarchical blocks, and such structure can have many equilibriums states. Some rock mass zones can accumulate and then release energy under certain conditions, i.e. they act as potential sources of energy [1–8]. This property has been observed in laboratory-scale experiments and in field studies [9, 10].

As a rule, energy is accumulated at boundaries of blocks or in fault zones when shear stress on the fault edges is balanced by friction between the edges. The balance is disrupted under certain conditions, and the accumulated energy releases. This process can either be stable relaxation or unstable catastrophe [11, 12]. Such events are particularly hazardous when take place nearby mined-out areas.

The problem of stresses and strains in rock mass adjacent to an underground opening is one of the key problems of rock mechanics. Many researchers have addressed different formulations of this problem, and its solution is in a varying degree used in all technologies of underground mineral mining. Special interest is provoked by formulations on redistribution of stresses in the adjacent rock mass under effect of various disjunctions [13–18]. Taking account of structural hierarchy and internal self-balancing stresses is possible in the framework of different approaches. In [19, 20] the mathematical apparatus and models with internal variables are developed; in [21, 22] the approach based on the method of non-Archimedean analysis is used while in [23, 24] the discrete element method finds application. In this paper, we take mathematical considerations from [8, 21] and analyze numerically deformation of rock mass with a disjunctive break nearby a working face.

2. Mathematical model

Let us consider a two-scale model of rock mass in the initial stress state [8, 21] (plane-strain deformation). At microscale elastic particles (grains) occur at the points of a square lattice. Pore space is filled with binding material with different elastic characteristics than the grains have (Figure 1). Plastic shearing between particles develops nonlinearly, with the stages of strengthening, strength loss and residual strength. The grain boundary sliding diagram relating tangential forces t_{12} and shears ε_{ij}^{R} (in a general case, it is assumed that $\varepsilon_{12}^{R} \neq \varepsilon_{21}^{R}$) is shown in figure 2 (a piecewise linear approximation). External loading causes deformation of grains and shearing between them. This, in turn, results in deformation of pore space. Elastic stresses in grains and in pore space are balanced by shearing forces at the grain contacts and, thus, the element of the medium has a certain energy margin. Internal self-balancing stresses can be considerable even in case of absence of external stresses.

The grain boundary sliding conditions are approximated by a piecewise linear function (figure 2). In this case, the increments in the microstresses and microshears are related as:

$$\Delta\varepsilon_{12}^{R} = \Delta t_{12} / G_1^s, \quad \Delta\varepsilon_{21}^{R} = \Delta t_{12} / G_2^s, \tag{1}$$

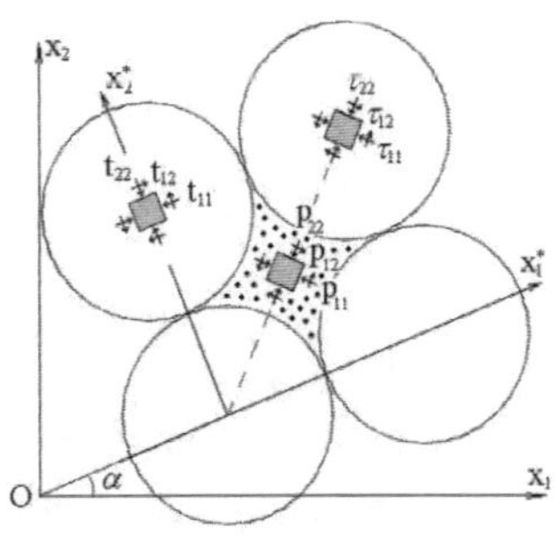

Figure 1. Internal microstructure of a medium.

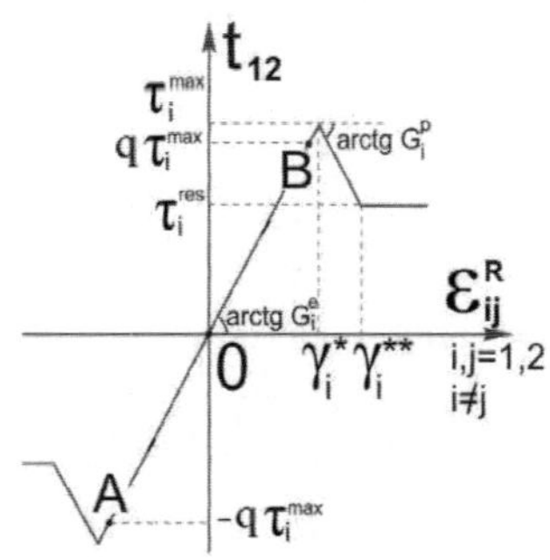

Figure 2. Grain boundary sliding diagram.

where G_1^s, G_2^s are the present moduli of grain boundary sliding for each family of boundaries. The values G_i^s are defined by the preset constants $\gamma_i^*, \gamma_i^{**}, \tau_i^{\max}, \tau_i^{res}$, where $i = 1,2$ is the number of a family of boundaries

$$G_i^s = \begin{cases} G_i^e, & 0 \le \gamma_i < \gamma_i^* \\ -G_i^p, & \gamma_i^* \le \gamma_i < \gamma_i^{**} \\ 0, & \gamma_i^{**} \le \gamma_i \end{cases}.$$

The constitutive equations of the model at macroscale relate the macrostress increments $\Delta\sigma_{ij}$ and macrostrain increments $\Delta\varepsilon_{ij}$

$$\begin{pmatrix} \Delta\varepsilon_{11} \\ \Delta\varepsilon_{22} \\ \Delta\varepsilon_{12} \end{pmatrix} = W \cdot \left((T+R)^{-1} + 2(T+P)^{-1} \right) \cdot W^{-1} \cdot \begin{pmatrix} \Delta\sigma_{11} \\ \Delta\sigma_{22} \\ \Delta\sigma_{12} \end{pmatrix}. \tag{2}$$

Coefficients in equation (2) depend on the microproperties of grains (matrix T), pore-filling material (matrix P), grain boundary sliding moduli (matrix R) and the angle α responsible for the orientation of regular packing of grains in the Cartesian coordinate system (matrix $W = W(\alpha)$). In turn, the macrostresses σ_{ij} and microstresses t_{ij} of grains and pore-filling material p_{ij} are not mutually independent and fulfill the condition of compatibility [21] given by:

$$\sigma_{ij}^{*} = t_{ij} + 2m \cdot p_{ij}, \qquad i, j = 1,2; \tag{3}$$

here, σ_{ij}^{*} are the macrostresses at the components of axes in the coordinate system of regular packing of grains (figure 1); the dimensionless value $0 < m < 1$ characterizes porosity [21].

Conditions (3) imply that one and the same macrostresses σ_{ij}^{*} can conform with many stress states of grains t_{ij} and pore space p_{ij}. Furthermore, the absence of macrostresses means no equal absence of internal microstresses which can be alternating and totally balancing each other.

The system of constitutive equations (2) and the strain–displacement relation is closed by the equilibrium equations (for increments)

$$\frac{\partial \Delta\sigma_{11}}{\partial x_1} + \frac{\partial \Delta\sigma_{12}}{\partial x_2} + \Delta X_1 = 0, \quad \frac{\partial \Delta\sigma_{12}}{\partial x_1} + \frac{\partial \Delta\sigma_{22}}{\partial x_2} + \Delta X_2 = 0. \tag{4}$$

where $\Delta X_1, \Delta X_2$ are the increments of mass forces.

In view of the nonlinearity of the grain boundary sliding diagram, the problem is formulated in terms of increments. The general solution is constructed by steps of loading with regard to the change in properties during deformation: $\sigma_{ij}^{k+1} = \sigma_{ij}^{k} + \Delta\sigma_{ij}^{k}$, $\varepsilon_{ij}^{k+1} = \varepsilon_{ij}^{k} + \Delta\varepsilon_{ij}^{k}$, where the superscript k is a number of iteration.

The analysis of the type of the system (2), (4) shows that for the descending branch of the grain boundary sliding diagram, there exists a critical slope G_i^p governed by shear moduli and Poisson's ratios of grains, μ^t, ν^t, and pore space, μ^p, ν^p. When G_i^p is lower than a critical value, the system is elliptical and deformation process is stable. In case that the slope exceeds the critical value, the process of deformation is unstable and jumps of strength loss can take place in the medium. In this study, we limit ourselves to the stable deformation parameters.

Based on the model (2), (4), the finite element-based algorithm and computer program have been developed, which allow solving quasi-static plane boundary value problems.

3. Numerical modeling results

The computational domain models a working face (Figure 3). We select linear distribution of macrostresses: $\sigma_{22}^0 = -\gamma(H - x_2)$, $\sigma_{11}^0 = \xi\sigma_{22}^0$, $\sigma_{12}^0 = 0$, where γ is the specific gravity of rocks mass, H is the depth of mining; ξ is the lateral earth pressure coefficient. Aside from the macrostresses, it is required to find initial microstress state to satisfy relations (3). We choose the distribution of microstresses depending on the parameter q: $t_{11}^0 = 0.5\sigma_{11}^0$, $t_{22}^0 = 0.5\sigma_{22}^0$, $t_{12}^0 = \pm q\tau_i^{\max}$, $p_{ij}^0 = (\sigma_{ij}^0 - t_{ij}^0)/2m$, $i, j = 1,2$ and analyze two problems. In the first problem, it is assumed that shear microstresses on grains and, consequently, in pore space are absent in the computational domain, i.e. $q = 0$ everywhere. In the sliding diagram in figure 2, this state conforms with the point O (origin of coordinates). In the second problem, a disjunction break in rock mass adjacent to the working face (colored grey in figure 3) is considered, and it is assumed that shear microstrains of grains in the fault make a certain part of the critical stress, e.g. $q = 0.9$, while it is still $q = 0$ in the adjacent rock mass. In the sliding diagram, this state is shown by the points A or B depending on the shear microstress sign.

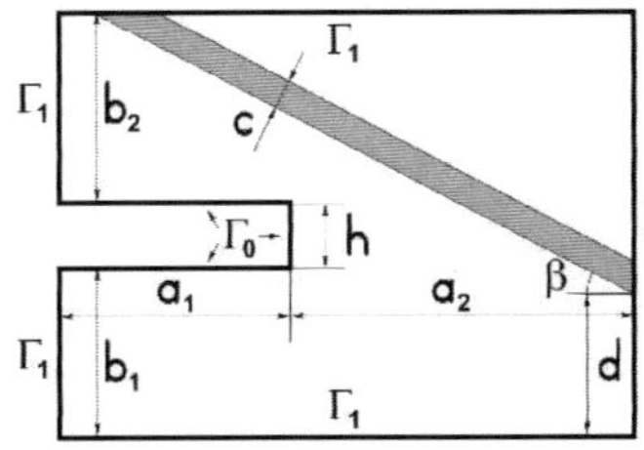

Figure 3. Computational domain, the disjunctive break is colored grey.

In this manner, we have that the same macrostresses are preset in both problems. The difference is distribution of microstresses of the domain elements.

The boundary conditions are set as the gradual decrease in the stresses in the roof and floor of the stoping void, as well as ahead of the working face: $\Delta\sigma_n\big|_{\Gamma_0} = -\Delta d^k \cdot \sigma_n^0$, $\Delta\tau_n\big|_{\Gamma_0} = -\Delta d^k \cdot \tau_n^0$, where σ_n^0, τ_n^0 are the normal and shear macrostresses at zero time (zero iteration); $\Delta\sigma_n, \Delta\tau_n$ are the increments in the normal and shear macrostresses; Γ_0 is the stoping void boundary; $0 \le \Delta d^k \le 1$ is the increment in the loading parameter at a k-th step. The condition $\sum_k \Delta d^k = 1$ means total unloading (absence of macrostresses) of the stoping void boundary. At the computational domain boundary Γ_1, it is assigned that $\Delta u_n\big|_{\Gamma_1} = 0$, $\Delta\tau_n\big|_{\Gamma_1} = 0$, where Δu_n is the increment in the normal component of displacement vector.

We assume non-dimensional values of parameters (all stress dimensions are related to the peak shear stress $\tau_1^{\max}$ in Figure 2; length dimensions — to the mineral bed thickness h in Figure 3): $\mu^t = 2.5 \cdot 10^3$, $\nu^t = 0.2$, $\mu^p = 0.7 \cdot 10^3$, $\nu^p = 0.3$, $\gamma_1^* = \gamma_2^* = 5 \cdot 10^{-4}$, $\gamma_1^{**} = \gamma_2^{**} = 10^{-3}$, $\tau_1^{\max} = \tau_2^{\max} = 1$, $\tau_1^{res} = \tau_2^{res} = 0.4$, $\gamma H = 32.5$, $\Delta X_1 = \Delta X_2 = 0$, $\xi = 0.4$, $m = 0.5$, $\alpha = 0$, $h = 1$, $a_1 = 3.3$, $a_2 = 5$, $b_1 = 2.3$, $b_2 = 2.7$, $c = 0.4$, $d = 2$, $\beta = 27^0$, $H = 300$ and perform calculations until total unloading of the stoping void boundary.

Figure 4 demonstrates the calculated results for the first problem without a fault zone. Figure 4a shows the evolution of the plastic deformation zones (light-grey color marks the local strength loss zones, which corresponds to the descending branch in the grain boundary sliding diagram; dark-grey — residual strength zones, which conforms with the horizontal branch in Figure 2). Figures 4b and 4c present, respectively, contour lines of the maximum shear stress $\tau_0 = 0.5\sqrt{(\sigma_{11} - \sigma_{22})^2 + 4\sigma_{12}^2}$ and maximum shear $\gamma_0 = \sqrt{(\varepsilon_{11} - \varepsilon_{22})^2 + 4\varepsilon_{12}^2}$. The contour lines of the macrostresses σ_{11}, σ_{22}, σ_{12} are shown in figures 4d–4f, respectively.

It is seen that plastic shears propagate from the natural stress raisers at the stoping void boundary depthward rock mass in the vertical direction predominantly.

Figure 5 presents the calculation results for the second problem with a disjunctive break. Apparently (Figure 5a), in the course of deformation, material inside the fault situated directly above the stoping front comparatively rapidly passes to plasticity and affects stress state of adjacent rock mass (figures 5b–5f). Sliding of the edges of the fault has the most pronounced influence on the distribution of the tectonic (σ_{11}) and shear (σ_{12}) stresses in rock mass.

IOP Conf. Series: Earth and Environmental Science **206** (2018) 012002 doi:10.1088/1755-1315/206/1/012002

In whole, under one and the same initial macrostresses, the release of the accumulated energy in the form of internal self-balancing microstresses in a relatively narrow zone of a disjunctive break under the influence of the stoping front advance results in the macroshear of the fault edges and changes the macrostress state of surrounding rock mass.

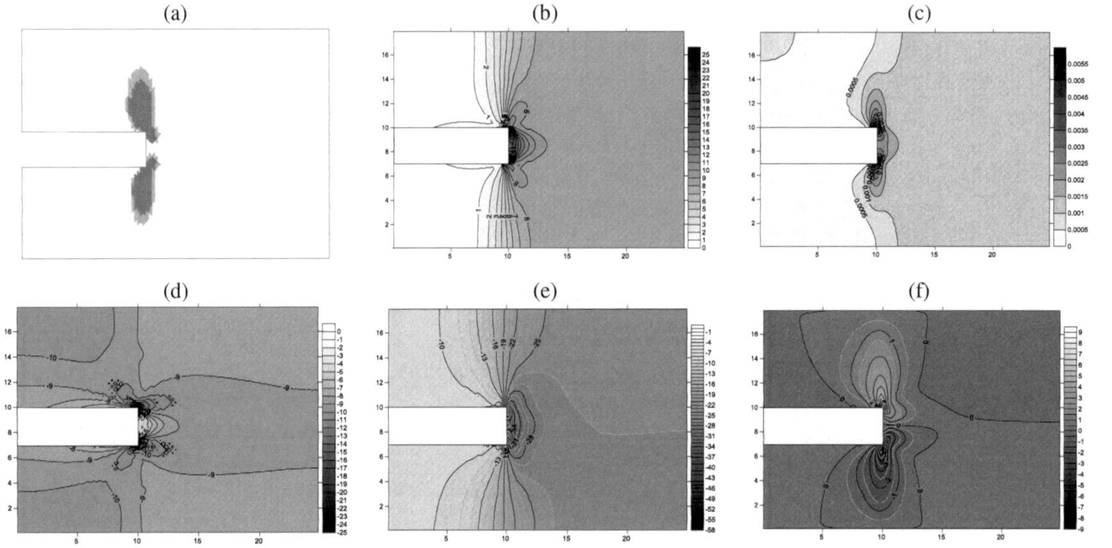

Figure 4. First problem calculation data without a fault: a) plastic deformation zones; b) contour lines of the maximum shear stress τ_0, c) contour lines of the maximum shear γ_0, d), e), f) contour lines of σ_{11}, σ_{22}, σ_{12}, respectively.

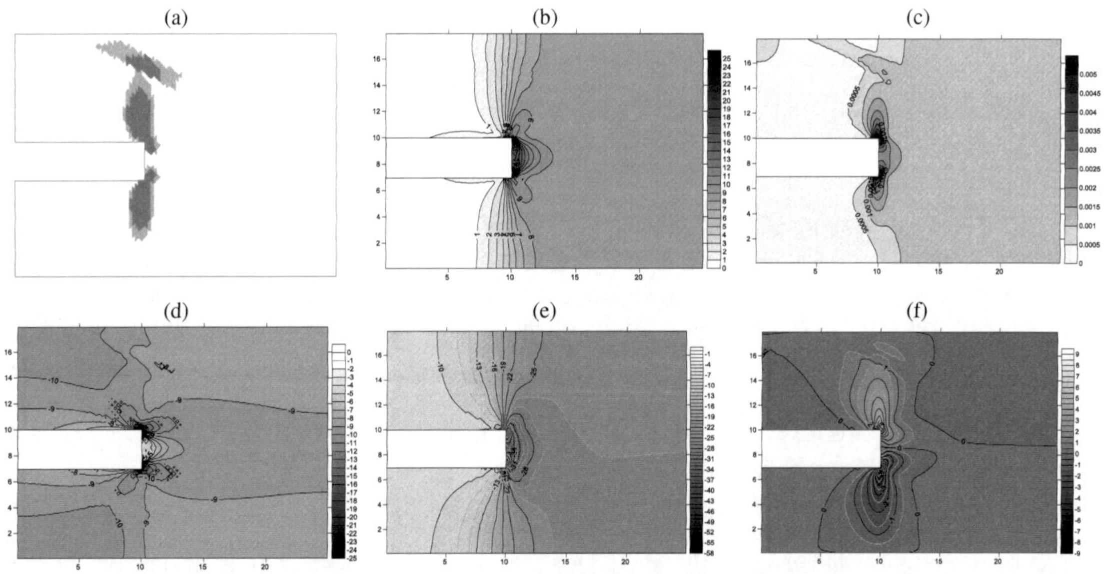

Figure 5. Second problem calculation data with a fault: a) plastic deformation zones; b) contour lines of the maximum shear stress τ_0, c) contour lines of the maximum shear γ_0, d), e), f) contour lines of σ_{11}, σ_{22}, σ_{12}, respectively.

It is expected that the influence of the disjunctive break will become even stronger in case of unstable deformation with the uncontrollable dynamic liberation of the accumulated elastic energy.

4. Conclusions

- The developed approach makes it possible to solve problems on deformation of structurally inhomogeneous geomaterials with regard to internal self-balancing stresses.
- Stoping near a disjunctive break can provoke release of internal self-balancing stresses and, thus, affect stress state of surrounding rock mass.

Acknowledgments
This study was supported by the Russian Science Foundation, Project No. 16-17-10121.

References

[1] Kocharyan G G 2016 *Geomechanics of Faults* (Moscow: Geos) p 424
[2] Melnikov N N (Ed) 2012 *Destruction and Self-Organization Processes in the Earth's Crust Areas under Strong Induced Impact* (Novosibirsk: SO RAN) p 631
[3] Ponomarev V S 2011 *Geotektonika* **2** pp 66–75
[4] Moroz A I 2004 *Self-Stress State of Rocks* (Moscow: MGU) p 288
[5] Peng Z and Gomberg J 2010 *Nature Geoscience* **3** 599–607
[6] Brune J N 1970 *J. of Geophysical Research* **75 (26)** 4997–5009
[7] Revuzhenko A F 1991 *Journal of Mining Science* **5** 391–396
[8] Lavrikov S V and Revuzhenko A F 2017 *Journal of Mining Science* **1** 12–20
[9] Goryainov P M and Davidenbko I V 1979 *Doklady Akademii Nauk SSSR* **247 (5)** 1212–15
[10] Stavrogin A N and Shirkes O A 1986 *Journal of Mining Science* **4** 235–244
[11] Aduskin A A, Gornov V V, Kurlenya M V, Oparin V N, Revuzhenko A F and Spivak A A 1992 *Doklady Akademii Nauk SSSR* **123 (2)** 263–9
[12] Sobolev G A and Ponomarev A V 2003 *Physics and Signs of Earthquakes* (Moscow: Nauka) p 270
[13] Kocharyan G G, Zolotukhin S R, Kalinin E V, Panasyan L L and Spungin V G 2018 *Journal of Mining Science* **1** 16–24
[14] Rib S V, Basov V V and Nikitina A M 2016 *Vestnik Sibirskogo Gosudarstvennogo Industrialnogo Universiteta* **(15)** 17–20.
[15] Bobryakov A P 2013 *Journal of Mining Science* **6** 873–880
[16] Fryanov V N, Pavlova L D and Gordeev S N 2015 *J. of Fundamental and Applied Minng Sciences* **2** 162–7
[17] Dordzhiev D Yu and Protosenya A G 2010 *Izvestiya TulGU. Nauki o Zemle* **2** 225–30
[18] Lavrikov S V 2003 *Journal of Mining Science* **5** 444–452
[19] Kolymbas D 2005 *Tunneling and Tunnel Mechanics.A Rational Approach to Tunneling* (Springer-Verlag Berlin Heidelberg) p 437
[20] Novozhilov V V and Kadashevich Yu I 1990 *Microstresses in Engineering Materials* (Leningrad: Mashinostroenie) p 223
[21] Revuzhenko A F 2000 *Mechancis of Elastoplastic Media and Nonstandard Analysis* (Novosirbisk: NGU) p 428
[22] Lavrikov S V, Mikenina O A and Revuzhenko A F 2011 *J. of Rock Mechanics and Geotechnical Engineering* **3 (2)** 153–60.
[23] Klishin S V, Lavrikov S V and Revuzhenko A F 2017 *AIP Conference Proceedings* **1909** 020086
[24] Lavrikov S V and Revuzhenko A F 2016 *Journal of Mining Science* **4** 632–637

Effect of compressive stresses on P-wave velocity in samples of equivalent geomaterials

V P Kosykh, P V Kosykh and A F Revuzhenko*

Chinakal Institute of Mining, Siberian Branch, Russian Academy of Sciences, 54 Krasny ave 54, Novosibirsk, 630091, Russia

E-mail: *revuzhenko@yandex.ru

Abstract. The authors have found experimentally that elastic wave velocity depends on compressive stresses in tubular samples made of equivalent geomaterials. The relation between the wave velocity and tangent modulus of elasticity is determined. The values of the dynamic modulus of elasticity as function of the compressive stress are obtained.

1. Introduction

In-situ rock masses experience constant or slowly varying static loads alongside multiple repeated weak variable loads (caused by tides, atmospheric variations, transport-induced vibrations, etc.) [1,2]. Rock is a medium having internal structure at different scales. Being subjected to deformation, rocks accumulate energy in the form of internal self-balancing stresses. Laboratory-scale and full-scale researches reveal that geomaterials can change their mechanical properties under the influence of cyclic loads [3–8] and release the accumulated energy under certain conditions [9,10].

Mechanical properties of rocks are currently determined using the static and dynamic methods. In the static tests, it is most often that a geomaterial sample is subjected to slowly varying compression by different loading programs, and the testing results are obtained as a set of deformation and strength characteristics of the material (static modulus of elasticity, modulus of deformation, limit strength, limit elasticity, etc.).

The dynamic tests most commonly use ultrasonic techniques. In this case, a sample is exposed to treatment by ultrasonic waves, and the velocities of the waves are used to determine dynamic modulus of elasticity and attenuation coefficient.

In a general case, the stress–strain curve is nonlinear, and the static modulus of elasticity E is, therefore, found from averaging. The dynamic modulus E_d is determined in an unloaded sample, as a rule. For this reason, E and E_d differ. Moreover, the difference between the elasticity moduli is influenced by an elastic after-effect.

There are a comparatively few known studies into the influence of cyclic long-term weak impact on the change of mechanical properties of geomaterials. The Institute of Mining, SB RAS, has designed a test bench for investigating effect of multiple weak impacts against static loading on evolution of properties of geomaterials; in particular, the test bench allows tracing the change in elastic wave velocity depending on loading conditions of samples [1].

This study focused on the elastic wave velocity in equivalent geomaterials depending on background level of static stresses of compression. The test bench enabled following up evolution of deformation and strength characteristics in geomaterial samples under weak impacts against a pre-set static load.

IOP Conf. Series: Earth and Environmental Science **206** (2018) 012003 doi:10.1088/1755-1315/206/1/012003

2. Research techniques

The test bench layout is shown in figure 1. A geomaterial sample is held in grippers 2 and 5. The right-hand gripper 5 is fixed on a vertical plate by means of membrane 4 and is aligned horizontally through its stem and a Teflon sleeve. The left-hand gripper 2 is also aligned horizontally by a Teflon sleeve and transmits compressive load from the screw 1 to the sample via a ball. The vertical plates are rigidly fixed on a solid base 7. Alongside the axial load, the sample is subjected to a dynamic force created by impacts of the known energy on the stem of the gripper 5. The measurement equipment of the test bench is the 8-channel LTR-EU-8-2 crate system manufactured by L-Card, Moscow, composed of 8 measurement modules. The system continuously takes data from 16 strain gauge transducers and records high-frequency (to 10 MHz) signals simultaneously in four channels. More details of the test bench are given in [11].

The hollow cylinder samples were made of the mixture of graded quartz sand particles 0.3 mm in size and a Plasticrete filler at a weight ratio of 1:1. The samples were 80 mm long and had external and internal diameters of 36 and 16 mm, respectively. After crystallizations the samples were dried at a temperature of 70°C for 6 hours. The prepared samples had density of 2.26 g/cm^3.

To measure longitudinal and circumferencial static deformation, constantan foil strain gages were glued on the samples and connected with the strain meter module of the crate system. The wave process was studied using four semi-conductor resistive-strain sensors with a spacing of 5 mm and strain conversion efficiency of 130. These sensors were glued on the sample surface along a line, at a distance of 21 mm from each other. The signals of the sensors, after amplification by wide-band amplifiers, were recorded in the form of oscillograms of axial dynamic strains.

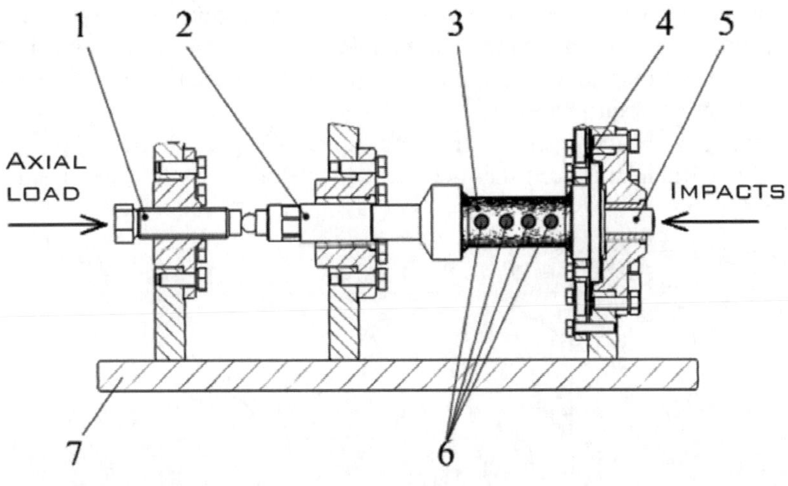

Figure 1. Test bench: 1 – compression load screw; 2 – left-hand gripper; 3 – geomaterial sample; 4 – membrane; 5 – movable gripper; 6 – strain sensors.

3. Results and discussion

During the tests, a sample was subjected to static step-wise loading up to a pre-set axial compressive force. At the same time, the stress–axial/lateral strain curves were plotted. After each step of loading, for a certain time, deformation of a sample increased while the loading was decreased. Figure 2 illustrates the change in the axial stress σ_1 in a sample at a step of loading.

It is seen in figure 2 that attenuation of stresses (and strains) lasts for 1.5–2 hours. Therefore, a sample was held for 2 h after each loading step up to complete stabilization of stresses. Then, the right-hand gripper stem was subjected to a series of 25 impacts, and oscillograms of dynamic strains were recorded.

Figure 3 shows a typical compressive stress–axial strain diagram. The values of stresses and strains at the moment of impacts are marked. The sample was not loaded to failure.

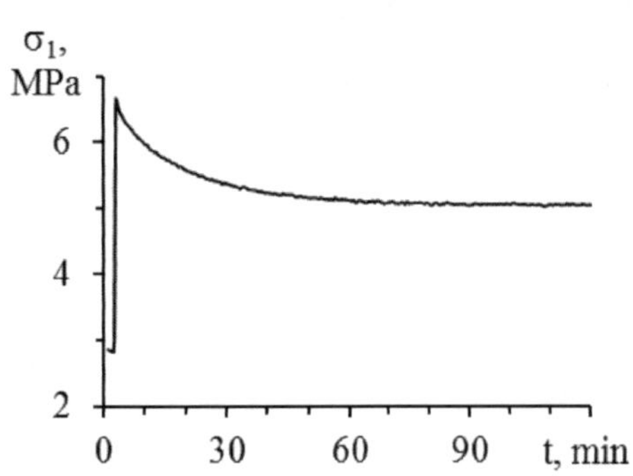

Figure 2. Stress variation in time after step-wise increase in loading of a sample.

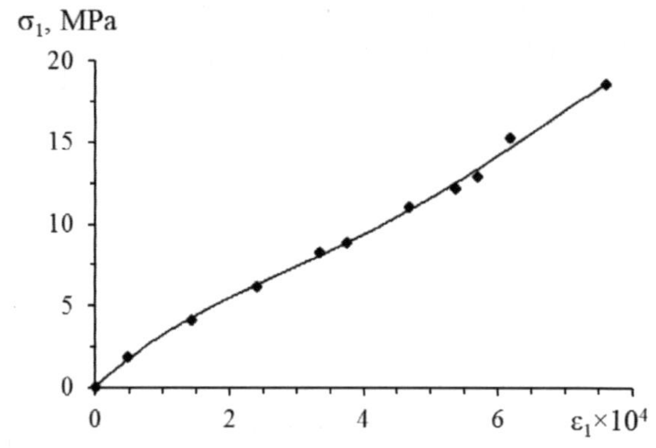

Figure 3. Compressive stress–axial strain curve in a sample made of equivalent geomaterial.

Apparently, the stress–strain curve is nonlinear. The average modulus of elasticity of the material was $2.31 \cdot 10^4$ MPa and Poisson's ratio was 0.35.

The typical oscillograms are presented in figure 4. The impacts excite vibrations in the bench–sample system, and the frequency and attenuation rate of the vibrations are governed by the sample properties. The oscillatory process in the bench–sample system involves movable parts of the test bench; moreover, free vibrations are excited in the sample. As a result, the test material is exposed to variable stresses of different frequencies.

The typical oscillograms of incremental strains $\Delta\varepsilon_1$ after an impact on the sample are recorded by two sensors spaced at 63 mm. It is seen that the attenuating strains are the superposition of different frequency oscillations.

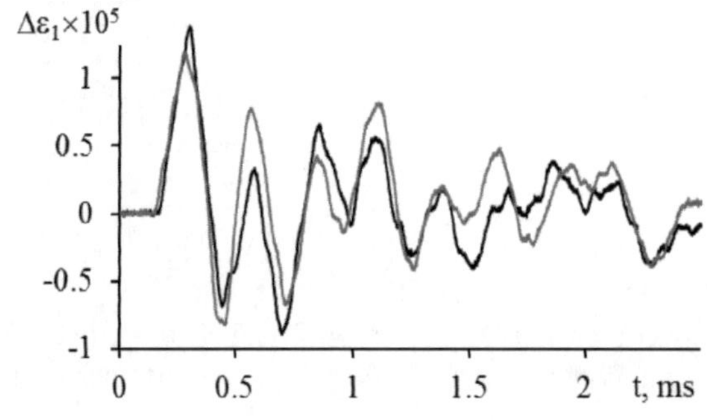

Figure 4. Oscillograms of incremental axial strains in a sample under impact.

Velocities of elastic waves in a sample were determined by correlation of oscillograms from different sensors. Figure 5 shows the start sections of oscillograms of incremental axial strains in a sample after impact after their normalization with respect to maximum amplitude.

It is seen in figure 5 that the second sensor signal is delayed by Δt relative to the first sensor. This delay is governed by the elastic wave velocity in the sample. Considering that the distance l between the first and second sensors is known, it is easy to find the elastic wave velocity $v=l/\Delta t$.

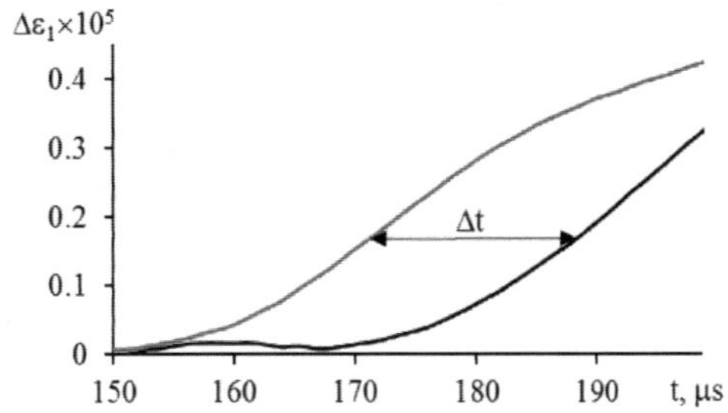

Figure 5. Time of delay of elastic wave in a sample.

Figure 6a demonstrates the determined elastic wave velocity depending on the compressive force.

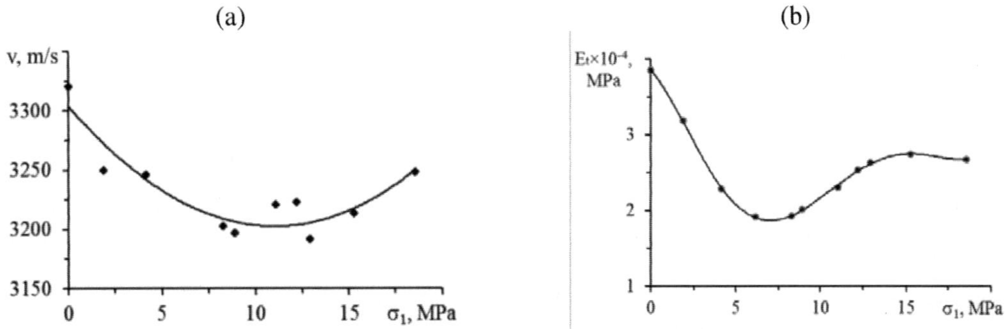

Figure 6. Relations of (a) elastic wave and (b) tangetn modulus of elasticity and copmpression stress in a sample.

As seen in figure 6a, the dependence of the elastic wave velocity on the compression stress is nonmonotonous. As the stress is increased, the elastic wave velocity decreases, passes a minimum and, then, grows. Such behavior correlates with the change in the tangent (local) modulus of elasticity E_t under the increase in the compression stress (figure 6b).

It is well-known that the velocity of elastic waves in rods depends on the local modulus of elasticity and on the density of a rod material in accordance with the expression $v = \sqrt{E_t/\rho}$ where E_t is the local (tangent) modulus of elasticity and ρ is the density of material.

From this formula, we find the dynamic modulus of elasticity E_d as function of the compression stress in a sample using the experimentally determined velocities of elastic waves (figure 6a). The results are depicted in figure 7 as the relation between the dynamic and tangent moduli of elasticity.

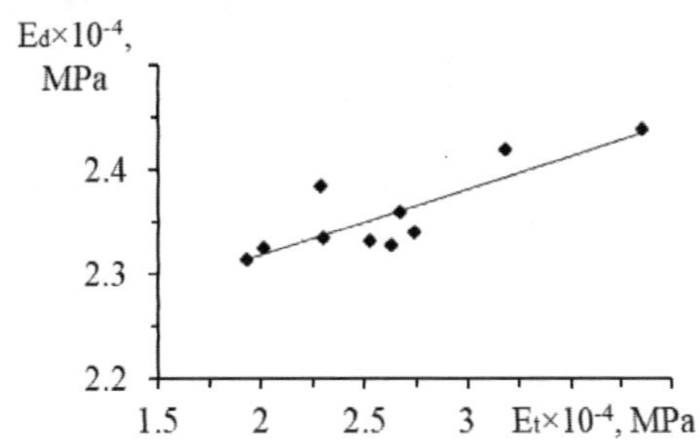

Figure 7. Relation of the dynamic and tangent moduli of elasticity.

Apparently, the experimental points group nearby the straight line. The coefficient of determination makes 0.7.

4. Conclusion

The authors have analyzed the influence of static stresses on the elastic wave velocities in samples made of equivalent geomaterials under uniaxial compression. It is found that the local deformation modulus and elastic wave velocities correlate. As the compressive stress is increased, the local modulus and elastic wave velocities nonmonotonously change. There exists a certain relation between the dynamic and tangent moduli of elasticity: as the tangent modulus increases so does the dynamic modulus of elasticity.

References

[1] Voznesensky E A 1999 *Dynamic Instability of Soil* (Moscow: Editorial URSS) p 264
[2] Sheinin V I, Blokhin D I, Maksimovich I B and Sarana E P 2016 *Journal of Mining Science* **52**(6) 1039–1046
[3] Mokhnachev M P 1979 *Fatigue of Rocks* (Moscow: Nauka) p 152
[4] Fuenkajorn K and Phueakphum D 2009 *Int. J. Rock Mech. &Min. Sci.* **46** 200–209
[5] Kocharyan G G, Kostyuchenko V N and Pavlov D V 2004 *Fizich. Mezomekhanika* **1** (7) 5–22
[6] Kurlenya M V, Adushkin V V, Garnov V V, Oparin V N, Revuzhenko A F and Spivak A A 1992 *Doklady Akademii Nauk* **323** (2) 263–265
[7] Kazantsev A A, Kosykh V P and Revuzhenko A F 2015 *IOP Conf. Series: Materials Science and Engineering* **91** DOI 10.1088/1757-899X/91/1/012089
[8] Klishin S V and Revuzhenko A F 2017 *IOP Conf. Series: Earth and Environmental Science* **84** 012003
[9] Lavrikov S V and Revuzkenko A F 2017 *Journal of Mining Science* **53** (1) 12–20
[10] Lavrikov S V and Revuzhenko A F 2015 *AIP Conference Proceedings* DOI 10.1063/1.4932811
[11] Kosykh V P, Kosykh P V and Revuzhenko A F 2017 *Journal of Mining Science* **53** (6) 1147–1151

Deformation of granular material flow in converging channels

A P Bobryakov, S V Klishin, V P Kosykh, S V Lavrikov, O A Mikenina and A F Revuzhenko*

Chinakal Institute of Mining, Siberian Branch, Russian Academy of Sciences, 54 Krasny ave, Novosibirsk, 630091, Russia

E-mail: *revuzhenko@yandex.ru

Abstract. The authors study deformations of a granular material flow from a converging channel using the method of Particle Image Velocimetry (PIV). It is experimentally proved that shears localize along two orthogonal families of slip lines. During material flow, shears periodically "switch" between the families. Furthermore, the flowing material is split into blocks of different densities, and the maximum rates of density variation are concentrated along the slip lines of one of the families. The results of numerical modeling by the discrete element method and the laboratory-scale experiments qualitatively agree.

1. Introduction

In mining, many processes are studied experimentally using physical models. Such studies, first of all, include rock pressure phenomena, explosion load on rock mass, processes of ore and coal discharge from stopes and bunkers, haulage, etc. Adequate modeling involves the analysis of similarity criteria, manufacture of models using equivalent materials and experimentation aimed to register kinematics of deformation failure and to measure stresses.

A convenient object for modeling processes in mining is granular materials and equivalent geomaterials made of them with addition of binders.

Granular materials possess two fundamental properties intrinsic to all types of rocks, namely, internal friction and dilatancy. At the same time, cohesive forces between particles are very low, and granular materials are incapable to resist tensile stresses. For this reason, it is very difficult to measure deformation of granular materials on lab-scale and, moreover, full-scale models.

The best suitable techniques for such studies are non-contact measurements one of which is the method of Particle Image Velocimetry (PIV). This method consists in comparison of successive images of deformation process and determination of velocities in the target area. As a result, it is possible to determine field of strains in a specimen [1, 2].

2. Flow of granular materials in converging channels (laboratory modeling)

The general view of the test slot-type bunker with the flat converging side walls 1 and the flap-gate discharge 2 is given in figure 1. On the inside of the transparent front wall 2, a thin pipe filled with a colored granular material was attached. As the main flow was let out of the bunker, the colored particles were entrained in it and visualized the flow trajectory.

Regarding granular materials, it is known that their mechanical properties differ in different packings. The cause is the filling conditions, first of all. For instance, rain-like fill process generates a dense packing of particles while stream-like fill results in a loose packing. It is possible to improve structure and stabilize mechanical parameters of a packing by subjecting it to alternating small-amplitude shearing [3, 4].

In this study, for stabilization of experimental results, the granular material was in stationary conditions. To that end, the material was continuously added while being discharged so that the bunker was always filled. Additional filling was carried out using a continuous batcher (not shown in figure 1). Fixed weight of the fill material ensured steady average pressure on the side walls of the bunker and constant discharge velocity.

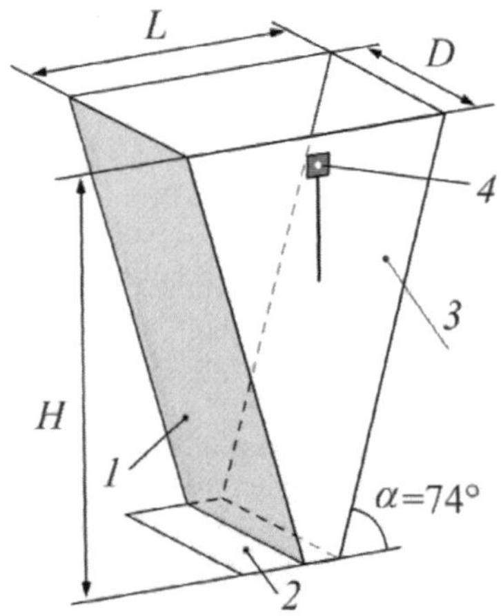

Figure 1. Slot-type bunker, general view: 1 – converging walls; 2 – flap-gate discharge; 3 – transparent front wall; 4 – nozzle of marker flow source. Bunker height $H = 310$ mm, width $D = 52$ mm, top length $L = 182$ mm.

The experiments showed that quartz sand flow velocity was constant with the continuous additional fill, while without the fill, during the period of complete discharge of the material from the bunker, the flow velocity was higher with the higher column of the material. To check adherence to the condition of time invariance, the discharge hole and the batcher were closed simultaneously, and the material in the bunker was weighed. The average weight of seven specimens was 2059 g at the deviation from the mean was no more than 6 %.

Thus, the adjustable continuous batcher provides stationary condition of the medium in the bunker in operation. The average density of the packing was 1.4 g/cm^3 and porosity was 45 %.

The flow process allows observing localization of deformations, clustering, initiation of new flows and their re-routing [5]. By way of illustration, figure 2 demonstrates the view of the marker belt during local flow. The images are taken at an interval of 20 s. It is seen that at a distance of 120 mm from the bunker top, the marker belt begins to crook and stretch. The flow path varies with time. The average flow velocity of the belt was 1.7 mm/s, standard deviation made 3 %, and maximum deviation totaled 6.5 %.

In this manner, the experiment shows that despite the stationary external conditions, the flow does not reach steady state.

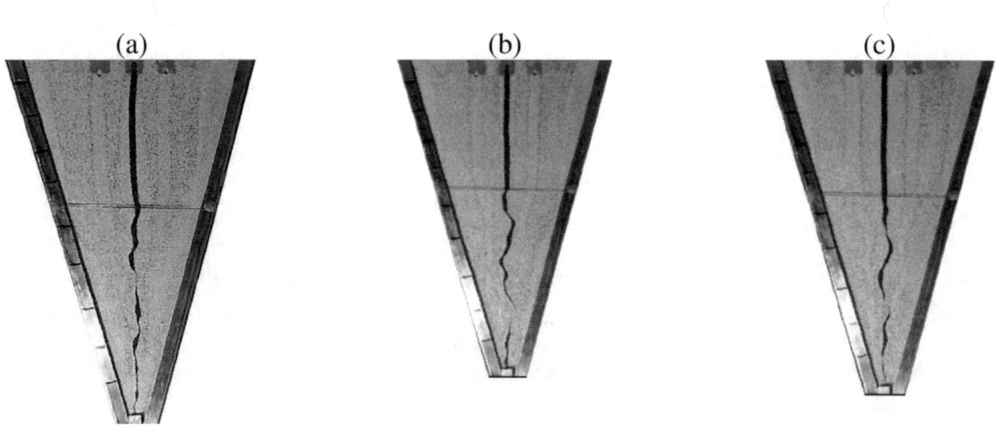

Figure 2. Marker belt deformation at different times of flow.

For the more detailed studies of kinematics, the flow process was continuously filmed on a digital camera at a speed of 25 frames per second. Each 12th frame was selected from the video file. The resultant series of images was processed in MatLab by the PIV method. Finally, fields of velocities (displacements) were obtained for the process of discharge at a step of 0.5 s; later on, they were additionally treated to calculate rates of shearing velocities and density variations.

Figure 3 demonstrates contour lines of shearing rates during gravity flow. The figures are the numbers of frames. It is seen that the maximum shear rates gradually localize in certain areas of the converging channel. Slip lines split the material into blocks. Splitting starts nearby the discharge hole and propagates upward as the material is discharged. Slip line progressively reach the top of the channel, which is clearly visible on frame 0181. Later on, the lines move downward and are slowly "healed". During this process, orthogonal families of slip lines are formed, and shearing "switches" from one family to the other. Successive switching of orthogonal families of localized slip lines proceeds during the whole experiment. For example, frames 0613–0733 show the sequence of slip lines, starting 25 s after discharge onset. Apparently, switch of sliding from one family to the other takes place every 4 s.

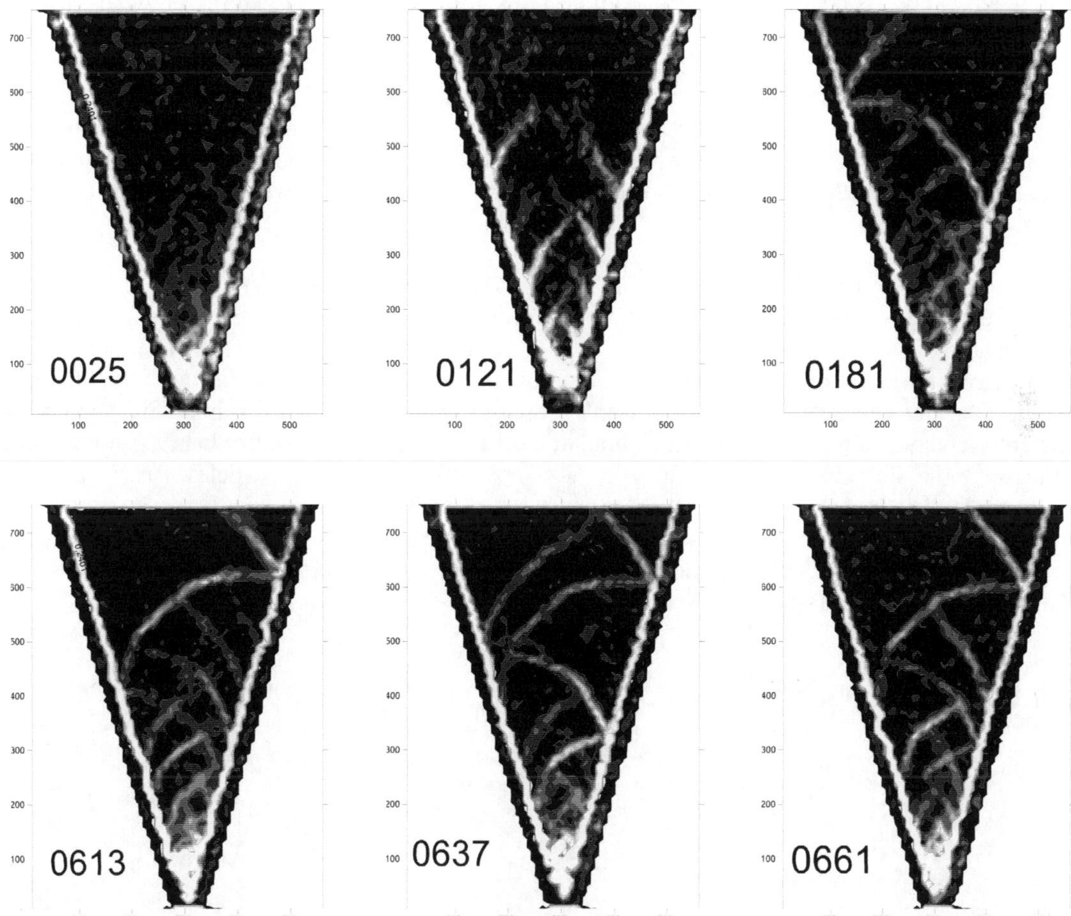

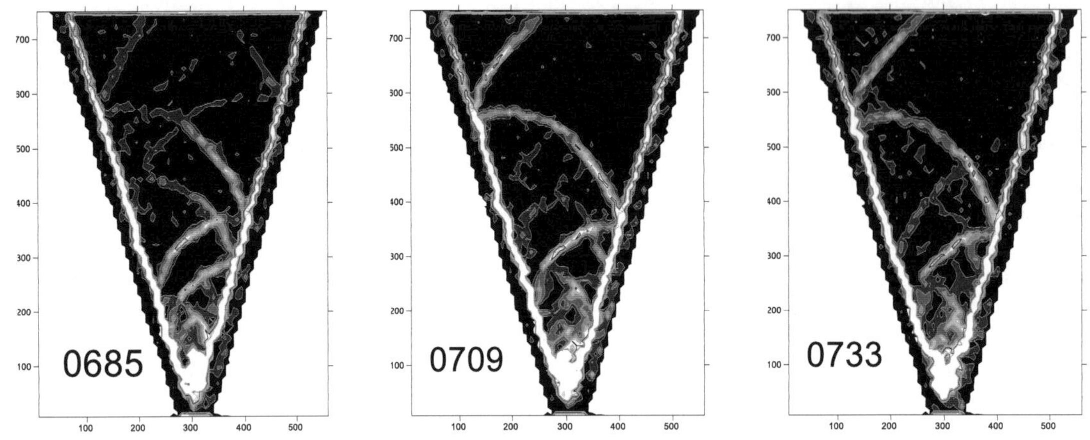

Figure 3. Rates of shearing in the flow of quartz sand in the converging channel.

Now, let us discuss contour lines of maximum rates of density variations in the granular material (volumetric strains) in figure 4. In this figure, the darker shades show higher rates of compaction. It is seen that the flowing material is divided into areas of different density. Along the localized deformation lines, the material experiences higher rate compaction. Furthermore, regardless the existence of two slip line families and their successive switching, the highest rates of compaction concentrate along one family, except for the onset of the flow process (frame 0121). This fact points at a certain asymmetry in the directions of shearing localization, the cause of which will be the object of the further research.

3. Flow of granular materials in converging channels (numerical modeling)

In the numerical investigations of kinematics of deformation in granular media, different variants of flow simulation models are very popular [6, 7]. In such models, a cell is set, and a material vanishes from it (leaves the study domain). Then, certain probabilities of the vacant cell filling by the material from the neighbor cells are defined. As a result, based on very simple rules, it is possible to obtain sufficiently complex flow patterns which simulate real processes. On the other hand, this approach fails to solve the problem of stresses. In order to calculate forces between particles, it is required to take closed models including stresses. It is the most convenient to carry out simulation using discrete element models [8–10]. Being the basic alternative to classical methods founded on the traditional continuum mechanics, the discrete element method currently finds wide application in studies of different modes of granular material flow.

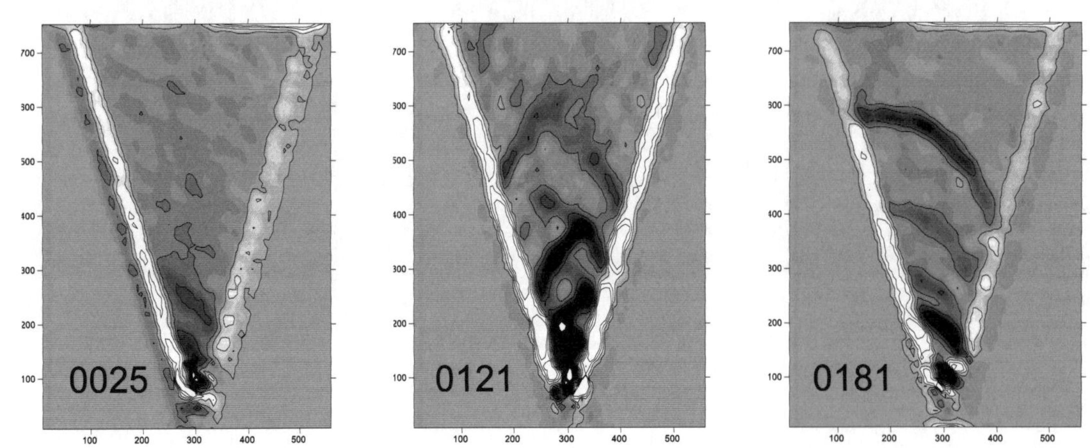

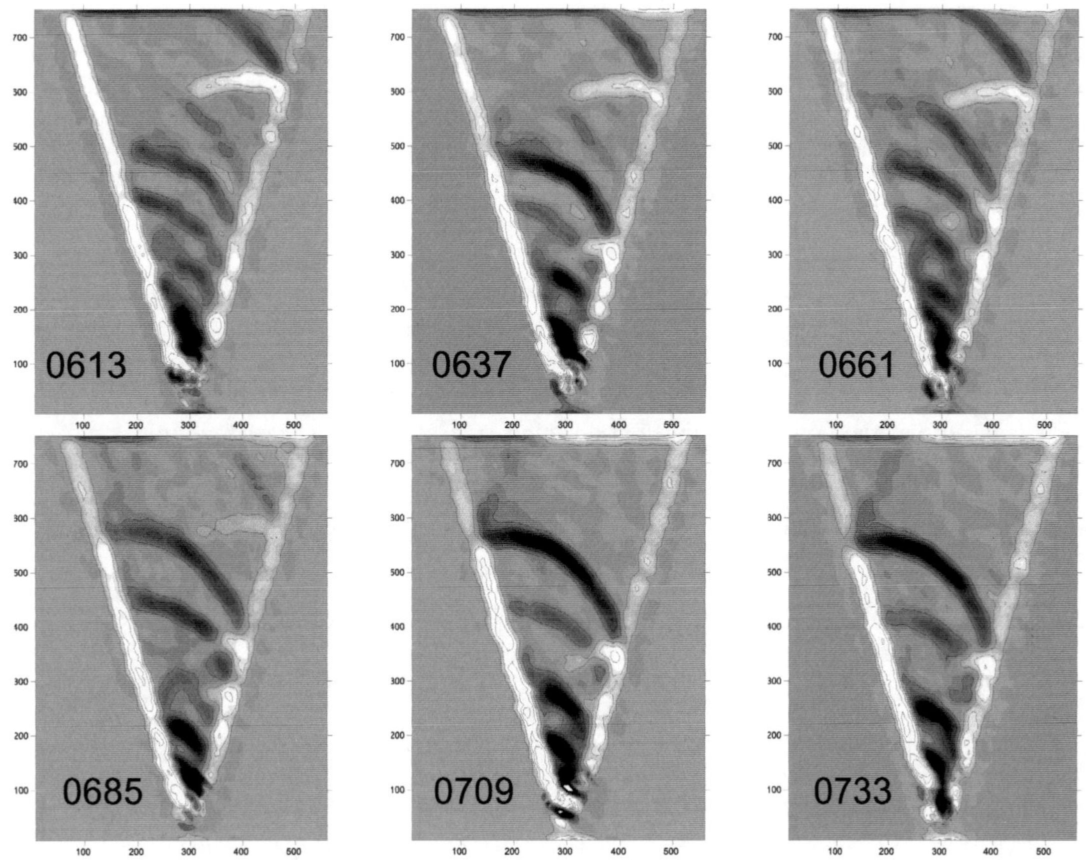

Figure 4. Contour lines of density variation rates in the flow of quartz sand in the converging channel.

Figure 5 demonstrates the result of 3D DEM-based calculation in the form of successive kinematic patterns of flow deformation at fixed times.

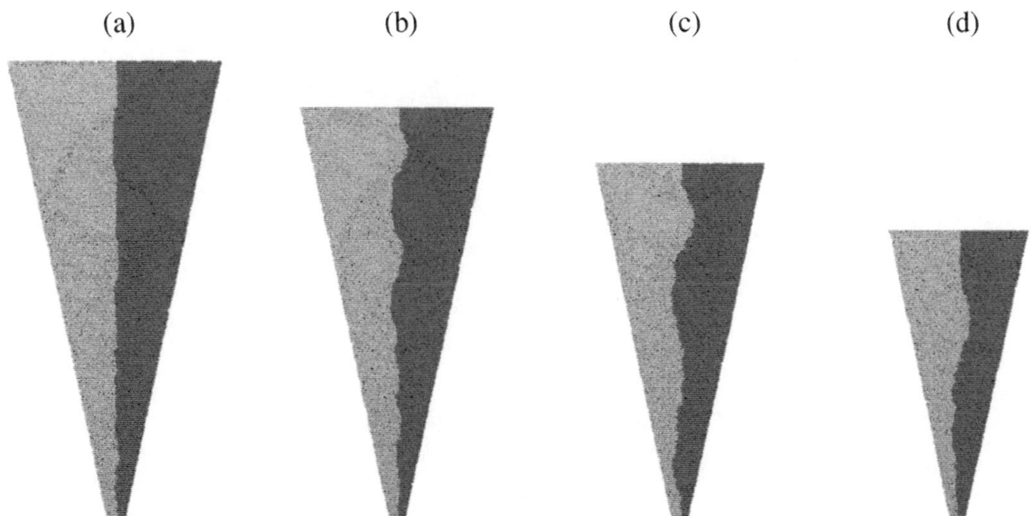

Figure 5. Successive kinematic patterns of granular material deformation calculated by DEM

The calculations show that deformations localize along separate slip lines which divide the flowing material into blocks. Then, deformation takes place as relative slip of these blocks (on each other and on the channel walls) as rigid integer bodies. After a certain time, these slip lines are "healed", and a new system of slip lines appears orthogonally to the previous one. On the whole, the experimental kinematics of flow and the DEM-calculated patterns correlate.

The numerically obtained stresses demonstrate quasi-periodic behavior also correlated with the sliding switches between the orthogonal families of slip lines.

Thus, the numerical modeling data agree qualitatively well with the experimental results.

4. Conclusions

1. Under stationary external action, gravity flow of granular materials in converging channels is steady-state. Shearing is localized along two orthogonal families of slip lines which function successively, one after the other.
2. The granular material flow is split into zones having different densities, and maximum compaction rates concentrate along one of the families of slip lines.
3. The data of the numerical modeling using the discrete element method agree qualitatively well with the laboratory-scale experiment on localized mode of granular material flow in converging channels.

Acknowledgements
This study has been carried out in the framework of the Fundamental Research Projects, Registration Nos. AAAA-A17-117121140065-7 and AAAA-A17-1171220900002-5.

References

[1] Hall S A, Muir Wood D, Ibraim E and Viggiani G 2010 *Granular Matter* vol 12 issue 1 1–14
[2] Słomiński C, Niedostatkiewicz M and Tejchman J 2006 *Archives of Hydro-Engineering and Environmental Mechanics* vol 53 **1** 71–94
[3] Bobryakov A P and Revuzhenko A F 1982 *Journal of Mining Science* vol 18 issue 5 373–379
[4] Bobryakov A P 2008 *Journal of Mining Science* vol 44 issue 2 115–122
[5] Revuzhenko A F, Stazhevskii S B and Shemyakin E I 1979 *Doklady Akademii Nauk SSSR* vol 246 **3**
[6] Osinov V A 1993 *Journal of Mining Science* vol 28 **5** 433–440
[7] Lavrikov S V and Revuzhenko A F 2000 *Journal of Mining Science* vol 36 **1** 8–16
[8] Lavrikov S V and Revuzhenko A F 2016 *Journal of Mining Science* vol 52 **4** 632–637
[9] Klishin S V and Revuzhenko A F 2017 *IOP Conf. Series: Earth and Environmental Science* **84** 012003
[10] Klishin S V and Revuzhenko A F 2015 *Journal of Mining Science* vol 51 **5** 951–956

IOP Conf. Series: Earth and Environmental Science **206** (2018) 012005 doi:10.1088/1755-1315/206/1/012005

Modelling the trajectory of a fracture that moves under the influence of the fluid pressure in hard rock roofs of in-seam working

N V Cherdantsev

Federal Research Centre of Coal and Coal Chemistry at the Siberian Branch of the Russian Academy of Sciences, 10 Leningradsky ave, Kemerovo city, 650065, Russian

E-mail: cherdancevn@bk.ru

Abstract. Presents the results of the solution to the problem of creation of a trajectory of movement in the roof rocks of the reservoir production artificially created cracks under the action of internal fluid pressure. The array is loaded equivalent gravitational field stresses and he is in condition of plane deformation. The emergence of cracks and admission thereto of fluid are not considered. Breeds of soil and roof of layer strong enough to deform elastically and upon reaching the limit state of collapse as the fragile materials. The problem is solved in the framework of model of geomechanical state of rock massif containing local zones in the ultimate state. The model is based on the main provisions of the solid mechanics, it is implemented by means of boundary element method and classical views of the state of the crack, its stable and unstable growth in an infinite plane of a brittle material based on the theories of Griffiths – Irwin.

1. Introduction

The issue of manifestation of different gas-geodynamic phenomena in deep depth while developing coal deposits appears to be of vital importance. Firstly, it is connected with the appearance of limit stress zones of a seam formed in selvages and in-seam workings adjacent to sidewalls and faces. The formation of such zones brings about significant incoming of methane into the workings and very often it is accompanied by sudden outbursts of fragmentary mined-rocks into workings [1–3]. Due to this, an important significant scientific and production task is in accurate predicting estimations of geo-mechanic state of an in-seam coal massif [4–6].

Depending on strength characteristics of a coal seam and adjoining country rocks their transition into limit state while being developed happens in different ways. In one case, limit stress of a roof country rock takes place in a small-sized goaf, in other cases it covers significant areas and, as a rule, causes the rock falls accompanied by dynamic phenomena [7–9].

To reduce the size of a poor caving of the roof the most effective way is to apply a method of directional hydraulic fracturing [9]. The idea of the method is in induced roof caving by means of artificial initiation of a fracture and its further development in rocks under the influence of the fluid injected into it under high pressure.

One of the first fundamental works that scientifically proved the effectiveness of hydraulic fracturing method is theoretical paper written by Khristianovitch S.A. and Zheltov Yu.P. On the bases of methods of deformable body mechanics the authors solved the issue of hydraulic fracture propagation in oil-bearing seams for the purpose of increasing the oil-well rate [10].

The basic goal for effective applying of the hydraulic fracturing method in colliery and ore undertakings is in defining optimal parameters for the initial (primordial) fracture with the purpose to obtain maximum effect for caving the roof by a growing fracture. The essential factor that influences on the trajectory of the fracture is in strata stress. Under real condition the fracture propagates near the working which changes it and this fact should be taken into the account.

By now the influence of a coal working on the trajectory of the fracture development has not been taken into account for solving similar issues [9].

2. Problem description and its solution

The results of solving the issue of fracture propagation under intrinsic pressure in a roof of in-seam working are introduced in the paper. In the process of solving the issue, the stress field in a coal massif is defined in the framework of a developed model for geo-mechanic state of a massif with in-seam working [4, 5].

The task is set as follows (figure 1). In coal rock massif under the condition of plane deformation and modeled weightless plane there is a working *1* of a rectangular cross-section with the size $b \times h$, driven at the depth H along a coal seam *2* with all its thickness. The strength characteristic of a seam is less than the one of a country rocks massif but more than the one where the seam contacts with an enclosing massif loaded by gravitational pressure on the top and on the bottom γH (γ – unit specific gravity of overlying strata), and at the walls $\lambda\gamma H$ (λ – coefficient of a lateral pressure). In a seam selvage, zones of inelastic deformation *3* with the width L are formed. In a working roof there is a narrow cut *4* that imitates the fracture with the length $2l$, loaded by pressure p on the section with the length of $2l_0$ (index 0 refers to the initial length of the crack).

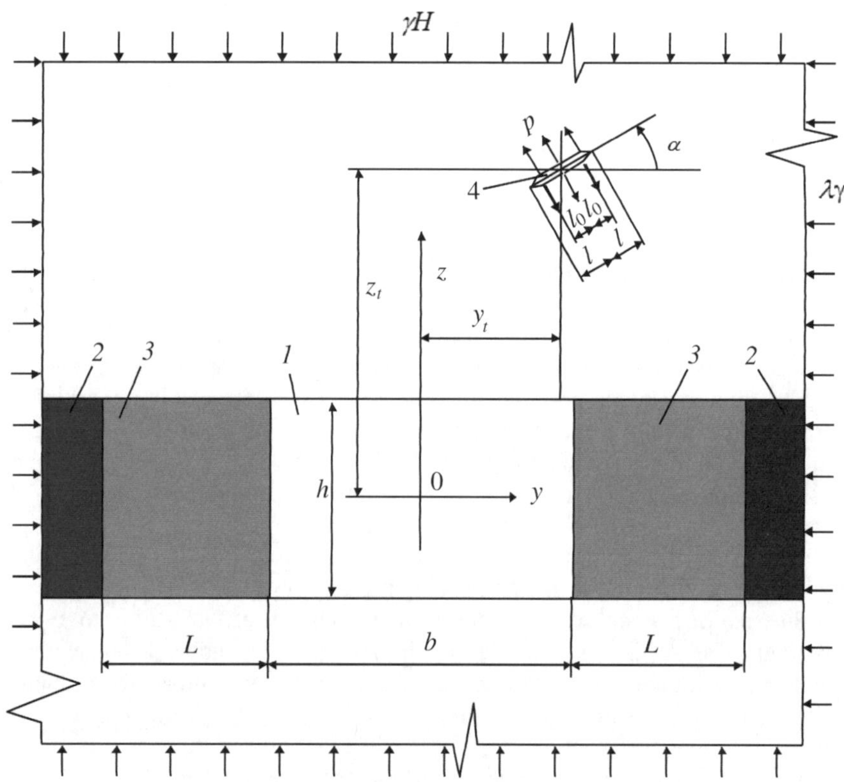

Figure 1. Analytical model of the problem about the fracture state in a working roof.

IOP Conf. Series: Earth and Environmental Science **206** (2018) 012005 doi:10.1088/1755-1315/206/1/012005

The cut is small compared to the output dimensions, and the coordinates of its middle in the y 0 z coordinate system, coinciding with the Central axes of the output, are indicated by y_t, z_t (index t denotes the cut). Cut inclined to the horizon at an angle α.

It is well-known that universal equation of thermodynamics is applied for theoretical analysis of the strength and large displacement discontinuity in solid deformable body problem. This equation can be applied to the body with a fracture, to its adjoined states: before and after the fracture propagation on a certain value Δl. In quasi-static process this equation looks like [11 – 13]:

$$k_I^2 + k_{II}^2 = \frac{E \cdot P}{1-\mu^2}, \tag{1}$$

where k_I – stress intensity factor due to the action of the first type of load p_I (normal wedging) acting on the banks of the crack, k_{II} – stress intensity factor from the action of the second type of shear load p_{II} acting on the banks of the crack, E – elastic modulus (Young's elasticity modulus), μ – Poison's ratio of country rock, P – density of surface energy that is necessary for the formation of the surface unit. Indices I, II – the type of load acting on the crack.

Stress intensity coefficients for the fractures are expressed by the following relationships [11 – 13]

$$k_I = \frac{1}{\sqrt{\pi l_0}} \int_{-l_0}^{l_0} p_I \sqrt{\frac{l_0+\xi}{l_0-\xi}} d\xi, \qquad k_{II} = \frac{1}{\sqrt{\pi l_0}} \int_{-l_0}^{l_0} p_{II} \sqrt{\frac{l_0+\xi}{l_0-\xi}} d\xi, \tag{2}$$

where ξ - the x coordinate of the point of the crack starting from its middle.

If the fracture is situated in a uniform stress field then p_I, p_{II} do not depend on variable ξ and coefficients k_I and k_{II} after being integrated according to Eq.(2) are determined as follows:

$$k_I = p_I \sqrt{\pi l_0}, \qquad k_{II} = p_{II} \sqrt{\pi l_0}, \tag{3}$$

in connection with hydrofracture angled at α towards the horizon in tight rocks the equations (3) are as follows:

$$k_I = (p - p_I)\sqrt{\pi l_0}, \qquad k_{II} = p_{II} \sqrt{\pi l_0}. \tag{4}$$

In equations (4) p_I, p_{II} are determined by the stresses at the "distant" edges and in this case, they are connected with stresses γH и $\lambda\gamma H$ as

$$p_I = \frac{\gamma H}{2}\left[1+\lambda+(1-\lambda)\cdot\cos 2\alpha\right], \ p_{II} = \frac{\gamma H}{2}\left[(1-\lambda)\cdot\sin 2\alpha\right]. \tag{5}$$

Substituting equations (4) into equation (1) allows us to determine the fluid pressure, called critical pressure, and it corresponds to the moment of fracture initiation (starting of the growth). We call it the first critical pressure.

$$p_1 = p_I + \sqrt{\frac{E \cdot P}{\pi l_0 (1-\mu^2)} - p_{II}^2}, \tag{6}$$

where index 1 denotes the number of the first critical pressure.

To determine the fluid pressure under which the fracture grows on a certain (given) value, it is important to take into account the fact that the fracture increases its length with the velocity closely equal to sound propagation in a given medium while its filling with fluid during the growth is significantly slow. Thus, while the fluid consumes a part of the fracture with the length $2l_0$, the length of the fracture itself is equal to $2l=2(\Delta l+l_0)$.

In this connection, to calculate the coefficients of stress intensity in Eq. (2) it is necessary to consider the above-mentioned condition and to set the integration limits correctly

$$k_I = \frac{1}{\sqrt{\pi l}}\int_{-l_0}^{l_0} p_t\sqrt{\frac{l+\xi}{l-\xi}}d\xi - \frac{1}{\sqrt{\pi l}}\int_{-l}^{l} p_I\sqrt{\frac{l+\xi}{l-\xi}}d\xi,\ k_{II} = \frac{1}{\sqrt{\pi l}}\int_{-l}^{l} p_{II}\sqrt{\frac{l+\xi}{l-\xi}}d\xi. \tag{7}$$

After integrating the relationship k_I, k_{II} appear to be the following:

$$k_I = \frac{2p\sqrt{l}}{\sqrt{\pi}}\operatorname{arctg}\frac{l_0}{\sqrt{l^2-l_0^2}} - p_I\sqrt{\pi l}, \qquad k_{II} = p_{II}\sqrt{\pi l}.$$

After substituting k_I and k_{II} into the equation (1) the value of the intrinsic pressure in a fracture corresponding to its propagation on a certain value Δl takes the following form:

$$p_2 = \frac{\pi}{2\operatorname{arctg}\frac{l_0}{\sqrt{l^2-l_0^2}}}\left(p_I + \sqrt{\frac{EP}{\pi l\left(1-\mu^2\right)} - p_{II}^2}\right), \tag{8}$$

The value p_2, determined by the relationship (8) is called the second critical value and it corresponds to the sustainable growth of the fracture. Index 2 denotes the number of the second critical pressure.

In Griffith's theory which is referred to purely elastic materials, the characteristics P has the same value in all directions for all types of fracture deformation and its deviation from the initial direction is expected if $k_{II}\neq0$. Thus, the direction of the fracture propagation takes place under the angle $\theta=\theta_c$ with respect to its initial position where the left part of the equation (1) has a maximal value out of all possible values under the given external load.

It is obvious to assume that the direction coincides with the area of maximal stress. It is known that there is no shear stress in this area and the normal stress is a principal stress and the area is called the main area.

According to this supposition the direction of the fracture development is angle θ_c with the initial (preceding) direction of the fracture. This angle is determined in the process of solving the following trigonometric equation with respect to the angle θ measured from the direction of the fracture at the moment of its initiation [13]

$$k_I\left(\sin\frac{\theta}{2} + \sin\frac{3\theta}{2}\right) + k_{II}\left(\cos\frac{\theta}{2} + 3\cos\frac{3\theta}{2}\right) = 0. \tag{9}$$

To solve the equation (9) it is necessary to know values k_I and k_{II}.

IOP Conf. Series: Earth and Environmental Science **206** (2018) 012005 doi:10.1088/1755-1315/206/1/012005

There are several methods for determining the coefficients of stress intensity at the ends of the fracture which is positioned in the area with "distant" edges and nonuniform stress field [13].

If the fracture is small comparing to the area dimensions and is positioned inside it then the equation (2) can be taken as one of the options for calculating the coefficient of stress intensity. Function p_I in it is equal to the stress which could be perpendicular to the fracture plane if it were shut; therefore the stress for the body without fracture is calculated analytically or numerically. Type *II* can be studied similarly.

In this case the stresses at the fracture edges together with its initial position can be determined in the framework of geo-mechanic model of the massif with in-seam working [5]. After that, normal and shear stresses acting at the edges of the fracture are determined according to the formula of stress component transformation under the rotation of coordinate system in the following way:

$$p_I = \frac{\sigma_z + \sigma_y}{2} + \left(\frac{\sigma_z - \sigma_y}{2}\right)\cos 2\theta + \tau_{yz}\sin 2\theta, \quad p_{II} = \left(\frac{\sigma_z - \sigma_y}{2}\right)\sin 2\theta + \tau_{yz}\cos 2\theta. \quad (10)$$

In equations (10) σ_y, σ_z, τ_{yz} are the stress components determined by the solving of a boundary problem about the state of a massif with in-seam working (*x*, *y*, *z* – indexes represent coordinate axes).

Thus, solving the problem it is necessary to carry out a computational procedure that consists of derived series of cycles. The fracture trajectory is a polygonal line of strait-line portions where in the limits of each the stresses are constant.

The mechanism of the fracture development is the following. The fracture is in unstable balance when the injected fluid reaches the pressure p_1 determined by the Eq. (6). For its further propagation to Δl it is necessary to increase the pressure till p_2 value which is calculated by the equation (8) and then the edge in a jump-like mode shifts into a nearly point. A new direction of the fracture together with the horizon forms an angle $\alpha+\theta$. In a jump-like mode for the length of the fracture development the fluid disseminates along the increased volume under reduced pressure. Thus, in its new position both critical pressures of the fracture are also reduced. For the following stage of the fracture propagation onto a given value (distance) to the first critical pressure which corresponds to a new position of a fracture, additional fluid pressure is required. The process of fracture propagation continues till its full propagation with the number of jumps *n*.

3. Computational experiment and analysis of the obtained results

The results of computational experiment are introduced below. The following parameters of the massif are taken as the initial data: γ=25 kHz/m^3, *H*=800 m, λ=0.7, *b*=5 m, *h*=3 m, seam hardness coefficient *f*=1, *P*=0.0087 MPa m, l_0=0.1 m, Δl=0.1 m, final half-length l_k=3.1m. Other data were being changed during the experiment.

According to the results of solving the elastoplastic problem on a seam marginal zone state it appeared that a maximum value of a bearing pressure zone is 1.657γH and the size of a limit stress zone *L* is equal 4.17 m.

In figure 2 the trajectories of the fracture movement for some initial positions relative to the workings are built. Sections *a*, *c* with number *1* – edge of a working, *2* – limit zone edges, *3* – initial fracture, *4* – fracture trajectory are marked.

In figure 2 *a*, *b* the calculations with the following parameters of a fracture are done: y_f=1.5 m, z_f=5 m, α=30°. Figure 2 *a* demonstrates that the trajectory of the fracture is a wavy line.

Figure 2 *c*, *d*, demonstrates the obtained results y_f=0.5 m, z_f=5 m, α=80°. As it can be seen from figure 2 *c* the fracture propagates linearly coinciding with the initial (primordial) fracture direction.

The critical pressure graphs corresponding to the growing fracture are built in figure 2 *b*, *d*. Curve 1 corresponds to p_1 and curve 2 is p_2.

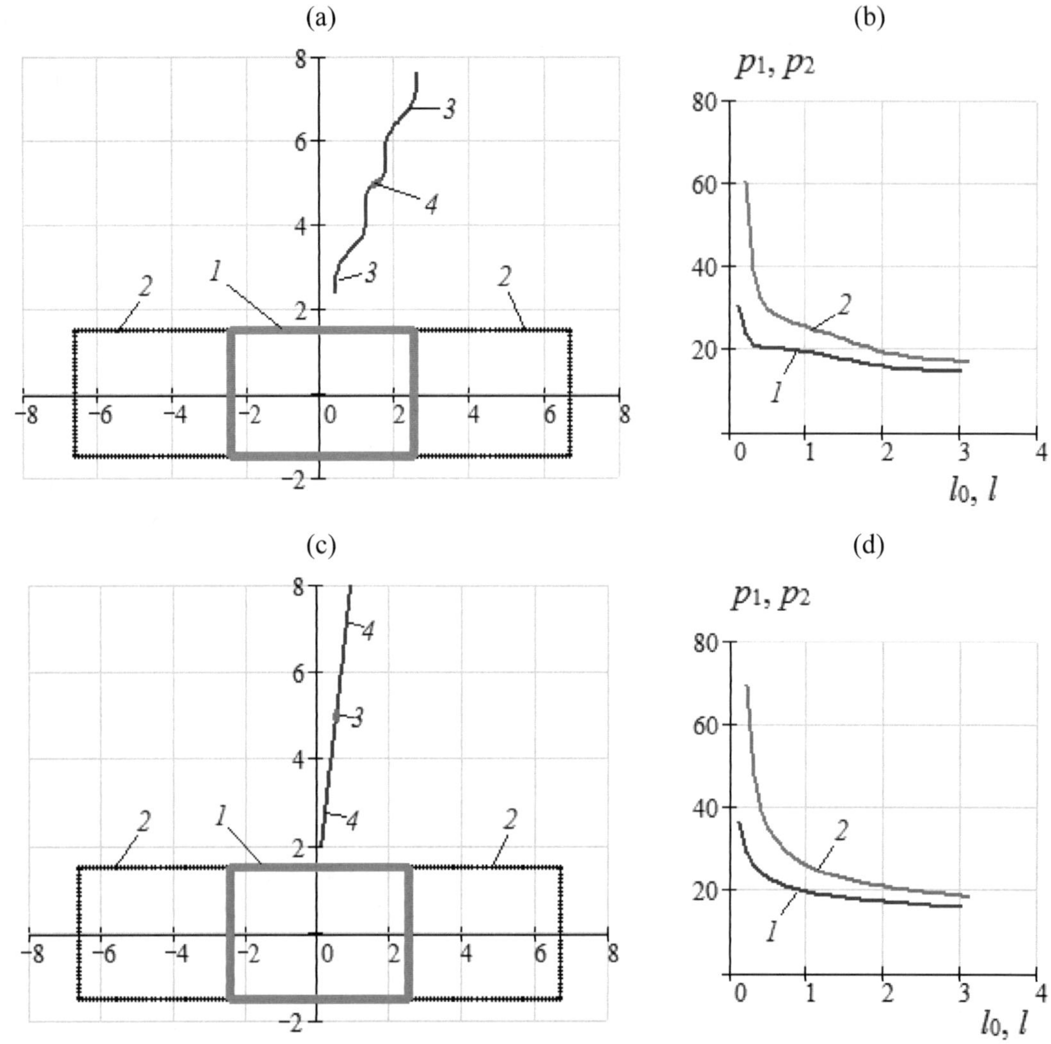

Figure 2. The trajectory of the fracture movement for its initial positions and corresponding graphs of critical pressures.

According to the obtained results of the research the orientation of the "primordial" fracture for its positions at the level z_t=5 m when the trajectory is close to the strait line are introduced.

The dependence graph of initial fracture angle of bend that corresponds to linear trajectory of its growth on initial abscissa is built in figure 3 *a*. The graph represents gently sloping curve without extreme points.

The dependence graphs of the first critical pressure (curve 1) and the second critical pressure on the angle of bend of the initial fracture are built in figure 3 *b*. The graph abscissa corresponds to graphs ordinates in figure 3 *a*. As it is seen from the figure the graphs are practically equidistant and each curve is bi-curved having their lows at $\alpha_t \approx 45°$ and the critical pressure aside of the working is higher than the one positioned directly above it.

The analysis of the results of the computational experiment and the graphs introduced in the figures show that the fracture trajectory has a wavy form if angle α_t is of arbitrary value. However, there are some values for the angle α_t where the fracture propagates along its initial direction only linearly.

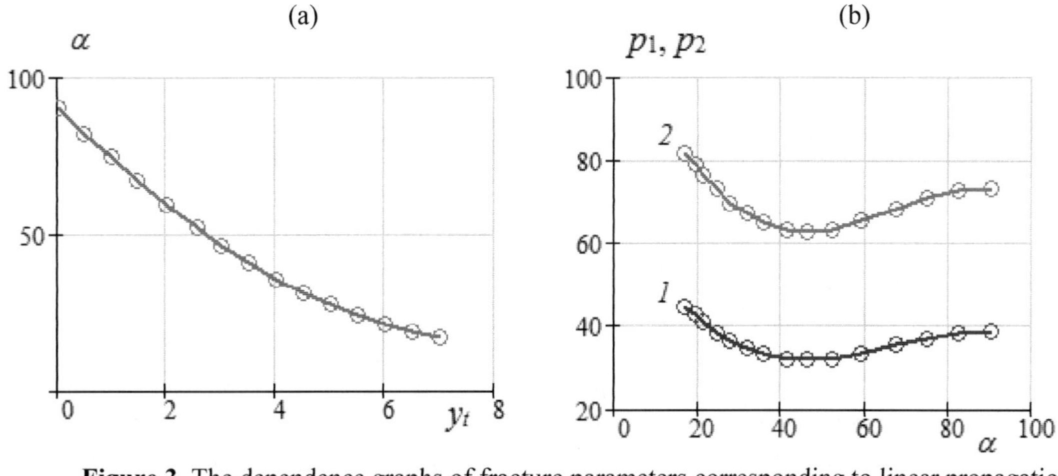

Figure 3. The dependence graphs of fracture parameters corresponding to linear propagation starting from its initial ordinate.

4. Conclusions

1. The model of geo-mechanic state of a coal massif with in-seam working provides the calculation of a fracture trajectory that grows in hard rock roof under the influence of intrinsic pressure under the given medium characteristics and initial parameters of the fracture.

2. In roof rocks of an in-seam working the fracture propagates generally in curvilinear way and its trajectory depends on the parameters of the initial fracture.

3. There are some positions of the initial fracture respective to in-seam working where it develops strictly in linear way.

4. In the working surrounding there are points and directions where the linear growth of initial fracture takes place under the minimal critical pressure.

Acknowledgments

The research was carried out due to the grant of Russian Science Foundation (project # 17-17-01143).

References

[1] Petukhov I M and Linkov A M 1983 *Mechanics of Rock Bumps and Outbursts* (Moscow: Nedra) p 280
[2] Tchernov O I and Puzyrev V N 1979 *Prediction of Sudden Gas and Coal Outbursts* (Moscow: Nedra) p 296
[3] Kozyreva Ye N and Shinkevich V V 2010 *Industrial Safety* **2** 28–35
[4] Fissenko G L 1976 *Limit State of a Rock Around Workings* (Moscow: Nedra) p 272
[5] Cherdanstev N V and Cherdantsev S V 2014 *BTP* **11** 41–45
[6] Cherdantsev N V 2017 *IOP Conference Series 1: Earth and Environmental Science* **84** doi:10.1088/1755-1315/84/1/012004.
[7] Cherdantsev N V 2017 *BTP* **5** 17–22
[8] Cherdantsev N V 2016 *BTP* **6** 46–50
[9] Klishin V I 2002 *Adapting Powered Support to the Condition of Dynamic Loading* (Novosibirsk: Nauka) p 200
[10] Khristianovich S A 1981 *Continuum Mechanics* (Moscow: Nauka) p 484
[11] Sedov L I 1984 *Continuum Mechanics* (Moscow: Nauka) p 560
[12] Kachanov L M 1974 *Basics of Fracture Mechanics* (Moscow: Nauka) p 312
[13] Hellan K 1988 *Introduction to Fracture Mechanics* (Moscow:Mir) p 364

Investigation of the temperature field in the areas of rock and coal concentration near the self-heating zone

S V Cherdantsev, P A Shlapakov, A Yu Erastov, S A Khaymin, K S Lebedev, V V Kolykhalov and E A Shlapakov

JSC "Scientific Center of VostNII on Industrial and Ecological Safety in the Mining Industry" (JSC "NC VostNII"), 3 Institutskaya street, Kemerovo, 650002, Russia

E-mail: svch01@yandex.ru

Abstract. The processes of coal self-heating, which, as a rule, have a local character, are the most acute problems for the coal industry. Experimental and theoretical studies show that self-heating is influenced by the processes of heat accumulation and temperature increase, which means that more heat is formed in the area of rock and coal concentration than released into the surrounding environment. Thus, self-heating can quickly and unexpectedly go into fire, and then into an endogenous fire. Any source of self-heating is a thermal "source", which is the cause for spreading thermal field that inevitably leads to an increase in the temperature of the rocks surrounding the faces and then the air atmosphere of the faces. A mixed boundary-value problem for a one-dimensional parabolic equation describing the temperature field in the areas of rock and coal concentration in the presence of self-heating zone was formulated in order to reveal the regularities of temperature field propagation in the vicinity of the self-heating zone. A formula is obtained that determines the temperature of the rocky-coal cluster. Calculations were performed on the basis of which the temperature dependences of the temperature in the rocky clusters were plotted against its thermophysical parameters and certain regularities of the temperature field in the cluster were revealed.

1. Introduction

It is proved that self-heating zones significantly increase the probability of inflammation and burning of dust and gas mixtures in the mine workings [1]. During the movement of the flame front, fine coal particles rise from the walls, roof and soils of workings, mix with gas and create critical combustion conditions that can transform into explosion, which is catastrophic in the conditions of the mine. The largest catastrophes with human casualties occurred in the mines of the Kuznetsk Coal Basin. In the catastrophe that occurred on March 19, 2007 110 people died at Ulyanovskaya mine (Novokuznetsk), and at Raspadskaya mine (Mezhdurechensk) on May 8-9, 2010 the disaster claimed the lives of 91 people.

In connection with the facts mentioned above, it seems to us actual to study the temperature field in coal pillars and the areas of rock and coal concentration in the presence of self-heating zones.

The change in the temperature field in the conditions of coal mines is a complex non-stationary physical process [2]. Firstly, the area containing coal is three-dimensional and, secondly, the area of rock and coal concentration and rock massif is an inhomogeneous and anisotropic medium, and therefore, the thermophysical coefficients are functions of coordinates. Thirdly, because of technological measures the rock massif is "cut" by workings of various purposes that significantly distort the temperature fields in the massif and cause difficulties in setting conditions on the boundaries of the considered areas.

2. Task setting

Based on the above it is not possible to construct an exact solution of the problem of a nonstationary temperature field. Therefore, we will discuss this problem with the following simplifying assumptions: 1) within the considered areas the rock and coal concentrations will be assumed to be homogeneous and isotropic, so that the thermophysical coefficients will be assumed constant and identical in all directions; 2) at the initial temperature of the rock and coal concentration we take the temperature T_r of the rocks in the self-heating zone, but its location, which is a heat source, will be assumed to be outside the calculation area; 3) we will consider the rock and coal concentration as a formation of "beams-strips", the process of propagation of the temperature field in which occurs only along their length, and the neighboring "beams-strips" exchange heat along their lateral surfaces.

On the basis of accepted assumptions, we consider the problem of the non-stationary temperature field in the rock and coal concentrations of face 21-1-9 in the mine "Olzherasskaya-Novaya" (figure 1). The problem is formulated as follows.

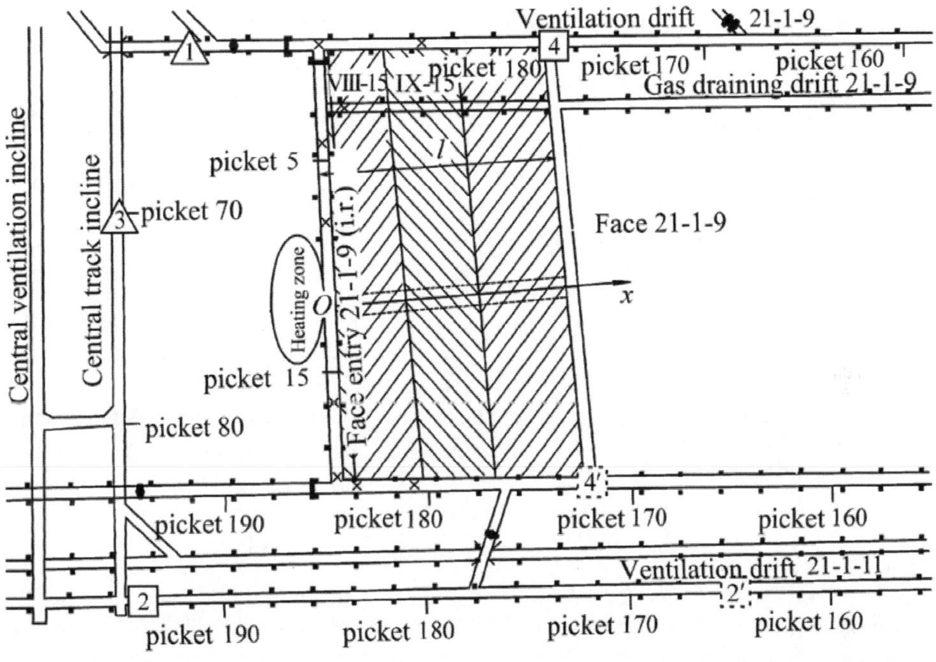

Figure 1. The plan of mining works in face 21-1-9 of mine "Olzherasskaya-Novaya".

There is a limited "beam-strip" (in figure 1 it is shown with a dotted line near axis x) with a length l and a square cross-section, the size of which b is equal to the thickness of the coal seam. A heat exchange takes place according to Newton's law between the lateral surface of the "beam-strip" and the surrounding medium. The temperature of the medium surrounding the side surface of the "beam-strip" is assumed to be constant and equal to its initial temperature T_s, and heat exchange with the surrounding medium occurs on the left and right ends of the "beam-strip". Moreover, its left end contacts with the self-heating zone, the temperature in which T_r, and the right one – with the face (figure 2), the wall temperature of which is T_w. It is required to find the temperature distribution along the length of the "beam-strip" at any point time.

Since the size b of the cross-section of the "beam-strip" is small in comparison with its length l, we can assume that the temperature difference in the cross-section of the "beam-strip" is insignificant [3, 4] and it can be neglected

$$\frac{\partial T}{\partial y} = \frac{\partial T}{\partial z} = 0,$$

thus, the problem under consideration degenerates into a one-dimensional problem.

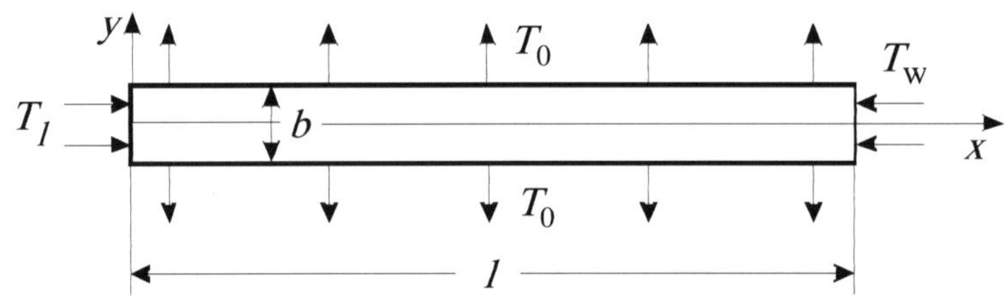

Figure 2. Heat transfer scheme of "beam-strip" with the environment.

Heat transfer from the lateral surface of the "beam-strip" to the environment is assumed to be a negative source of heat q, the value of which, in accordance with Newton's law, is defined as follows [3, 4]

$$q = -\frac{\alpha_1}{Sl}\left[T(x,t) - T_0\right]Pl = -\frac{\alpha_1}{h}\left[T(x,t) - T_0\right], \tag{1}$$

where α_1 is the coefficient of heat exchange between the side surface of the "beam-strip" and the surrounding rocky clusters, $h = S/P$ is the ratio of the cross-sectional area of the "beam-strip" to the perimeter of the cross section, which for the square cross section is $h = 1/4b$.

Thus, the heat equation in the one-dimensional problem under consideration is represented in the form

$$c_p\rho\frac{\partial T(x,t)}{\partial t} = \lambda\frac{\partial^2 T(x,t)}{\partial x^2} - \frac{\alpha_1}{h}\left[T(x,t) - T_0\right]. \tag{2}$$

where c_p is the specific heat capacity of the rock coal concentration, determined at a constant pressure p; ρ, λ are the density and thermal conductivity coefficient of the rock and coal concentration, respectively.

Let us divide equation (2) by $c_p\rho$ and, first, make in it a replacement

$$W = T(x,t) - T_0, \quad \nu = \frac{\alpha_1}{c_p\rho h}, \tag{3}$$

which allows us to rewrite equation (2) as follows

$$\frac{\partial W}{\partial t} = a\frac{\partial^2 W}{\partial x^2} - \nu W, \tag{4}$$

and the second replacement

$$W = e^{-\nu t} u \,, \tag{5}$$

leads the inhomogeneous equation (4) to the classical heat equation

$$\frac{\partial u}{\partial t} = a \frac{\partial^2 u}{\partial x^2} \,, \tag{6}$$

belonging to the parabolic type [5], in which the quantity $a = \lambda/(c_p\rho)$ is the coefficient of thermal diffusivity.

Adding to the equation (6) the initial condition

$$u\big|_{t=0} = \psi(x) \,. \tag{7}$$

and two boundary conditions

$$\left. \frac{\partial u}{\partial x} - \frac{\alpha_2}{\lambda} u \right|_{x=0} = 0 \,, \quad \left. \frac{\partial u}{\partial x} + \frac{\alpha_2}{\lambda} u \right|_{x=l} = 0 \,, \tag{8}$$

we obtain the boundary value problem, where α_2 is the coefficient of heat transfer between the environment and the end cross-sections of the "beam-strip", which is different from α_1.

The boundary value problem (6) – (8) describes the process of temperature propagation in the "ball-strip" provided that it is placed in the medium with an initial temperature T_s and exchanges heat with the medium along its entire length, and at the ends of the "beam-stripes" the heat transfer of the self-heating zone with the "beam-strip" takes place at one of its end and the atmosphere of the face – at the other.

We need to note that the boundary conditions (8) contain not only the desired function u, but also its partial derivative, so that the boundary value problem (6) – (8) is a mixed-type problem.

3. Construction of the problem solution

To construct a mixed problem, we use the Fourier method and represent the particular solutions of (6) as the product of two functions [5-7]

$$u(x,t) = u_n(x) u_n(t) \,. \tag{9}$$

and after placing (9) into (6) we obtain two ordinary differential equations

$$\frac{du_n(t)}{dt} = -a\Theta_n^2 u_n(t) \,, \quad \frac{d^2 u_n(x)}{dx^2} + \Theta_n^2 u_n(x) = 0 \,,$$

integrating which, we find their particular solutions [8, 9]

$$u_n(t) = a_n \mathrm{e}^{-\Theta_n^2 at} \,, \tag{10}$$

$$u_n(x) = c_{1n} \cos(\Theta_n x) + c_{2n} \sin(\Theta_n x) \,, \tag{11}$$

where the values of the index n in formulas (9) – (11) belong to the interval $n \in 1, 2, \ldots, \infty$.

The eigenvalues Θn and the integration constants a_n, c_{1n}, c_{2n}, containing in equations (10) and (11), determine from the initial condition (7) and two boundary conditions (8). Placing the boundary conditions at the beginning into (11), we obtain a homogeneous system of equations

$$Hc_{1n} - \Theta_n c_{2n} = 0,$$
$$[H\cos(\Theta_n l) - \Theta_n \sin(\Theta_n l)]c_{1n} + [H\sin(\Theta_n l) + \Theta_n \cos(\Theta_n l)]c_{2n} = 0, \tag{12}$$

which, according to Kronecker-Kapeli theorem [10], has at least one non-zero solution if the determinant of the system is zero, i.e.

$$\begin{vmatrix} H & -\Theta_n \\ H\cos(\Theta_n l) - \Theta_n \sin(\Theta_n l) & H\sin(\Theta_n l) + \Theta_n \cos(\Theta_n l) \end{vmatrix} = 0,$$

where $H = \alpha_2/\lambda$.

Expanding the determinant and performing the transformations, we obtain the characteristic equation

$$2\mathrm{ctg}(\mu_n) = \frac{\mu_n}{p} - \frac{p}{\mu_n}, \tag{13}$$

in which

$$\mu_n = \Theta_n l, \quad p = \frac{\alpha_2}{\lambda} l. \tag{14}$$

Equation (13) is transcendental and therefore has an innumerable set of roots, the first ten of which were found by with the help of MathCAD.

After that, from the first formula (14) we find the eigenvalues

$$\Theta_n = \frac{\mu_n}{l}, \tag{15}$$

then, taking $c_{1n} = 1$, from the first equation of system (12) we find $c_{2n} = H/\Theta_n$ and, using relations (14), we obtain $c_{2n} = p/\mu_n$. Therefore, an eigenfunction corresponds to each eigenvalue

$$u_n(x) = \cos\frac{\mu_n x}{l} + \frac{p}{\mu_n}\sin\frac{\mu_n x}{l}, \tag{16}$$

which follows from (11).

Inserting formulas (16) and (10) taking into account (15) into equation (9), we find particular solutions of the problem under consideration

$$u_n(x,t) = a_n \mathrm{e}^{\frac{-\mu_n^2 a}{l^2}t}\left(\cos\frac{\mu_n x}{l} + \frac{p}{\mu_n}\sin\frac{\mu_n x}{l}\right)$$

and we build a general solution

$$u(x,t) = \sum_{n=1}^{\infty} a_n \mathrm{e}^{\frac{-\mu_n^2 a}{l^2}t}\left(\cos\frac{\mu_n x}{l} + \frac{p}{\mu_n}\sin\frac{\mu_n x}{l}\right). \tag{17}$$

Then inserting (17) into the initial condition (7), we obtain the function

$$\psi(x)=\sum_{n=1}^{\infty}a_n\left(\cos\frac{\mu_n x}{l}+\frac{p}{\mu_n}\sin\frac{\mu_n x}{l}\right),\tag{18}$$

multiplying which by the function $u_m(x)$, where $m\in 1, 2,\ldots,\infty$, and then integrating and taking into account that the functions $u_n(x)$, $u_m(x)$ are orthogonal [5 - 9], we find a_n

$$a_n=\frac{1}{B_n}\int_0^l\psi(x)u_n(x)dx,\tag{19}$$

where

$$B_n=\int_0^l u_n^2(x)dx=\int_0^l\left(\cos\frac{\mu_n x}{l}+\frac{p}{\mu_n}\sin\frac{\mu_n x}{l}\right)^2dx=$$

$$=\frac{l}{2}\frac{\left[(\mu_n^2-p^2)\cos\mu_n\sin\mu_n+p(p+2)\mu_n+(\mu_n^2-2p^2\cos^2\mu_n)\mu_n\right]}{\mu_n^3}.\tag{20}$$

We find the integral in (19) by taking function $\psi(x)$ as a constant value equal to T_s

$$I_n=\int_0^l\psi(x)u_n(x)dx=\int_0^l T_0\left(\cos\frac{\mu_n x}{l}+\frac{p}{\mu_n}\sin\frac{\mu_n x}{l}\right)dx=\frac{T_0 l}{\mu_n^2}(\mu_n\sin\mu_n-p\cos\mu_n+p).\tag{21}$$

Taking into account the trigonometric relations

$$\sin\mu_n=\frac{1}{\sqrt{1+\text{ctg}^2\mu_n}},\quad \cos\mu_n=\frac{\text{ctg}\,\mu_n}{\sqrt{1+\text{ctg}^2\mu_n}}$$

and formula (13), we find the relations

$$\sin\mu_n=\frac{2p\mu_n}{\mu_n^2+p^2},\quad \cos\mu_n=\frac{\mu_n^2-p^2}{\mu_n^2+p^2},$$

inserting which into formulas (20) and (21), we first find

$$B_n=\frac{l}{2}\frac{p(p+2)+\mu_n^2}{\mu_n^2},\quad I_n=\frac{2pl}{\mu_n^2}T_0,$$

then by the formula (19) we find

$$a_n=\frac{2T_0 p}{p(p+2)+\mu_n^2}$$

and inserting a_n into formula (17), we obtain the function

$$T(x,t)=4T_0\sum_{n=1}^{\infty}\frac{p}{p(p+2)+\mu_n^2}\left(\cos\frac{\mu_n x}{l}+\frac{p}{\mu_n}\sin\frac{\mu_n x}{l}\right)e^{\frac{-\mu_n^2 a}{l^2}t}.$$

Finally, returning using formulas (3), (5) to the required variable $T(x,t)$, we obtain the formula

$$T(x,t)=T_0\left[1+4e^{-\nu t}\sum_{n=1}^{\infty}\frac{p}{p(p+2)+\mu_n^2}\left(\cos\frac{\mu_n x}{l}+\frac{p}{\mu_n}\sin\frac{\mu_n x}{l}\right)e^{\frac{-\mu_n^2 a}{l^2}t}\right], \tag{22}$$

which is a solution of the mixed boundary-value problem under consideration.

4. Results and discussion

On the basis of the solution obtained, calculations and graphs are made (figures 3-6), in which value $\bar{T}$ represents the required temperature related to the initial temperature T_s. The calculations were performed with the following initial data of the rock and coal concentration [11]: T_0 = 287 K; ρ = 1400 kg/m^3; c_p = 0.835·10^3 J/(kg K); λ = 46.5·10^{-2} W/(m·K); α_1 = 1.5·W/(m^2·K); α_2 = 2·W/(m^2·K); l = 130 m; b = 3 m. The remaining parameters were calculated from the previously obtained formulas h = 0.75 m; p = 0.559; ν = 17.11·10^{-9} sec^{-1}; a = 0.39·10^{-6} m^2/sec. Since ten eigenvalues μ_n were found during calculation of the required temperature, the first ten eigenvalues were retained, then in formula (22) the difference in the temperature values found with the nine and ten retained members was only 0.17%.

Analyzing the graph in figure 3, we note that as the distance from the location of self-heating increases, the temperature of the rock and coal concentration decreases practically linearly.

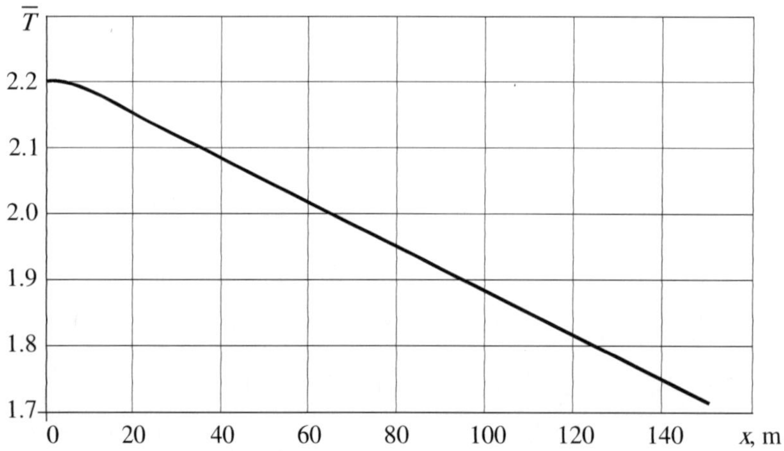

Figure 3. Dependence of the relative temperature of the rock and coal concentration on the distance to the zone of self-heating.

Figures 4 and 5 show graphs of the dependences of the relative temperature of a rock and coal concentration on its thermophysical parameters.

As it follows from the presented graphs, when the coefficient of heat transfer of rock and coal concentration increases from 1.5 W/(m^2·K) to 4 W/(m^2·K) in the direction from the location of self-heating to the face, the relative temperature of rock and coal concentration decreases by 12%, and in the perpendicular direction – only by 1%.

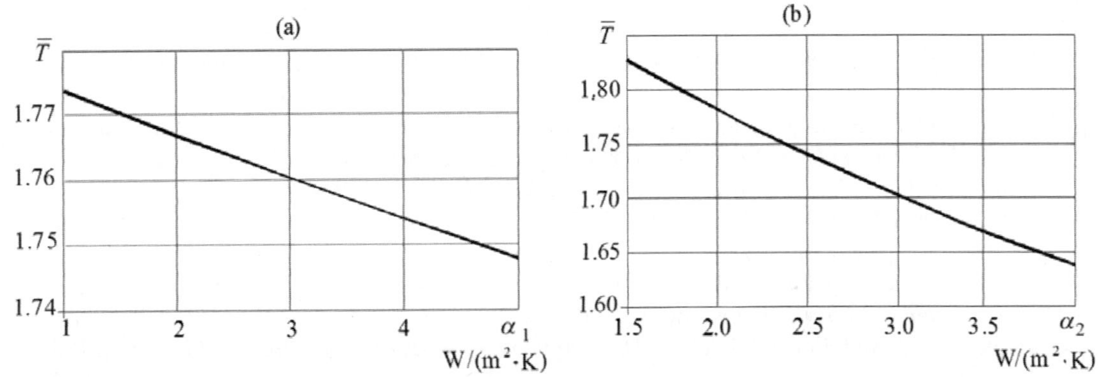

Figure 4. Dependences of the relative temperature of the rock and coal concentration on the heat-transfer coefficients α_1 (figure 4, *a*) and α_2 (figure 4, *b*).

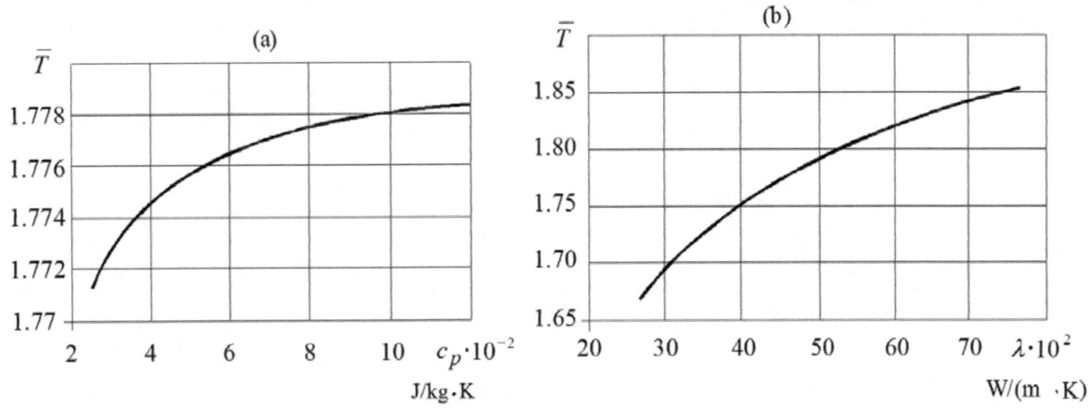

Figure 5. Dependences of the relative temperature of rock and coal concentration on the specific heat (figure 5a) and thermal conductivity coefficient (figure 5b).

With an increase in the thermal conductivity in the interval $\lambda \in [0.25; 0.75]$ Вт/(m·K) the relative temperature increases by 11%, and with the change in the specific heat capacity in the interval $c_p \in [250; 1200]$ J/(kg·K) – only by 0.4 %.

5. Conclusions

1. A mixed boundary-value problem is formulated for a one-dimensional parabolic equation describing the temperature field in the areas with rock and coal concentrations near self-heating zones.

2. Using the method of separation of variables, a solution of the mixed problem is constructed, on the basis of which a formula is obtained that makes it possible to calculate the temperature of the rock and coal concentration at any distance from the self-heating zone at any point of time.

3. During the analysis of the received formula it is established:

a) when the coefficient of heat exchange of the rock and coal concentration increases from 1.5 Вт/(m²·K) to 4 W/(m²·K) in the direction from the zone of self-heating to the face, the relative concentration temperature decreases by 12 %, and in the perpendicular direction – only by 1%;

b) with the increase in the thermal conductivity within the interval [0.25; 0.75] W/(m·K), the relative temperature increases by 11%, and with a change in the specific heat capacity in the interval [250; 1200] J/(kg·K) – only by 0.4 %.

IOP Conf. Series: Earth and Environmental Science **206** (2018) 012006 doi:10.1088/1755-1315/206/1/012006

References

[1] Amelchugov S P, Bykov V I and Tsybenova S B 2002 *Physics of Combustion and Explosion* **3** 48–54
[2] Kremnev O A and Zhuravlenko V Ya 1986 *Heat and Mass Transfer in the Mountain Massif and Underground Structures* (Kiev: Naukova Dumka) p 344
[3] Lykov A V 1978 *Heat and Mass Transfer* (Moscow: Energia) p 480
[4] Kutateladze S S 1979 *Fundamentals of the Theory of Heat Transfer* (Moscow: Atomizdat) p 416
[5] Koshlyakov N S, Gliner E B and Smirnov M M 1970 *Equations in Partial Derivatives of Mathematical Physics* (Moscow: Vyssh. Shkola) p 712
[6] Tikhonov A N and Samarskii A A 1999 *Equations of Mathematical Physics* (Moscow: MGU) p 798
[7] Vladimirov V S and Zharinov V V 2004 *The Equations of Mathematical Physics* (Moscow: Fizmatlit) p 400
[8] Stepanov V V 2016 *Course of Differential Equations* (Moscow: URSS) p 512
[9] Pontryagin L S 2001 *Ordinary Differential Equations* (Izhevsk: Regular and Chaotic Dynamics) p 400
[10] Ilyin V A and Poznyak E G 2005 *Linear Algebra* (Moscow: Fizmatlit) p 280
[11] Rzhevsky V V and Novik G Ya 1984 *Fundamentals of Rock Physics* (Moscow: Nedra) p 359

IOP Conf. Series: Earth and Environmental Science **206** (2018) 012007 doi:10.1088/1755-1315/206/1/012007

Discrete element modeling of gravity flow of broken rocks in the technology of longwall top coal caving

S V Klishin[1]*, V I Klishin[2] and G Yu Opruk[2]

[1]Chinakal Institute of Mining, Siberian Branch of the Russian Academy of Sciences, 54 Krasny ave 54, Novosibirsk, 630091, Russia
[2]Institute of Coal, Federal Research Center of Coal and Coal Chemistry, Siberian Branch of the Russian Academy of Sciences, 10 Leningradsky ave, Kemerovo, 650065, Russia

E-mail: *sv.klishin@gmail.com

Abstract: The authors perform 3D discrete element modeling of pre-broken coal and dirt discharge in underground mining with sublevel caving of thick coal seams. The process of force interaction between rocks and powered roof support is analyzed. The difference between the discharge regimes with feeders equipped with smooth and corrugated surfaces is demonstrated. The mass flow of coal during discharge is shown.

1. Introduction

New technologies of thick gently and steeply dipping coal bed mining with the in-built effect of physical destruction are based on the use of rock pressure, which makes it possible to give extra options to mechanized mining equipment. As a consequence, new prospects open for development of technologies and structures of powered roof support. The scientific research and engineering evaluation of the technology of pre-broken top coal caving with the controlled discharge in mining thick coal seams are carried out in many countries [1–6]. The current-level study of mechanical means for longwall top coal caving in thick gently and steeply dipping coal beds shows that the improvement of the related technologies and powered roof supports consists in the search and implementation of new engineering solutions on discharged coal flow regulation. These technologies are united by the common idea of controlling coal flow to face conveyor using feeders installed in the powered roof support units. For the implementation of such technologies, the optimal directions of top coal treatment to ensure the wanted coal flow mode have earlier been substantiated: vibroseism treatment and hydraulic fracturing of coal seams using wells drilled from intermediate or sublevel drifts.

Previously proposed configuration of powered roof support with adjustable coal flow to face conveyor is developed with regard to geomechanical processes in coal and roof rocks, includes benefits of the known alternatives and eliminates their shortcomings (figure 1) [7, 8].

Introduction of feeders in the structure of powered roof support for discharge and load of caved top coal is a new trend in high-productive technologies in coal mining. Thus, the studies into the basic relations and mechanisms that govern rational designs of powered roof support with the forced-and-adjustable coal discharge are of the fundamental and applied relevance.

This paper describes the numerical experiment results on simulation of gravity flow of pre-broken rock mass aimed to find regular patterns in interaction between powered roof support and granular material in the course of discharge.

Figure 1. Top coal caving technology in thick gently dipping coal seam (a); general view of powered roof support with adjustable coal flow to longwall conveyor.

The most promising method to model deformation of geomaterials is the discrete element method DEM [9]. Efficiency of this method is connected with the focusing of the modern geomechanical research on the role of the internal structure of a medium. Granular materials possess such fundamental properties as dilatancy, internal friction and cohesion, which result in the nonlinear behavior and anisotropy. Within DEM a study domain is modeled by a set of discrete elements (particles) having spherical shape, as a rule. Particles are assumed non-deformable, and the rate of their deformation is represented by a value of their overlap. Movement of particles in space is described using Newton's law of motion. Forces arising on collision of particles are calculated from models of the contact mechanics with regard to the normal and tangential components. In the framework of such approach with selected particle interaction potential, it is possible to describe movement of solid particles in fluid, heat exchange processes, adhesion, polydisperse media, etc. [10–13].

2. Discrete element modeling

In accordance with DEM, a granular material is represented by a set of N spherical particles (discrete elements). Each i-th element has a radius r_i, the known physical properties of density, elasticity and plasticity moduli, and contact properties of friction, cohesion, etc. ($i = 1, \ldots, N$). Movement of each i-th discrete element with the gravity center radius $\boldsymbol{x}_i$ consists of translational and rotation motions and is given by:

$$m_i \frac{d^2 \boldsymbol{x}_i}{dt^2} = \sum_j \boldsymbol{f}_{ij} + m_i \boldsymbol{g} \tag{1}$$

$$I_i \frac{d^2 \theta_i}{dt^2} = \sum_j (\boldsymbol{r}_{ic} \times \boldsymbol{f}_{ij} + \boldsymbol{M}_{r,ij}) \tag{2}$$

where $\boldsymbol{\theta}_i$ is the rotation of an i-th particle relative to coordinate axes; m_i is the mass; I_i is the moment of inertia; $\boldsymbol{g}$ is the acceleration of gravity; $\boldsymbol{r}_{ic}$ is the vector from the particle center to a contact point. The contact force $\boldsymbol{f}_{ij}$ affects the i-th particle from the side (or boundary) of a j-th particle, and depends on the overlap of the particles, as well as on the elasticity and plasticity moduli; $\boldsymbol{M}_{r,ij}$ is the moment of rolling resistance of particles. Summing in (1) and (2) is carried out over all elements and boundaries which contact the i-th particle. As the shape of discrete elements is assumed unaltered over the time of contact, the rate of deformation is described by the value of overlap of the contacting particles.

The force interaction at the contact point between the spherical particles i and j is presented in the vector form based on the Kelvin–Voigt viscoelastic model with regard to the normal and tangential components (figure 2):

$$\boldsymbol{f}_{ij} = f_{n,ij}\boldsymbol{n}_{ij} + f_{t,ij}\boldsymbol{t}_{ij} \tag{3}$$

Here, $f_{n,ij}$ and $f_{t,ij}$ are the viscoelastic normal and tangential components of the contact force, respectively; $\boldsymbol{n}_{ij}$ is the unit vector governing the plane of contact between two spheres (vector of normal to the intersection plant of spheres, directed along the line connecting the centers of the spheres); $\boldsymbol{t}_{ij}$ is the unit vector belonging to the contact plane.

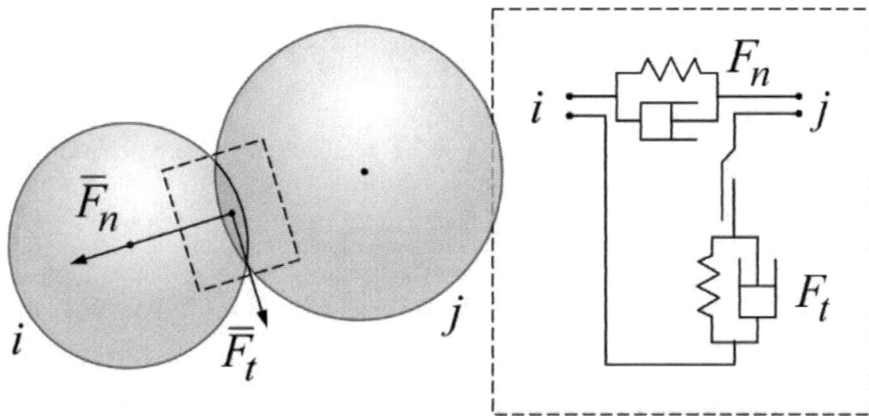

Figure 2. Schematic of the contact interaction of spherical particles based on the Kelvin–Voigt rheological model.

It is chosen to determine the normal and tangential forces in (3) in terms of particle interaction based on the Hertz law and from the Mindlin Deresevich scheme, respectively [14–16]. As the normal and tangential components of the contact force have been determined at each contact, we go to the next step of loading, and the calculations are repeated for the existing and new contacts.

Making an original equilibrium packing of solid spherical particles is an important initial step in discrete element modeling and is studied as a separate problem by many authors [17–19]. Currently, a number of approaches to generation of an original dense packing are available and differ by the way of achieving the final configuration of the domain composed of separate particles. In this paper, we use a dynamic scheme of simulation of particle motion under the action of a preset external force (gravity). Here, a set of initially noncontacting particles of preset size move under gravity in an assigned bounded domain, and the final equilibrium state is determined with regard to the particle–particle and particle–boundary contact interactions.

3. Numerical experiment

The 3D modeling of powered roof support unit (figure 3) is carried out using the widely applicable STL File Format (StereoLithography). In this format, information about an object is stored as a list of triangular faces which describe the object surface and the normal to the faces. Such representation only shows the geometry of the object surface without its color, texture or other model attributes included in the automated design systems.

The numerical calculations disregard the force interaction between the elements of powered roof support (when they contact or intersect). The forces between particles and powered roof support are calculated from formula (3).

IOP Conf. Series: Earth and Environmental Science **206** (2018) 012007 doi:10.1088/1755-1315/206/1/012007

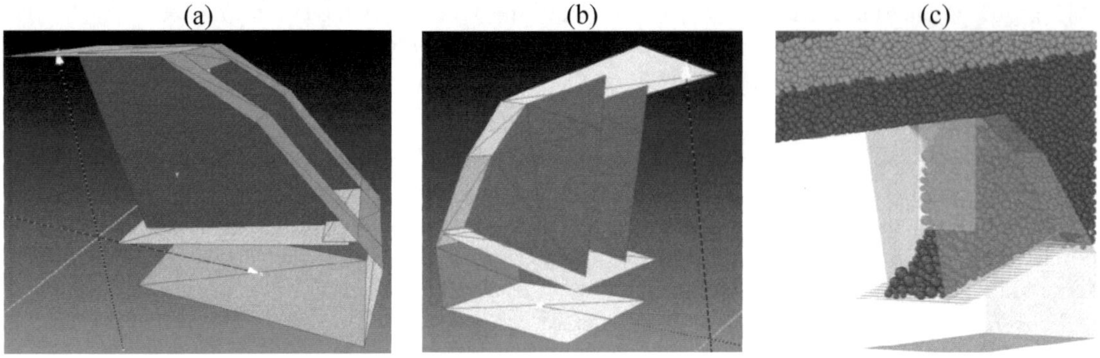

Figure 3. 3D model of powered roof support with discharge (a), (b); equilibrium state of material when the feeder is switched off (c).

Let us discuss 3D problem formulation. Let in the space *Oxyz* a domain (container for granular material) be set as a parallelepiped bounded by planes oriented along the coordinate axes; the upper boundary is free of stresses (figure 4). The container has a length of 10 m and a width of 2 m; the material fill height is 12.5 m. The vector of gravity $\boldsymbol{g} = (0; 0; -9.81)$ is directed along the axis *Oz* vertically downward. The front and back walls are perfectly smooth, i.e., the condition close to plane deformation state in the *Oy*-direction is fulfilled. The physical parameters of particles conform with the properties of coal (table 1). The coal seam thickness is assumed as 3 m, the height of the powered roof support unit is 3.5 m.

Table 1. Physical and geometrical parameters of particles.

	Coal	Dirt
Elasticity modulus, MPa	$5.4 \cdot 10^3$	$1.5 \cdot 10^4$
Poisson's ratio	0.16	0.21
Density, kg/m^3	1350	2500
Diameter, m	0.08–0.16	0.08–0.16

Figure 4 shows such elements of powered roof support as canopy 1 with a length of 2.5 m, vertical flow shield 2 with a height of 1.75 m to limit velocity and volume of granular material flow; feeder 2 with a length of 3.575 m arranged at an angle of 12° relative to a horizontal plane; discharge 4 with a width of 1 m (initially closed); side guards 5 with a spacing of 1 m.

After creation of equilibrium packing, granular material was represented by a set of 127 000 discrete elements composed of 58 t of coal particles and 195 t of dirt particles. Then, the sliding friction angle $\varphi_{ij} = 30°$ and the rolling resistance angle $\psi_{ij} = 30°$ between particles were fixed. For the visualization of the discharge kinematics, discrete elements were colored. In the course of discharge, particles with the gravity center coordinate meeting condition that $z_i < 0$ were removed. Thus, bulk flow of coal and dirt was calculated.

The scope of the numerical experiment embraced simulation of broken rock flow by gravity. In this case, the shield limiting flow rate and volume was withdrawn and the feeder was inoperative. Figure 5 shows the deformation patterns of coal and dirt flow by gravity at certain moments of time. It is seen that without the limitation of the flow velocity and volume from the side of the powered roof support, the particles of dirt, which have higher density than the particles of coal, enter the flow and cause dilution and high loss of the valuable component.

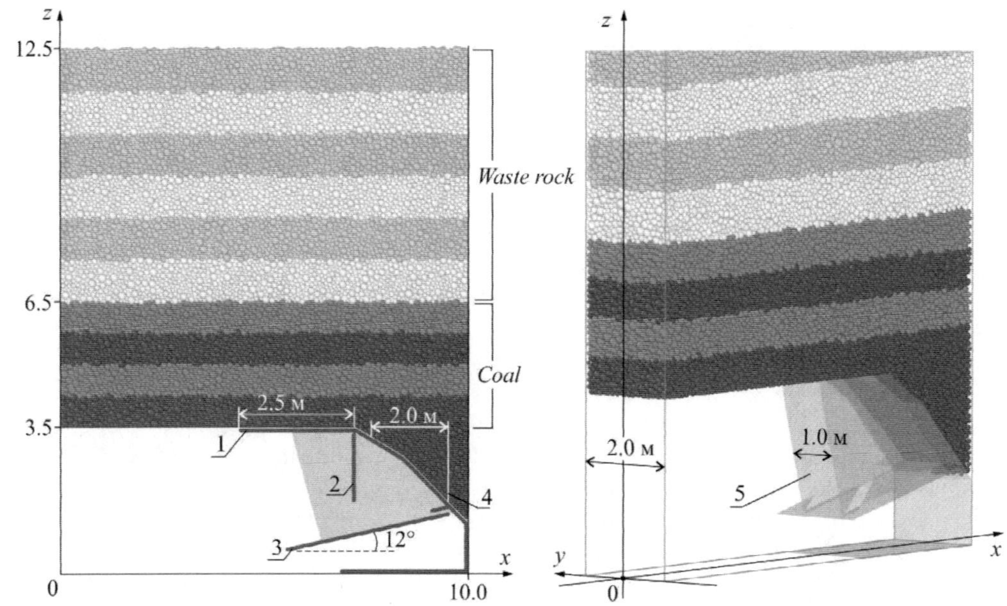

Figure 4. Schematic of numerical experiment on coal discharge.

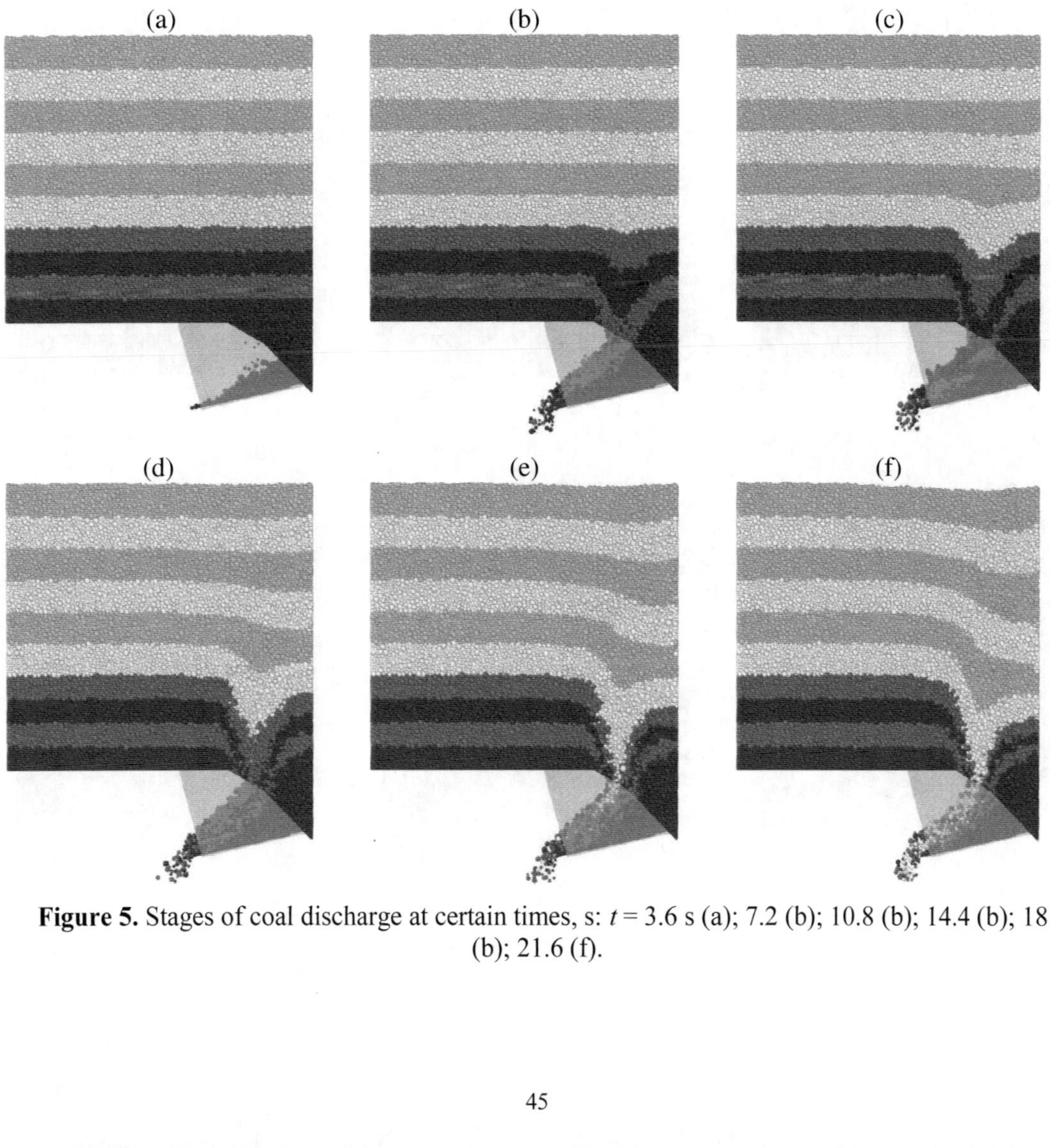

Figure 5. Stages of coal discharge at certain times, s: $t = 3.6$ s (a); 7.2 (b); 10.8 (b); 14.4 (b); 18.0 (b); 21.6 (f).

The mentioned fact is confirmed by the plots of the bulk flows of coal M_c and dirt rock M_r in figure 6. The bulk flow is found as the total mass of certain material particles passing the plane $z = 0$ per unit time.

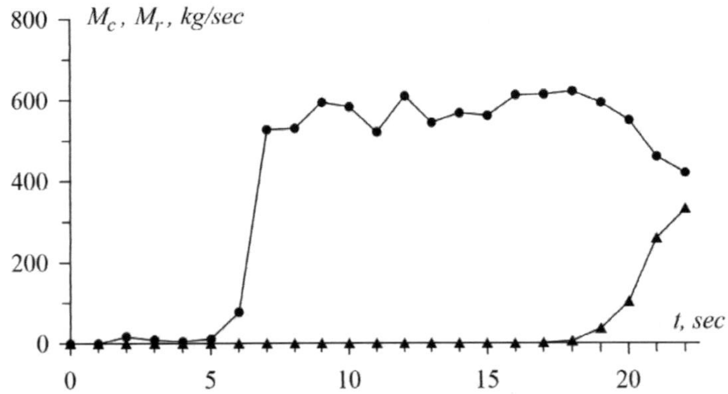

Figure 6. Bulk flows of coal (•) and dirt (▲) in the course of discharge.

The process of elliptical flow zone formation [20] above the discharge is illustrated in figure 7. Here, the kinematic patterns of discharge at fixed times, namely, distributions of moduli $|\boldsymbol{v}_i|$ of velocity vectors of each particle, are colored. The dark shade are the particles with low velocity (close to zero), the light shade are the particles with high velocity (to 5 m/s). It is evident in figures 5 and 7 that the discharge configuration appears above the outlet and propagates up to the free surface.

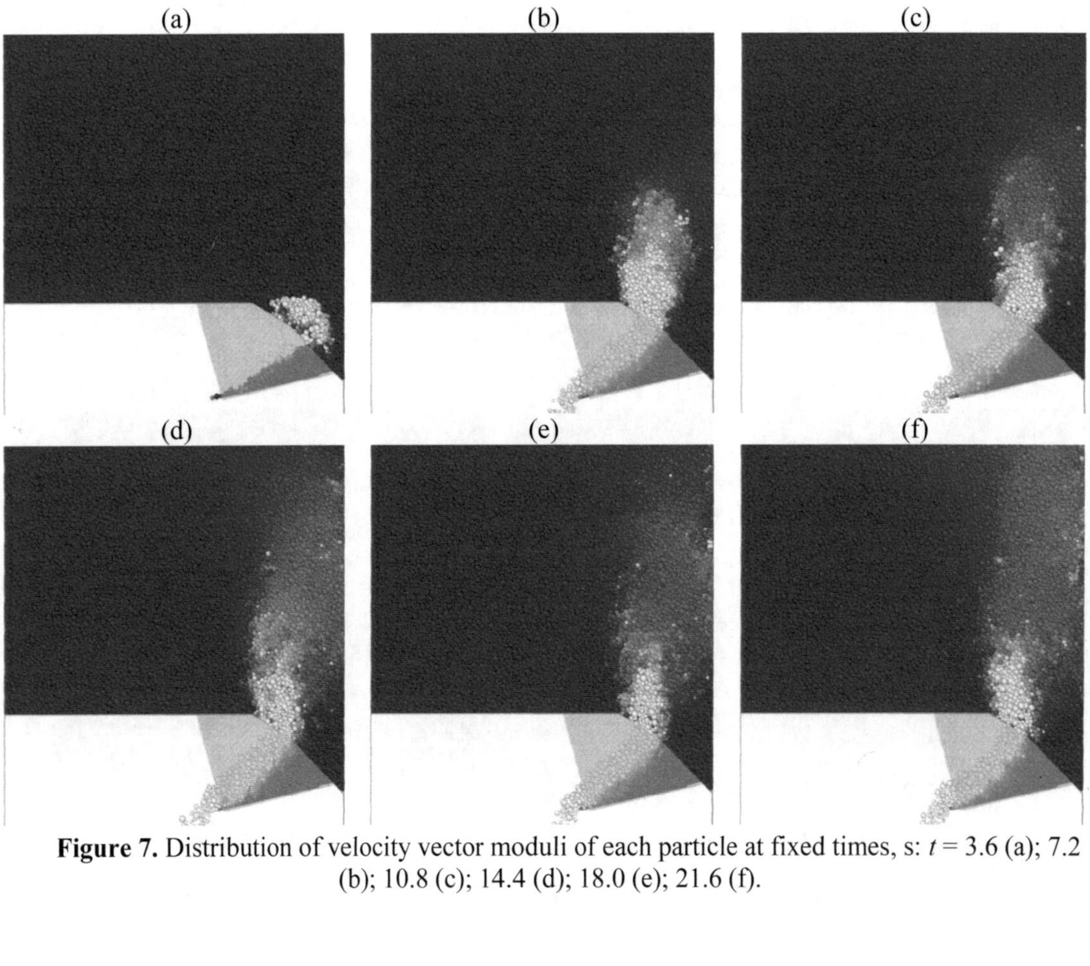

Figure 7. Distribution of velocity vector moduli of each particle at fixed times, s: $t = 3.6$ (a); 7.2 (b); 10.8 (c); 14.4 (d); 18.0 (e); 21.6 (f).

Two experiments were meant to analyze influence of the feeder structure on the discharge flow behavior. To this effect, the feeder surface was set smooth in the first experiment and as corrugation in the form of steps 0.15 m long and parallel to a horizontal plane. The feeder tilt was 12° in both experiments. The 3D model of the powered roof support with the corrugated feeder is demonstrated in figure 3c. As soon as broken rock mass reaches equilibrium state, the discharge was opened and the material flew until the next stage of equilibrium state. Such state is provided by the immobile vertical shield during deactivation of the feeder. In the experiments, the pressures on the shield and the feeder, P_d and P_f, were recorded (figure 8).

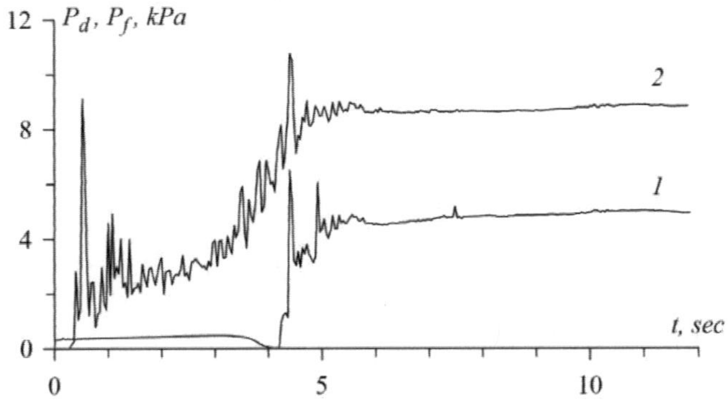

Figure 8. Pressure of granular material on shield (*1*) and deactivated feeder (*2*).

Later on, the feeder was set into back-and-forth motion by the law:

$$\boldsymbol{X}_f = \boldsymbol{A}(1 - cos\omega t) \tag{4}$$

where $\boldsymbol{X}_f = (x_f, y_f, z_f)$ is the radius-vector of the feeder center of gravity; $\boldsymbol{A} = -0.15\cdot(\cos 12°; 0; \sin 12°)$ is the amplitude of vibrations; $\omega = 0.5$ is the frequency of vibrations. In this manner, the total travel of the feeder is 0.3 m and its vibration period is 2 s.

Owing to contact friction between the particles and feeder, the material was involved into motion. In some time, the flow gained steady state. The coal bulk flow rate M_c in the described mode of discharge is shown in figure 9. In case of the smooth feeder, average M_c is 47 kg/s at the total discharge mass of 2193 kg of coal (figure 9a). With the corrugated feeder, these indexes are 32 kg/s and 1500 kg, respectively, which is an unexpected and surprising result (figure 9b).

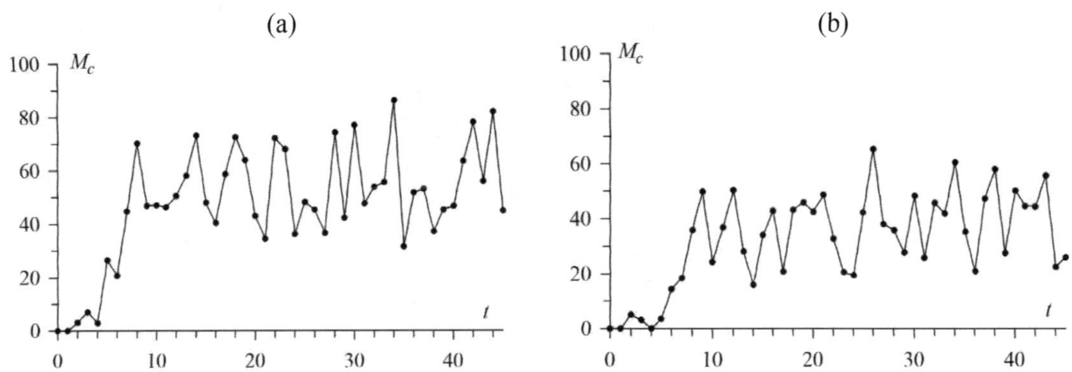

Figure 9. Bulk flow rate of coal with smooth feeder (a) and corrugated feeder (b).

Unlike in situ tests, the numerical experiment readily allows recording forces applied by the granular material on the powered roof support elements. The force interaction between the flow and the powered roof support is described in figure 10. The variation in the pressure P_d (kPa) on the immobile vertical shield in case of the smooth and corrugated feeders is shown in figures 10a and 10b, respectively. It is seen that the change in the pressure of the granular material flow has periodic behavior and reaches extremums at the points of the end positions of the feeder. The peak pressure on the shield is higher in case of the corrugated feeder. The horizontal component (figures 10c and 10d) and vertical component (figures 10e and 10f) of the total force exerted by the flow on the feeder changes periodically, as well. It is worthy of noticing that the peak vertical load on the smooth feeder is higher by an order of magnitude as against the corrugated feeder.

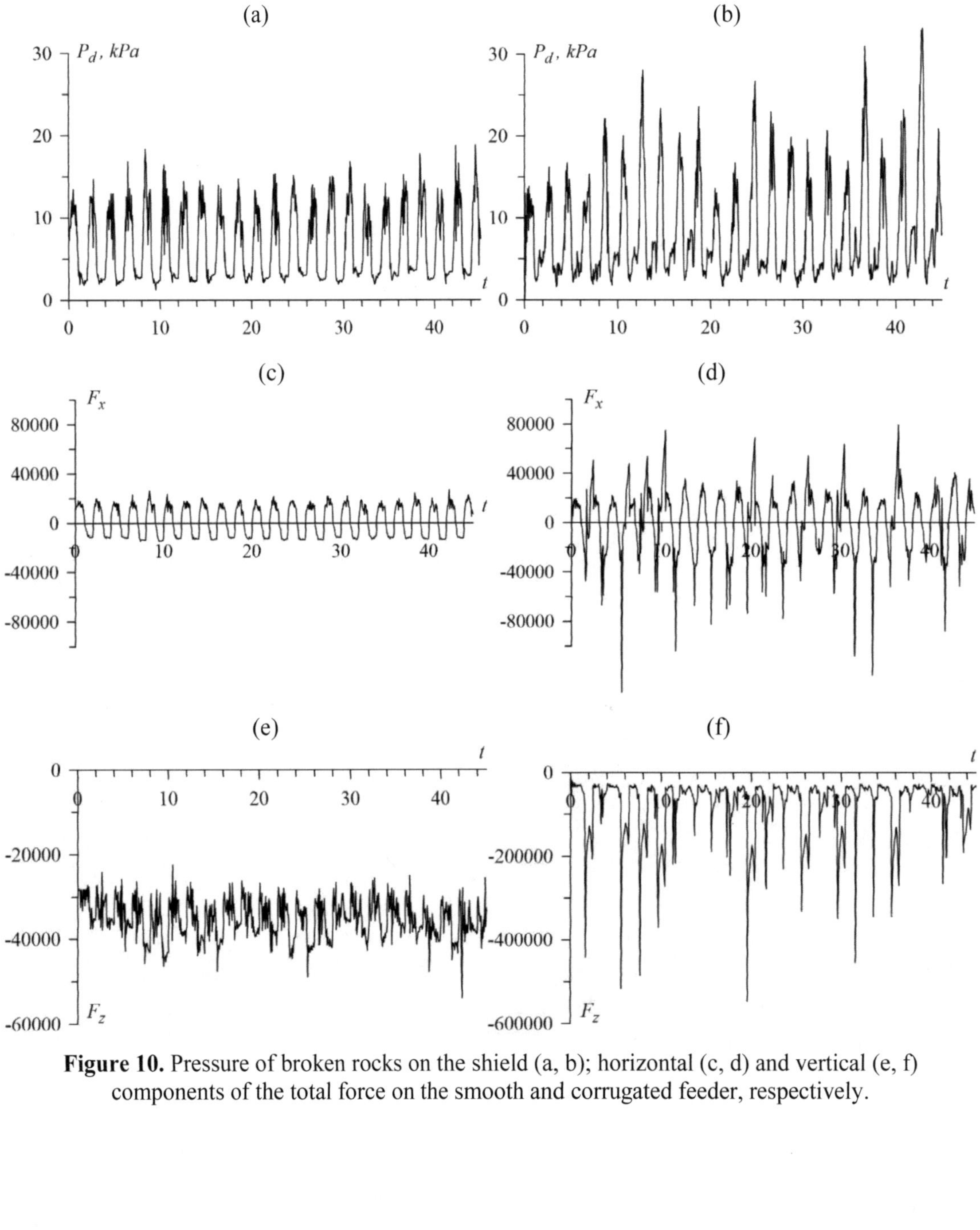

Figure 10. Pressure of broken rocks on the shield (a, b); horizontal (c, d) and vertical (e, f) components of the total force on the smooth and corrugated feeder, respectively.

IOP Conf. Series: Earth and Environmental Science **206** (2018) 012007 doi:10.1088/1755-1315/206/1/012007

4. Conclusions

The described discrete element model of gravity flow of broken rocks in the technology of longwall top coal caving includes all basic process stages: creation of initial equilibrium state of rocks; discharge of the flow to the feeder of powered roof support; functioning of the feeder and further discharge in accordance with the flow chart of the powered roof support operation.

The study demonstrates periodic nature of variation in the material bulk flow rate and in the pressure exerted by the flow on the elements of the powered roof support in the course of discharge. The influence of the feeder surface configuration on the study characteristics is illustrated.

Acknowledgments

The study has been supported by the Ministry of Education and Science of the Russian Federation in the framework of the Federal Targeted Program on R&D in Priority Areas of Advancement in the Science and Technology of Russia for 2014–2020, Topic: Development of Efficient Robotic Longwall Top Coal Caving Technology and Equipment (Unique Identifier RFMEF160417X0173).

References

[1] Melnik V V and Sushchev R A 2009 *Mining Informational and Analytical Bulletin* **5** 198–210

[2] Jabinpoura A, Bafghib A Y and Gholamnejad J 2016 *International Academic Journal of Science and Engineering* **3(2)** 102–109

[3] Kumar R, Singh A K, Mishra A K and Singh R 2015 *International Journal of Mining Science and Technology* **25(6)** 885–896

[4] Hebblewhite B K 2005 *The 19th International Mining Congress and Fair of Turkey IMCET2005* (Izmir, Turkey) pp 169–178

[5] Guo J, Ma L, Wang Ye and Wang F 2017 *Energies* **10 (9)** 1371

[6] Unver B and Yasitli N E 2006 *International Journal of Coal Geology* **66(4)** 227–252

[7] Klishin V I, Shundulidi I A., Ermakov A Yu and Soloviev A S 2013 *Technology of Excavation of Reserves of Thick Flat Seams With the Release of Coal* (Novosibirsk: Nauka) p 248

[8] Klishin V I, Fokin Yu S, Kokoulin D I and Kubanychbek B. *Thick Coal Mining With Powered Support with Adjustable Coal Discharge* 2007 (Novosibirsk: Nauka).

[9] Cundall P A and Strack O D L 1979 *Géotechnique* **29(1)** 47–65

[10] Lavrikov S V and Revuzhenko A F 2016 *Journal of Mining Science* **52(4)** 632–637

[11] Klishin S V, Lavrikov S V, Mikenina O A and Revuzhenko A F 2018 *Journal of Physics: Conference Series* **973(1)** 012008

[12] Klishin S V, Lavrikov S V and Revuzhenko A F 2017 *AIP Conference Proceedings* **1909** 020086

[13] Lavrikov S V and Revuzhenko A F 2017 *AIP Conference Proceedings* **1893** 030122

[14] Johnson K L 1985 *Contact Mechanics* (Cambridge University Press)

[15] Mindlin R D 1949 *J. Appl. Mech.* **16** 259–268

[16] Mindlin R D and Deresiewicz H 1953 *J. Appl. Mech.* **(20)** 327–344

[17] Guises R, Xiang J, Latham J-P and Munjiza A 2009 *Granular Matter* **11(5)** 281–292

[18] Makse H A, Johnson D L and Schwartz L M 2000 *Physical Review Letters* **84** 4160–4163

[19] Revuzhenko A F, Klishin S V and Mikenina O A 2015 *Physical Mesomechanics* 18(3) 244–249

[20] Hustrulid W and Kvapil R 2008 *5th International Conference and Ex-hibition on Mass Mining* (Luleå Sweden) pp 107–132

Approach to construction of mathematical models of geo-massif stress-deformed state under the influence of natural and technogenic forces

A B Tsvetkov and L D Pavlova

Siberian State Industrial University, 42 Kirova street, Novokuznetsk, 654007, Russia

E-mail: ld_pavlova@mail.ru

Abstract. The paper presents the developed approach to the construction of mathematical models of the stress-strain state of a geo-massif under the influence of natural and technogenic forces. According to the proposed concept, the geological environment is considered as a nonlinear hierarchically organized multiscale system. The natural state of the geo-massif is formed as a result of the influence of gravitational, geotectonic forces and internal pressure of fluids, and the man-made impacts lead to a change in the natural stress field as a result of formation of disturbances in the integrity of rocks in the unloaded zones and stress concentrators around the areas. Description of the stressed-deformed state of the geo-massif is carried out through a hierarchical family of mathematical models, which are constructed in the form of boundary value problems of the theory of deformable solid. First, at the lower hierarchical level, the determining relationships are developed, in which the influence of natural forces is taken into account, and then the generalization of the obtained models for determining technogenic impacts is carried out at the upper level. The proposed approach allows the links in the resulting hierarchical structure of models to be built, which makes it possible to carry out studies of both the disturbed and the natural state of the geo-massif.

1. Introduction

Geological environment is a natural object with a complex structure, therefore its description is an actual task. Geomaterials and geo-environments demonstrate all the specific features of a complex structure and are considered as non-linear hierarchically organized multiscale systems, which include not only extended geo-environments, but all loaded solid bodies [1].

Technogenic impacts on the geo-environment lead to a change in its state, the description of which must be carried out not only taking into account the influence of natural forces, but also the conditions of rocks deformation that have changed as a result of a wide range of mining operations.

The construction of underground structures, such as mines, tunnels, bunkers, underground storages of water, oil, gas, requires the performance of mining works and subsequent arrangement of free space in accordance with the designation of the projected facility.

2. Methods of research

The methodology of describing complex systems was considered by H. Haken [2], who introduces three levels of description: micro-, meso- and macroscopic.

A continual, phenomenological description of the geo-environment is used at a mesoscopic and macroscopic level. The microscopic level corresponds to the description of the system at the level of elementary particles and atoms. However, it is difficult to describe the geo-environments on this scale due to the huge number of particles and the difference in scales. Therefore, at the micro level, it becomes necessary to introduce parameters that would allow the most significant features of behavior to be described at the meso level.

At the meso- and macrolevel the relationship between parameters is ensured by equations that take into account at the micro level not the processes themselves, but their manifestation and influence on the geo-environment under the influence of natural and technogenic forces. The meso- and macroscopic state of solid media and geomaterials has its own characteristic features, manifested in the deformations of geomaterials, which, depending on the acting stresses, can be elastic or plastic while maintaining continuity of the medium.

To describe the geomechanical state of geomaterials, a concept is proposed, according to which the geo-environment is considered as a hierarchical nonlinear system, subjected to the impact of natural and technogenic forces. It is proposed to implement this concept on the basis of a hierarchical approach, which is as follows:

- the object of limited sizes is allocated at the macro level;
- its decomposition into subjects of meso- and micro-level is carried out depending on the research tasks;
- the integral parameters characterizing the state of the natural object are determined;
- connections are established between the set parameters.

The description of the geo-environment through the hierarchical system allows general scenarios of the geomassive behavior to be formed, and also is the basis for determining of its stress-strain state by means of mathematical modeling.

The specific solution of the choice of the research object within the offered hierarchical approach depends on the considered subject area. If the task is to study the stress-strain state of the geo-massif under the impact of natural and man-made forces, we can confine ourselves to the meso level, considering it representative. In this case, the result of the decomposition of the geo-environment macro-volume of the at the meso-level is a natural or disrupted due to technogenic impacts geo-massif, consisting of blocks rock layers and coal layers deformed under the action of applied forces. The state of the geomassif rocks at the micro level depends on the forces acting in it (figure 1).

The natural state, typical for the geomassif not affected by technogenic impacts, is formed during the geological time scales under natural conditions of occurrence of rocks (figure 1). There are theories in which, under certain conditions, the natural state of the undisturbed geomassif is described by the relations of the elasticity theory [3, 4].

In the disturbed state, the geomassif is under the influence of technogenic influences, which lead to the localization of deformations and the formation of the unloaded zones and stress concentrators around the areas of disturbance of rocks integrity, that significantly alter the natural state of the massif. In this case, irreversible deformations and the transition of rocks from the elastic to the elastoplastic, limit or out-of-limit states are observed, the description of which must be carried out by means of mathematical relation constructed on the basis of laws corresponding to the nonlinear character of rock deformation.

Due to the complexity in describing the geomechanical state of the rock massif formed in the result of combination of natural and man-made forces, the modeling of geo-massif stressed-deformed state should be based on the family of mathematical models that, because of the generality of connections and the hierarchical structure, should provide an opportunity to study both the disturbed and the natural state of the geo-massif.

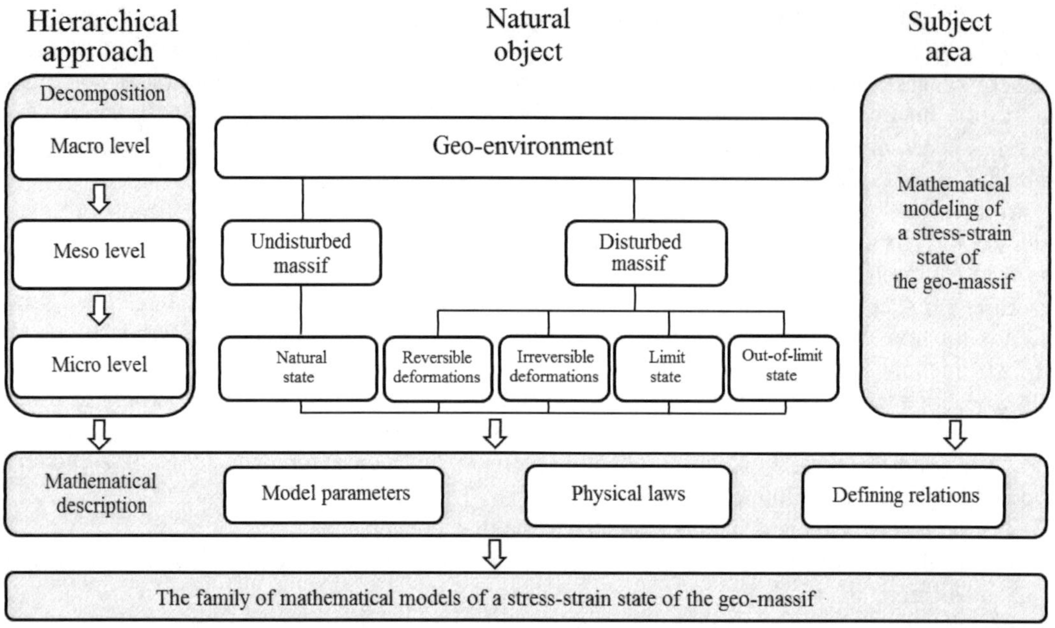

Figure 1. The concept of mapping the research object on the technology of mathematical modeling.

The resolving system of nonlinear equations must, on the one hand, reflect the state of the rocks at the microlevel, and, on the other hand, at the mesolevel take into account the result of the work of the bulk forces typical to the macrolevel.

On the basis of the proposed concept, it is required to determine the main parameters of the geomassif characterizing its stress-strain state under the joint influence of natural and technogenic factors and to establish connections between them by means of equations.

3. Results and discussion

According to the results of studies [5, 6, etc.], it is established that the natural state of geomassif is formed as a result of the influence of gravitational, geotectonic forces and fluid pressures, which, depending on the physical and mechanical properties of the rocks, are balanced by stresses arising

As a result of technogenic influences, the change of the natural stress field in the zones of influence of mine workings is accompanied by the change in the laws of rocks deformation of and their transition from elastic to elastic-plastic, limit and out-of-limit states [4, 6, etc.]. Rocks have unequal tensile strengths under uniaxial compression or tension, so their resistance to applied forces must be taken into account when determining the stress-strain state of the geomassif [7, 8].

The technogenic impact on the geomassif leads to a redistribution of the stress field in relation to its natural state. Therefore, in order to build a family of mathematical models of the stress-strain state of geomassif under joint influence of natural and man-made forces, it is first necessary to develop determining relationships that take into account the effect of natural forces, and then generalize the resulting models to determine the technogenic impacts.

When constructing a family of mathematical models of the massif stress-strain state, a hierarchical approach is proposed, which provides technology for developing models according to the principle from the bottom to the top in the order of taking into account natural and technogenic factors.

Mathematical models of the geomassif stress-strain state are constructed in the form of boundary-value problems of the theory of deformable solid. To establish relations between the parameters of models, while preserving the continuity of the geomassif, it is possible to use the equilibrium equations, Cauchy relations, the boundary conditions that determine the relations.

At the lower hierarchical level, the mathematical models are designed to investigate the deformation of geomassif rocks, the stress-strain state of which is determined by the action of natural forces – gravitational, geotectonic and internal pressure of natural gas. The hierarchical structure of the family of developed mathematical models of the geomassif stress-strain state is shown in figure 2.

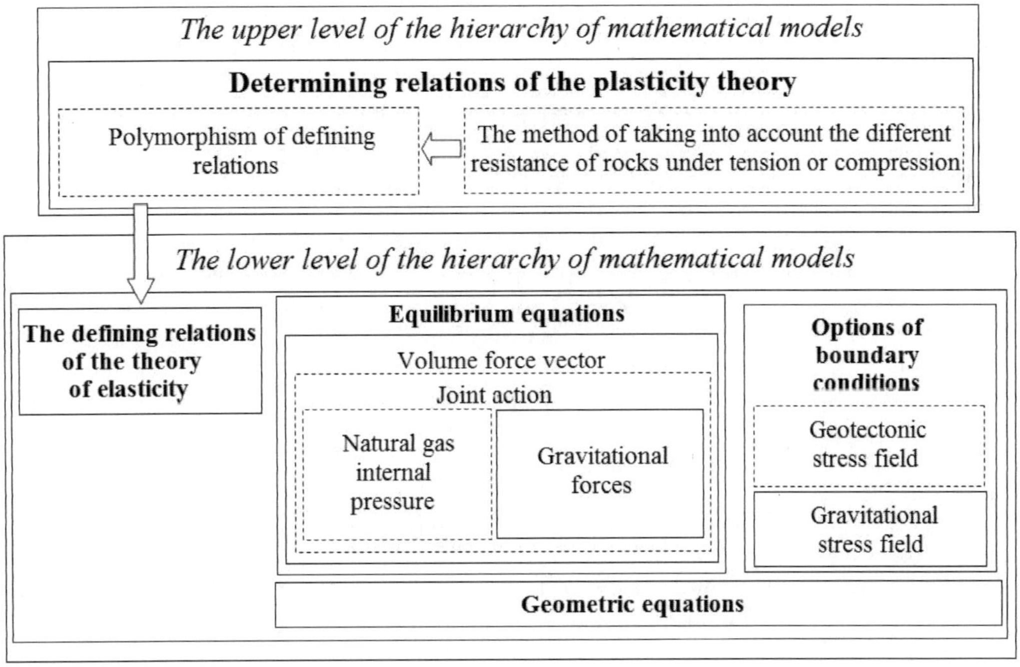

Figure 2. Hierarchical structure of the family of mathematical models of the geomassif stress-strain state.

The law of rocks deformation under the combined effect of gravity forces and natural gas pressure must be taken into account in the defining relations of the mathematical model, and the acting forces – in the equations of equilibrium.

Since the effect of the surrounding strata on the studied section of geomassif leads to the formation of a gravitational or geotectonic stress field in it, it is necessary in the mathematical model of the lower level to provide the possibility to determine the ratio of the vertical and horizontal components of the stress tensor.

In order to take into account the external influences on the section of geomassif under consideration, it is required to make a transition from external forces acting on the surface to internal forces near the boundary, by developing or modifying the boundary conditions of the boundary-value problem that connect the effects of external and internal forces. Therefore, it is necessary to define boundary conditions for the boundary value problem that correspond to the gravitational or geotectonic stress fields (figure 2).

Thus, mathematical models of the geomassif stress-strain state of the lower-level hierarchy should provide for the mechanisms of accounting the effect of the joint action of gravitational, geotectonic forces and internal gas pressure in one boundary-value problem.

The combined effect of natural and man-made forces on the rock massif leads to the formation of a new geomechanical state, for the study of which the mathematical models of the geo-massif stress-strain state of the upper hierarchical level are developed.

Numerous studies [4, 6, 9, etc.] have established that under technogenic impact on the geo-massif in the mining zone, redistribution of stresses relative to the natural state of the geo-massif and formation of zones of tensile and contraction of rocks around the gob are observed. Therefore, in the models of the upper hierarchical level it is necessary to provide in the determining ratios the mechanism of taking into account different rocks resistance to the tensile and contraction forces (figure 2).

It should be noted that underground mining of a coal seam leads to non-linear deformation of rocks, which under the influence of natural and man-made forces transfer from elastic to elastic-plastic, limit or out-of-limit states. Therefore, in the top-level models it is necessary to take into account the laws of rock deformation, which are characteristic for the corresponding zones of mining work, and to develop a method for their connection by developing an approach to joint use of defining relations of theories of elasticity and plasticity in one boundary-value problem.

To construct such a class of inhomogeneous boundary value problems with a system of nonlinear equations, it is necessary that the defining relations of mathematical models of deformation of upper-level rocks have polymorphism for transition to the defining relations of low-level models. Since static equations and geometric equations in the theory of plasticity and elasticity have the same form, the inheritance of the properties of models between levels of hierarchy is ensured by the general structure of differential equations of the developed mathematical models that can be constructed both in two and in a three-dimensional setting.

The developed concept is implemented in the family of mathematical models of the geo-massif stress-strain state presented in [10-15].

4. Conclusion

Thus, the concept of investigation of the geo-massive stress-strain state under the influence of natural and technogenic forces was proposed. According to it the parameters determining the geomechanical state of the rock massif and the mechanisms of their interconnection are established in the process of decomposition of the geo-environment through a hierarchical approach.

The description of the geo-massive stress-strain state is carried out using different laws of rock deformation in the zones of influence of mine workings, which requires the integration of the theory of elasticity describing reversible deformations in the undisturbed geo-massif and the theory of plasticity allowing inelastic deformations arising from technogenic impacts to be taken into account.

References

[1] Makarov P V, Smolin I Yu, Stefanov Yu P, Kuznetsov P V, Trubitsyn AA, Trubitsyna N V, Voroshilov S P and Voroshilov Ya S 2007 *Nonlinear Mechanics of Geomaterials and Geomedium* (Novosibirsk: Geo) p 235

[2] Hacken G 1991 *Information and Self-Organization. Macroscopic Approach to Complex Systems* (Moscow: Mir) p 240

[3] Dinik A N, Morgaevsky A V and Savin G N 1938 *Proceedings of the Meeting on the Management of Rock Pressure* (Moscow: Izd. AN SSSR) pp 176–85

[4] Borisov A A 1980 *Mechanics of Rocks and Massifs* (Moscow: Nedra) p 360

[5] Zhukov V V, Chernov E V and Dovchenko G N 1973 *Stress-Deformed State of the Layered Massif* (Leningrad: Nauka) p 132

[6] Turchaninov I A, Iofis M A and Kasparyan E V 1989 *Fundamentals of Rock Mechanics* (Leningrad: Nedra) p 488

[7] Ilnitskaya E I, Teder R I, Vatolin E S and Kuntysh M F 1969 *Properties of Rocks and Methods for their Determination* (Moscow: Nedra) p 392

[8] Shtumpf G G, Ryzhkov Yu A, Shalamanov V A and Petrov A I 1994 *Physico-Technical Properties of Rocks and Coals of the Kuznetsk Basin: A Reference Book* (Moscow: Nedra) p 447

[9] Pevzner M E, Iofis M A and Popov V N 2005 *Geomechanics* (Moscow: MSU) p 438

IOP Conf. Series: Earth and Environmental Science **206** (2018) 012008 doi:10.1088/1755-1315/206/1/012008

[10] Tsvetkov A B and Pavlova L D 2013 *Scientific and Technical Bulletin of the Volga Region* **6** 485–8
[11] Tsvetkov A B 2014 *Scientific Review* vol 8 (Part 2) 599–603
[12] Tsvetkov A B, Pavlova L D and Fryanov V N 2015 *Mining Informational and Analytical Bulletin* **1** 365–70
[13] Tsvetkov A B, Pavlova L D and Fryanov V N 2016 *IOP Conference Series: Earth and Environmental Science* **45** 012009
[14] Tsvetkov A B, Pavlova L D and Fryanov V N 2017 *IOP Conference Series: Earth and Environmental Science* **84** 012005
[15] Tsvetkov A. B., Pavlova L D and Fryanov V N 2018 *IOP Conference Series: Earth and Environmental Science* **134** 012067

The choice of the site for inclined shaft and parameters of its stage-by-stage construction in the conditions of the Tersinsky coal region

V A Voloshin[1], S V Rib[1], O A Petrova[1] and A V Belyaev[2]

1Siberian State Industrial University, 42 Kirova street, Novokuznetsk, 654007, Russia
2JSC "UK Sibirskaya", 5 Ermakova ave, Novokuznetsk, 654007, Russia

E-mail: voloshinva1966@gmail.com

Abstract. The problem of the construction of inclined trunks in the conditions of coal mines in the Tersinsky geological and economic region of Kuzbass is discussed in the paper. The feasibility study for the technological option of the inclined boundary shaft construction in seam 67 in the complex mining and geological conditions of the mine was carried out. The recommendations on the monitoring of the workings state and development of measures ensuring the reduction of negative manifestations of rock pressure during the construction and operation of the inclined shaft are given taking into account the cumulative influence of the system of the previously mined workings and seasonal increased water inflow.

1. Introduction

The construction and maintenance of opening workings in difficult mining and geological conditions is a very urgent task for mine-designers [1,2]. Annually in Kuzbass the construction of more than 7 km of opening inclined workings is carried out, about 4 km of permanent mine openings need to be repaired. Strengthening of the support and repair for the entire lifetime is required in almost 35% of cases during the lifetime of the workings.

In the course of the work on the construction of the boundary inclined ventilation shaft in seam 67 of the brunch "Mine "Uvalnaya" of JSC "UK Sibirskaya" in difficult mining and geological conditions, the results of engineering geological, hydrogeological and geophysical surveys in the construction area were analyzed.

Based on the results of analysis of the mining and geological situation and the spatial location of previously mined areas the following alternative options and technologies for conducting the boundary and inclined ventilation shaft are justified (figure 1):

1) Performance of stripping works, building the frame of the boundary ventilation shaft, concreting the working and filling up it to the existing level of the earth's surface.

2) Construction of a boundary inclined ventilation shaft under the previously sealed working to the vertical shaft.

3) Construction of the boundary inclined ventilation shaft from the junction with the previously driven working at a distance of 8 m from the sealed part horizontally and 10 m vertically.

4) Construction of the boundary inclined ventilation shaft from the junction with the previously driven working at a distance of 8 m from the sealed part horizontally and 18 m vertically (below the level of seam 66).

The four options for choosing the location for the construction of the boundary ventilation shaft in seam 67 are shown in figure 1.

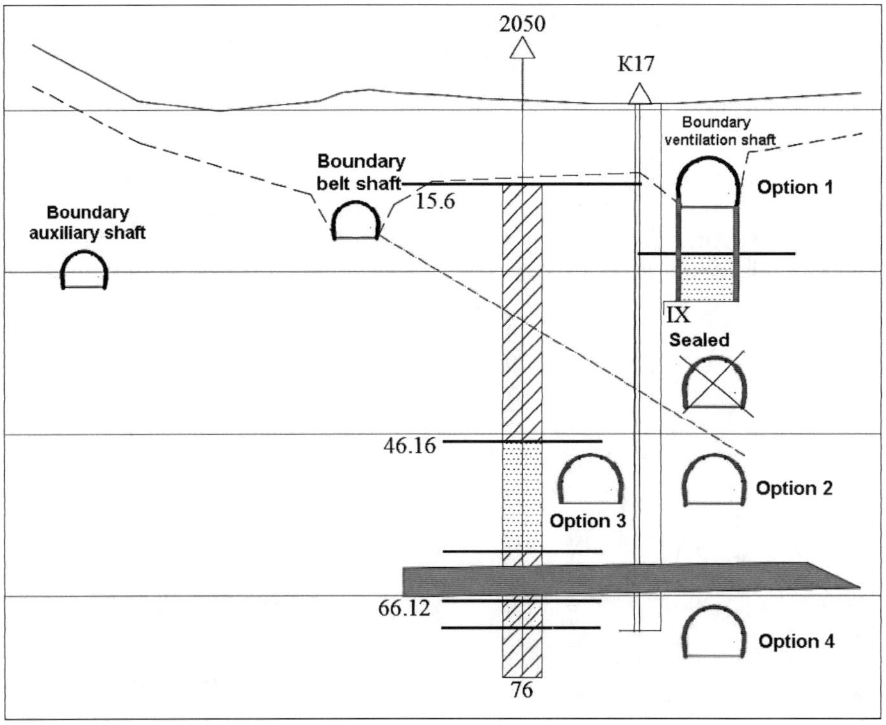

Figure 1. Spatial position of options for the boundary inclined ventilation shaft.

2. Methods of research

The calculation of the parameters of the stress-strain state (SSS) of the coal-bearing massif near the newly driven boundary inclined ventilation shaft is presented on the basis of numerical solution by the finite element method of differential equations of continuum mechanics using a package of computer programs [3-5].

For substantiating the shape and size of the zones of coal and rocks destruction near the ventilation shaft, the results of mine observations and laboratory studies of the properties of rocks and their deformation diagrams were used. The following sources of information were used as input parameters for the mathematical model:

- geological sections along the wells;
- the sizes of the zone of rock massif movement according to the forecast;
- the height of the dome of natural equilibrium and the load-bearing capacity of the frame support elements, calculation of the support setting parameters of permanent workings with frame and concrete support in accordance with the current instructions.

The following geometric model of the coal-bearing massif is adopted for modeling:

- the length of the vertical section along the axis of the ventilation shaft in seam 67 is assumed equal to 300 m;
- the mining depth of seam 67 in the area of ventilation drift construction in seam 67 is 62 m;
- the width of the opening workings of the ventilation, belt and auxiliary shafts was adopted taking into account the design recommendations and actual side rocks sloughing according to field observations 7.0-8.0 m;

- the height of the workings is taken equal to the maximum extracted volume of the extracted rock mass – 5.0-5.5 m;
- the origin of coordinates on the vertical section is taken at the intersection of the roof in seam 67 (vertical axis of ordinates) and the left side of the previously driven and "sealed" section of the boundary ventilation shaft in seam 67 (horizontal axis of abscissas).

3. Results and discussion

According to the analysis of the results of rock displacements distribution in the vicinity of the ventilation shaft the nature of deformation and the vector directions of rock displacement did not change significantly in comparison with the displacements during performing mining operations under the previously driven working (figure 2).

The values of the displacement of roof rocks decreased by almost 15%, in the soil of the working the probability of swelling increased by 20%. Horizontal displacements in the soil of the working increased, which indicates an increased probability of destruction of rocks at the contact of the soil and the wall of the working, especially in the case of increased water inflow in rocks highly prone to soaking.

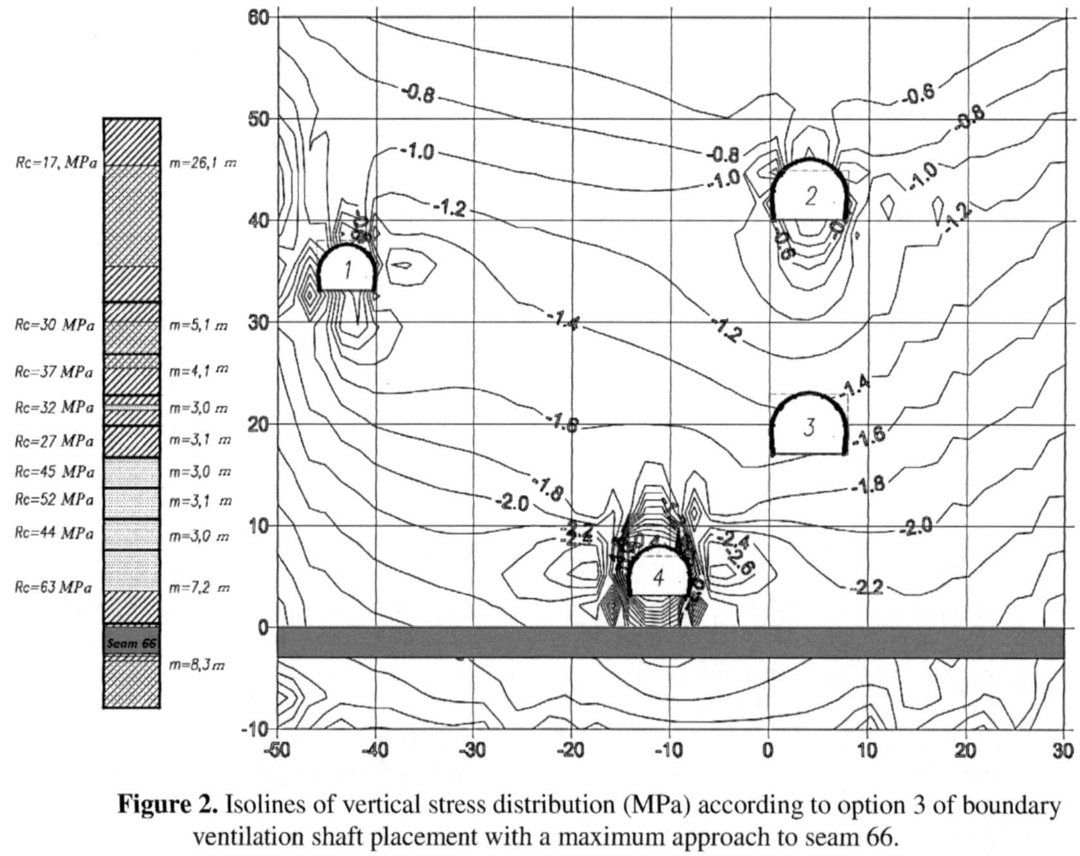

Figure 2. Isolines of vertical stress distribution (MPa) according to option 3 of boundary ventilation shaft placement with a maximum approach to seam 66.

It should be noted that when constructing the boundary ventilation shaft in the area with no contiguous previously driven working, the vertical stresses and the expected rock displacements of the roof and walls decreased significantly, on average by 25%.

The analysis of the simulation results of the options for placement of the boundary inclined ventilation shaft in relation to the previously driven workings of the technological site made it possible to draw conclusions about the advisability of adopting the third construction option.

It is necessary to take into account a number of factors to ensure the stability of the boundary inclined ventilation shaft during its construction and maintenance during the entire period of its operation.

First, it is the working support or to be more exact the stages of support setting. Obviously, when driving the boundary inclined ventilation shaft, the greatest stresses are observed in the places near the previously driven and sealed working. The stability of the wall rocks will be reduced, therefore at the stage of the boundary inclined ventilation shaft construction near the previously driven working, it is recommended to use back laths before installing the frame metal support in the right wall and roof of the working.

The *second* factor to ensure the stability of the opening working is the technology of conducting (selection of the cross-section during the driving and subsequent expansion). Simulation of different options for working driving shows a significant advantage of the working construction in two stages. Initial driving of a working of narrow cross-section with subsequent expansion to the design cross-section.

Driving the working with a narrow cross-section will reduce the probability of increased stresses on the part of the rock massif and ensure a reliable working support. Expansion of the output will be accompanied by a redistribution of the rock massif SSS and will significantly reduce the probability of soil heaving and deformation of support elements in the inclined ventilation shaft.

The revealed regularities in the isolines distribution of host rocks residual strength according to simulation results showed that when driving a working with a full cross-section, the zone of heaving rocks in the soil of the working will amount to 6 m^2 in the cross-section, that is, an active soil pressure up to 5-7 m^3 per running meter in the working should be expected. The destruction of rocks in the roof is expected at a height of more than 2 m, the greatest pressure is expected from the walls of the mine. To ensure the sustainability of the working, it is recommended to develop measures for the installation of back laths into the walls.

When driving the working with a partial cross-section, the SSS changes substantially in the vicinity of the ventilation shaft: the zone of the heaving rocks in the soil of the mine will be less than 1 m^3 of rock. Rock destruction in the roof is expected at a height up to 1.5 m, the greatest pressure is expected from the walls of the mine, which is significantly less than when driving a working with a wide cross-section. The most dangerous is the zone near the roof, in places of stress concentration. It should be noted that the choice of the domed shape of the working is correct, however, it is possible to recommend measures for the installation of back laths into the right walls of the working during driving immediately after extraction of the rock mass.

In connection with the expected large displacements and destruction of rocks in the walls of the mine, it is proposed to consider the option of a stage-by-stage mining operations. At the first stage, the working is driven with a narrow cross-section 16 m^2, at the second stage it extends to a design cross-section of 22 m^2.

In figure 3 and 4 the distribution graphs of residual rock strength isolines during driving the boundary inclined ventilation shaft in seam 67 with a full and narrow cross-section are given.

Reduction of the intensity of roof and soil rocks convergence by 30% significantly increases the stability of the working and reduces the risk of negative impact on the conditions of driving and maintenance of the boundary ventilation shaft of seam 67. The expected displacements during the driving of working with a small cross-section are compensated in full by the strength characteristics of the erected support.

From the comparison of isolines in figure 4 it follows that the vertical load by the action of the roof rocks is reduced by 30%. The rock heaving during the formation of the fracture zone in weak soil rocks with intensive water inflow in the spring period is recommended to be avoided by increasing the working cross-section due to active dinting of the working soil. The option of initial working arrangement of smaller cross-section, followed by the increase in height to the design cross-section by the method of dinting, will substantially increase the working stability during its maintenance.

Negative manifestations of increased watering of the field, intensive pressure of coal and rocks from the walls and roof of the mine, and rock heaving were the objective prerequisites for the development of a set of measures to monitor the condition of the support and surrounding rocks around the inclined workings.

To ensure the safety and effectiveness of mining operations, as well as reduce the risk of rock inrushes and collapses while driving inclined shafts, a comprehensive methodology for studying the state of the rock massif and the support elements has been developed.

The *methodology* considers the following provisions:

- visual inspection of the contour massif of mine workings;
- visual inspection of the condition of the roof support and the walls of workings;
- visual and instrumental survey of the roof rocks for the presence of rock lamination processes;
- instrumental control of the host rocks convergence in the vicinity of mine workings;
- observation over the movement of rocks and the earth surface;
- physical modeling on equivalent materials;
- numerical finite element modeling (FEM).

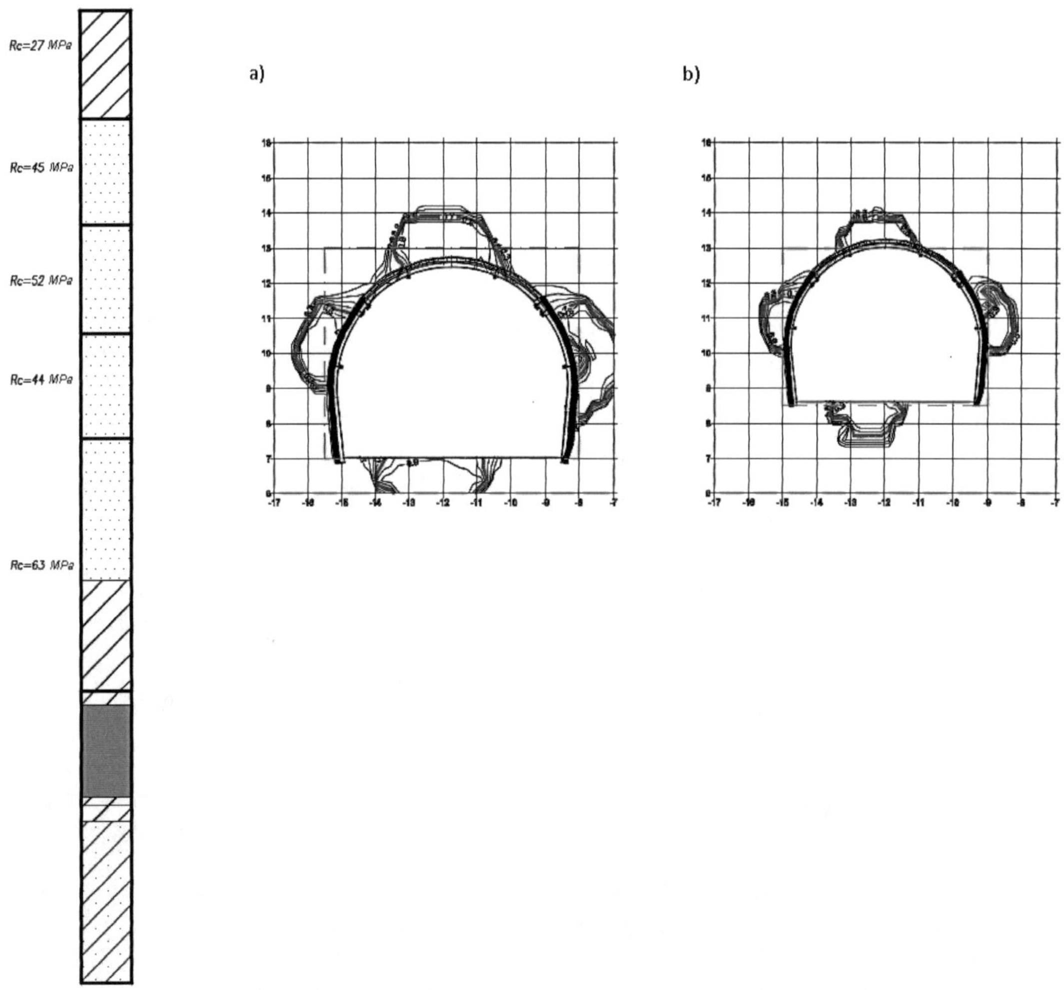

Figure 3. The distribution isolines of the ratio of residual rock strength to the initial one when driving the boundary inclined ventilation shaft with a full cross-section (a) and narrow cross-section (b).

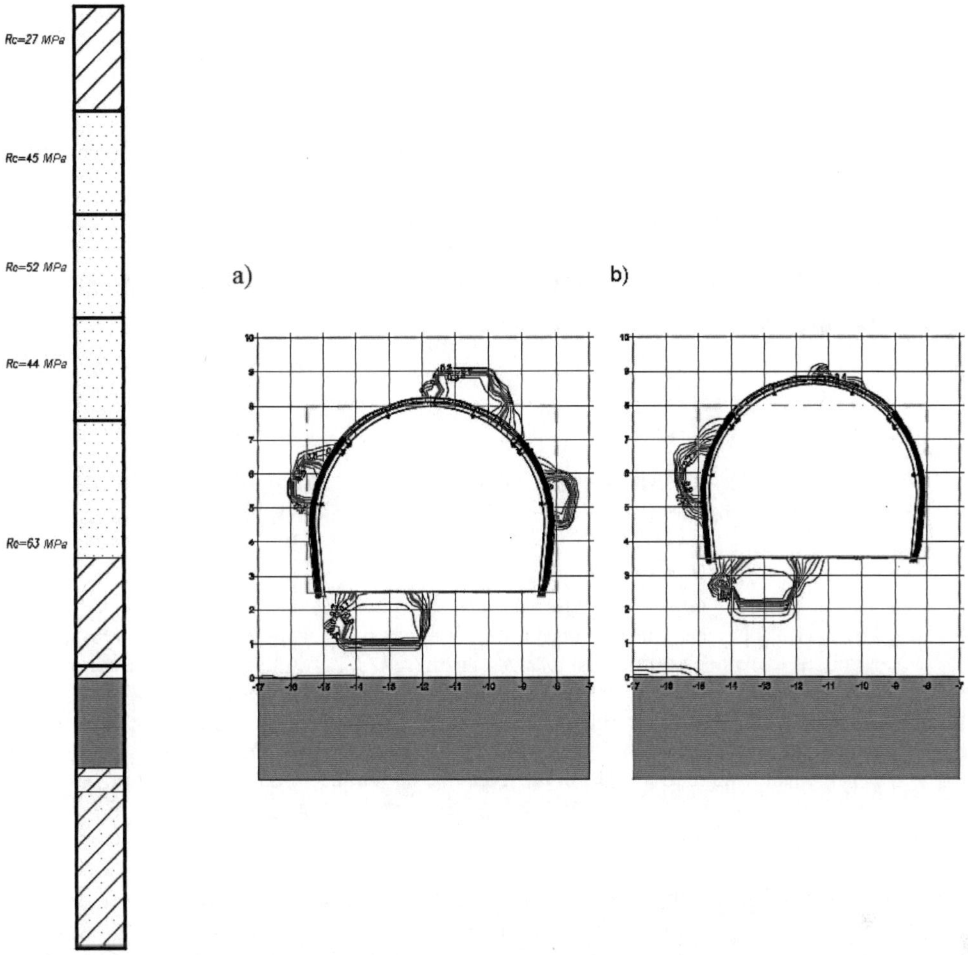

Figure 4. The distribution isolines of the ratio of the residual rock strength to the initial one during driving the boundary inclined ventilation shaft in the area maximally close to seam 66; a) with a full cross-section; b) with a narrow cross-section.

4. Conclusions

Thus, the developed technological solutions for choosing the location of the inclined shaft and the technology of its step-by-step driving and supporting with the use of observations over the condition of the support and the surrounding rocks in the maintained working allowed the costs for the working support while driving to be significantly reduced. The monitoring of the state of the opening inclined working carried out over the period of two years showed a satisfactory condition and the absence of the necessity for any additional repair works in the working.

References

[1] Kosarev N F et al 2004 *Construction of Inclined Mine Workings* (Kemerovo: Kuzbassvuzizdat) p 347
[2] Pershin V V et al 2004 *Bulletin of KuzSTU* **4** 24–26
[3] Fadeev A B 1987 The *Finite Element Method in Geomechanics* (Moscow: Nedra) p 221
[4] Rib S V et al 2015 *Mining Information Analytical Bulletin* **7** 23–29
[5] Rib S V et al 2015 *Mountain Information and Analytical Bulletin* **3** 414–422

Geomechanical substantiation of parameters of workings intersection taking into account the influence of dynamic bearing pressure

V V Basov[1] and A G Serg[2]

[1]Siberian State Industrial University, 42 Kirova street, Novokuznetsk, 654007, Russia
[2]JSC "Mine "Bolshevik", 27 Tsentralnaya street, 654235, Novokuznetsk, Russia

E-mail: vadimbasov@yahoo.com

Abstract. The technique of initial data preparation for geomechanical substantiation of parameters of mine workings interfaces is offered. The results of the mine experiment on a stressed-deformed state of a rock massif in the intersection of a belt road with a diagonal connection are presented taking into account the spatio-temporal location of the mining face. A comparative evaluation of the results of instrumental measurements and calculated displacements of the roof rocks in the intersection of mine workings was carried out.

1. Introduction

The relevance of studies of geomechanical processes in the vicinity of the interfaces of mine workings is confirmed by the difficulty of maintaining them near the faces. This is due to the dynamic change in the parameters of the bearing rock pressure, which depend on the distance between the workings intersection and the face, as well as the the rock overhang length of active roof over the gob. Therefore, studies that take into account the spatio-temporal position of the workings intersection, the face and the state of the underworked roof rocks are relevant.

Objective of the study. To carry out a mine experiment and determine the parameters of the strained-deformed state (SDS) of the rock massif in the vicinity of the intersection of the faces and development workings taking into account the effect of the dynamic bearing pressure.

Tasks of the study:

1) to develop a program and methodology for the mine experiment.

2) to make visual observations of the change in the shape, deformation and destruction of the roof and wall rocks in the intersection of the mine workings when the face advances to the site under consideration.

3) to carry out instrumental measurements of coal convergence and sloughing from the walls of the workings in the junction zone with a different spatial location of the face.

4) to conduct an assessment of the geomechanical parameters of the rock massif in the vicinity of the intersections of the mining faces and developing workings taking into account the effect of the dynamic bearing pressure.

5) to compare the results of instrumental measurements and calculated displacements of the roof rocks in the intersection of the workings.

2. Research methodology

The mine studies were conducted in accordance with the developed methodology, which included:

V – visual observations of the rock massif condition in the vicinity of the workings intersections;

- inspection of the state of the support in the intersection of mine workings;
- monitoring of mixing depth benchmark stations (DBS) in the investigated area;
- measurement of height and width of the workings in the intersection with the help of a range finder;
- sketching;
- taking photographs.

A mine experiment was carried out at one of the mines of the Kuznetsk coal basin at the time of mining working area 1-1 (stage I), 1-2 (stage II). The mining works on the development of a flat coal seam 1 with a thickness of 3.6 m was made at a depth of 385 m in difficult mining and geological conditions.

The investigated site was located at the intersection of the best road 1-2 with a connection 1-5 and a diagonal connection 1-2-2 (then, the drift, connection, diagonal connection) prepared for the next working area 1-2 (figure 1).

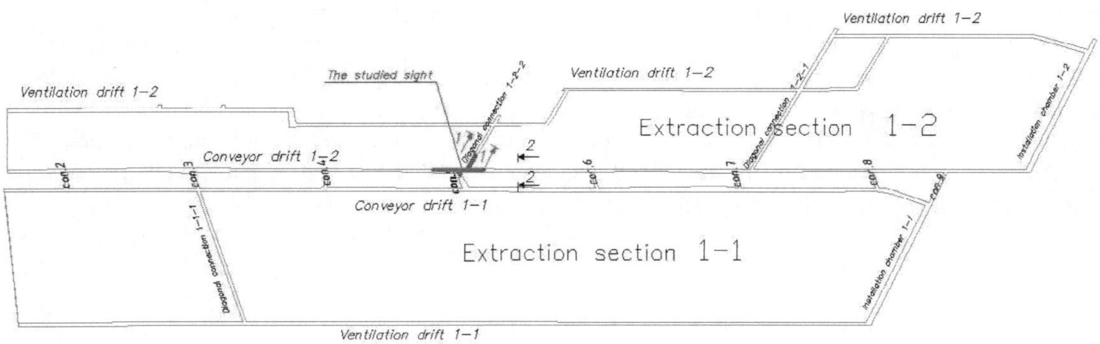

Figure 1. Fragment from the mining plan.

The length of the face 1-1 was 200 m. The length of the face 1-2 to the diagonal connection 1-2-1 was 200 m, after 120 m and in the area of the diagonal connection 1-2-2 70 m. The drift, connection and diagonal connection were driven with a width of 4.9 m, 4.6 m, 4.6 m, respectively, in the coal seam with an anchorage (figure 2).

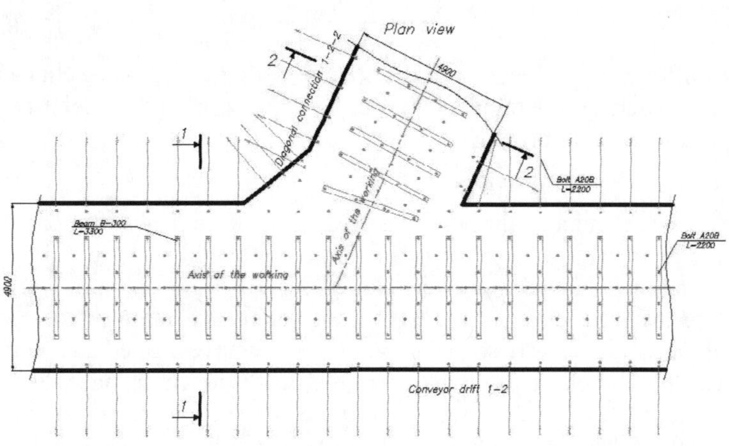

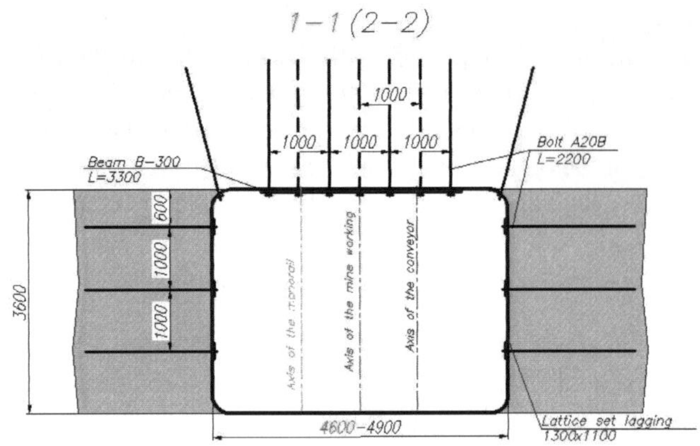

Figure 2. Cross section of a diagonal connection 1-2-2, conveyor drift 1-2, connection 1-5.

The instrumental measurements of the rock deformation in the intersection of mine workings were carried out using a mine range finder according to the proposed scheme (figure 3).

Monitoring of rock displacements in the dome of natural equilibrium of the investigated area of intersections was carried with the help of depth benchmark stations. The scheme of the stations includes the mine working, the vertical borehole, the benchmarks fixed in the borehole, the system of change in the displacement of the benchmarks (figure 3, c). For the upper benchmarks $R_{ПО}$, the displacements were assumed to be zero. The displacements of the lower benchmarks $R_{П1}$, $R_{П2}$ were determined with respect to the upper benchmark $R_{ПО}$.

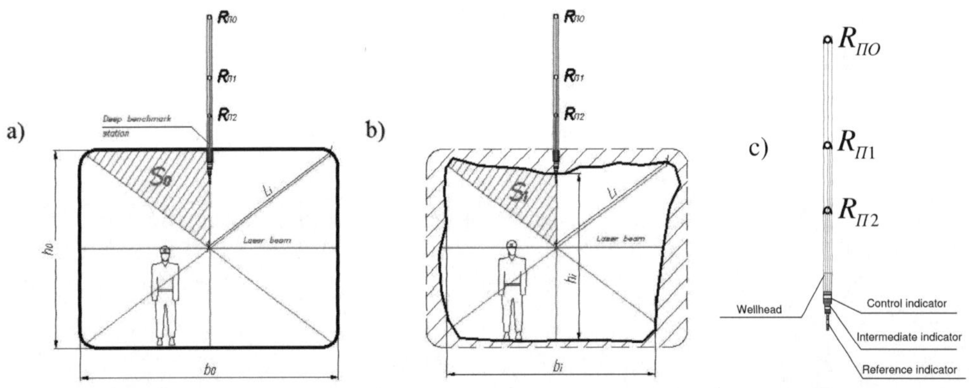

Figure 3. Scheme of instrumental measurement of deformation in the intersection of workings with the help of a range finder: a – the initial cross-section of the working; b – deformed section of the working; c – construction of the DBS.

At the first stage, when mining the working area 1-1, a visual observation of the rock massif condition in the vicinity of the investigated area and the examination of the state of the roof support and the walls of the workings intersections were carried out.

It was established that at a distance of 100-150 m behind the mining face, deformation of the bearing elements of the support was observed at 40 m from the investigated area in the direction of the gob of face 1-1. On the section of the junction of the drift with the connection the following was noted: intensive development of cracks in the roof at the intersection, delamination of the roof rocks, coal sloughing with the subsequent breach of the latticed set lagging from the side of the gob of working area 1-1 (figure 4, a).

As a result, in order to bring drift 1-2 into operational condition, the muck draw was performed with the subsequent cleaning, installation of the metal support of KMPT-27 type with a step of 1 m and baking the voids with timber (figure 4, b). Connection 1-5 and the diagonal connection were strengthened with wooden pit-prop in two rows along the walls of the working.

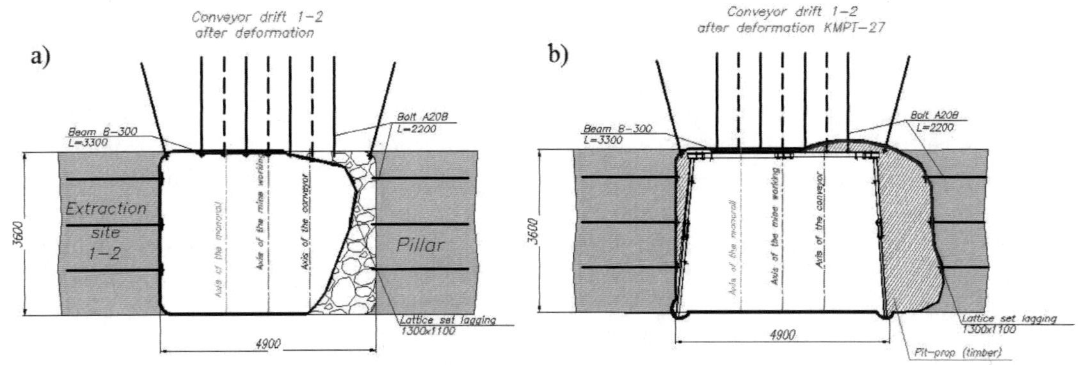

Figure 4. Conveyor drift 1-2 in the intersection between the connections: a – after deformation of the working; b – after strengthening with a metal support of KMPT-27 type.

At the second stage of the mine experiment, the visual observations and instrumental measurements of rock convergence at the intersection of the drift with a diagonal connection were performed while mining the working area 1-2.

3. Results and discussion

Based on the results of processing the mine convergence measurements in the investigated area and analyzing the data of roof rock displacements according to the DBS, a dependency graph of the vertical displacements distribution on the distance to the face 1-2 was plotted (figure 5).

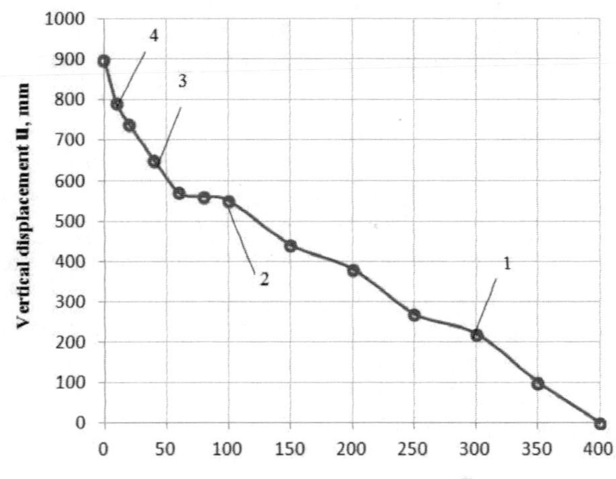

Figure 5. The dependence of the of vertical displacements distribution on the distance to the face 1-2.

According to the results of the survey of the support on the mating of the excavations, a repeated deformation of the anchorage in the form of extrusion of the supporting elements of the anchors on the sides was recorded at a distance of 300 m from the face of the lava (Fig. 5, section 1). At the junction with the drift, coal was extruded and subsequently exposed.

With the approach of the stoping at a distance of 100 m from the face in the roof of the drift at the intersection and for 20 m, the SVP deflection was observed in the locking part of KMPT-27 support (figure 5, section 2).

At a distance of 70 m from the face, a re-convergence of the working roof at the intersection is fixed (figure 5, section 3).

At a distance of 50-30 m from the face in the zone of influence of the bearing pressure of face 1-2, the props breach of the metal supports were observed in the places at the welding seams in 5 m from the intersection of the draft with a diagonal connection. As a result, at the walls of the drift the ejection of the coal pillar started from the finished extraction pillar 1-1 (figure 5, section 4).

In 10 m from the studied area, the intensive lowering of the drift roof was recorded during the face advancement (figure 5, section 5).

At the time of the intersection location in the alignment with the face, a critical condition of the drift support was noted (figure 6). The rock taking-down was performed at the intersection of the workings. Polymer resins were used to fill the voids of the domes in the roof.

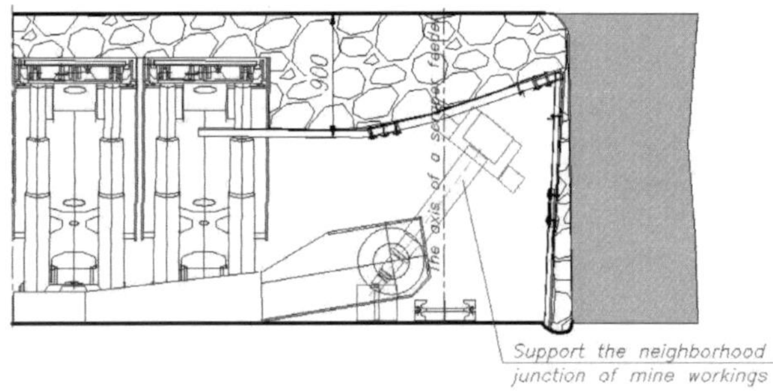

Figure 6. Intersection of the belt road and a diagonal connection in-line with the face.

In accordance with the objectives of the study, the displacement of the roof rocks at the intersection of the mine workings was calculated on the basis of normative documents [1, 2] and instructions [3, 4] for the following options of mining engineering and geological conditions:

1) favorable geological conditions;

2) near a discontinuous geological disturbance;

3) in the zone of increased rock pressure, near the breaking disruption and water cut of rocks;

4) in the zone of increased rock pressure, near the faulting and water content of the rocks after additional exploration.

The core was taken at the site of influence of the face bearing pressure 1-2 to carry out supplementary exploration. According to the results of these studies, it was revealed that at a distance of 50 m from the face in the roof of the workings intersection, an increased fracturing of rocks was registered, which is the reason for the redistribution of stresses and as a consequence the decrease in the physical and mechanical parameters of the roof rocks, which were taken into account during obtaining calculated displacements.

Figure 7 shows the results of calculated rock displacements in the intersection roof of the drift and a diagonal connection for 3 stages of coal-face work (spatial and temporal location of the face):

1) in the zone of the bearing rock pressure (70 m);

2) for the section of workings intersection 10 m ahead of the face;

3) for the section of workings intersection in a line with a cleaning face.

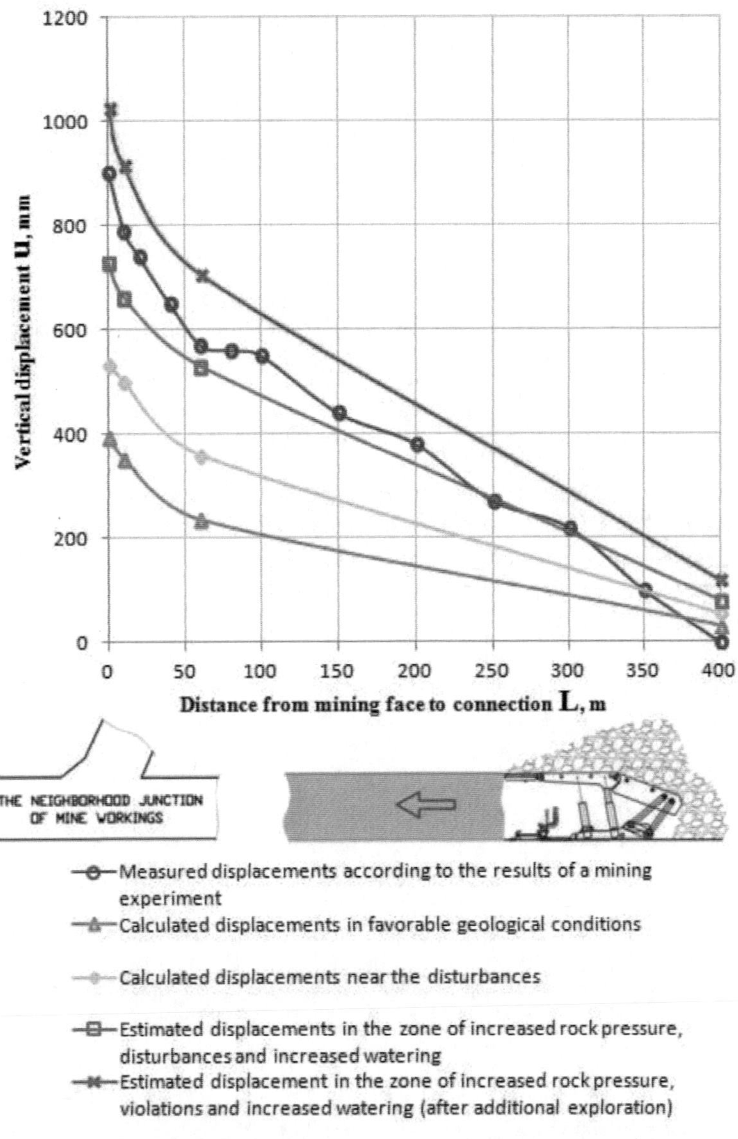

Figure 7. Graphs of vertical displacements distribution in the intersection of mine workings: 1-4 according to the instruction, 5 – instrumental measurements.

The analysis of the mine measurements makes it possible to state that the calculated displacements in the investigated area in the bearing pressure zone of 10 m and in in a line with the face have a discrepancy of about 20%.

Graph 3 of displacements with calculated values according to the instruction has a similar character with the actual data up to the zone of influence of the face; however, it does not fully ensure the correctness of the forecast of the SDS geomechanical parameters of the rock massif in the zone of influence of the bearing pressure. This is the cause for unreasonable selection of the support parameters at the intersection of mining and development workings, since in most cases the calculated values are either too high or less than the actual ones.

4. Conclusions

The obtained results of the calculated displacements of the roof rocks do not fully ensure the correctness of the choice of the support parameters at the workings intersection under the influence of the dynamic bearing pressure. To obtain a more complete volume of information on rock displacements, it is proposed to perform numerical simulation of geomechanical processes and determine the parameters of the SDS rock massif in the vicinity of the intersections of faxes and development workings taking into account the influence of dynamic bearing pressure with subsequent mathematical model calibration based on the results of mine measurements.

References

[1] *Federal norms and regulations in the field of industrial safety "Instruction for the calculation and use of anchor support in coal mines"* (approved by the Order of Rostekhnadzor No. 610 from December 17, 2013)

[2] *Federal rules and regulations in the field of industrial safety "Safety Rules in Coal Mines"* (approved by Rostechnadzor Order No. 550 of November 19, 2013. Registered in the Ministry of Justice of Russia on December 31, 2013, No. 30961)

[3] *Guidelines on the rational location, protection and maintenance of mine workings in the coal mines of the USSR 1986* (Leningrad: Ministry of Coal Industry) p 222

[4] *Method of calculation and selection of support parameters at the intersections of workings for single and paired preparation of extraction pillars 2004* (St. Petersburg: Department of Coal Industry of the Ministry of Energy of Russia) p 84

IOP Conf. Series: Earth and Environmental Science **206** (2018) 012011 doi:10.1088/1755-1315/206/1/012011

The influence of rock interlayer location on the stress-strain state of the rock massif near the underground mine

S V Rib

Siberian State Industrial University, 42 Kirova street, Novokuznetsk, 654007, Russia

E-mail: seregarib@yandex.ru

Abstract. The results of a numerical study of rock deformation near the preparatory mine working during underground mining of coal seams of a complex structure are presented. The evaluation of the stress-strain state of rocks with the location of rock interlayer within the coal seam is performed: in its upper, middle and lower parts.

1. Introduction

A diverse structure of coal seams prevails within the coal-bearing areas of the mines in the south of Kuzbass, [1]. There is a significant variability in the conditions of seams position even within the same extraction site of a mine field. The mechanical properties of coal and rock interlayers in the seams, fracturing, deformation and strength characteristics of rocks in the immediate roof and seams soil change drastically.

For example, at mine "Osinnikovskaya" (Kuzbass) during mining seam E_5 the interlayer of hard rock is observed, which increases the stability of the edge part of the seam. Under the influence of mining works the strained rocks in the roof, soil and coal in the seam are subjected to additional deformation, which can lead to a negative manifestation of rock pressure.

2. Methods of research

Laboratory experiments with the use of equivalent materials were carried out to establish the influence of rock interlayers on the rocks deformability [2]. According to the experiments results it can be stated that the values of the vertical displacements near the working with a rock interlayer decreased with respect to the displacement value near the working with no interlayer [3].

It is proposed to evaluate the influence of the rock interlayer in the seam on the parameters of the stress-strain state (SSS) with the help of CoalPillar computer software [4] intended for modeling the processes of change in displacements, stresses and deformations under the influence of structural seam heterogeneity, natural stress fields and mine workings in the coal and rock massif.

The calculations of SSS parameters of the rock massif in the vicinity of the mine workings are performed in conditions of plane deformation. The dimensions of the calculated area along the width are assumed equal to 600 m, in height – 485 m; the working depth in the mined seam is 350 m. The fragment of the calculation scheme of the problem is shown in figure 1.

On the contour of the model the following boundary conditions are accepted:

- on the left and right boundaries of the calculation area the horizontal displacements equal to zero;
- at the lower boundary of the model the vertical displacements equal to zero;
- the upper limit of the calculation area is free from external load.

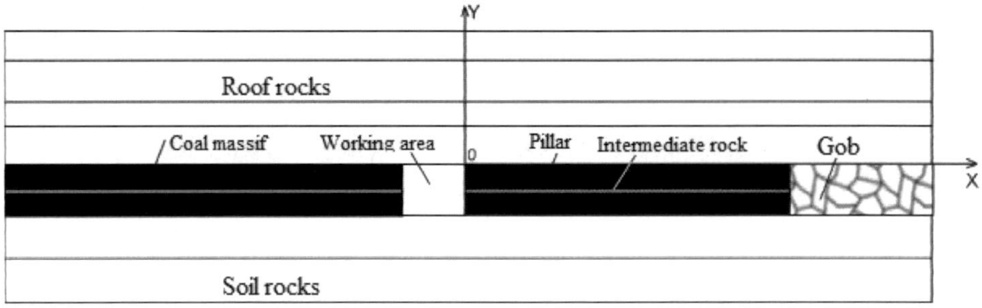

Figure 1. Fragment of the calculation scheme.

For computational experiments the mining and geological conditions of the mine “Osinnikovskaya” (seam E_5) were used. The seam thickness was 3.25 m. Its position was horizontal. The coal strength in the seam was 11 MPa, aleurolite interlayer was 0.5 m with a compressive strength of 30 MPa. The origin of the coordinates is the point of intersection from the axis of the right wall of the working with the roof of the developed seam.

The working with a width of 5 m, height 3.25 m is guarded by the coal pillar 30 m in width from the influence of the gob 210 m in length. SSS modeling of the massif was performed, and the location of the rock interlayer relative to the seam roof and soil was varied. Four options of the model are considered (table 1).

Table 1. Calculation options for modeling.

Option No.	The presence of a rock interlayer in the seam	Location of the interlayer	Distance from the seam roof to the rock seam, m
1	absent	-	-
2	rock interlayer of aleurolite with thickness 0.5 m	in the upper part of the seam	0.5
3	rock interlayer of aleurolite with thickness 0.5 m	in the middle part of the seam	1.5
4	rock interlayer of aleurolite with thickness 0.5 m	in the lower part of the seam	2.25

To establish the influence of the rock interlayer on the SSS of the rock massif in the vicinity of the preparatory working, the characteristic points (figure 2) are indicated, the coordinates of which are given in table 2.

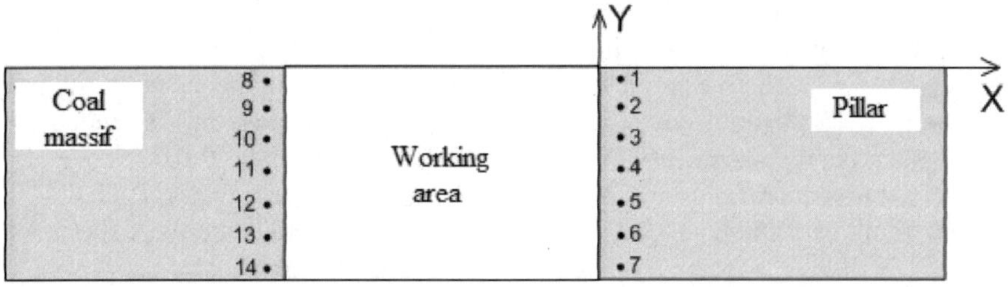

Figure 2. Arrangement of the characteristic points.

Table 2. Coordinates of the characteristic points.

Coordinates of points	Number of the characteristic point													
	1	2	3	4	5	6	7	8	9	10	11	12	13	14
x, m	0.2	0.2	0.2	0.2	0.2	0.2	0.2	-5.2	-5.2	-5.2	-5.2	-5.2	-5.2	-5.2
y, м	-0.1	-0.6	-1.1	-1.6	-2.1	-2.6	-3.1	-0.1	-0.6	-1.1	-1.6	-2.1	-2.6	-3.1

3. Results and discussion

Qualitative differences were found in the distribution of horizontal stresses while solving the problem in the elastic formulation (figure 3, table 3) in the boundary sections of the pillar, i.e. in the walls of the working and in the boundary part of the seam from the side of the gob.

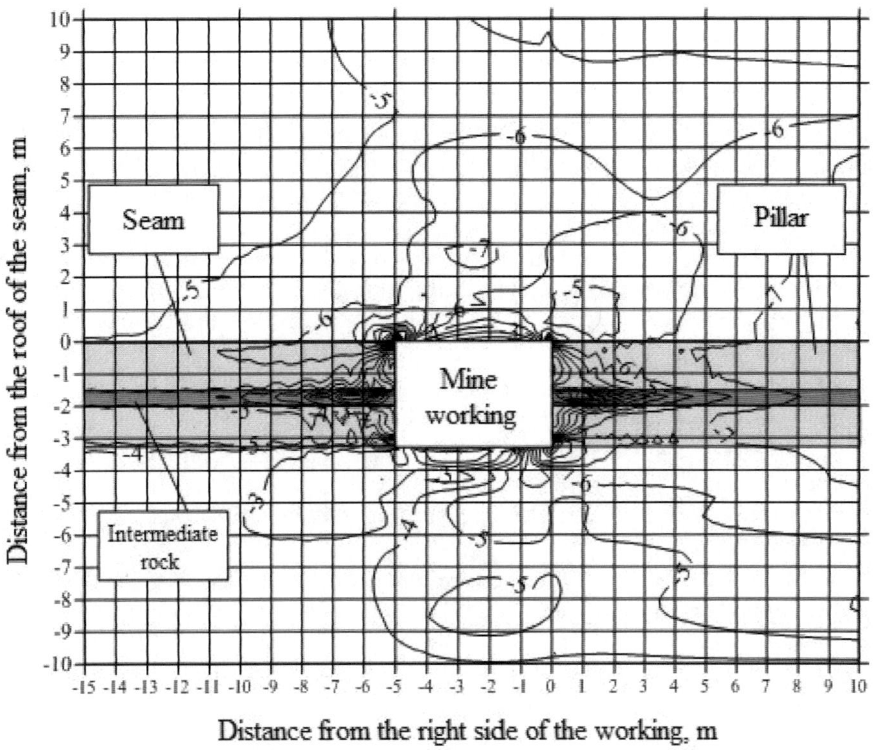

Figure3. The isolines of horizontal stresses distributions, MPa (option 3 with the location of the rock interlayer in the middle of the seam).

Table 3. Calculated parameters of the stressed state of the rock massif in the vicinity of the mine working at different locations of the rock interlayer in the seam.

Position of the characteristic point of calculation	Parameter name	Point No.	Model No. 1	Model No. 2	Model No. 3	Model No. 4
in the right wall of the mine	horizontal stresses, MPa	No.1	-4.8	-4.9	-4.7	-4.7
		No.2	-0.68	1.7	-0.68	-0.65
		No.3	-0.18	-1.5	-0.72	-0.19
		No.4	-0.13	-0.26	0.3	-0.22
		No.5	-0.28	-0.3	-0.79	1.2
		No.6	-1.3	-1.4	-1.4	-1.2
		No.7	-12	-11	-11	-12
in the left wall of the mine	horizontal stresses MPa	No.8	-7.5	-7.7	-7.5	-7.6
		No.9	-0.082	2.5	-0.26	-0.18
		No.10	-0.1	-1.8	-0.87	-0.11

No.11	-0.037	-0.19	0.31	-0.21
No.12	-0.19	-0.22	-0.35	0.72
No.13	-0.85	-0.89	-0.93	-0.63
No.14	-6.6	-6.5	-6.6	-6.8

The figure shows that the presence of the rock interlayer in the seam clearly changes the pattern of stress isolines in comparison with the operation of the working in the seam of simple structure.

4. Conclusions

The proposed methodology for estimating the stress-strain state of the rock massif in the vicinity of the mine working, protected entirely by the pillar in seams of complex structure taking into account the influence of mining works, allows the expected convergence of the roof and soil to be predicted, as well as the working walls, depending on the geological conditions of their operation, and gives the opportunity to use the results of SSS research for the geomechanical justification of the parameters of mining systems in the undeveloped deposits at the design stage.

References

[1] Kuntysh M F and Baronskaya E I 1980 *Methods for Evaluating the Seam Properties of complex Structure* (Moscow: Science) p 144
[2] Rib S V 2017 *Science-intensive Technologies for the Development and Use of Mineral Resources* **3** 155–157
[3] Rib S V and Basov V V 2017 *Bulletin of Siberian State University on Economics and Finance* **4(22)** 45–51
[4] Rib S V et al 2014 *Certificate of Registration of an Electronic Resource "Software Complex for Forecasting Geomechanical and Technological Parameters of Underground Mine Workings and Heterogeneous Coal Pillars with Testing of a Numerical Model Based on the Results of Point Monitoring"* (State Ac. of Sci, INIPI RAO) No. 20629; date of reg. 09.12.2014

IOP Conf. Series: Earth and Environmental Science **206** (2018) 012012 doi:10.1088/1755-1315/206/1/012012

Creation of a competitive production on the basis of electrothermal and plasma technologies of combined coal and carbonate mineral raw materials processing with obtaining fuel and non-fuel products

A V Zhukov, M I Zvonarev, Yu A Zhukova and A A Shmelev

Far Eastern Federal University, 8 Sukhanova street, Vladivostok, 690091, Russia

E-mail: Yul25juk@mail.ru

Abstract. Physicotechnical and organizational-technological models of technologies and technical means were developed, an integrated feasibility study of the organization of production of competitive products was performed, and the main technical and economic parameters of the newly created production were calculated for obtaining fuel and non-fuel products. When implementing the organizational and technological model of innovative production at the first stage, the following combined directions for processing mineral raw materials and new products are planned: 1) processing of carbonic acid mineral raw materials: calcium carbide, carbon dioxide (in gaseous, liquid or solid state); 2) acetylene, plant growth regulators (REGROST), plant protection products (TAKAR). Fuel products in the second stage: 1) synthetic ethyl alcohol (ethanol), antifreeze, ethylene glycol, dichloroethane, synthetic drying oils, acetone, etc.; 2) Carbamide (urea), ammonia, nitrogen in gaseous and liquid state, methanol, gasoline, etc. The previously developed economic-mathematical model (EMM) of production organization has a high degree of universality. It is advisable to implement the EMM differentially for alternative options at coal-mining, energy, gas producing and fuel-energy enterprises.

1. Introduction

In the Strategy for the Development of Fuel and Energy Resources in the Far Eastern Economic Region until 2020" the following tasks were set: "... to explore ways of overcoming the crisis situation by developing Far Eastern oil and gas fields on the basis of progressive methods of extraction and chemical processing of primary raw materials, and also by introduction of a policy of economy…".

In this connection, it was necessary: 1) to develop a physico-technical and economic-technological model of technologies and technical means for deep chemical stage-by-stage processing of carbon-carbonate mineral raw materials; 2) to apply electrothermal and to develop plasma ecologically safe resource and energy saving technologies for processing carbonate and coal mineral raw materials using several modules (subsystems) to produce competitive fuel and non-fuel products [1, 7-9].

The developed and patented technologies provide for a complex waste-free processing of carbon-carbonate mineral raw material using several modules that at the first stage ensure the production of calcium carbide, calcium oxide, carbon dioxide, at the second stage – processing products of calcium carbide: synthetic and gaseous liquid fuel and non-fuel products.

In the new production complex for the production of calcium carbide (CaC_2), preliminary calcination of limestone ($CaCO_3$) is carried out to obtain calcium oxide (CaO) and carbon dioxide (CO_2); batch from calcium oxide (CaO) and coal (C) are used to produce calcium carbide (CaC_2).

Carbon dioxide (CO_2) formed during calcination of limestone is captured and carbon dioxide (H_2SO_3) is produced from it by means of equipment for trapping off-gases (figure 1).

During the production of acetylene hydrated lime – $Ca(OH)_2$ is formed from calcium carbide, which is used in construction industry, in agriculture for the production of superphosphate – $Ca(H_2PO_4)$, in the water purification systems, etc.

Synthesis of (carbamide) urea – $CO(NH_2)_2$ is possible with the participation of carbon dioxide (CO_2) by addition of ammonia (NH_3) or carbamide can also be produced from calcium carbide using nitrogen (N_2). In this case, to obtain nitrogen (N_2) the separation of the air mixture is used – the rectification of the liquid air or by the oxidation of ammonia: $4NH_3 + 3O_2 = 2N_2 + 6H_2O$.

Carbamide (urea) is used to increase the efficiency of plant growth regulators (REGROST).

Figure 1. Installation for the capture of gases and the production of carbon dioxide ("dry ice").

In Primorsky Territory the demand for calcium carbide is about 1000-1500 tonnes/year, carbon dioxide: 2000-2500 and more tonnes per year. At present, these products are imported from Eastern Siberia and partly from China, Temir-Tau (Kazakhstan).

The investment attractiveness of this project is very high, since its implementation allows local, abundant, inexpensive sources of raw materials to be used and valuable products to be produced on its basis with high added value and with sufficiently high demand stability for different sectors of the economy and a wide range of applications, which provides increased economic stability of the enterprise during changes in market conditions and its different segments.

The main profit comes from the sale of agricultural products (REGROST and TAKAR), the price for which is set at 1.5-2 times lower than for imported analogues. Given the novelty of the products, preliminary measures on informing consumers about the advantages of REGROST and TAKAR in comparison with the known analogues [3-5] are planned in parallel with the construction of the plant.

2. Methods of research

Calcite carbide, calcium oxide, acetylene, methanol, carbamide, fertilizers REGROST and TAKAR are produced as the main products after processing of coal and carbonate mineral raw material; as well carbon dioxide, drinking or calcined water, calcium hydrate are produced as additional products.

Whereas, processing of acetylene makes it possible to obtain: synthetic ethylene and ethyl alcohol, naphthalene, hydrogen: synthetic drying oils (oil varnish), chloroprene rubber. In accordance with the polymerization reaction, the following products are also created: ethylene, polyethylene (polymer), propylene, polypropylene in the following directions of acetylene (C_2H_2) processing:

1) Acetic aldehyde (CH_3CHO) → synthetic ethyl alcohol (C_2H_5OH);
2) Benzene (C_6H_6) → naphthalene ($C_{10}H_5$) → hydrogen (H_2);
3) Vinyl-acetylene ($CH_2 : CH \cdot C : CH$) → divinyl-acetylene ($CH_2 : CH \cdot C : C \cdot CH : CH_2$) → synthetic drying oils (oil varnish);
4) Vinyl-acetylene ($CH_2 : CH \cdot C : CH$) → Chloroprene ($CH_2 : CH \cdot C\ Cl : CH_2$) → chloroprene rubber;
5) In accordance with the polymerization reaction, the following products are created:
6) T, R, cat
7) ethylene ($nCH_2 = CH_2$) → $(- CH_2 - CH_2 -)$ n – polyethylene (polymer);
8)
9) T, R, Cat

$$10)\ \text{propylene } nCH_2 = CH - CH_3 \longrightarrow \left[\begin{array}{c} -CH_2 - CH \\ \downarrow \\ CH_3 \end{array} \right] n - \text{polypropylene.}$$

Figure 2 shows an example of a physical-technological model of an installation for the processing of coal and carbonate mineral raw materials. Similar physical and technological models are developed and protected by inventions for most of the products discussed above and published author's certificates [7,8,9].

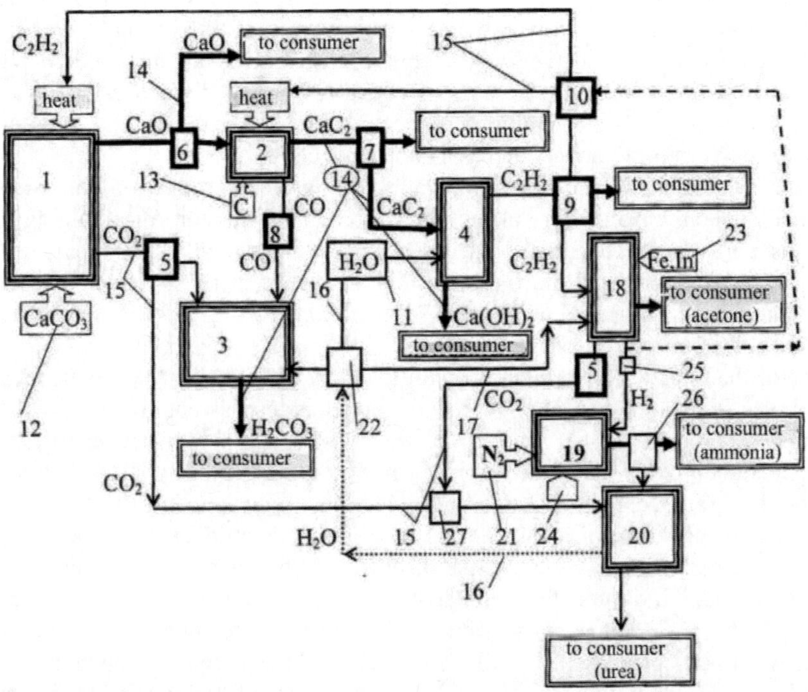

Figure 2. Physical and technological model of the unit for the processing coal and carbonate mineral raw materials: reactor of electrothermal furnace RKZ-2.5 (1), unit for calcining limestone and obtaining calcium carbide (2), unit for gas capture and production of carbon dioxide (3), unit for the production of acetylene (4), unit for capturing carbon dioxide (5), unit for producing calcium oxide (6), a calcium carbide storage bunker (7), acetylene station (9), electrothermal unit for melting mineral charge (10), a calcium hydrate storage (11), acetone production unit (18), unit for the production of ammonia (19), a device for the production of carbamide (20).

The invention can be used to produce calcium carbide, acetylene, quicklime and hydrated lime and carbon dioxide. The plant for processing carbonate mineral raw materials comprises a lime kiln 1, a furnace for calcium carbonate production 2, a carbon dioxide synthesis reactor 3, a gas generator 4, the first 5 and the second 8 gas sampling units, dispensers 6, 7, an acetone synthesis reactor 18, an ammonia synthesis reactor 19, a carbamide synthesis reactor 22. The first gas distribution unit 9 is connected to the first inlet of the acetone synthesis reactor 18, the second inlet of which is connected to the vapor source 22.

The invention makes it possible to expand the spectrum of the obtained commercial products of deep acetylene processing while ensuring a high level of diversification of production and exclusion of the generation of man-made wastes [7,8]. Similarly, the following products can be obtained from calcium carbide: acetylene – synthetic ethanol, ethylene glycol, dichloroethane, antifreeze, acetone, synthetic drying oils, etc. [9].

3. Results and discussion

The developed economic-mathematical model (EMM) of the organization of innovative high-tech production, built for coal-mining, power and industrial enterprises engaged in both coal, gas, oil production and the production of natural and synthetic gaseous and liquid fuels, is possible during the reconstruction and diversification of this production. Such an economic and mathematical model of the organization of production has a high degree of universality and it is expedient to implement the EMM differentially for alternative options for coal-mining, energy and gas producing enterprises.

Economic-mathematical and organizational-technological models are planned to be applied for development of territorial investment programs of construction and diversification of enterprises in fuel and energy industries that use and process coal and carbonate mineral raw materials into traditional and alternative energy carriers and non-fuel products.

In the integrated feasibility study of the organization of high-tech production, the calculation of the main technical and economic parameters of the newly created production for obtaining fuel and non-fuel products was performed. The paper discusses a new ecologically clean patented method of combined processing of coal and carbonate mineral raw materials, which is the scientific, methodological and organizational development of the construction of the first-due mining and chemical complex in Primorsky Teritory.

The relevance of the topic of this project is conditioned by the fact that thermochemical processing of coal and carbonates in newly created or reconstructed industrial enterprises provides the most valuable materials and products for the production sector, construction industry, energy engineering in this pilot project: calcium oxide, calcium carbide, carbon dioxides in gaseous, liquid and solid state ("dry ice"), acetylene, growth regulators and plant protection products (GRPPP).

Table 1 presents complex data of technical and economic indicators of investment projects, where column 4 indicates the performance of JSC "Spasskcement" in production of cement of various grades in 2012. The main initial indicator for comparison is the volume of mineral raw material used throughout the year: in line 3.1 it is shown that in the compared options for calculations – 990990 tonnes of limestone is accepted as ~ 1 million tonnes. For the second component – coal for the production of calcium carbide it will be required by approximately 100 thousand tonnes more, but this is completely compensated by the exception in this production of other mineral resources: clay, gypsum components, etc. [6, 10, 11].

Table 1. Integrated table of technical and economic indicators of investment projects (before the ruble devaluation in 2014).

No.	Name of the TEP	Units	Cement	Type of product					
				RKZ – 2.5			Plasma reactor		
				Calcium carbide	Carbon dioxide	Total	Calcium carbide	Carbon dioxide	Total
1	The volume of output	T	1415700	2500	3500	—	1200	1650/ 1319	—
2	The volume of mineral raw materials:								
2.1	Limestone	T	990990	5500	—	5500	2640	—	2640
2.2	Coal	T	218018	1750	—	1750	840	—	840
3	Cost of production	rub./t	1826	18220	4998	—	13990 (14419)	4172	—
4	Sale price	rub./t	3286	30000 – 36000	15000 – 25000	—	30000 – 36000	15000 – 25000	—
5	Number of IPP	person	1785	30	15	45	9	4	13

IOP Conf. Series: Earth and Environmental Science **206** (2018) 012012 doi:10.1088/1755-1315/206/1/012012

6	Product profitability	%	21.42	29.36 – 37.80	51.67 – 63.00	38.55 – 50.22	41.65 – 48.05 (40.22 – 46.85)	56.64 – 65.98	46.97 – 55.81 (46.04 – 55.13)
7	Revenues	million rub.	4652	75 – 90	52.5 – 87.5	127.5 – 177.5	36.00 – 43.20	19.79 – 32.98	55.79 – 76.18
8	Profit	million rub.	996.26	22.02 – 34.02	27.13 – 55.13	49.15 – 89.15	15.00 – 20.76 (14.48 – 20.23)	11.21 – 21.76	26.20 – 42.51 (25.68 – 42.00)
9	Revenue per person per year	million rub.	2.61	2.50 – 3.00	3.50 – 5.83	2.83 – 3.94	4.00 – 4.80	4.95 – 8.25	4.29 – 5.86
10	Volume of limestone per 1 tonne	th/t	0.70	2.20	—	2.20	2.20	—	2.20
11	Volume of coal per tonne	th/t	0.15	0.70	—	0.70	0.70	—	0.70
12	Energy costs per 1 ton of products	kWh	182.30	4000	1000	5000	2700	1000	3700
13	Pay-off period	year	—	3.17 – 2.06	1.48 – 0.73	2.24 – 1.24	1.13 – 0.82 (1.52 – 1.09)	0.89 – 0.46	1.03 – 0.64 (1.24 – 0.76)

The above data on the project for the processing of carbonate and coal mineral raw materials with the use of resource and energy-saving technologies only on the first subsystem (production of calcium carbide and carbon dioxide) shows their high efficiency and, equally importantly, environmental safety.

Technical and economic calculations of the application of a newly created plant for the processing of carbonate and coal mineral raw materials using an experimental industrial plasma reactor with a power of 500 kW throughout the above technological chain of products: calcium carbide, carbon dioxide, acetylene. REGROST and TAKAR with the same annual volumes of consumed mineral raw materials – **990990** tonnes of limestone and **315315** tonnes of coal show the following integrated production parameters:

1) volume of investments, mln. rub. – **21375**;
2) volume of output, t:
3) calcium carbide – **450450/656250**;
4) carbonic acid – **492244/616360**;
5) acetylene – **54000**; REGROST – **90000**; TAKAR – **144000**;
6) number of production personnel, person – **10425**;
7) profitability of the products sold, % - **68.9-69.3**;
8) revenue, million rubles – **87043- 95359**;
9) profit, million rubles – **60912 -65053**;
10) revenue per 1 person per year, million rubles – 8.87.

Thus, the proposed investment project, which ensures the rational combination of innovative (combined) technologies in a single technological complex, allows the extremely cost-effective and integrated use of local mineral raw materials and energy to be performed on the basis of highly profitable production with high profit margins and added value with a minimal for production projects payback period [10].

4. Conclusions

1) The parameters of physical-technological and economic-mathematical modeling of processes makes it possible to consider the configuration and parameters of the systems of restructuring and diversification of coal-mining and energy enterprises in more detail, to agree and determine the operating modes of control subsystems and production organization, to forecast technological and economic indicators of newly created and diversified production complexes

2) When processing coal and carbonates, it is planned to develop a physico-technical and economic-technological model for newly created technologies; the selection and justification of the scheme for the production of synthetic gaseous and liquid fuels is considered based on the specific conditions, the cost and quality of coal, energy supply, market conditions and other factors.

3) In the territorial investment programs it is advisable to envisage the application of electrothermal technologies and to continue the development of plasma ecologically safe resource and energy-saving complexes using the principles of modular design of a new production.

References

[1] Sorokin A P, Avdeyko G P, Alekseev A V, Baklanov P Ya, Zhukov A V and Podolyan V I et al 2001 *Dal'nauka* **1** 12
[2] Zhukov A V, Zhukova Yu A, Zvonarev M I and Umarov M S 2017 *Proc. of Int. Sci. Conf. on Science-intensive Technologies of Development and Use of Mineral Resources* (Novokuznetsk) **3** pp 21–26
[3] Zhukov A V and Zhukova Yu A 2014 *Basic Research* **9** 13–7
[4] Zhukov A V 2014 *FGBNU NII RINKTSE Innovation and Expertise* **2(13)** 235–42
[5] Zhukov A V, Zhukova Yu A, Umarov M S and Mikhalkov A V 2015 *Development of Geo-resources in the Asia-Pacific Region. Mining Information and Analytical Bulletin* **4** 51–9
[6] Zhukov A V, Kusraeva D E, Zhukova Yu A, Shmelev A A and Umarov M S 2016 *Advances in Modern Natural Science* **12(1)** 154–61
[7] Zhukov A V, Zvonarev M I and Zhukova Yu A 2008 Patent 74912 Bul 20
[8] Zhukov A V, Zvonarev M I and Zhukova Yu A 2009 Patent *RU* 2362735 C1 Bul 21
[9] Zhukov A V, Zvonarev M I and Zhukova Yu A 2009 Patent RU 2373178 C2 Bul 32
[10] Kutasov M L 2012 *GTh of Training Specialists in the direction 080502.65 – Economics and Management at the Enterprise*. Supervisor of the GTh – Doctor of Tech. Sci., Prof. Zhukov A V (Branch of Far Eastern State Technical University Spassk-Dalniy of Primorsky Territory) 102
[11] Zhukov A V, Zhukova Yu A and Umarov M S 2017 *Proc. of Sci.-Prac. Conf. on Geosystems of Northeast Asia: Territorial Organization and Dynamics* (Vladivostok, Pacific Institute of Geography, FEB RAS) pp 431–435

Mineral resources base of the south of Kemerovo region (Mezhdurechensky administrative district)

Ja M Gutak and O Ya Gutak

Siberian State Industrial University, 42 Kirova street, Novokuznetsk, 654007, Russia

E-mail: GutakJaroslav@yandex.ru

Abstract. The article considers current state of mineral resources base of Mezhdurechensky administrative district of Kemerovo region. The main mineral resource of the region for the present and the near future is coal, explored reserves of which ensure operation of coal-mining enterprises for a period of 15 to 85 years. Another relevant mineral resource in the district is placer gold, which is mined by small diggers at several placers in Kuznetsk Alatau and Mountain Shoria. The district has significant resources of metallic (manganese, ore gold) and non-metallic (talc, tremolite, vermiculite, etc.) minerals.

1. Introduction

Mezhdurechensky administrative district is located in the southeast of Kemerovo region. It covers the valley of the river Tom in its latitudinal part and adjacent areas of Kuznetsk Alatau (in the north) and Mountain Shoria (in the south). Geologically, area is located at junction of the Caledonian structures of Kuznetsk Alatau and Mountain Shoria (border is drawn along the latitudinal section of the Tom river) with the Hercynian Kuznetsk marginal trough (large coal basin is formed in the trough in the Late Paleozoic). This defines metallogenic specialization of the area and its minerals, of which only coal and gold are currently relevant. Analysis of district mineral resources base starts with coal, then resource potential for metallic and non-metallic minerals is examined.

2. Mineral resources overview

2.1. Coal

Resources of coal in Mezhdurechensky district are genetically related to the Permian deposits of Kuznetsk coal basin (Tom-Usinsky geological and economic region). Within the latter, four sites with prospects of coal extraction are distinguished (the sites are often referred to as the deposits): Cheksinskoye, Raspadskoye, Olzherasskoye and Tomskoye. Commercial coal mineralisation is associated with deposits of Balakhon (Upper Balakhon sub-series) and Kolchuginskaya series of the Permian system. Systematic geological prospecting and exploration for coal in the district had begun in the late twenties of the last century and is closely associated with V.I. Yavorsky, outstanding Russian geologist. At present, coal mining is conducted by Raspadskaya PJSC: Raspadskaya Mine, Raspadskaya Coke Mine, Raspadsky Open Pit Mine (43.9% of coal production in the region); South Kuzbass PJSC: Olzherasskaya Novaya Mine, V.I. Lenin Mine, Krasnogorsky, Olzherassky, Tomusinsky open pit mines (29% of coal production), Sibuglemet Holding of Mezhdurechye JSC: Mezhdurechensky Open Pit Mine (25.1% of production). Coal mining is conducted on Raspadskoye, Olzherasskoye and Tomskoye coal deposits. Reserves life of enterprises varies from 15 to 85 years.

Cheksinskoye coal field is considered as a long term reserve (due to remoteness and inaccessibility). Tom-Usinsky geological and economic area is characterized by a wide range of coal grades, such as: G, GZH, KZh, K, KO, OS, T and A. Total inferred resources of the area are estimated as 2.7 billion tons. Of these, by the beginning of our century, explored by A, B and C_1 categories are 10527 million tons [5]. Immediate industrial prospects of the district are associated with refinement of reserves of the Tomskoye, Olzherasskoye, and Raspadskoye fields upper horizons and commissioning of their deep horizons.

2.2. Ferrous metals

2.2.1. Manganes. Usinskoye deposit. Located 60 km to the north of Mezhdurechensk in the Usa river basin in the middle part of Kuznetsk Alatau mountain range. Discovered by K.V. Radugin in 1939. Manganese reserves are 98.5 million tons (including 69339 thousand tons by A + B + C_1 categories) [1,3].

Ores include rhodochrosite and marganocalcite. Stratiform deposit (the Lower Cambrian limestones) is confined to the western steeply dipping wing (70-90 °) of synclinal fold and is formed by structural and lenticular accumulations (several hundred meters long, and 20-65m thick) extending 4-6km to the north-west. Ore bodies are traced to depths of more than 500 meters in patches. In northern and central areas, rhodochrosite and manganocalcite ores are predominant, interbedded with manganese limestone (Mn content in ore is 16.7-19.6%). Total thickness of layers exceeds 150m. Carbonate manganese ores from the surface are oxidized: oxidation zone is up to 40m in the north and 50-75m in the south. Oxidized ores are composed of psilomelane, vernadite, less often of pyrolusite (Mn content is 26.5-27.7%). Operational work at the deposit had been started several times, but each time it ended with nothing. The last attempt was made in this century by Cheksu Co (Republic of Khakassia). The reason is carbonate mineral form of manganese ores, which is inapplicable in metallurgy, low average manganese content in ores and inaccessibility of the site.

2.3. Precious metals. Gold

2.3.1. Placers. Alluvial gold of Mezhdurechensky district is known since the mid-thirties of the nineteenth century, when the party of Aboltin, second lieutenant of the corps of mining engineers had discovered a rich placer on the river Petropavlovka (the right tributary of the river Bazas). In 1836, the placer along the river Fedorovka was discovered, and in 1837 its operation had started. In 1842, the placer of the river Veselaya and in 1843 the placer of the river Bazas were discovered. From those times to the present, mining operations are periodically carried out at alluvial deposits. There is no exact data on the amount of gold produced at the placers. Obviously, each new development of placers, decreases gold content in sands. Currently, the minimum industrial gold content in placers of Mezhdurechensky district is stated at 125 mg/m^3.

Alluvial placer in the middle flow of the river Usa and its tributaries - Berezovaya, Shatay, Murta. Mining operations are currently being conducted by the "Golden Pole" CJSC.

Tebinskaya placer originates 10km above the mouth of the river Teba and extends to its headwaters, and from the mouth of the river Koksu (the right tributary) to its headwaters. The placer had been operated by Altai gold-mine of Zapsibzoloto industrial association using gold dredgers since 1977 to 1985.

Placer on the river Zaslonka includes placer of the river Zaslonka itself (the right tributary of the river Mrassu) in its middle flow and placer of its right tributary, the river Ploskaya. Placer of the river Zaslonka is explored on its valley segment with the length of 8.5km, the placer of the river Ploskaya is explored from its mouth 0.5km up the valley. Pay-Cher 2 prospectors association has obtained the license to develop the placer.

The Orton-Fedorovsky placer cluster (three alluvial gold deposits along the rivers Fedorovka, Orton, Maly Orton, Bazas and their tributaries, which are currently in operation). Placer of the river

IOP Conf. Series: Earth and Environmental Science **206** (2018) 012013 doi:10.1088/1755-1315/206/1/012013

Fedorovka is considered the largest in the Mountain Shoria. Reserves by C_1 + C_2 categories here are 1.187 tons, and inferred resources (P_1) are 2.01 tons. The placer is alluvial, valley, its length is 12.1km with a width of 50m. Peat capacity is 3.1m, sands capacity is 1.3m; average metal content in sands is 3 g/m^3, and in total mass is 0.183 g/m^3. Gold is coarse, mostly lamellar, during mining nuggets up to 20kg in weight were encountered. The placers of the Orton and Bazas rivers (with tributaries) are significantly inferior to those of the Fedorovka river in terms of the amount of mined, in-place and prospected gold. The main indigenous sources of the Orton-Fedorovsky alluvial gold placer cluster are Fedorovskaya, Orton-Magyzinskaya and Asanovskaya ore zones belonging to placer-generating gold ore quartz formation. Inferred resources of alluvial gold in cluster are 2.81 tons by P_1 category, and 0.11 tons by P category [6]. Since 1993, development of placer gold along the rivers Orton and Small Orton, as well as in three blocks of placer adjacent to the bottom of the placer along the rivers Orton and Fedorovka, is provided by Kuznetskaya prospectors association (since 2006 - Aurum prospectors association). Bazas prospectors association conducts mining along the Bolshoi Orton and the Maly Orton rivers.

2.3.2. Primary deposits. Fedorovskoe 1 (Kashkatskoye) deposit. Located on the left side of the river Orton (headstream of the left tributary of the Fedorovka river). It was discovered by the Orton party of Krasnoyarsk geological department in 1959. It is a series of vein-stockwork zones in the intensively deployed Late Precambrian and Lower Cambrian volcanogenic- sediments and magmatic associations penetrating in them. Gold content in traces is up to 248g/t. Gold is free, mostly coarse with gold grain size of 1 to 6mm. Distribution of metal is uneven with rich nests alternating by empty intervals. Total resources of the deposit by P_2 category are 8.1 tons of gold. Part of ore of the deposit has been mined from the surface to a depth of 10-15m [6].

2.4. Non-metallic minerals

2.4.1. Talc. The Alguyskoye deposit of low-iron powder talc is located in the headstreams of the Alguy river, 9km from Luzhba station. Discovered in 1960 by geologists of Tom-Usinskaya exploration expedition. Talc reserves amount to 11 million tons. High quality talc. In recent years, there have been several attempts to launch exploitation of the deposit. The last attempt was made by Rusotalk CJSC in 1998. However, due to the deposit inaccessibility, cost of talc does not ensure profitability of its extraction.

2.4.2. Asbestos. Topchulskoye deposit. Located 3.0 km to the south of the village Luzhba. It is confined to massif of hyperbasites of the Konchik mountain. It is explored from the surface by ditches and pits. Two asbestos zones with length of 4 and 1km were traced with power of 370m and 200m, respectively. Within the zones, 15 deposits of chrysotile asbestos with thickness of 10-22mm with length of fibers up to 12mm are allocated. Fiber content is 1.5-11%, average content over the deposit - 4.1% (content of grade I and II fiber is 1.03%). Inferred resources by P_1 category is 1.0 million tons [2,4]. It should be noted that this type of minerals is considered carcinogenic overseas and its development is prohibited by law.

2.4.3. Tremolite rocks. Alguyskoye deposit. Located in the same ore field with the Alguyskoye talc deposit. Evaluation of the deposit was carried out to a depth of 100m. Four main tremolite deposits lying on the folds wings were identified and outlined. The deposits are in form of consonant lenses and layers with capacity of up to 150m and length of up to 800m. The main components of tremolite rocks are tremolite (65-75%) and calcite (20-30%). Tremolite reserves by C_2 category are about 12 million tons. Rocks are suitable for production of household semi-porcelain and construction ceramics.

Degtyarnoye deposit. Located on the right side of the river Orton, along the river Degtyarnaya 2.5 km from its mouth. It is represented by large reservoir with capacity of 80–100m of intensively tremolitized dolomites of Mrasskaya formation of the Lower Cambrian. The strike is sublatitudinal, transferring to the south-west on its western flank. Established length of reservoir is 1.5km. Ores are composed of tremolite (50.6%), talcum (13.73%), dolomite (16.94%), calcite (11.68%) and quartz (6.1%). In terms of quality, tremolites correspond to that of Alguyskoye deposit. Inferred resources by P_1 category are estimated at 50 million tons, by P_2 category at 25 million tons, which is 3-4 times higher than reserves of the Alguyskoye deposit.

2.4.4. Graphite. Konyukhovskoye deposit. Located 9km to the NNW from the station Teba (the Konyukhovsky stream, left tributary of the river Izrass). It is localized in metamorphic formations of the Yenisei series (biotite gneisses, slates with marbles and small bodies of amphibolites). Stratiform-type deposit. Thickness of the graphitized stratum is 160–200m. Stratum is traced 1.5km along the strike. Content of graphite in rich ores is 4-20.5%, in poor ores - 1-3%. Graphite concentrate belongs to clearly crystalline natural graphite. Estimated reserves of the deposit at depth of 200m are defined in the amount of 4.5 million tons.

2.4.5. Vermiculite. Tebinskoye deposit. Located 3-4km north from the station Teba. It is localized in weathering crust with thickness of 6-17m (area 1km^2) along gorglendites, amphibolites, and amphibole-feldspathic gneiso-granites of the Tomsk association. Vermiculite is formed due to hydration in the zone of hypergenese phlogopite and biotite. Content of vermiculite in the deposit varies from 5 to 70%. C_1 reserves are 425.9 thousand tons, - C_2 reserves are 280.1 thousand tons. Vermiculite can be used to manufacture thermal insulating products, lightweight concrete and mortar, as a filler in production of plastics, rubber, enamels and paints.

2.4.6. Marshallite. As part of Alguyskoye talc deposit. Marshallite here forms independent deposits and is present in talc ores. In the northeast of the deposits, marshallite contain talc mineralization. Identified reserves of raw materials amount to about 2.1 million tons. It can be used in foundry and ceramic production.

2.5. Stone-cast materials

2.5.1. Diabase. Usinskoye field. Located 8km from the mouth of the river Usa. It is associated with the Syrkashevsky site of the lower Triassic, localized in the Intermediate and Ishanovskaya suite of the Permian system. Industrial stone reserves were approved in 1956 by categories A + B in the amount of 35,812 thousand m^3. Rocks can be used as a quarrystone, paving stone and in facing products manufacturing. Production tests had shown good foundry ability of diabase.

2.6. Brick and tile clay and loam

Mezhdurechenskoye deposit. On the right board of the Usa river 2.5km north from Mezhdurechensk. Eluvial-deluvial loams of the Upper Quaternary age. Average capacity is 4.8m. Subject to introduction of mineral filler into the mixture, it is suitable for production of grade "100" solid brick. Reserves of loam by C_1 category is 227.0 thousand m^3.

Olzherasskoye deposit. Located in confluence of the rivers Olzheras and Usa (3.0km eastward from Mezhdurechensk). Brownish heavy loam. Thickness of productive layer is 3.0 - 7.0m. With introduction of mineral fillers, it is possible to use in production of "125" and "150" grades construction bricks. Category A reserves are 487.9 thousand m^3.

2.7. Carbonate mineral materials for finish lime production

Belsynskoe deposit. Located 1.5km from the mouth of the river Belsu. Deposit is localized in the Lower Carboniferous sediments and is represented by limestone stratum of 17.5m. Limestone is suitable as a mineral for production of air lime of 1 and 2 grades, hydraulic fluid and silica Portland cement. Limestone reserves by A + B + C_1 category are 11924.0 thousand tons. Large stripping ratio is indicated as a negative feature of the deposit structure.

IOP Conf. Series: Earth and Environmental Science **206** (2018) 012013 doi:10.1088/1755-1315/206/1/012013

2.8. Sand-gravel mix

Tom-Usinskoye deposit. Located within the boundaries of Mezhdurechensk on two neighbouring islands. It is composed of channel alluvial deposits of the river Tom. Capacity of the deposit body is from 5.0 to 7.0m. Reserves by B + C_1 categories were approved in 1956, and registered in State Reserve in the amount of 2,618 thousand m^3, including 1027 thousand m^3 of B category resources. Deposit is located in water protection zone of the river Tom.

Chulzhanskoye deposit. Located along the Novokuznetsk-Abakan railroad from the village Karai to the village Kameshok. Localized in sediments of the first floodplain terrace of the river Tom. Sand-gravel mix forms a bedding plain body with thickness of 0.4 to 1.2m. Sediments are water-flooded. Estimated gravel reserves by C_2 category are 3917.0 thousand m^3; reserves of sand are 3123.0 thousand m^3.

2.9. Building stones

2.9.1. Sandstones. Usinskoye field. Located 2km from the mouth of the river Olzheras upstream the river Usa along its right bank. Sandstone layer of 70 to 77m thickness is placed in sediments of the Permian system (Ishanovskaya suite). Sandstone reserves by A + B category are 4635.0 thousand m^3.

2.9.2. Granites. The Elk Stone deposit. Located on the right bank of the river Tom 6.0km from the Teba station. Intrusive body is composed of fine crystalline granites, granite-porphyries, crystalline diabase porphyrites. Granites prevail. Minerals are applied as crushed stone in pavements foundation, heavy concrete filling, railway ballast, asphalt concrete roadway aggregates. Reserves by C_1 category are 6,318.0 thousand m^3, by C_2 category - 65,388.0 thousand m^3. The facility is located in water protection zone of the river Tom.

2.10. Facing and ornamental stones

Mezhdurechensky district has a significant number of promising occurrence of this type of minerals. Unfortunately, absence of stone processing production in the district and underdeveloped market leave this type of mineral resources outside of business interest. On promising occurrences, only preliminary assessment was carried out.

2.10.1. Marble. Chernousovskaya group of occurrences are on the left bank of the river Chernaya Usa; Belousovskoye occurrence is in the upstream of the river Belaya Usa; Usinskoye occurrence is in the Usinskoye manganese deposit area; Belsynskoye occurrence is 15km north from Luzhba station; Izrasskoye occurrence is Belsu-Izrass interfluve area; Ust-Zaslonkovskoe occurrence is in the estuary of the river Zaslonka (right tributary of the river Mrassu); Orton occurrence is in the basin of the river Orton and its left tributary - river Bazas; Kolosovskoye occurrence in the upstream of the river Kolos (right tributary of the Orton river); Ust-Fedorovskoye occurrence above the mouth of the river Fedorovka (left tributary of the Orton river); Degtyarnoye occurrence along the Degtyarny stream (right tributary of the Orton river) in the area of tremolite deposite of the same name; Fedorovskoe occurrence in the upstream of the Koreysky stream (left tributary of the Fedorovka river); Upper Orton occurrence in the left side of the river Bolshoy Orton above the mouth of the Maly Orton river.

2.10.2. Dolomites and dolomitic marbles. Shorskoye occurrence in the left side of the river Orton above the mouth of the river Fedorovka; Zaslonkovskoye occurrence at the mouth of the river Sukhoy Log (right side of the river Zaslonka); Zaslonkovskoe occurrence in the upstream of the river of the same name. Ophicalcites (serpentinized calcite-dolomite and dolomite marbles): Luzhbinskoe occurrence in the left side of the river Alguy (right tributary of the Tom river); Zaslonkovskoye occurrence in the valley of the river of the same name 2km from its mouth.

2.10.3. Magmatic intrusive rocks. Tebinskoye deposit (granitoids) on the left bank of the river Tom in the area of Teba station, Izygashskoye occurrence (granites) on the right board of Mrassu (valley of the river Izygash); Zaslonka occurrence (Ekaterininskoe) (pegmatite) at the estuary of the Yekaterinovka stream (right tributary of the river Zaslonka).

2.10.4. Volcanic and subvolcanic rocks. Nazasskoye occurrence (plagioclase porphyrites, metatrachites tuffs) in the left side of the river Nazas (left tributary of the Usa river). The rocks of these occurrence have good polishing ability and are suitable for facing slabs production.

3. Conclusions

Despite the significant resource potential of Mezhdurechensky district, only coal and alluvial gold are currently of practical impact, with overwhelming predominance of coal. This disproportion reflects regional geological and economic conditions, and indicates general dependence of economy on only one type of mineral. The way to improve current situation is seen in development of regional road network and restart of exploration of a wide range of mineral resources. Absence of good roads in most parts of the district leads to unreasonably high cost of transportation and greatly increases cost of extracted minerals. Plans to build the Abakan-Tashtagol highway have been announced recently. In case of such project fulfillment, many of the described minerals may become profitable for mining. It also seems necessary to complexify mining process, especially in the areas with close location of deposits of several types of minerals. In this sense, the Alguysky site looks very promising, where significant reserves of high-quality talc, tremolite rocks, marshallite, and ophicalcite are concentrated almost in one deposit. It probably makes sense to study ores of the Usinskoye manganese deposit in more details in order to develop modern technologies for their application. Development of small mining enterprises seems to be expedient (for example, in stone-cutting). Such an enterprises do not require significant stocks of minerals. It can promptly change types of processing raw materials and supply necessary products from local natural facing stone.

References

[1] *Geology of the USSR. Volume XIV. Western Siberia (Kemerovo, Novosibirsk, Omsk, Tomsk regions, Altai Territory). Minerals. Book 1* 1982 (Moscow) p 319
[2] *Geology of the USSR. Volume. XIV. Western Siberia (Kemerovo, Novosibirsk, Omsk, Tomsk regions, Altai Territory). Minerals. Book 2* 1982 (Moscow) p 196
[3] Kondakov A N and Voznaya A A 2013 *Mineral Resources of the Kemerovo Region. Book 1. Metallic Minerals.* (Kemerovo) p 290
[4] Kondakov A N and Vvoznaya A A *2016 Mineral Resources of Kemerovo Region. Book 2. Nonmetallic Solid Minerals* (Kemerovo) p 496
[5] *Coal base of Russia. Volume II. Coal basins and deposits of Western Siberia (Kuznetsky, Gorlovsky, West Siberian basins; deposits of the Altai Krai and the Altai Republic)* 2003 (Moscow) p 604
[6] Yuriev A A, Rynda G S, Zeyfert L L, et al 2002 *State Geological Map of the Russian Federation* scale 1: 200000. 2nd edition. Kuzbass series. Novokuznetsk N-45-XXIX

IOP Conf. Series: Earth and Environmental Science **206** (2018) 012014 doi:10.1088/1755-1315/206/1/012014

Risk-oriented provision of mining operations safety at the enterprises of mineral resources sector in Russia

A Z Vartanov[1], I V Petrov[1,2] and A V Fedash[1]

[1]Institute of Comprehensive Exploitation of Mineral Resources Russian Academy of Sciences, 4 Kryukovsky Tupik, Moscow, 111020, Russia
[2]Financial University under the Government of the Russian Federation, 49 Leningradsky ave, Moscow, 125993, Russia

E-mail: direct.coal@mail.ru

Abstract. The development of the Russian economy for many years will be based on the effective use of mineral resources. At the same time the development of deposits characterized by the worst mining and geological conditions, while increasing the intensity of mining operations, leads to an increase in mining risks. In the paper, recommendations are given on a set of measures aimed at ensuring the safety of mining operations at the enterprises of the mineral resources sector in Russia based on the analysis of the current regulatory and legal framework and the assessment of factors that affect the growth of the probability of emergency situations. The obtained conclusions are drawn from the situation in the Russian coal industry. The proposed recommendations are based on a risk-oriented approach illustrated by the example of the coal industry. In the paper the directions for improvement of the mechanism of organizational economic provision of mining operations safety at the enterprises of the coal industry are presented.

1. Introduction

Despite the significant growth of non-resource-based industries, the state cannot refuse from involvement of mineral and raw materials into economic circulation as the basis for the development of the Russian economy. According to the assessment of the Ministry of Natural Resources of Russia for 2017 the gross value of Russian mineral resources was 28 trillion dollars and had the following structure (table 1). At the same time, the state sets ambitious goals for the development of the Arctic in order to involve its natural resource and mineral resources potential in economic circulation [1].

To date 37% of the explored deposits have been put into industrial development, but it should be noted that from 1/3 up to 2/3 of all balance reserves with modern technologies of extraction and processing are unprofitable.

Analyzing the emerging situation, it is possible to identify the main problems of the development of the mineral resources sector, which have been formed for many years:

- Long operating time of most working mines (often beyond the established time limit).
- Depletion of large near-surface deposits easy to develop.
- Use of obsolete equipment and norms of technological design by mining enterprises in new mining and geological conditions.
- Lack of interest of mining companies in financing pilot and basic research.

Table 1. Structure of the gross value of mineral resources in the Russian Federation (for 2017).

Type of mineral resources	Share in the total cost
natural gas	32 %
coal	23 %
oil	16 %
ore mineral	14 %
other mineral resources	15 %

At the same time, the current development of mineral resources sector in Russia is characterized by a deterioration in mining and geological conditions with a simultaneous intensification of mining works, which ensures sufficient profitability of mining. These two factors, as well as a lack of qualification of employees and motivation for unconditional compliance with the requirements of mining safety rules and increase in labor productivity lead to an increase in mining [2, 3].

Despite the considerable attention paid by the state to the issues of mining safety through the tightening the requirements and support of relevant scientific research, the level of accidents at the enterprises of the mineral resources sector remains rather high, which is evident from the comparative assessment of the mining operations safety in the coal industry of Russia and the USA (table 2). At the same time, mine operators bear significant costs both implementing measures to prevent accidents and eliminating accidents.

Table 2. Comparative assessment of the mining operations safety in the coal industry in Russia and the USA (1991-2009).

Indicator	Russia	USA
Average annual production (billion tons)	0.26-0.30	1.05-1.10
Fatal injuries (person/year)	115-180	25-49
Fatal injuries (person/100 million tonnes)	54+/- 12	3.36+/- 0.55
Dead (person/10 thousand workers)	7.9+/- 1	3.8 (5.1 underground)

In the period from 2000 to 2016 at the mines of the Kuznetsk and Pechora coal basins, 9 major accidents occurred which resulted in the death of 385 miners.

The facts indicated above proves the need to search for new approaches to ensuring the mining operations safety at the enterprises of mineral resources sector in Russia.

2. Methods of research

A systematic analysis of the factors affecting the development of the mineral resources sector in Russia was carried out. The following groups of factors were accepted for consideration [4]:

- Infrastructural
- Environmental
- Market conditions
- Political
- Internal

Investigating the internal factors of the development of mineral resources sector in Russia, it is necessary to pay attention to the factors directly influencing the level of safety of mining operations:

- Regulatory framework and law enforcement practice
- Mountain-geological and natural-climatic conditions
- The need to ensure competitiveness
- Existence of deviations from technical and technological norms
- Degree of absence of demand on the part of the state and business for application of high-tech research in the mining industry

- Technological support of mining processes
- Staffing of modern mining industry.

3. Results and discussion

The conducted integrated analysis of the industry operating conditions made it possible to identify the main contemporary problems in the development of mineral resources sector in Russia:

- Depletion of favorable, well-explored reserves not far from the main consumers of raw materials and infrastructure;
- The occurrence of reserves at great depths in complex geological, gas and hydrodynamic conditions, as well as in nature protection zones;
- The need to develop the mineral and raw materials base of the northern and far eastern territories of Russia;
- Accumulation in large volumes of technogenic wastes leading to high pollution of territories;
- Incompleteness of information on the state of the subsoil, increased risk of explosive destruction processes, hard-to-predict rock bumps, emissions of gases, saturated brines and waters into mine workings in the developed fields;
- Dependence of mining enterprises on the supply of imported mining equipment and spare parts;
- Increase in energy costs, production costs and processing of mineral raw materials;
- Growth of volumes of difficult to enrich raw material;
- Volumes of extraction and processing of some types of raw materials are insufficient for domestic perspective demands with their simultaneous exporting.

All this is exacerbated by the problems accumulated in the mineral resources sector in Russia:

- Long operating time of most mines;
- Increase in the number of hazardous production facilities with close to the maximum or fully exhausted technical and technological resources;
- Use of out-dated equipment and norms of technological design by mining enterprises in new mining and geological conditions;
- Increased intensity of mining operations, new technologies do not meet the safety requirements;
- Lack of detailed study of the risks of negative industry-related processes and support of mining geotechnologies;
- Lack of modern regulatory and legal framework;
- Lack of interest of mining companies in financing pilot and basic research.

According to Rostechnadzor, the main sources of danger and the reasons for the increase in accidents are:

- Poor quality of design and technical solutions;
- Critical level of wear and tear on plant and equipment;
- Poor or untimely performance of maintenance and repair works;
- Low motivation to ensure work safety and personnel qualification;
- Low technological and labor discipline;
- Lack of effective incentives to upgrade the production facilities and equipment of enterprises involved in industrial safety activities.

Noting the increase in the number of hazardous production facilities, it is possible to present the ranked structure of the operating coal mines of the Russian Federation in terms of the degree of hazard in mining (table 3). At the same time, it should be said that at present five coal mines are in operation in Russia, at which all types of hazards are present simultaneously, and in general there are two or more hazards at 33 mines [5].

Table 3. Ranked structure of operating coal mines of the Russian Federation in terms of the degree of hazard of mining operations (2017).

Type of hazard	Number of mines
Dust explosion	54
Rock bumps	33
Coal self-ignition	32
Sudden rock, coal and gases outbursts	30
Water irruption	25

Speaking about the most dangerous factor and type of emergency – explosion of methane-air mixture, it should be noted that often methane explosions in coal mines are caused by organizational and technological violations, which include:

- Selective mining of the most favorable layers in the formation;
- Operation of very gassy mines with a methane content more than 15 m^3/t without their preliminary gas drainage;
- Absence of preliminary and low efficiency of current gas drainage of methane-bearing coal seams;
- Use of long-pillar method along the strike with the collapse of the roof without application any methods of its weakening and caving;
- Use of exhaust method for the ventilation of mines and combined schemes for ventilating the faces during the mining coal seams prone to spontaneous self-ignition;
- Contract bonus system.

It is necessary to give an example of the Australian coal industry, which occupies the first place in the world for its mining operations safety (fatal injuries 3.0 person/100 million tonnes), where the law prohibits the exploitation of coal mines with methane-bearing seams of more than 9 m^3/t without preliminary gas drainage. In Russia very gassy mines with a methane content more than 20 m^3/t are operated without pre-degassing.

Separately, we should touch on the environmental consequences of mining and mining activities. The impact of accumulated and newly generated wastes on the environment, loss of resources through waste accumulation, emissions and dumping, as well as the unsettled relationships between mine owners and environmental authorities (forest and water resources, land use, specially protected areas) not only reduce the competitiveness of the industry, but also leads to global environmental problems that can be solved by switching to the best available technologies [6, 7].

The facts mentioned above proves the need to switch to a new system of regulatory and legal support for mining safety, based on a risk-oriented approach with a preliminary assessment of the field by constructing a digital model with a forecast of mining risks and an economic evaluation of the investment project throughout the life cycle of the deposit [8, 9].

These problems require the improvement, creation and wide introduction of highly efficient exploration technologies, low-waste, resource-saving technologies for extraction and processing of solid minerals. It is necessary to develop combined technologies for extraction and deep processing of raw materials as a unified technological complex for the implementation of a complete, closed cycle of combined physical, technical and physical-chemical geotechnologies, optimizing the timing and order of phased commissioning of production facilities and mandatory waste disposal in the mined-out space. These tasks can be performed only through the development of domestic mining machinery to ensure import-independence, including through the localization of the most significant productions [10].

To stimulate mining companies to implement modern approaches ensuring the mining operations safety, it is necessary to use the experience accumulated in the framework of Eurasian technical and economic cooperation in the field of exploration, extraction and processing of solid mineral resources for the creation of appropriate trust funds financed by subsoil users [11].

4. Conclusions

Thus, the following main directions for ensuring the safety of mining operations at the enterprises of the mineral resources sector in Russia should be indicated:

- Modern scientific support for the operation of mines with the assessment of risks of manifestation of negative technogenic processes throughout the life cycle of the field;
- Improvement of legal and regulatory risk-based regulation of integrated mining safety on the basis of scientifically-based approaches;
- Full-scale complex examination of large mining projects;
- Financing of pilot and fundamental research focused on ensuring the mining operations safety;
- Development of mechanisms for involvement of the results of scientific research in the real sector of the Russian economy, creating a regulatory and legal framework and principles for supervisory activities that ensure the mining operations safety.

At the same time, the following indicators should serve as targets (indicators) for ensuring the safety of mining operations: decrease in the number of workers in the zone of high mining risk (for example, by 25% every five years, starting from 2020). This can be done either by refusing to develop high risk deposits or by robotizing production processes with the transition to fundamentally new closed geotechnologies.

References

[1] Vartanov A Z, Galchenko Yu P, Kalabin G V, Petrov I V and Fedash A V 2018 *Vestnik MGSU* vol 13 **2(113)** 240–48

[2] Petrov I V 2013 *Proc. of Xth All-Russian Sci. and Prac. Internet Conf. on Supply and Demand in the Labor Market and the Market of Educational Services in the Regions of Russia* pp 185–94

[3] Kazakov V B, Kalacheva L V, Petrov I V and Surat I L 2017 *Coal* **6(1095)** 48–50

[4] Petrov I V and Stoyanova I A 2014 *Monitoring. Science and Technology* **2** 54–8

[5] Zharikov I F, Zakharov V N and Norel B K 2015 *Mechanics of Solids* vol 50 **6** 41–49

[6] Goncharenko S N, Le B Z, Stoyanova I A and Petrov I V 2014 *Mining Journal* **9** 143–46

[7] Vartanov A Z, Zakharov V N, Fedash A V and Petrov I V 2015 *Mining Informational and Analytical Bulletin* **P** 7

[8] Malinnikov V A, Zakharov V N, Malinnikova O N and Uchaev D V 2017 *Global Challenges and Data-Driven Science* (Book of Abstracts of the International Conference) p 115

[9] Zakharov V N, Chernikov A G, Malinnikova O N and Isaev Y S 2016 *Research in geoinformatics: the works of the Geophysical Center RAS* vol 4 **2** p 42

[10] Petrov I V, Fedash A V and MSc E J D 2014 *Science-intensive Technologies for the Development and Use of Mineral Resources* **1** 214–16

[11] Vartanov A Z, Petrov I V and Fedash A V 2017 *Mining Journal* **11** 14–7

IOP Conf. Series: Earth and Environmental Science **206** (2018) 012015 doi:10.1088/1755-1315/206/1/012015

Problems of innovative development of the fuel and energy industry in Russia

A V Fedash[1], A Z Vartanov[1] and I V Petrov[1,2]

[1]Institute of Comprehensive Exploitation of Mineral Resources Russian Academy of Sciences, 4 Kryukovsky Tupik, Moscow, 111020, Russia
[2]Financial University under the Government of the Russian Federation, 49 Leningradsky ave, Moscow, 125993, Russia

E-mail: direct.coal@mail.ru

Abstract. The fuel and energy industry in Russia is the basis for the functioning of the national economy, on the one hand, providing energy and energy resources for all types of economic activities, as well as for the nonmaterial sphere, on the other hand, forming the greater part of budget revenues from energy exports. In modern economic conditions, and also taking into account the stage of the economic cycle, these problems can be solved through innovations. The prerequisites for the implementation of innovative activities in the fuel and energy industry in Russia are considered. The authors described the level of innovative development at the enterprises of fuel and energy sector giving reasons for its relatively low importance. Based on the technical and economic characteristics of the products and production processes of the companies of the fuel and energy industry, specific features of the organization and implementation of innovative activities in it are determined. Studying the experience of implementing innovative ideas in the Russian fuel and energy sector allowed the factors hampering the success in this direction to be formulated. To implement innovation activities effectively in the Russian fuel and energy industry, it is necessary to carry out a number of transformations both at the level of industries and in individual enterprises in order to reduce the negative impact of the identified factors.

1. Introduction

The Russian fuel and energy industry is currently in stagnation. In order to overstep this period we need to create incentives for the development of the national economy. The worsening conditions at the external markets, which negatively affect the volumes of Russian energy exports due to the combination of factors, including the reduction in demand with the slowdown in the growth of the world economy and the economies of the largest consumers of fuel and energy resources (FER), are negatively affect the development of the fuel and energy industry (FEI) [5].

In addition, there is an increase in the supply of FER from other countries, which also worsens the competitive position of Russian energy resources.

2. Methods of research

Given the conditions that prevailed in 2016, the main challenges for the sustainability of Russia's energy industry and ensuring the FEI growth were formed on the basis of the following factors:

- Preservation of technological and financial sanctions of Western states;
- Domestic demand for energy, associated with low expected growth rates of Russian GDP;

- Lack of free or credit funds available for investment projects with long payback periods;
- Depreciation of fixed assets of FEI enterprises and increase in the probability of accidents.

The financial and technological sanctions adopted by the countries of the European Union, the United States and other countries manifested themselves among others in closure of the European capital market for commercial companies and state financial institutions of the Russian Federation, which led to the unstable situation in FEI. Especially these sanctions negatively affected the technical and technological support of FEI.

3. Results and discussions

As the analysis of exports of different types FER shows, the most sensitive to changes in the external market conditions is the export of gas, the volumes of which may drop by 20% by the year 2025. The volume of coal exports is also quite sensitive (decrease by 13-17% by the year 2025). The dynamics of exports of oil and oil products is controversial: if in the mild scenario a small (about 3.5%) growth in volumes of export is expected by 2035, then in the pessimistic scenario they fall by 16%. [9]

Alongside with the reduction in domestic demand for FER, this challenge is also connected with the tightening of the maximum requirements for investment and price load on the part of FEI, the risks of reducing the opportunities and profitability of investments in the energy sector in Russia.

The situation in the economy in 2016-2017 was formed under the impact of the complication of external economic situations, primarily the oil price fall, the extension of economic sanctions by the EU and the USA, and a stable tendency to investment activity reduction. At the same time, as the calculations show, in general, the situation in the energy sector of the Russian Federation in 2016-2017 can be characterized as stable.

Despite the decline in the average annual prices for URALS crude oil, the amount of oil produced with gas condensate increased by 2.5% compared to 2015, coal production – by 3.1%, electricity generation increased by 2.1%, gas production – by 0.7%. In 2016, the Russian export of energy resources increased by 1.5% compared to 2015, including oil – by 4.2%, natural gas – by 6.7%, coal – by 8.8%. As a result, there was a tendency to stabilization of the ruble exchange rate with its lower dependence on world oil prices. Ruble in 2016 strengthened against the dollar by more than 16% [10, 11, 12].

In 2016, the coal mining enterprises of Russia produced 385.7 million tonnes (+ 3.1% to 2015) [12]. At the same time, according to the Federal Agency for Subsoil Use "Rosnedra", in 2016 the balance reserves of coal increased by 1238.5 million tonnes [12]. Taking into account the facts discussed above, as well as the growth of total coal supplies by 11.5% in relation to 2015, the situation in the reporting period was positive. The coal industry was characterized by challenges associated with the continued fall in investment in the industry and caused by the uncertain development of the external and domestic markets for coal products. At the same time, the coal industry was much less exposed to threats related to dependence on imported equipment, which was due to the high degree of diversification of supplies. [3]

The total supply of Russian coal in 2016 amounted to 338.2 million tonnes (101.6% by 2015). Volumes of coal supply to the domestic market amounted to 168.6 million tonnes instead of 178.0 million tonnes planned for 2016 (95.6 percent by 2015), including 86.7 million tonnes for the needs of the electric power industry (92.5%); for coking – 39.9 million tonnes (99.2%); residential energy consumption and agroindustrial complex – 20.3 million tonnes (86.8%) [12]. The decrease in the amount of coal shipped to the domestic market was due to the evolving demand in coal. The obstacle for the development of domestic coal market was the macroeconomic situation in the country and the world, the transition of generating capacities to gas fuel and the continued implementation of gasification programs of the regions in housing and communal sector. [8]

It should be noted that, given the domestic energy market demand for coal products, the Russian coal industry fully provides consumers with coal in the domestic market.

The main challenge that determines the technogenic and natural threats affecting the fuel and energy complex in Russia is the high level of wear and tear on the majority of production assets in the FEI.

In 2016, according to the data of Situational and Analytical Centre of the Ministry of Energy of the Russian Federation 41 accidents (incident) were registered at the coal industry enterprises which is 41% lower than in 2015, among them 8 major accidents (the same in 2015), including: 6 accidents (87.5%) – in the mines; 2 accidents (12.5%) – on the open pit mines. [11, 12]

Causes of accident in 2016 were: explosion of gas and coal dust – 4 accidents (in 2015 there were no accidents); fires – 1 accident (by 4 accidents fewer than in 2015); flooding of mine workings, water or clay breakthrough – 1 accident (in 2015 1 accident); rock or support collapse – 1 accident (1 accident fewer than in 2015); other types (fall of the excavator) – 1 accident (in 2015 there were no accidents) [11, 12].

According to the updated information of Rostekhnadzor, 56 people got fatal injuries at the enterprises of the industry in 2016 (by 36 people more than in 2015) [11, 12].

In 2016, specific fatal injuries amounted to 0.145 cases per million tonne of production (0.05 per million tonne of production in 2015), including: 54 people (over 96%) got fatal injuries under the ground (+ 41 to 2015); at the open-pit mines – 2 people (less than 4%), (by 5 fewer in comparison with 2015) [11, 12].

A sharp increase in deaths is associated with the technogenic accident that occurred on February 25, 2016 at the mine "Severnaya" of JSC "Vorkutaugol", when as a result of the explosion of methane and coal dust 30 miners were killed. Assessing the state of energy safety in respect to man-caused and natural challenges, it should be noted that due to the high level of physical deterioration of fixed assets in the field of coal mining and non-compliance with safety rules in mining operations, which led to an increase in accidents in 2016 in comparison with 2015, including accidents with fatal accidents, safety measures of conducting mining operations were strengthened at the mines and open-pit mines.

In 2016, all enterprises of coal industry of the Russian Federation transited to the elements of a multifunctional safety system, namely: determination of the personnel location in mine workings by the system, the systems of air-gas monitoring, search and detection of people in emergency situations. At all mines and large open-pit mines, auxiliary mine rescue teams were created that included the workers of operating organizations.

To improve the energy security in relation to man-made and natural challenges, in the near future it is necessary to complete the modernization of FEI technological base. In the beginning, technological improvement of the existing production facilities will be carried out due to the limited investments. Then, their main transformation and formation of new production capacities will take place on the basis of domestic and foreign technologies that meet our requirements. To reduce and prevent the consequences of accidents, natural disasters and manifestations of natural anomalies, it is necessary to carry out work on improvement of quality of fuel and energy supply system of the country and its regions during emergencies. These works should be carried out in the following main areas: normative consolidation of the necessary volume of (estimated) sparing main and auxiliary equipment of heat supply sources and heat networks, as well as emergency stock of materials and spare parts, etc. [1, 2, 4, 6, 7]

Financial and technological sanctions on the part of the countries of the European Union, the United States and other countries negatively affected the work of coal industry enterprises. The analysis showed by 2016 that in the coal industry the share of mechanized support of foreign production used in coal mines reached almost 45%, coal-mining and tunneling combines 79% and 47%, respectively [11, 12].

In this regard, the Ministry of Industry and Trade of the Russian Federation adopted the import substitution program, the goal of which is to reduce this dependence to 43% by 2020.

Starting from 2014 there was a change in the dynamics of coal consumption, it began to decline leading to the reduction in production [8]. At the same time, the main consumer of coal is China (about 50% of world consumption), therefore, the development strategy of its economy significantly influences the dynamics of world prices for coal. The main competitors in this market for Russia are China, the USA, India, Australia and Indonesia. The main risk of market development is the global environmental initiatives for the decarbonization of economies, limiting production and import of products characterized by a significant "carbon footprint".

4. Conclusions

Thus, in the next few years, the growth of world coal consumption is unlikely, at the same time the positive dynamics of coal prices is possible this year [8].

In 2016, the coal export from Russia amounted to 169.6 million tonnes (+ 7.8% by 2015), instead of 147 million tonnes provided by the plan for 2016. 82.5 million tonnes of coal were shipped in the Atlantic direction. Most of the exports in this direction (76% or 62.5 million tonnes) account for port shipment, including the Baltic ports (14.4 million tonnes). Through the border points 20 million tonnes are exported, mainly to the countries of Eastern Europe and the CIS [11, 12].

In the conditions of aggravation of a competition between the countries – exporters of coal production, the growth of Russian coal supplies for export is taking place, primarily to the countries of the Asia-Pacific region. The coal market supports demand in the industrial centers – Japan, South Korea, India, Vietnam and the Philippines. The main consumers of Russian coal in the Atlantic market are the countries of the European Union (54 million tonnes or 65.5% of the export volume in this direction and 32.7% of the total volume of Russian exports), the CIS countries that have concluded agreements with the EU (Moldova – 0.9 million tonnes), the Middle East (15 million tonnes), Africa (2.3 million tonnes) and South America (1.0 million tonnes) [11, 12].

The largest consumers of the Russian coal in Europe are the UK (10.7 million tonnes in 2016), Germany (8.6 million tonnes) and Poland (5.4 million tonnes) [12]. At the same time, there is an increase in world prices for coal, which improves the financial condition of coal companies. In 2016, the pre-tax profit in the industry amounted to 60 billion rubles, the share of unprofitable enterprises in total coal output reduced to 8%, debt on loans and credits decreased. An increase in the share of exports in the supply of coal products indicates the preservation of the positive dynamics of coal exports while ensuring the full demand for coal in all sectors of the Russian economy.

The factor, that has negatively affected the modern development of the coal industry, is the growth of transportation tariffs. The consequences of this factor are the complication of the conditions of competition in the external market and the deterioration of the economic situation of the Russian organizations.

Due to the growth in world prices for coal and the increase of share of coal export in 2016-2017 compared to 2015, the dynamics of the level of growth of this energy carrier is at a good level of the projected indicators for the country's budget.

The modern innovative development of FEI in relation to foreign economic and foreign policy challenges is characterized by the fact that the development of the Russian economy will take place, at least in the next 2-3 years, in the conditions of continuing geopolitical instability, the continued imposition of economic and technological sanctions on Russia by the EU and the USA.

As a priority measure to ensure the FEI sustainability in the unstable macroeconomic situation, it is necessary to transit to a taxation system based on the financial result, improve the pricing system for electricity and natural gas on the domestic market, and implement mechanisms for eliminating non-payments. To minimize the sanctions risks, it is necessary to continue implementation of the policy of import substitution in FEI, including the expansion of state support measures for projects that aim to create domestic technologies, equipment and software for the needs of FEI. To attract investment into FEI, liberalization of investors involvement into large projects is necessary.

References

[1] Zakharov V N 2013 *Mining Informational and Analytical Bulletin* **S1** 57–69
[2] Petrov I V 2013 *Proc. of Xth All-Russian Sci. and Pract. Internet Conf. on Supply and Demand in the Labor Market and the Market of Educational Services in the Regions of Russia* pp

185–94
[3] Dobrokhotova M V and Petrov I V 2015 *Best Available Technologies. Application in Various Industries. Collection of papers* (Moscow: Pero) **2** pp 163–69
[4] Kalacheva L V 2015 *Coal* **3** 51–5
[5] Romanov S M and Petukhov P P 2014 *Mining Informational Analytical Bulletin* **S1** 148–56
[6] Fedash A V 2011 *Coal* **5** p 10
[7] Fedash A V 2011 *Mining Industry* **3(97)** p 38–42
[8] Linnik Yu N, Afanasyev V Ya and Linnik V Yu 2014 *Coal of Russia: State and Prospects* (Moscow: INFRA-M) p 271
[9] *Russia's energy strategy for the period untill 2035* http://www.energystrategy.ru/ab_ins/source/ES-2035_09_2015.pdf.
[10] *Russian Statistical Yearbook 2015* (Moscow: Rosstat) p 728
[11] *Russian Statistical Yearbook 2016* (Moscow: Rosstat) p 725
[12] *Russian Statistical Yearbook 2017* (Moscow: Rosstat) p 686

IOP Conf. Series: Earth and Environmental Science **206** (2018) 012016 doi:10.1088/1755-1315/206/1/012016

Modern forms of project interaction of science, business, and government in the coal industry

E V Goosen[1,2], S M Nikitenko[1] and V I Klishin[1]

[1]Federal Research Center of Coal and Coal Chemistry, Siberian Branch of the Russian Academy of Sciences, 18 Sovetskiy ave, Kemerovo, 650000, Russia
[2]Kemerovo State University, 6 Krasnaya street, Kemerovo, 650043, Russia

E-mail: egoosen@yandex.ru

Abstract. The fuel and energy complex (FEC) of Russia is going through hard times. The problems of development and investment are particularly acute in the coal sector, where demand and prices have fallen particularly hard due to the financial crises of 2008 and 2014, unstable energy prices, sanctions and the transition of a number of countries to "carbon-free energy development". In these circumstances, only the transition to flexible technologies, which are able to quickly adapt to changes and extinguish the external negative shocks, within the framework of close cooperation of science, government and business can save the Russian coal industry. The purpose of this article is to analyze the currently available project forms of cooperation between science, business and government for the benefit of the innovative development of the coal industry and to summarize the experience of implementing such project forms of cooperation at the Federal Research Centre of Coal and Coal Chemistry of Siberian Branch of the Russian Academy of Sciences (FRC CCC SB RAS), Kemerovo region.

1. Introduction

The fuel and energy complex (FEC) of Russia has always been and still is a key sector of the Russian economy. The characteristic feature of this sector is capital intensity and a long period of return on investment. This hinders the attraction of investment in the fuel and energy sector and limits the possibilities for innovative development. These problems are particularly acute in the coal industry, where demand and prices have fallen particularly hard due to the financial crises of 2008 and 2014, unstable energy prices, sanctions and the transition of a number of countries to "carbon-free energy development". Unfavorable conditions for development and investment in related industries (coal engineering, transport and coal generation) also had a negative impact on the coal industry. While until 2008 infrastructure projects for the coal industry were actively financed from the budget, state funds, special programs of direct state support, etc., during the crisis, these sources of direct and indirect support for the coal industry decreased considerably.

The coal industry is seriously affected by the worldwide transition to new technologies within Industry 4.0: digitalization, robotization, modern big data processing systems, and the development of the industrial Internet, which displace people from all stages of the production process, including transportation and logistics, and further reduce the investment cycle in all sectors, making long-term investments in coal mining and coal energy even less attractive. In these circumstances, only the transition to flexible technologies, which are able to quickly adapt to changes and extinguish the external negative shocks, within the framework of close cooperation of science, government and business can save the Russian coal industry.

The purpose of this article is to analyze the currently available project forms of cooperation between science, business and government for the benefit of the innovative development of the coal industry and to summarize the experience of implementing such project forms of cooperation at the Federal Research Centre of Coal and Coal Chemistry of Siberian Branch of the Russian Academy of Sciences (FRC CCC SB RAS), Kemerovo region.

2. Data and method

When preparing the article, the authors used the data of the Ministry of Energy of the Russian Federation and the Federal State Statistics Service. The principle research methods were the descriptive method and case-study.

3. Discussion of the problem

3.1. Forms of project interaction of science, business, and government in the coal industry and in related industries

As of the beginning of 2017, there were 181 coal mining enterprises in Russia, including 66 mines and 115 open-cuts. Coal processing was carried out at 65 coal washing plants, as well as at the sorting facilities existing in the majority of coal companies (table 1).

Table 1. The number of coal enterprises in the Russian Federation in 2016–2018.

Year / indicator	Total number of coal enterprises	Number of coal mining enterprises		Number of coal washing plants
		mines	open-cuts	
2016	191	60	131	63
2017	181	66	115	65

Currently both in Russia and in the world much attention is paid to project forms of interaction of science, government and business in the fuel and energy sector and related industries. The analysis of the world energy sector allowed the authors to conclude that the priority directions of joint projects in the energy sector are: (1) the development of modern technologies for the production of complex and hard-to-recover geo-resources in small fields with the use of low-waste technologies; (2) the support of independent companies (IC) and the promotion of national (local) markets; (3) the development of flexible transport systems and logistics; (4) the support of extensive cooperation of the independent oil and gas companies (IOGC) with the companies of related industries, service and engineering companies, created research organizations and major machine-building companies; (5) the formation of open branched fragmented value chains within large vertically integrated companies (VICs); (6) the creation of conditions for the localization of supporting and related industries, providing a comprehensive and balanced development of the territories where VICs and ICs are located.

To date, in the Russian coal industry and the related industries there is a fairly well-defined range of available project forms of cooperation between government, business and science. The most common forms include:

- self-supporting infrastructure projects financed on the basis of self-sufficiency at the expense of the National Welfare Fund (infrastructure projects) and the Russian Direct Investment Fund,
- concessions,
- ublic-private partnership agreements (PPP projects),
- Federal target programs (FTP),
- projects implemented in the framework of the Decree of the RF Government No. 218 dated 9 April 2010 (Government Decree No. 218),
- special investment contracts (SPIC).

These project forms of cooperation differ in the scale of projects, the degree of innovation, the nature of management and the ways of involving the parties in cooperation, etc. It is important to emphasize that all these forms of cooperation are aimed directly at the development of the coal industry, their task is to remove barriers to its development through developing related industries (transport, energy, coal engineering, etc.). This practice is not exclusively Russian. All over the world, the companies operating in the industries related to FEC receive state support. Table 2 gives a brief description of these forms and a list of related industries in which they are most common.

Table 2. The number of coal enterprises in the Russian Federation in 2016–2018 [1-6].

Form	Definition	Relation to FEC	Specifics of interaction between parties, and degree of innovation development
Infra-structural project	A major infrastructure project of particular importance for the development of the industry (region), included in the list of Federal (regional) projects financed by the National Welfare Fund.	Development of transport and energy infrastructure, development of new deposits	The state provides the funding on a return basis, the business implements the project The stage of mass production The scientific organization can be involved at the stage of development and implementation of the contract
Concessions	Agreements permitting, under certain conditions, to perform prospecting, exploration and (or) exploitation of mineral deposits owned by the state (or local authorities). Concessions are legally formalized through the contracts for the provision of rights, licenses, or lease.	Development of transport infrastructure	The business provides the funding and implements the project, the state grants exclusive rights to use the created or reconstructed objects The stage of mass production The scientific organization can be involved at the stage of development and implementation of the contract
PPP project	Cooperation of the public partner, on the one hand, and the private partner, on the other hand, which is legally formalized for a certain period of time and based on the pooling of resources and the distribution of risks	Attracting investments to infrastructure projects and programs for the development of new territories, including resource regions	The business creates or reconstructs the object, manages the object, and after the return of the invested funds returns the object to the state The stage of mass production The scientific organization can be involved at the stage of development and implementation of the contract
Federal target program	The program of budget funding of research, giving access to specific developments and products in the technological areas that are a priority for the Russian economy and contribute to improving its competitiveness through cooperation between business and scientific organizations.	Possibility of financing high-tech innovations for the benefit of fuel and energy complex and related industries on the basis of converting the results of scientific research into a market product.	The state provides funding to the scientific organization at the stage of R&D with the obligatory participation of business as an industrial partner, which undertakes the duty to introduce innovation and receives rights to intellectual property created during the project.

Government Decree No. 218	The program of development of cooperation between Russian higher education institutions, state scientific institutions and organizations implementing complex projects for the creation of high-tech production.	. Possibility of financing high-tech innovations for the benefit of the fuel and energy complex and related industries on the basis of converting the results of scientific research into a market product.	The state provides funding to the business at the stage of experimental industrial production with the obligatory participation of the scientific and educational organization
SPIC	The agreement between the investor and the state, which fixes the obligations of the initiator of the project to develop a new production of industrial products within the prescribed period, as well as the obligations of the Russian Federation and (or) its subject in terms of guarantees of stability of tax and regulatory conditions and the provision of support measures.	Possibility of financing the introduction of high-tech innovations for the benefit of the fuel and energy complex and related industries on the basis of converting the results of research into a market product and the creation (expansion) of markets for innovative products	The private investor undertakes to create or modernize and (or) master the production of industrial products in the territory of the Russian Federation, and the state (subject of the RF) undertakes to guarantee the sales market and tax benefits for the entire duration of the project. The stage of mass production The scientific organization can be involved at the stage of development and implementation of the contract

3.2. EPC contracts with the engineering firms

As foreign experience shows, the optimal choice of the project operator – the organization that cooperates with the regional authorities, business and scientific structures – plays a crucial role in the success of the project. The operator can be a private manufacturing company, a public authority or a private intermediary agency. However, specialized engineering firms are the best operators in fuel and energy projects. The modern market of engineering services is characterized by a complex, multistage structure, complex forms of interaction between the customer and the contractor, so EPC contracts (engineering, procurement, construction contracts) are often used in such projects. Figure 1 shows a detailed scheme of the EPC contract.

The EPC contract is a contract for the turn-key construction (reconstruction) of an industrial facility. The engineering firm acts as an EPC contractor that performs the bulk of the investment and construction project and assumes all the risks of its implementation from the moment of design to the moment of transfer of the finished object to the customer (including the fulfillment of warranty obligations), for which the contractor is financially responsible to the customer [8]. The advantage of EPC contracts is the possibility to reduce the project implementation period due to the possibility of parallel launching of works (the start of the next stage before the finish of the previous one, as it is not necessary to complete the design work to start the construction). Despite the fact that EPC contracts are usually signed for turn-key construction projects where licensed proven technologies are used, technology developers are still actively involved in those projects to take into account the specific conditions of the project. The advantages of EPC contracts include the reconciliation of a fixed cost, which protects both the investor and the customer from many investment and construction risks [9].

Currently, the major international contractors are companies from industrialized countries, although in the past 20 years companies from developing countries, primarily China, have become increasingly involved (see table 3).

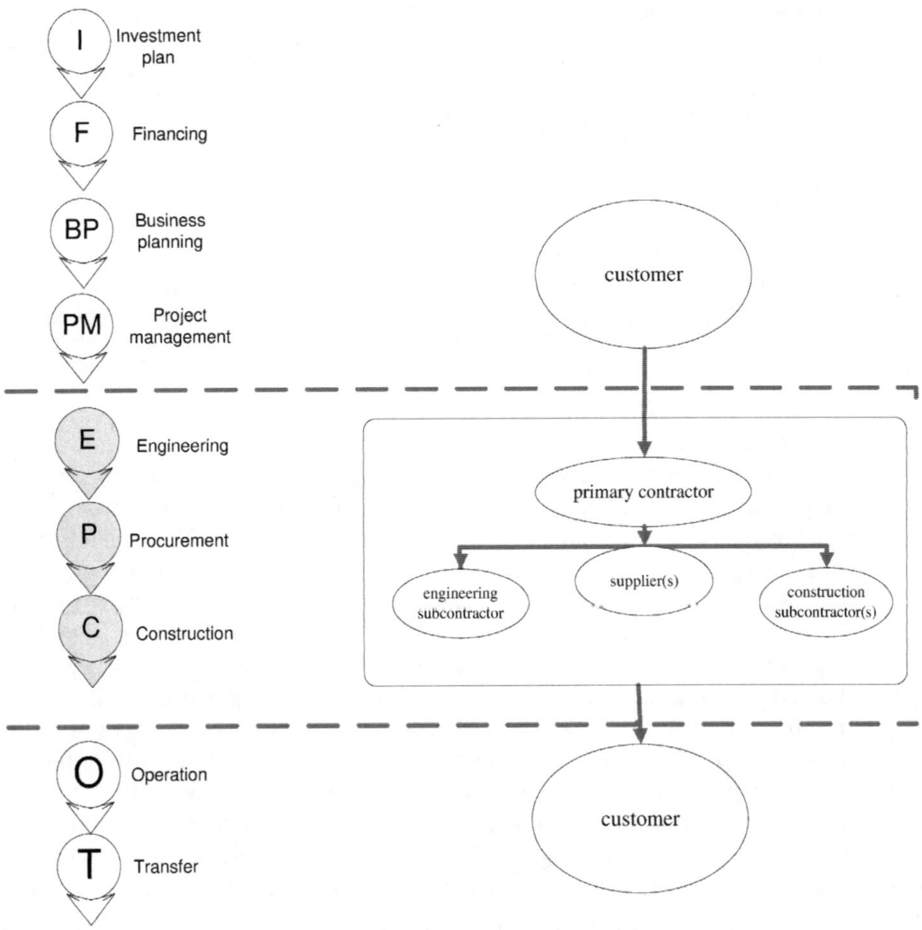

Figure 1. Scheme of the EPC contract [7].

Table 3. Countries with the most developed contract markets [10-12].

	Country	Cumulative turnover of the companies, billion US dollars	Number of companies
1	China	124.050	62
2	Spain	78.896	13
3	USA	71.069	32
4	France	50.744	5
5	Germany	46.547	5
6	South Korea	42.422	13
7	Brazil	41.877	4
8	Italy	28.899	16
9	Japan	22.246	14
10	Turkey	20.409	42

The main areas of EPC contract cooperation in the world include the construction of buildings (25 – 26.5%), transport infrastructure (26–28%), and oil and gas complex (18 – 23%). Energy and mining facilities account for less than 7% [10]. The structure of EPC projects in the fuel and energy sector is presented in table 4.

Table 4. Countries with the most developed contract markets [10-12].

	Sphere of activity	Number of projects
1	Oil and gas sector	1 135
2	Extraction of mineral resources	627
3	Electric power industry	561
4	Construction and infrastructure	78

Engineering contracts in the coal industry are widely used in the following activities: 1) exploration and development of deposits; 2) the construction and maintenance of facilities for production, processing, transportation and storage of coal; 3) the construction and maintenance of processing facilities [13].

3.3. Experience of cooperation between science, government and business in Russia with the participation of engineering companies

In Russia, EPC contracts are still rarely used, and engineering is understood in a very narrow sense as equipment installation and commissioning, which does not correspond to the modern world trends, the engineering services market is concentrated in the oil and gas sector, which accounts for more than 70% of the revenue of engineering services (figure 2). A small number of EPC contracts have been signed in the energy sector, the first contracts appeared in the sphere of coal mine construction. The demand for the services of engineering companies in Russia is formed mainly due to government orders and public corporations, which helps to attract the world's largest providers of EPC services to the Russian market. At the same time, one of the main drawbacks of attracting foreign companies is the high price for the services provided under the contract. This, in turn, gave impetus to the emergence and development of domestic engineering companies in Russia, which were formed basing on the pre-existing structures. However, in most cases, foreign engineering firms with licenses for modern technologies act as primary EPC contractors in the fuel and energy sector in Russia. Domestic companies mainly act as intermediary agencies. The share of large completed projects is still very small. All this leads to the conclusion that in Russia the EPC contracts market is at the stage of formation.

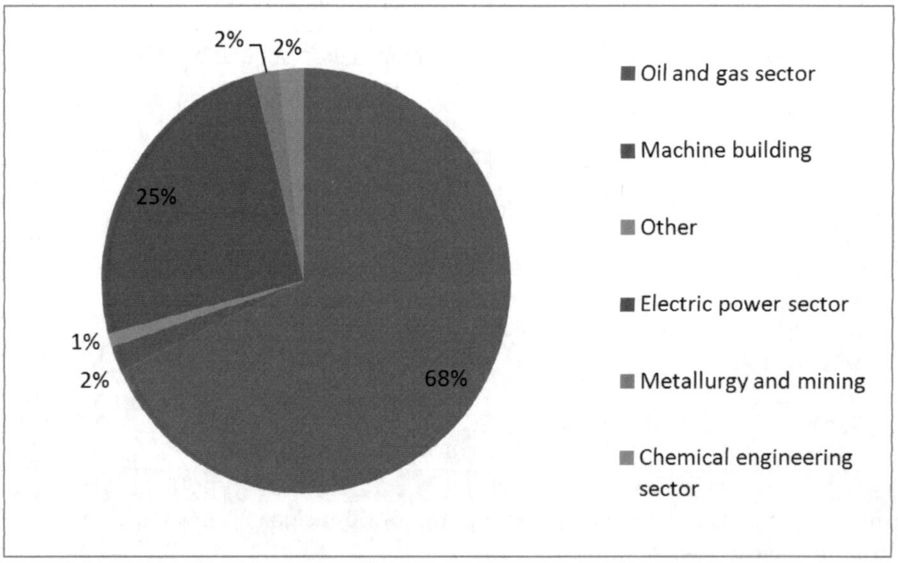

Figure 2. Structure of the Russian engineering services market [15].

To understand the key success factors of project implementation, especially in the coal industry, it is important to analyze all of the project forms of cooperation of government, business and science, and the projects involving engineering firms deserve particular interest.

In 2014, the engineering company "InTechPromEngineering" was established in Kemerovo region for preparing and implementing the major regional projects in the mining sector. The founders of the company were the Institute of Coal of the Siberian Branch of the Russian Academy of Sciences, Kuzbass Center of Welding and Control (the industrial partner of joint projects) and the non-profit organization (public professional association) "Association of machine builders of Kuzbass". This created the conditions for long-term cooperation of government, business and science representatives in the coal mining region [16]. OOO "InTechPromEngineering", basing on the results of the intellectual activity of the FRC CCC SB RAS, is involved in the implementation of several innovative projects for the benefit of the coal industry: the first stage of the regional project "Creation of new types of tools based on superhard composite materials for the effective destruction of rocks" has been implemented in the framework of the Federal target program; currently, several projects focusing on efficient and safe extraction of coal from thick steep and shallow seams using the elements of robotization in the main technological operations are being implemented [17]. Specialists from India, Georgia, and Vietnam have shown interest in the first results of the projects, which confirms their innovative character and allows developing the export component. In order to stabilize the implementation of these and future projects, the regional association of employers "Kuzbass Union of Employers" has signed the three-way Agreement with the Autonomous non-commercial organization "National Center for Development of Public-Private Partnership" (Moscow) and the Board of Administration of Kemerovo region addressing the development of PPP projects in the region [18]. In order to create a methodological base and monitor the implementation of PPP projects in the coal industry and related industries, in 2018, a specialized project office was created on the basis of the Laboratory for monitoring and forecasting of PPP projects in the field of integrated subsoil development of the FRC CCC SB RAS; the Office involved the specialists from Kemerovo region and neighboring regions.

This experience is interesting it shows a practical example of uniting and formalizing the interests of the authorities, business structures, and scientific organizations in the development of the leading industries (mining and engineering) in a certain region.

4. Conclusion

The analysis showed that Russia has created sufficient organizational and legal conditions for fruitful project cooperation between government, business and science in the coal industry in the field of innovation. However, despite the fact that there is a significant demand for such projects, modern forms of interaction are being implemented very slowly. This is especially noticeable on the example of the projects involving engineering firms as the main operator. In these circumstances, the experience of particular small regional engineering companies is extremely important, as in the long term they can become the 'seedlings' for the formation of new effective forms of cooperation in this field.

Acknowledgements

This research was supported by the Russian Science Foundation, Project No. 16-18-10182 "Formation of the organizational and economic mechanisms of integrated development of mineral resources based on the partnership between science, business and authorities in the resource-type regions".

References

[1] *Federal Law No 224-FZ dated 13 July 2015 "On public-private partnership and municipal-private partnership in the Russian Federation and on amendments to certain legal acts of the Russian Federation"*

[2] *Federal Law No 488-FZ dated 31 December 2014 "On the industrial policy in the Russian*

IOP Conf. Series: Earth and Environmental Science **206** (2018) 012016 doi:10.1088/1755-1315/206/1/012016

Federation"
[3] *Federal Law No 115-FZ dated 21 July 2005 (edited on 03 August 2018) "On concessional agreements"*
[4] *Decree of the RF Government No.708 dated 16 July 2015 (edited on 16 December 2017) "On special investment contracts for certain branches of industry*
[5] *Decree of the RF Government No.218 dated 9 April 2010 (edited on 17 August December 2018) "On the measures of state support of the cooperation of Russian higher education institutions, state academic and research institutions, and the organizations implementing complex projects on establishing high-tech manufacturing within the subprogram "Institutional development of the R&D sector" of the State program of the Russian Federation "Development of science and technologies for 2013–2020"*
[6] *Decree of the RF Government No.1204-p dated 16 June 2018 "On amendments to the Decree of the RF Government No.449-p dated 18 March 2016"*
[7] Loots P and Henchie N 2007 *Worlds Apart: EPC and EPCM Contracts: Risk Issues and Allocation* (London: Mayer Brown)
[8] Mikhailov R B 2014 *Territory of Science* **4** 133–7
[9] Broome J C and Hayes R W 1997 *International Journal of Project Management* **15** 255–261
[10] *Handbook of Top 250 Contractors 2015* 2016 (Brussels: EU, ENR)
[11] *Reviewing ENR's top 250 Global Contractors: China Takes the Lead* 2016 (EYGM Limited)
[12] *Industry Report 2016* 2016 (N.Y.: USA, IBIS World)
[13] Mishin S A 2013 *Project Management – Technology of Success: Proceedings of the 12th Int. Conf. on Project Management* (Moscow)
[14] Manturov D V 2013 *Herald of the Bauman Moscow State Technical University. Series "Mechanical Engineering"* **2** 3–17
[15] Smetana V V 2014 *Enginneering in the concept of the RF industrial policy* report of the work group of the Federation Coincil interim comission of RF regulation "On engineering and engineering-related activity" (Ekaterinburg)
[16] Nikitenko S M, Goosen E V and et al 2017 *IOP Conf. Series: Earth and Environmental Science* **53** 012018
[17] Nikitenko S M and Goosen E V 2017 *IOP Conf. Series: Earth and Environmental Science* **53** 012018
[18] Pakhomova E O, Goosen E V and Nikitenko S M 2017 *IOP Conf. Series: Earth and Environmental Science* **53** 012018

Typology of resource regions by the level of innovative potential: fuzzy approach

E S Kagan [1,2], S G Gutova[2] and E S Chernova[2]

[1]Russia Federal Research Center of Coal and Coal Chemistry, Siberian Branch of the Russian Academy of Sciences, 18 Sovetskiy ave, Kemerovo, 650000, Russia
[2]Kemerovo State University, 6 Krasnaya street, Kemerovo, 650000,

E-mal: kaganes@mail.ru

Abstract. The paper discusses a new approach to developing a typology of resource regions based on their readiness for comprehensive exploitation of mineral resources and the level of innovative development. The use of fuzzy clustering apparatus is explained by the fuzzy boundaries for defining classes as it allows improving the quality of the typology.

1. Introduction

The mineral resource complex of Russia is the main driver of economic development and a source of budget revenues, which determines many socio-economic indicators of the country's development, including the current and potential level of people's welfare. At present, such "resource" development model is gradually exhausting itself and there is an urgent need to change the development paradigm of the resource complex of Russia and its mining regions. The basis of the new paradigm is the mining regions' ability to move from extensive use of resources to comprehensive exploitation of mineral resources, which, according to academician K. N. Trubetskoy, assumes full and low-waste high-performance and economically justified extraction of mineral resources. The transition to comprehensive exploitation of mineral resources is impossible without active innovation. Comprehensive exploitation of mineral resources involves the use of resource-saving technologies; involvement of all associated natural resources in the process of extraction and processing, while increasing their useful qualities; respect for the subsoil and the environment. Comprehensive exploitation of mineral resources will focus on the combined technological solutions, allowing to create a closed technological cycle and to maintain the ecological balance in the territories of mineral resource exploitation [3, 9, 14, 16].

Despite the fact that in all Russian mining regions the resource specialization is steadily being developed, they are extremely heterogeneous in terms of development, the share of extractive industries in the regional GRP, innovation potential, income level, etc. But most importantly, the regions have different degrees of readiness to move to comprehensive exploitation of mineral resources. This fact should be taken into account in the development of their strategic and long-term development programs. Under these conditions, it is extremely important to develop new approaches to the qualitative assessment of regional differences, while the typology of regions (the division of regions into groups according to the set of qualitative characteristics) becomes an important tool for analysis [1, 2, 4, 7, 8, 10, 11].

IOP Conf. Series: Earth and Environmental Science **206** (2018) 012017 doi:10.1088/1755-1315/206/1/012017

2. Data and methods

The complexity of the typology of mining regions is explained by two factors: the lack of generally accepted approaches (qualitative and quantitative features and criteria) to the allocation of mining (resource) regions, and the lack of a universal approach to the typology of Russian regions as a whole. This is manifested in the predominance of the rating approach based on a small number of criteria.

In the present study, the selection of criteria for the typology of regions was carried out from the position of assessing their readiness for comprehensive exploitation of mineral resources. At the initial stage of this study, a group of 36 regions specializing in oil, gas and coal production was identified from the entire list of Russian regions. Therefore, the first criterion for clustering was the level of the region's resource dependence. To assess resource dependence, the share of extractive industries in the GRP of the region was determined [12]. The indicator characterizing resource dependence was presented as a coefficient calculated by the ratio of the share of extractive industries in the GRP of the region to the share of extractive industries in the GDP of the Russian Federation. For the studied group of regions, the value of this indicator ranged from 0.107 (Republic of Ingushetia) to 6.05 for (Khanty-Mansiysk Autonomous Okrug) [5, 18].

The implementation of comprehensive exploitation of mineral resources requires a qualitative leap, a sharp increase in the level of the regions' innovative development. The lag in innovation is seen as one of the main threats to national security. Innovative development is a determining factor that affects the formation of the level of the regions' competitiveness. Innovations should cover all stages of the production and technological cycle, from exploration and production of raw materials to processing and transportation of its products and processing [17]. There-fore, the second criterion characterizing the readiness of the region for the comprehensive exploitation of mineral resources is the level of innovative development. The level of the regions' innovative development was estimated through the value of an integral indicator calculated on the basis of the approach developed by the National Research Institute Higher School of Economics [15]. In the calculation of this indicator requires the information about the values of the four indices, characterizing different aspects of innovation development: "Socio-economic conditions for innovative activity", "Scientific and technical potential", "Innovation activity", and "Quality of innovation policies". For the studied group of regions, the value of the innovation development index ranged from 0.173 (Nenets Autonomous Okrug) to 0.575 (Republic of Tatarstan).

Traditionally, the problem of dividing objects into classes basing on a group of features can be solved in two ways.

1. Splitting objects into groups using cluster analysis methods. Yet, the cluster analysis for objects whose values differ significantly from the others even during the preliminary procedure of standardization leads to the fact that the cluster analysis does not identify the differences be-tween the objects, considering them insignificant compared to the abnormal results. The initial attempt to use the cluster analysis procedure to classify regions did not show positive results. Thus, for example, in the course of this procedure, regions with a low level of innovative development and medium resource dependence, as well as regions with an average level of innovative development and low resource dependence, fell into one cluster.

2. The typology of regions is carried out on the basis of the set boundaries (levels) of indicators. The disadvantage of this approach is the lack of a common view on the setting of such boundaries. In a number of works [15] such boundaries are set by the results of the previously conducted cluster analysis, the shortcomings of which were described above. At the same time, the boundaries of indicators belonging to different clusters can overlap.

In our opinion, a more productive approach to dividing regions into groups reflecting their readiness to comprehensive exploitation of mineral resources is the approach based on the use of fuzzy clustering algorithms. The use of fuzzy intervals and overlapping ranges will reduce uncertainty in setting cluster boundaries and improve the quality of classification.

3. Results and discussion

At the initial stage, the criteria were presented in the form of linguistic variables (LVs) with the corresponding term sets: "level of the region's innovative development" {T1-low, T2-medium, T3-high} and "level of the region's resource dependence" {T1-weak, T2-medium, T3-high} [13, 6]. Table 1 presents the expressions for the membership functions (MFs) of the corresponding terms of these two criteria.

Table 1. Membership functions of the terms of linguistic variables.

LV "Level of the region's innovative development"	LV "Level of the region's resource dependence"
$\mu(T1)=\begin{cases}1, & x\le 0,2\\ 1-\dfrac{x-0,2}{0,2}, & 0,2<x\le 0,4\\ 0, & x>0,4\end{cases}$	$\mu(T1)=\begin{cases}1, & x\le 0,15\\ 1-\dfrac{x-0,15}{0,65}, & 0,15<x\le 0,8\\ 0, & x>0,8\end{cases}$
$\mu(T2)=\begin{cases}0, & x\le 0,2\\ \dfrac{x-0,2}{0,1}, & 0,2<x\le 0,4\\ 1, & 0,4<x\le 0,5\\ 1-\dfrac{x-0,5}{0,2}, & 0,5<x\le 0,7\\ 0, & x>0,7\end{cases}$	$\mu(T2)=\begin{cases}0, & x\le 0,15\\ \dfrac{x-0,5}{0,65}, & 0,15<x\le 0,8\\ 1, & 0,8<x\le 1,5\\ 1-\dfrac{x-1,5}{1,5}, & 1,5<x\le 3\\ 0, & x>3\end{cases}$
$\mu(T3)=\begin{cases}0, & x\le 0,5\\ \dfrac{x-0,5}{0,2}, & 0,5<x\le 0,7\\ 1, & x>0,7\end{cases}$	$\mu(T3)=\begin{cases}0, & x\le 1,5\\ \dfrac{x-1,5}{1,5}, & 1,5<x\le 3\\ 1, & x>3\end{cases}$

Table 2 presents a fragment of the table with the initial (crisp) and fuzzy values of the criteria for the studied group of regions.

The fuzzy clustering procedure included two stages. Firstly, the possible number of clusters was determined. Since the criteria have three levels, the maximum possible number of clusters is nine. Then a reference point was defined for each cluster. For example, for a cluster that characterizes the regions with the medium level of resource dependence and medium level of innovative development, the reference point will have fuzzy coordinates: $\{(0,1,0);(0,1,0)\}$. In fuzzy clustering, three options are possible.

1. The region belongs to only one cluster if the coordinates of the region match the reference point of the cluster (the region's MF of each indicator for a certain term equals one). For example, Nenets Autonomous Okrug has a low level of innovative development (MF of LV terms equals (1;0;0) and a high level of resource dependence (MF of LV terms equals (0;0;1), so this region will belong only to the cluster, which is characterized by a low level of innovative development and high resource dependence, and will be the reference region for this cluster.

2. The region belongs to two adjacent clusters if the region's MF of only one indicator for a certain term equals one. For example, Tomsk Region has a medium level of innovative development (MF of LV terms equal (0;1;0)). However, MFs of LV terms "Level of the region's resource dependence are: T1=0; T2=0.246; for T3=0.754. Therefore, the degree of Tomsk Region's belonging to two clusters will be determined: 1) with the medium level of innovative development and a medium level of

resource dependence ((0;1;0), (0;1;0)); 2) with a medium level of innovative development and high level of resource dependence ((0;1;0), (0;0;1)).

3. The region belongs to four adjacent clusters, if none of the region's MF indicators equals one. For example, for the Republic of Tatarstan the level of innovative development is estimated as "medium" with the MF value equaling 0.724, and as "high" with the MF value equaling 0.376. At the same time, resource dependence for this region is estimated as "medium" with MF equaling 0.716 and "high" with MF equaling 0.274. Therefore, the extent of this region's membership in four cluster-frames will be evaluated: 1) with the medium level of innovative development and the medium level of resource dependence; 2) with the high level of innovative development and the medium level of resource dependence; 3) with the medium level of innovative development and the high level of resource dependence; 4) with the high level of innovative development and the high levels of resource dependence.

Table 2. Fragment of the table with crisp and fuzzy values of the criteria.

Region	Level of the region's innovative development				Level of the region's resource dependence			
	Crisp value	T1	T2	T3	Crisp value	T1	T2	T3
Amur Region	0.237	0.816	0.185	0.000	1.467	0.000	1.000	0.000
....								
Kemerovo Region	0.334	0.332	0.668	0.000	2.284	0.000	0.477	0.523
Krasnoyarsk Kray	0.461	0.000	1.000	0.000	1.562	0.000	0.959	0.041
Nenets Autonomous Okrug	**0.173**	**1.000**	**0.000**	**0.000**	**6.008**	**0.000**	**0.000**	**1.000**
Republic of Bashkortostan	0.485	0.000	1.000	0.000	0.334	0.717	0.283	0.000
Republic of Buryatia	0.309	0.455	0.546	0.000	0.379	0.648	0.352	0.000
Republic of Tatarstan	**0.575**	**0.000**	**0.624**	**0.376**	**1.927**	**0.000**	**0.716**	**0.284**
...								
Tomsk Region	**0.464**	**0.000**	**1.000**	**0.000**	**2.631**	**0.000**	**0.246**	**0.754**

The basic concept of cluster analysis is that of the distance between objects. In the fuzzy clustering the concept of fuzzy similarity between objects in some parameter can be used for calculating the distance between objects. The convergence between two objects A and B is determined by the formula: $d_{AB} = \frac{1}{2}\sum_{i=1}^{k}|\mu_{iA}(x) - \mu_{iB}(x)|$, where k is the number of LV terms. Then the distance between two objects by n criteria is counted as: $r_{AB} = \frac{1}{n}\sum_{i=1}^{n} d_{AB}$. The degree of belonging to a cluster is defined as $Cl_j = 1 - r_j$.

Thus, the typology of regions according to two selected criteria is developed according to the following algorithm. The distance from each region to the reference point of the cluster and the degree of its belonging to all clusters are determined. If the cluster does not contain any element, it can be excluded from further consideration (in this study, such a cluster is the cluster with the high level of innovative development and the low resource dependence). If the fuzzy coordinates of the region coincide with the coordinates of the reference point, this region is considered a reference for this cluster, which is another advantage of this approach. Table 3 shows a fragment of the results of the dividing the regions into clusters with indication of the degree of their belonging to some cluster.

Table 3. Fragment of the results of dividing regions into groups.

		Level of the region's resource dependence		
		T1-low	T2-medium	T3-high
Level of the region's innovative development		…	…	…
	T2-medium	**((0;1;0),(1;0;0))** Arkhangelsk Region (0.6), Volgograd Region (0.6), Kaliningrad Region (0.47), Kursk Region (0.35), Republic of Bashkortostan (0.86), Republic of Buryatia (0.6), Republic of Karelia (0.31), Khabarovsk Kray (0.74), Chechen Republic (0.53)	**((0;1;0),(0;1;0))** Amur Region (0.59), Arkhangelsk Region (0.33), Astrakhan Region (0.53), ***Belgorod Region (1),*** Volgograd Region (0.57), Transbaikal Kray (0.6), Irkutsk Region (0.65), Kaliningrad Region (0.22), Kemerovo Region (0.57), Krasnoyarsk Kray (0.98), Kursk Region (0.8), Magadan Region (0.4), Murmansk Region (0.88), ***Perm Kray (1),*** Republic of Bashkortostan (0.64), Republic of Buryatia (0.45), Republic of Karelia (0.71), Republic of Tatarstan (0.67), Republic of Tyva (0.6), Republic of Khakassia (0.59), Samara Region (0.99), Tomsk Region (0.62), Tyumen Region (0.97), Republic of Udmurtia (0.44), Khabarovsk Kray (0.76)	**((0;1;0),(0;0;1))** Astrakhan Region (0.53), Irkutsk Region (0.58), Kemerovo Region (0.6), Krasnoyarsk Kray (0.52), Magadan Region (0.62), Orenburg Region (0.76), Komi Republic (0.72), Republic of Sakha (0.85), Sakhalin Region (0.85), Republic of Tatarstan (0.45), Tomsk Region (0.88), Tyumen Region (0.47), Republic of Udmurtia (0.44), Khanty-Mansiysk Autonomous Okrug (0.79), Chukotka Autonomous Okrug (0.64), Yamalo-Nenets Autonomous Okrug (0.89)
	T3-high	**((0;0;1),(1;0;0))**	**((0;0;1),(0;1;0))** Republic of Tatarstan (0.55)	**((0;0;1),(0;0;1))** Republic of Tatarstan (0.33)

Yet, researchers at National Research Institution Higher School of Economics argue that in 2015, compared to the previous years, the difference between regions regarding "Socio-economic conditions" and "Scientific and technical potential" indices was smaller than the difference regarding the "Innovation activity" and "Quality of innovation policies" indices [15]. Table 4 presents the descriptive statistics for the four innovative indices calculated based on 2015 data.

At the second stage of the study the typology of regions was developed based on the modified Level of innovative development calculated basing on only two indices: "Innovation activity" and "Quality of innovation policies". The descriptive characteristics for the "Level of innovative development (modified)" are presented in the last line of table 4.

Table 4. Descriptive statistics of the indices reflecting the regions' innovative development in 2015.

	Descriptive statistics (innovative development)					
	Medium	Minimum	Maximum	Lower Quartile	Higher Quartile	Statistical deviation
Level of innovative development	0.322	0.173	0.575	0.249	0.371	0.090
"Socio-economic conditions" index	0.383	0.225	0.557	0.320	0.446	0.077
"Innovation activity" index	0.232	0.000	0.590	0.149	0.307	0.124
"Quality of innovation policies" index	0.389	0.000	0.811	0.310	0.487	0.180

"Scientific and technical potential" index	0.295	0.133	0.464	0.234	0.338	0.076
Level of innovative development (modified)	0.311	0.038	0.7	0.218	0.371	0.133

The analysis of the results presented in table 5 leads to the conclusion that the use of the modified coefficient significantly changes the typology of the regions, especially for classes with a high level of innovative development.

Table 5. Division of regions into classes based on the Level of innovative development (modified).

		Level of the region's resource dependence		
		T1-low	T2-medium	T3-high
Modified Level of the region's innovative development (modified)	T1-low	**Cluster 1: ((1;0;0),(1;0;0))** Arkhangelsk Region (0.74), Volgograd Region (0.36), Kaliningrad Region (0.88), Kursk Region (0.14), Republic of Buryatia (0.58) ***Republic of Ingushetia (1),*** Republic of Kalmykia (0.98), Republic of Karelia (0.41) ***Chechen Republic (1)***	**Cluster 2: ((1;0;0),(0;1;0))** Amur Region (0.98), Arkhangelsk Region (0.47), Astrakhan Region (0.49), Volgograd Region (0.33), Transbaikal Kray (0.96), Irkutsk Region (0.51), Kaliningrad Region (0.63), Kemerovo Region (0.4), Kursk Region (0.59), Magadan Region (0.25), Murmansk Region (0.68), Republic of Buryatia (0.43), Republic of Kalmykia (0.52), Republic of Karelia (0.81), Republic of Tyva (0.84), Republic of Khakassia (0.95), Samara Region (0.55), Tyumen Region (0.62), Republic of Udmurtia (0.64)	**Cluster 3: ((1;0;0),(0;0;1)** Astrakhan Region (0.49), Irkutsk Region (0.45), Kemerovo Region (0.42), Magadan Region (0.47), ***Nenets Autonomous Okrug (1),*** Orenburg Region (0.72), Komi Republic (0.92), Republic of Sakha (0.55), ***Sakhalin Region (1),*** Tyumen Region (0.12), Republic of Udmurtia (0.65), Khanty-Mansiysk Autonomous Okrug (0.78), Chukotka Autonomous Okrug (0.96), Yamalo-Nenets Autonomous Okrug (0.66)
	T2-medium	**Cluster 4: ((0;1;0),(1;0;0))** Arkhangelsk Region (0.53), Volgograd Region (0.67), Kursk Region (0.41), Republic of Bashkortostan (0.82), Republic of Buryatia (0.57), Republic of Karelia (0.19), Khabarovsk Kray (0.72)	**Cluster 5: ((0;1;0),(0;1;0))** Amur Region (0.52), Arkhangelsk Region (0.26), Astrakhan Region (0.51), ***Belgorod Region (1),*** Volgograd Region (0.64), Transbaikal Kray (0.54), Irkutsk Region (0.55), Kemerovo Region (0.58), Krasnoyarsk Kray (0.95), Kursk Region (0.86), Magadan Region (0.53), Murmansk Region (0.82), ***Perm Kray (1),*** Republic of Bashkortostan (0.61), Republic of Buryatia (0.42), Republic of Karelia (0.59), Republic of Tyva (0.66), Republic of Khakassia (0.55), Samara Region (0.95), Tomsk Region (0.62), Tyumen Region (0.88), Republic of Udmurtia (0.35), Khabarovsk Kray (0.73)	**Cluster 6: ((0;1;0),(0;0;1)** Astrakhan Region (0.53), Irkutsk Region (0.49), Kemerovo Region (0.6), Krasnoyarsk Kray (0.49), Magadan Region (0.75), Orenburg Region (0.78), Komi Republic (0.58), Republic of Sakha (0.95), Tomsk Region (0.88), Republic of Udmurtia (0.36), Khanty-Mansiysk Autonomous Okrug (0.72), Chukotka Autonomous Okrug (0.54), Yamalo-Nenets Autonomous Okrug (0.84)

T3-high	**Cluster 7: ((0;0;1),(1;0;0))** Republic of Bashkortostan (0.39), Khabarovsk Kray (0.27)	**Cluster 8: ((0;0;1),(0;1;0))** Krasnoyarsk Kray (0.51), Republic of Bashkortostan (0.18), Republic of Tatarstan (0.86), Khabarovsk Kray (0.28)	**Cluster 9: ((0;0;1),(0;0;1))** Krasnoyarsk Kray (0.05), Republic of Tatarstan (0.64)

4. Conclusion

The approach to the typology of regions based on the application of fuzzy sets, considered in the article, has the following advantages. Firstly, it makes attribution the region to a particular class more grounded, which reduces the uncertainty in the setting of their boundaries of clusters. Secondly, in most classes the most typical regions (reference points) can be distinguished, and the degree of similarity of the other group of regions with the reference regions can be estimated through the distance to the reference point. This approach allows changing the structure of clusters, specifying the qualitative and quantitative criteria for the classification of regions to a particular type. In this study, it allowed changing the composition of the clusters showing a high level of innovative development. Finally, the developed approach to the typology of regions opens up quite interesting prospects for further research and evaluation. So, at the next stage the given typology can be used to create an expert system for the formation of a fuzzy base of rules for assessing the resource regions' readiness for comprehensive exploitation of mineral resources.

Acknowledgements

The article was prepared with the financial support of the Russian Science Foundation, under the agreement No. 16-18-10182 "Formation of organizational and economic mechanisms for comprehensive exploitation of mineral resources in resource-type regions on the basis of the partnership of science, government and business".

References

[1] Boots B et al. 2002 *Typology of Russian Regions* (Moscow) p 130
[2] Golyashev A V and Grigoriev L M 2014 *Types of Russian Regions: Sustainability and Shifts in 2003-2013* (Moscow: Analytical Center under the President of the Russian Federation) p 47
[3] Guriev S, Plekhanov A and Sonin K 2010 *Issues of Economics* **3** 4–23
[4] Ilyina I N 2013 *Public Administration Issues* **2** 91–102
[5] Kagan E S and Goosen E V 2017 *High Tech. of Dev. and Use of Mineral Res.* **3** 163–170
[6] Kagan E S 2012 *Izvestiya of Altai State University Journal* **1(1)** 160–3
[7] Kondratyev V 2015 *Theoretical and Practical Aspects of Management* **6** 20–7
[8] Lapaev S P 2014 *Vestnik of Orenburg State University* **8(169)** 100–5
[9] Levin S N, Kagan E S and Sablin K S 2015 *Journal of Institutional Studies* **3(7)** 92–101
[10] Mikheeva N N 2009 *Region: Economics and Sociology* **2** 23–42
[11] Orlov V P 2007 *Mineral Recourses of Russia. Economics and Management* **2** 3–5
[12] Official website of the Federal Service of State Statistics http://www.gks.ru
[13] Pegat A 2013 *Fuzzy Modelling and Management* (Moscow: Binom. Laboratory of Knowledge) p 798
[14] *Development of Resource Saving and Resource Reproducing Geotechnologies for Complex Exploitation of Mineral resource Deposits* 2014 Trubetskoy K N ed. (Moscow: Institute for Comprehensive Exploitation of Mineral Reources of the RAS) p 196
[15] Abdrakhmanova G et al *National Research University Higher School of Economics* Gokhberg L ed. (Moscow: HSE) p 260
[16] *Resource Regions of Russia in the "New Reality"* 2017 Kuleshov V V ed. (Novosibirsk: IEOIP SB RAS) p 308
[17] Eder L et al 2017 *Drilling and Oil* **5** 20–9
[18] Kagan E S and Goosen E V *IOP Conf. Series: Earth and Environmental Science* **84** 012016

IOP Conf. Series: Earth and Environmental Science **206** (2018) 012018 doi:10.1088/1755-1315/206/1/012018

Assessment of resource regions' readiness for comprehensive exploitation of mineral resources based on shift-share analysis

K S Sablin[1], E S Kagan[2] and A A Sharov[2]

[1]Kemerovo State University, Department of Economics and Public Administration, 6 Krasnaya street, Kemerovo, 650043, Russia
[2]Kemerovo State University, Department of Applied Mathematics, 6 Krasnaya street, Kemerovo, 650043, Russia

E-mail: sablin_ks@mail.ru

Abstract. The paper discusses the use of shift-share analysis method for assessing resource regions' readiness for comprehensive exploitation of mineral resources. The relevance of this study is explained by the fact that the choice of the factors affecting regional processes and the quantitative assessment of their impact are limited by the available statistical information. The lack of necessary statistical data significantly reduces the possibility of using econometric methods. The shift-share analysis method allows assessing the sectors of resource regions' economy, which serve as keys to identifying their readiness for comprehensive exploitation of mineral resources. The study focuses on the group of 36 resource regions specializing in the extraction of oil, gas and coal.

1. Introduction

Much attention has recently been paid to the development of resource economies and resource regions. This is due to the fact that most countries which have significant reserves of natural resources, including Russia, lag behind the countries that do not have such natural resources in terms of the rate and quality of economic growth, social development, and political and economic institutions. The need to implement the Russian economy's transition to the innovative development path (scenario) has recently been a key element in the political agenda. The peculiarity is that Russia consists of regions that are extremely different in their social and economic characteristics, which is reflected, for example, in their uneven economic development, the widening of the gap in the most important indicators of regional production, income and poverty, the quality of life of the population. The gap in the level of indicators that evaluate the innovative activity is one of the most significant compared to the other socio-economic indicators. Thus, the gap in the share of internal research and development costs in the gross domestic product has grown to 143-148 times, and in terms of the number of advanced production technologies used the gap has grown to more than a thousand times [1]. At the same time, it is the export-oriented raw materials sector of the Russian economy and resource-type regions that determines the country's place in the international division of labor, serves as the most important source of income of the budget system, and largely marks the direction of development and growth of the domestic economy. In this paper we understand resource regions (resource-type regions) as the regions whose basic sectors of economy are export-oriented extractive industries, and/or primary processing industries, producing raw materials and/or intermediate products.

The special relevance of this problem is given by the fact that, as the analysis [2] shows, the Russian resource regions are characterized by the development of an enclave dual economy. This development path is associated with the formation of a high-performance export-oriented sector isolated from the rest of the economy and represented by enterprises of extracting and intermediate industries (oil and mining industries, ferrous and non-ferrous metallurgy, etc.). The development of this sector does not provide significant incentives for the modernization of the regional economy, its integration into the all-Russian economic space; it does not create sufficient incentives for the development of the manufacturing industry, the innovation sector and other sectors of the economy. At the same time, export-oriented extractive industries and primary processing industries remain the "locomotives" of the Russian economy and determine its place in the global division of labor.

J. Stiglitz claimed that if the standard economic science focuses on the deformation of initiatives as a result of the issuance of certain privileges and protections, they must have a more harmful aspect since these privileges are usually obtained through corruption and bribery of public officials [3]. The consolidation of privileges for the representatives of wealth enclaves is a source of negative externalities associated with the deformation of incentives for entrepreneurs from other sectors of the economy, which are reoriented from the creation of a new value to rent-oriented behavior associated with participation in the redistribution of value in the high-performance export-oriented sector of the regional economy. The core of the problem is the situation when the "wealth enclaves" remain steady and hinder (or block) the transition to the development path of forming an integrated regional economy, which would provide modernization of manufacturing industries, stimulate innovation activity of small and medium businesses in the region and investment in the sphere of health care and education, research and development. Sustainable preservation and stable "reproduction" of the "wealth enclaves" in the regions rich in mineral resources increase their resource dependence, limit investment flows to a narrow range of sectors producing raw materials and intermediate products, and do not allow breaking short vertical value chains associated with the export of these products to world markets.

The problem of changing the extensive model of Russia's economic development is directly related to the assessment of the resource regions' readiness to move to the comprehensive exploitation of mineral resources, which entails significant positive external effects, i.e. the benefits received by non-privileged actors of the regional economy, located outside the "wealth enclaves", including entrepreneurs from other sectors of the resource-type economy. Under the influence of positive externalities, it is possible to increase the number of entrepreneurs capable of implementing productive activities, which in turn leads to an increase in the level of education and has a positive effect on the growth of labour productivity [4]. In other words, a productive entrepreneur is a pillar of growth and a source of Schumpeterian innovations in the economy, and the positive externalities associated with their activities are the basis for a broad modernization of the regional economy and the elongation of horizontal value chains.

2. Data and methods

This study presents the group of 36 resource regions specializing in oil, gas and coal production. These regions were selected on the basis of the coefficient calculated by E. Kagan and E. Goosen [5]. The indicator is the ratio of the share of extractive industries in the GRP of the region to the value of the share of extractive industries in GDP of the Russian Federation. To assess the readiness of the studied group of regions for comprehensive exploitation of mineral resources, the data of the official website of the Federal State Statistics Service for 2010 and 2015 [6] were used.

The choice of factors affecting regional processes, including resource regions' readiness for comprehensive exploitation of mineral resources, and the quantitative assessment of this impact are limited by available statistical information. The absence of the necessary statistical data significantly reduces the possibility of using econometric methods. The shift-share analysis method (SS), which has not yet been widely used in Russian studies, is used in foreign literature for developing typologies of regions and performing economic diagnostics of regional processes. Shift-share analysis is a method of analyzing regional growth through quantitative measurement of the contribution of structural factors to the economic dynamics of the region [7, 8, 9, 10, 11, 12]. The main idea of the method is to decompose the growth of the regional indicator into three components: National Share (NS) reflecting the influence of national growth factors; Industry Mix (IM) reflecting the influence of industry growth factors, and Regional Shift (RS) reflecting the influence of regional growth factors on the economic dynamics of the region.

The National Share NS_{ij}^{t} for sector '*i*' of the economy in the region '*j*' in the year '*t*' is determined as the share of the regional indicator (y_r) growth proportional to the growth of the national indicator (Y_N): $NS_{ij}^{t} = y_{ij}^{t-1} \cdot (\frac{Y_N^t}{Y_N^{t-1}} - 1)$, where $y_{ij}^{t-1}, Y_N^{t-1}, Y_N^t$ stand for the value of the indicators in the sector '*i*' and the region '*j*' in the year(*t*-1), and the values of the national indicator in the years (*t*-1) (basic period) and *t* (current period) respectively.

The Industry Mix for sector '*i*' of the economy in the region '*j*' in the year '*t*' reflects the value of the growth of the regional indicator determined by the change in the region's industry structure. Its value shows how much the growth of the regional indicator would amount if the growth rate of sector '*i*' of the regional economy equaled the national rate: $IM_{ij}^{t} = y_{ij}^{t-1} \cdot (\frac{Y_{iN}^t}{Y_{iN}^{t-1}} - \frac{Y_N^t}{Y_N^{t-1}})$, where Y_{iN}^{t-1}, Y_{iN}^t are the values of this indicator in the sector '*i*' of the national economy in the current and the basic periods. IM is used for assessing the contribution of the regional features of the studied indicator's structure into its cumulative growth. The positive value of this indicator means that the structure of this industry in the region is more effective compared to the national level, i.e. the share of the fast-growing industry in the region is higher than in the national economy.

The Regional Shift is frequently defined as competition effect. It reflects the difference between the real and the estimated growth in the case the growth rate of the considered sector of the economy equaled the national rate for this sector: $RS_{ij}^{t} = y_{ij}^{t-1} \cdot (\frac{y_{ij}^t}{y_{ij}^{t-1}} \frac{Y_{iN}^t}{Y_{iN}^{t-1}})$. This component characterizes the contribution of all the regional factors except the structural factor into the cumulative growth of this indicator. It serves as the basis for assessing the contribution of the endogenous factors, including human potential, investment climate, institutional conditions and other factors determining the specifics of the region, into its economic growth [13]. The total growth of the RS of the sector '*i*' in the region '*j*' is calculated as the sum of three components: $y_{ij}^t - y_{ij}^{t-1} = NS_{ij}^t + IM_{ij}^t + RS_{ij}^t$.

The shift-share analysis allows identifying the industries that are under the influence of a group of growth factors. Regional economic clusters are based on the industries for which RS values will be positive, which indicates the presence of special conditions in the region, providing these industries with competitive advantages and, as a result, the most successful development path. Based on the IM and RS values components and the ratio between them, 6 types of industries were identified [14] and presented in Table 1.

This typology allows us to assess the sectors of the resource regions' economy which are key for assessing their readiness to implement comprehensive exploitation of mineral resources. Such sectors are extractive industries, revealing the regions' resource potential, branches of manufacturing industry, defining the ground for the regions' transition to the intensive way of development, and infrastructure (transport and communications), showing the extent of the regions' development and their readiness for comprehensive exploitation of mineral resources.

Table 1. Typology of industries based on IM and RS ratio.

Type of industry	Ratio of indicators	Interpretation of indicators

1	IM>0, RS>0	Both regional and sectoral development conditions are favourable
2	IM<0, RS>0, $\lvert IM\rvert<\lvert RS\rvert$	Favourable regional development conditions outweigh partially unfavourable sectoral development conditions
3	IM>0, RS<0, $\lvert IM\rvert>\lvert RS\rvert$	Favourable sectoral development conditions outweigh partially unfavourable regional development conditions
4	IM<0, RS>0, $\lvert IM\rvert>\lvert RS\rvert$	Favourable regional conditions fail to outweigh unfavourable sectoral development conditions
5	IM>0, RS<0, $\lvert IM\rvert<\lvert RS\rvert$	Favourable sectoral development conditions fail to outweigh unfavourable regional development conditions
6	IM<0, RS<0	Both regional and sectoral development conditions are unfavourable

The most commonly used regional indicators are employment rate, turnover, value added, profit, etc. They should reflect either the scale of activities of enterprises and industries, or their effectiveness. In the first case, the indicator helps select the leading industries in the region, characterized by the fastest and most sustainable growth and high competitiveness, capable of extending the horizontal value chain. In the latter case, the most promising enterprises and industries can be selected [13].

In foreign studies, the employment rate is most often used as the regional indicator. The advantage of this indicator is that it accurately characterizes the general economic dynamics and does not depend on cost indicators. However, some researchers [15] argue that the employment rate does not adequately reflect the changes in the scale of production activities due to the specifics of the Russian labour market and low inter-regional and inter-sectoral mobility of labour resources. The employment rate indicators do not reflect the real economic dynamics, the growth of the output of raw materials and intermediate products is not necessarily accompanied by the growth of employment, and the modernization of mining and processing industries, as a rule, is associated with the Russian resource companies purchasing imported equipment and reducing the number of employees at their enterprises [see: 16]. In this regard, the most appropriate indicator may be the volume of shipped products. Within the framework of this , three sectors of the economy were selected for analysis: extractive and manufacturing sectors of industry and infrastructure, represented by transport and communication. With the aim of assessing the resource regions' readiness for comprehensive exploitation of mineral resources two regional indicators were chosen: the volume of shipped products and fixed capital investment in differents sectors of economic activities. On the one hand, fixed capital investments show resource companies' readiness to update fixed assets in order to maintain the existing level of development of extractive and manufacturing industries, on the other hand, they reflect the companies' priorities in terms of the importance of certain industries for them.

3. Results and discussion

Table 2 presents the typology of resource regions based on the analysis of the dynamcs of the amount of shipped products.

Table 2. Typology of resource regions based on the analysis of the dynamcs of the amount of shipped products in three sectors of the economy.

Sectors of the economy and classes				Transport and communications	
				4	6
Extractive	5	Manufacturing	5	(5,5,4) Khanty-Mansiysk Autonomous Okrug – Yugra Krasnoyarsk Kray	(5,5,6) Kursk Region Republic of Karelia Komi Republic Republic of Kalmykia Volgograd Region Kemerovo Region

1	1	(5,1,4) Nenets Autonomous Okrug Bashkortostan Republic	(5,1,6) Belgorod Region Arkhangelsk Region Kaliningrad Region Chechen Republic Trans-Baikal Kray
	5	(1,5,4) Murmansk Region Tatarstan Republic Perm Kray Samara Region Tyumen Region Irkutsk Region Tomsk Region Sakhalin Region	(1,5,6) Astrakhan Region Orenburg Region Republic of Sakha (Yakutia) Khabarovsk Kray Amur Region Magadan Region Chukotka Autonomous Okrug
	1	(1,1,4) Republic of Udmurtia	(1,1,6) Yamalo-Nenets Autonomous Okrug Republic of Buryatia Republic of Tyva Republic of Khakassia

Two classes were identified in the extractive and manufacturing sectors of the economy: Type 1 (both regional and sectoral development conditions are favourable) and Type 5 (favourable sectoral development conditions fail to outweigh unfavourable regional development conditions). In the sector defining transport and communications Type 4 (favourable regional conditions fail to outweigh unfavourable sectoral development conditions) and Type 6 (both regional and sectoral development conditions are unfavourable) were identified. The Republic of Ingushetia was excluded from this classification as its amount of shipped products in the manufacturing sector of industry equals zero.

The analysis showed that the most effective structural shifts occurred in the Republic of Udmurtia, which is characterized by favourable regional and sectoral development conditions in the extractive and processing sectors of the economy. Yet, in its sector of transport and communications favourable regional conditions fail to outweigh unfavourable sectoral development conditions. The major mineral resource in Udmurtia is oil, while significant peat reserves were found there as well, and both brown and black coal were extracted. At the same time, the Republic has advanced manufacturing industries and diversified agriculture. The major branches of industry are machine building, metal industry, ferrous metallurgy and wood processing industry.

The group of regions that are characterized by favourable regional and sectoral conditions in the extractive and manufacturing sectors and unfavourable regional and sectoral conditions for transport and communications presents certain interest as well. This group includes Yamalo-Nenets Autonomous Okrug, the Republic of Buryatia, the Republic of Tyva, and the Republic of Khakassia. For instance, Yamalo-Nenets Autonomous Okrug ranks among the leading Russian regions in terms of hydrocarbons, especially natural gas and oil. So, its economy is based on oil and gas extraction.

The mentioned national republics are referred to the regions of "new resource exploitation", which implies active development of both extractive (for example, coal mining) and manufacturing industries. At the same time, these subjects of the Russian Federation are characterized by generally weak development of internal communication infrastructure. In the Republic of Buryatia and the Republic of Tuva there is no railway connection, which weakens their competitive position, and levels favorable regional and sectoral conditions of development in the extractive and manufacturing sectors of the economy.

To identify the priorities of resource regions' development in terms of the renewal of fixed assets in the extractive and manufacturing sectors of the economy, and in the transport and communications sector, we present the results of the analysis of the dynamics of fixed capital investments in these sectors (table 3).

Table 3. Typology of resource regions based on the analysis of the dynamcs of the amount of fixed capital investments in three sectors of the economy.

Sectors of the economy and classes		Transport and communications		
Extractive	Manufacturing	2	4	6
5	5	(5,5,2) Republic of Kalmykia	(5,5,4) Krasnoyarsk Kray	(5,5,6) Republic of Karelia Khabarovsk Kray
5	1	(5,1,2) Murmansk Region Kemerovo Region	(5,1,4) Kursk Region	(5,1,6) Trans-Baikal Kray Republic of Sakha (Yakutia) Amur Region Magadan Region
3	1	(3,1,2) Volgograd Region	(3,1,4) Khanty-Mansiysk Autonomous Okrug - Yugra	(3,1,6) Tatarstan Republic Tyumen Region
1	5	(1,5,2) Astrakhan Region Sakhalin Region	(1,5,4) Kaliningrad Region	(1,5,6) Belgorod Region Komi Republic Chechen Republic Orenburg Region Yamalo-Nenets Autonomous Okrug Republic of Khakassia Chukotka Autonomous Okrug
1	1	(1,1,2) Bashkortostan Republic Republic of Udmurtia Perm Kray Samara Region Republic of Buryatia Republic of Tyva Irkutsk Region Tomsk Region		(1,1,6) Arkhangelsk Region Nenets Autonomous Okrug Republic of Ingushetia

In the extractive sector of the economy three classes (Type 1, Type 3, and Type 5) were identified. In the manufacturing sector of the economy two classes were identified: Type 1 and Type 5. In the transport and communications sector three classes (Type 2, Type 4, and Type 6) were allocated. Let us consider the resource regions whose economy saw structural changes which allowed them to become leaders in terms of investment in fixed assets in the mentioned sectors of the economy.

The group of the leading regions in the analysis of the dynamics of fixed capital investments is characterized by favorable regional and sectoral conditions of development in the extractive and manufacturing sectors of the economy, and favorable regional conditions for development which outweigh partially unfavorable sectoral conditions for development in the transport and communications sector. For this group of resource regions, this is due to a number of different reasons. For example, the Samara Region is one of the leaders among the subjects of the Russian Federation in the development of public-private partnership projects, which determines the need to invest resources, both from the private and public sectors of the economy in the renewal of fixed capital. Tomsk Region is leader in innovations not only within the Siberian Federal District, but also in Russia as a whole, which means a diversified economy of the region. The Republic of Tyva and the Republic of Buryatia are the regions of "new resource exploitation", which predetermines the necessity of investing resources in the extractive and manufacturing sectors and transport infrastructure. Another group of regions (Arkhangelsk Region, Nenets Autonomous Okrug, and the Republic of Ingushetia) are characterized by "failure" in terms of the volume of investments in the infrastructure sector of the economy, with unfavorable regional and sectoral development conditions.

4. Conclusion

The shift-share analysis of the resource regions in terms of the volume of shipped products showed that they fall into classes that are characterized either by favorable regional and sectoral conditions for development, or favorable sectoral development conditions which fail to outweigh the unfavorable regional conditions for the development of the extractive and manufacturing sectors of the economy, while transport and communication are characterized by Types 4 and 6. On the other hand, the typology of resource regions obtained on the basis of the analysis of the dynamics of fixed capital investments showed that a significant amount of these regions belong to Type 1 (favorable regional and sectoral development conditions) concerning the investment in the extractive and manufacturing sectors. In this sense, resource regions have a significant potential for the transition to the intensive development path in the context of their readiness for comprehensive exploitation of mineral resources.

Acknowledgements

The paper was supported with the Grant of the President of the Russian Federation, project No. MK-5941.2018.6

References

[1] Ignatov V G 2009 *Terra Economicus* **7 (2)** 132–138
[2] Levin S N, Kagan E S and Sablin K S 2015 *J. of Institutional Studies* **7 (3)** 92–101
[3] Stiglitz J 2003 *Globalization and its Discontents* (Moscow: National Public and Scientific Fund)
[4] Baumol W J 2013 *The Microtheory of Innovative Entrepreneurship* (Moscow: Gaidar Institute Publishing)
[5] Kagan E S and Goosen E V 2017 *IOP Conf. Series: Earth and Environ. Sci.* **84 (1)** 163–170
[6] *Official site of Federal Service of State Statistics* http://www.gks.ru
[7] Barff R and Knight P L III 1988 *Growth and Change* **19 (2)** 1–10
[8] Esteban J A 1972 *Regional and Urban Economics* 1972 **2 (3)** 249–255
[9] Herzog H W and Olsen R J 1979 *J. of Regional Science* **19** 393–395
[10] Houston D 1967 *Southern Economic Journal* **33(4)** 577–581
[11] Knudsen D 2000 *Socio-Economic Planning Sciences* **34 (3)** 177–198
[12] Stevens B and Craig L 1980 *J. of Regional Science* **20(4)** 419–437
[13] Karlina T V 2011 *Herald of Perm University* **4 (11)** 18–29
[14] Papadopoulou E 2010 *Applying Shift-Share Analysis (SSA) on LEADER. Initiative Local Action Groups in Greece* (Aristotle University of Thessaloniki) http://www.rudi-europe.net/uploads/media/Case-study_Greece_2_01.pdf
[15] Pritvorova T P and Kuttybaeva N B 2016 Proc. *of the VI Int. Sci. Conf. on New Approaches in Economy and Management* (Prague: Vмdecko Vydavatelskй Centrum "Sociosfira-CZ") pp 90–94
[16] *Annual Report of SUEK* 2016 http://www.ar2016ru.suek.ru

IOP Conf. Series: Earth and Environmental Science **206** (2018) 012019 doi:10.1088/1755-1315/206/1/012019

Intellectual property as an active resource for sustainable development of the regional economy

S M Nikitenko, M A Mesyats and A P Protsenko

Kemerovo Institute (branch) of Plekhanov Russian University of Economics, 39 Kuznetsky ave, Kemerovo, 650992, Russia

E-mail: nsm.nis@mail.ru

Abstract. This article is devoted to the study of the importance of the intellectual component in the sustainable development of the economy of the Russian regions, taking into account foreign practices. The article outlines the factors constraining the development of pledge transactions with rights to intellectual property objects (IPO) in Russia, identifies key problems that require a solution for turning IPO into an active resource for sustainable development of the regional economy. The accents of organizational, legal, economic and methodical character are made. The article is prepared with the support of the Department of Education and Science of the Kemerovo Region within the framework of the Agreement on Grants from the Regional Budget for the Implementation of Research Projects (the project "Intellectual Property as the Basis for Sustainable Development of the Coal-mining Region (case study Kemerovo Region)").

1. Introduction

In modern conditions of commodity production, the significant importance of intellectual property is due to the possibility of using it as a tool for creating added value, a means of capitalizing the assets of enterprises and organizations, and also as an investment resource [1]. Considering the orientation of the Russian economy towards innovative development and modernization, it is obvious that there is a need to form a civilized market for intellectual property, and, accordingly, to create favorable conditions for its functioning. At the same time, not only the legal basis (regulation) and the operation of the mechanism of mutual relations between subjects of the intellectual property market are implied, but also the organizational and economic instruments that can be used in this segment of the economy. Under such conditions, intellectual property will undoubtedly act as an active resource for the sustainable development of the national and regional economy.

2. Data and methods

This research is based on the results of scientific research of domestic and foreign specialists, practical materials, as well as legal documents related to lending issues secured by rights to intellectual property. In addition, in carrying out this study, the authors used official statistical and other materials of foreign and Russian organizations (in particular, the World Bank on Intellectual Property Rights and Rospatent), as well as regional enterprises.

3. Results and discussion

In foreign studies, the use of organizational and economic instruments based on pledge schemes for IPO was highlighted by the following authors: J. Marshall, R. Caldwell, B. Cain, L. Esposizione, A. Kukrus, V. Antonova, J. Chen and Z. Yuan, B. Amablea, J.B. Chatelaina, K. Ralf [2; 3-5; 6]. The works of Russian scientists also reflect the aspects of the pledge of IPO exclusive rights from the standpoint of organizational, legal and economic aspects [7; 8; 9]. Among the works of Russian scientists, mention should be made of the study of A.A. Kravtsov, focused on the analysis and systematization of problems relating to patent statistics [10].

Issues related to the sustainable development of the economy in its various manifestations are revealed in their works by a number of foreign scientists, such as L. Brown, E. Lovins, H. Lovins, T. Page, J. Rawls, K. Flwein, P. Hocken, et al. Problems of sustainable development of regions (including resource-producing regions) are covered by domestic scientists V.A. Vasilenko, G.I. Gritsko, A.F. Shupletsova, A.V. Samarukha and others. The connection of the production and technological chains, specific to the mining industry, with the economic development of a region is demonstrated in the studies of S.M. Nikitenko, E.V. Goosen, et al. [11-13].

The sustainable development of the regional economy is largely due to the investment climate, the development of which can facilitate pledge transactions with IPO. Authors have identified country specific features of legal regulation and registration of collateral in some EU countries (table 1). In the course of the study of foreign experience, it was found that lending transactions secured by IPO rights in European countries are governed by both the general (for example, the Dutch Civil Code) and the special legislation (for example, the UK Patent Law, the Italian Intellectual Property Code, the French Intellectual Property Code, the French Trade Code). In many countries, it is possible to register transactions with IPO in the state register, which guarantees the rights of parties involved in the transaction [14].

Table 1. Objects of intellectual property and registration authorities for pledge transactions for the purpose of lending in certain EU countries.[a]

IPO and features of its use in pledge transactions	Britain	Spain	Italy	Holland	France
Patent	+	+	+	+	+
Patent application	+	+	N/A	N/A	N/A
Trademark	N/A	+	+	+	+
Database	N/A	N/A	N/A	+	N/A
Industrial Designs	N/A	N/A	N/A	+	N/A
Pledge Registration Authority	UK Intellectual Property Office	Spanish Patent and Trade marks Register	Italian Registrar for Trademarks and Patents (UIBM)	- Tax authorities of the Netherlands; - IP registries of the Netherlands	National Institute for Industrial Property (INPI)

[a]Source: compiled by the authors on the basis of [5; 14; 6; 15; 16].

In particular, the US practice shows that for many companies, intellectual property rights are the most valuable assets. Loans secured by patents increase the availability of external funds and return on shareholders' capital. The US Patent and Trademark Office registers transactions related to the transfer of patents to other owners, in addition to IP inventors (the so-called transferring documents) in the Patent Transfer Database, but do not keep records of patent pledge assignments provided non-payment of a loan by a borrower, creating the prerequisites for increasing uncertainty regarding the transfer of intellectual property rights. In this regard, the authors, representing in their work the analysis of the American experience of pledge transactions, indicate: "... it should be done so that the transfer of patent income property rights could be enforced with low costs for not only the debtor but also competing creditors. Creditors should be protected against the borrower's ability to transfer, refuse or "license" the security in the form of a patent, as well as, from the termination of the borrower's maintenance of the patent in force, prosecution and exploitation. This legal improvement is the way to increase the aggregate debt ceiling and the growth of innovations" [2].

Zhou Li's research shows that China has formed three models of lending to high-tech small and medium-sized enterprises (SMEs) on the basis of IP rights pledge:

1) Beijing pattern which is a direct secured loan (Patent transaction directly between the bank and the enterprise / Copyright protection / Copyright guarantee);

2) Pudong Pattern which is an indirect collateral credit (a bank and state fund guarantee and a patent counter-guarantee);

3) Wuhan Pattern which is a combination of a direct and an indirect collateral loan.

Li Long analyzed the characteristics of loans secured by IPO rights, the cases of granting guarantees to them, and ways to assess the IPO in Japan. He believes that the main characteristic of lending on the security of IP rights in Japan is that it is implemented mainly by Japanese "policy investment banks" and local banks, and is coordinated with other commercial banks. Few commercial banks lend on the security of rights to IPO alone, mainly because of the difficulty of assessing the rights to IPO and lack of relevant experience.

The general regulation of the pledge of IP rights in Estonia is established by PL (Public Law). The pledge of an unregistered IPO is regulated in section 3 of PL "Pledge of rights", according to which the following conditions are specified in § 314: 1) the right to be pledged (for example, to IPO) must be valued; 2) the right to be pledged must be separable, and normative acts of pledge with transfer of possession to the creditor must be applied [4].

India has the developed pledge legislation. Most of the loans provided to date in India are secured loans [17]. In December 2012, the World Intellectual Property Rights (IPR) Bank was established in India to assist the fast-growing transfer of technology and intellectual property rights between the United States, the European Union, the United Kingdom, Russia, South Africa, Japan, China and other parts of the world. The World IPR Bank offers a set of services for manufacturers of original equipment and companies involved in the transfer of technology and IPO. The key services of the bank include: transfer of technology and IPO; technology and IP security; financial and licensing transactions; technology and IP banking; technology and IPR assessment.

The attitude of Russian commercial banks to the use of IPO as collateral for loans is ambiguous. This is explained by the fact that the regulatory acts of the Central Bank of the Russian Federation (Instructions, Regulations, Directives) regulating the procedure for creating reserves for possible losses on loans neither assume nor include intellectual property in the list of collateral. Nevertheless, a number of large Russian banks, based on the provisions of the Civil Code of the Russian Federation, have conducted such operations in recent years.

Despite the fact that there are many problems, the ways of resolving them are seen in the change in the position of the regulator (the Central Bank of the Russian Federation), legislative bodies, commercial banks themselves in the role and place of IPO in lending for the accelerated implementation of new scientific ideas and technologies.

Since the adoption of the 4th part of the Civil Code in Russia in 2006, the processes related to the conclusion of agreements on the alienation of rights to IPO have become more active. By now, some experience has already accumulated in the country of carrying out pledge transactions involving the rights to IPO (table 2). Any contracts for the disposal of the exclusive right to IPO and trademarks (including pledge transactions) in Russia are subject to registration in Rospatent [18-23].

The analysis of the Russian practice has shown that about 20 Russian and foreign banks have participated in pledge transactions with IPO rights registered in Rospatent since 2013. It is known that this pledge practice in Russia is often inherent in the activity of head banks. At the same time, the Russian practice of lending against the security of IPO rights shows that many transactions are carried out with the involvement of foreign banks branches and foreign appraisers, and are structured according to English law [7].

Table 2. Number of IPO pledge agreements registered by Rospatent in 2009-2017. [a]

	2009	2010	2011	2012	2013	2014	2015	2016	2017
Pledge agreements over exclusive rights to the results of intellectual activity	-	8	16	17	20	15	13	9	13
Pledge agreements over exclusive rights to trademarks	16	62	52	24	60	40	58	72	98
Number of trademarks referring to which the pledge agreements are concluded	N/A	195	280	82	191	258	362	689	750
Total number of pledge agreements over exclusive rights to IPO	16	70	68	41	80	55	71	81	111

[a]Source: compiled by the authors on the basis of [3-4].

In the course of the study conducted by the authors it was established that not all banks participating in contracts of pledge of IP rights in Russia have a regular nature of concluding such transactions. All banks participating in pledge contracts registered in the period from 2013 to 2016 by Rospatent can be divided into the three categories:

- banks that regularly participate in pledge transactions (for example, Sberbank of Russia PJSC, Moscow Credit Bank OJSC, International Financial Club OJSC AKB and other banks that are registered as pledgees in repeated annual pledge transactions);
- banks that periodically participate in pledge transactions (for example, Far Eastern Bank OJSC participated once in the pledge agreements in 2013 and 2014, Discovery Financial Corporation Bank PJSC in 2013, 2015 and 2016, Promsvyazbank OJSC in 2013 and as Promsvyazbank PJSC in 2016, Russian Agricultural Bank OJSC in 2013 and 2016, etc.);
- banks which were one-time participants in pledge transactions (for example, the European Bank for Reconstruction and Development, London, Great Britain, Center-Invest Commercial Bank OJSC, Absolut Bank Commercial Bank JSC, VESTA Investment Bank LLC), Amsterdam Trade Bank NV, Amsterdam, Netherlands and other banks).

4. Conclusion

At present, the main constraining factors for the development of the regional economy in the direction aimed at the active use of IP, according to the authors, are the following:

- the discontinuity of the production and technological chains, the absence of a model (conditions, respectively) of the sustainable development of a mining region;
- the lack of a methodology for identifying, selecting and preliminary assessing the latest promising technologies ensuring sustainable economic development of the mining regions (taking into account the sectoral specifics of the economy and the existing production and technological potential of the regions) on the basis of making production and technology chains with the production of high added value products;
- inadequate incentives for the development of IPO and the lack of adequate organizational and economic instruments with IPO participation in the mining regions;
- insufficiently active involvement of small and medium-sized businesses in the process of providing IPO access to mining regions.

The identified factors serve as the basis for conducting a study aimed at determining the significance of IP as an active resource for sustainable development of the economy of mining regions. Accordingly, the number of research tasks should include the following:

- research of patent-licensing activity of organizations of mining regions;
- identification of criteria affecting the intellectual activity of a region, and the main factors hindering the achievement of sustainable development of mining regions;
- development of the methodology for identifying, selecting and preliminary assessing the most effective technologies in mining, enrichment and deep processing of minerals based on patent studies, expert assessments and formalized mathematical models;
- development of regulations governing the procedures for the use of IPO as the basis for sustainable development of the mining regions.

The authors consider it expedient to direct the research towards the following fundamental scientific task: to search for the essential interconnection of the categories “knowledge economy” “global technological leadership” and “technological sovereignty” as the basis for the paradigm of the postindustrial model of economic growth of the national economy, as well as the identification of the systemic role of IP as the main productive force in a post-industrial society, forming a new technological mode [27].

Acknowledgements
The article is prepared with the support of the Department of Education and Science of the Kemerovo Region within the framework of the Agreement on Grants from the Regional Budget for the Implementation of Research Projects (the project “Intellectual Property as the Basis for Sustainable Development of the Coal-mining Region (case study Kemerovo Region)”). The Agreement No. 21 of August 14, 2018.

References

[1] Nikitenko S M, Patrakova L P and Goosen E V 2009 *Regional Economy: Theory and Practice* http://techlaw.org/wp-content/uploads/2011/05/110426_IP-rights-and-loan-financing_1.pdf?x1365424 17–26
[2] *Code de la propriété intellectuelle (version consolidée au 23 février 2015)* http://www.wipo.int/wipolex/ru/text.jsp?file_id=363403
[3] *Annual report of Rospatent for 2012* http://old.rupto.ru/about/reports/2012
[4] *Annual report of Rospatent for 2013* http://old.rupto.ru/about/reports/2013
[5] *Annual report of Rospatent for 2014* http://old.rupto.ru/about/reports/2014
[6] *Annual report of Rospatent for 2016* http://old.rupto.ru/about/reports/2016/otchet_2016_ru.pdf
[7] *Commercialization of intellectual property* http://rniiis.ru/kommertsializatsiya- is.html
[8] Valletdinova E N 2014 *Intellectual property right* **2 (34)** 23–29
[9] *Annual report of Rospatent for 2017* https://rupto.ru/content/uploadfiles/docs/otchet_2017_ru.pdf
[10] Livio Esposizione *IP rights and loan financing: European perspective* http://techlaw.org/wp-content/uploads/2011/05/110426_IP-rights-and-loan-financing_1.pdf?x13654
[11] Nikitenko S M and Mesyats M A 2017 *IOP Conf. Ser.: Mater. Sci. Eng.* **253** 012024
[12] Kravtsov A A 2017 *Journal of the New Economic Association* **3 (35)** 144–167
[13] Nikitenko S M 2010 *Innovations* **6** 15–21
[14] *Annual report of Rospatent for 2015* http://old.rupto.ru/about/reports/2015
[15] *Dutch Civil Code* Book 3 URL: http://www.dutchcivillaw.com (accessed: 08.02.2017).
[16] *World Intellectual Property Rights Bank (World IPR Bank)* http://www.wiprbank.com/#
[17] Kurakova N G and Zinov V G 2016 *Innovations* **6 (212)** 48–57
[18] *We invite you to participate in a pilot project on lending against intellectual property* http://fasie.ru/press/industry/priglashaem-k-uchastiyu-v-pilotnom-proekte-kreditovaniya-pod-zalog-intellektualnoy-sobstvennosti/

[19] *The Ministry of Economic Development proposes to finance innovative companies on the security of patents* https://www.rupto.ru/ru/news/minekproposition
[20] *The advisor to the Head of Rospatent, Grigory Senchenya, spoke about possible directions for the activation of state policy in the sphere of intellectual property* http://old.rupto.ru/press/news_archive/inform2016/spbcongress
[21] Bruno Amable, Jean-Bernard Chatelain and Kirsten Ralf June 2010 *Journal of Economic Dynamics and Control* vol 34 issue 6 1092–1104
[22] Jie Chen and Zeming Yuan 2012 *Proc. of the 2nd International Conference on Industrial Technology and Management* (Singapor: IACSIT Press) vol 49
[23] Kukrus A and Antonova V 2005 *Proc. of the 13th Scientific Conference of Economic Policy* (Berlin, Tallinn) pp 248–256
[24] Marshall J, Caldwell R and Cain B *Taking Security Over Patents* https://united-kingdom.taylorwessing.com/synapse/march14.html

Resource regions of Russia: socio-economic indicators and innovative development

I V Filimonova[1,2,3], L V Eder[1,2,3], A V Komarova[1,2], I V Provornaya[1,2] and V Yu Nemov[1,2]

[1]Trofimuk Institute of Petroleum Geology and Geophysics, Siberian Branch of the Russian Academy of Sciences, 3 Koptyug ave, Novosibirsk, 630090, Russia
[2]Novosibirsk State University, 2 Pirogova street, Novosibirsk, 630090, Russia
[3]Federal Research Center of Coal and Coal Chemistry, Siberian Branch of the Russian Academy of Sciences, 18 Sovetskiy ave, Kemerovo, 650000, Russia

E-mail: FilimonovaIV@list.ru

Abstract. The article is devoted to the study of the influence of the natural resources located in the regions of Russia on the level of socio-economic and innovative development. At the first stage of the study, a cluster analysis was applied to the database of indicators of the level of extraction specialization in the regions of Russia. During the next stage of the study, hypotheses on the connection between socio-economic indicators and specialization of the region in extractive industries were tested via multifactor regressions. Positive dependence of socio-economic variables on resource specialization variables with a high level of significance was revealed. However, most of Russia's resource regions have low level of innovative production, which can have a negative influence on country's economic growths

1. Introduction

The extractive industries of the Russian Federation play a key role in the development of the country. However, current development of the Russian economy follows the resource-inertial trajectory [1, 2].

The aim of the work is to identify the dependence of the level of socio-economic indicators and innovative development on the level of resource specialization of Russian regions. To achieve this goal, the following tasks were accomplished (1) empirical analysis of indicators of socio-economic and innovative development, as well as the level of specialization in extractive industries for Russian regions; (2) the allocation of resource regions and their classification in accordance with the indicators of resource specialization; (3) analysis of the relationship between the degree of resource specialization and the level of socio-economic and innovative development.

Recent studies indicate a negative relationship between resource dependence and variables, explaining the growth of productivity at the macro level [3]. A wide range of explanatory variables for such analysis includes: the development of human capital (Gylfason, 2001, Stijns, 2006, Shao and Yang, 2014), the growth of export of processing industries (Wood and Berge, 1997), investment, education and transparency (Papyrakis and Gerlagh, 2007) and institutional development (Mehlum, 2006; Boschini et al., 2013) [4-6].

There is a number of studies that challenge both the negative relationship between resources and development in general, and the choice of specific indicators for analysis [7]. Another critical approach relates to the time series used for the analysis of the "resource curse" (Alexeev and Conrad,

2009; Stijns, 2005; et al., 2011; James, 2015). It is important to note that most of the researchers reviewed the time interval of 1960-1990, when most of the developed countries moved from the industrial to the postindustrial type of economy, or, in the case of countries with a lower level of development - from mostly pre-industrial to industrial [8].

Most of the research on the influence of the natural resource on the economy of the territory was carried out at the macro level. However, in the situation of the heterogeneous structure of the regional economy and the level of socio-economic development within a single country, a more detailed analysis of the relationship of specialization in extraction and economic indicators is needed [9, 10].

2. Methods of research

At the first stage of the study, a cluster analysis was applied to the database of indicators of the level of extraction specialization in the regions of Russia. The Ward Method was chosen as a clustering algorithm, because it provides compact and well-separated clusters [11]. During the next stage of the study, hypotheses on the connection between socio-economic indicators and specialization of the region in extractive industries were considered. Econometric tools of multifactor regressions were used to test possible dependencies.

An important issue in such analysis is the choice of indicators and factors, as well as their calculation method [12, 13]. Independent variables, such as the share of the mining industry in the GRP, the share of employment in extractive industries, the proportion of mineral extraction tax in total tax revenue, were calculated as a percentage of the overall performance of the regions. This allowed not only to conduct inter-regional comparisons, but also to assess the production, fiscal and social structure within the regions. The dependent variables, such as income per capita, the level of investment per capita and GDP per capita were calculated as the relation of the absolute values and the population of the region. Per capita indicators allow more accurate assessment of the level of development of the territories under consideration.

3. Results and discussion

As a result of cluster analysis all 85 regions of Russia were divided in 4 groups. Group 1 “Monoresource regions” consist of Khanty-Mansiysk Autonomous District, Yamalo-Nenets Autonomous District, Kemerovo Region, which specialize in extraction of one particular natural resource. Group 2 “Hydrocarbon resource region” has Nenets Autonomous District, Komi Republic, Astrakhan Region, Republic of Bashkortostan, Republic of Tatarstan, Udmurt republic, Perm Region, Orenburg region, Samara Region, Tyumen region, Krasnoyarsk region, Irkutsk region, Tomsk Region, The Republic of Sakha (Yakutia) and Sakhalin region. Group 3 ‘Non-hydrocarbon resource regions” included Magadan Region, Chukotsky Autonomous District, Belgorod region, The Republic of Karelia, Murmansk region, Amur region, Arhangelsk region, The Republic of Khakassia, Transbaikal region. Group 4 “Non-resource regions” consists of all the other regions.

The first, second and third groups, composing the list of resource regions of Russia, have relatively high indicators of socio-economic development. Such indicators as monthly income per capita, investment per capita, GRP per capita are at a level above the Russian average for most of the considered territories. However, in terms of output of innovative products, this indicator exceeds the average Russian level only for 6 resource regions out of 27 (table 1).

Table 1. Main indicators of socio-economic and innovative development of Russia’s resource regions in 2016, thousand rubles.

Region	Per capita income per month	Investments per capita	GRP per capita	Share of innovative goods and services in GDP, %	Production of innovative goods and services per capita

Sakhalin region	49.6	508.8	1701.8	11.16	189.8
Republic of Tatarstan	32.6	165.4	473.9	20.36	96.0
Samara Region	26.8	80.1	386.9	18.78	72.7
Magadan region	50.8	267.1	851.1	7.19	61.5
Perm Region	28.4	90.0	397.8	9.19	36.6
Republic of Bashkortostan	28.1	88.7	323.6	10.58	34.3
Tyumen region	28.0	177.0	619.2	3.58	21.8
Krasnoyarsk region	28.0	145.7	564.5	3.64	20.5
Komi Republic	31.5	235.7	610.7	3.24	19.9
Belgorod region	29.6	92.6	442.8	4.28	18.9
Tomsk region	24.3	94.1	439.9	3.78	16.6
Kemerovo Region	21.3	57.9	310.1	3.85	12.0
Udmurt republic	23.9	56.3	328.0	3.48	11.4
Astrakhan Region	22.8	116.4	314.9	3.44	10.8
Transbaikal region	22.8	77.9	229.8	4.19	9.7
Irkutsk region	22.3	107.3	420.1	2.27	9.5
Orenburg region	22.0	82.9	388.5	1.89	7.4
Amur region	29.7	160.2	343.6	1.98	6.8
Khanty-Mansi AO - Yugra	44.2	488.5	1928.2	0.33	6.2
Murmansk region	36.1	133.4	512.2	1.12	5.8
Yamalo-Nenets AO	67.5	2046.9	3395.2	0.15	4.9
Republic of Sakha (Yakutia)	38.9	285.9	781.5	0.38	2.9
Chukotka AO	63.9	195.7	1273.1	0.17	2.2
Republic of Karelia	25.7	56.4	335.2	0.09	0.3
Republic of Khakassia	21.2	49.8	319.8	0.06	0.2
Nenets AO	70.0	1938.9	4956.8	0.00	0.1
Tyva Republic	14.1	26.9	149.8	0.05	0.1
Russian Federation	**30.7**	**99.7**	**443.5**	**5.91**	**26.2**

Three multifactorial regressions of testing influence of resource specialization were built. The models differed in dependent variables, such as per capita incomes, per capita investment and per capita GRP.

All regressions under consideration were significant, with the R-square and the normalized R-square having high levels, which indicates the high quality of the built model (table 2).

Table 2. Multifactorial regression results for GRP per capita.

Regression statistics	
Multiple R	0,861515
R-square	0,742208
Normalized R-square	0,708583
Standard Error	578,1163

Dispersion analysis

	df	*SS*	*MS*	*F*	*Significance F*
Regression	3	22131707	7377236	22,0731	5,85E-07
Residual	23	7687024	334218,4		
Total	26	29818731			

The share of production in GRP was a significant factor for all the regressions tested; the share of mineral extraction tax in tax revenues was significant for explaining per capita investment and per capita GRP; employment in the extractive industry has an impact on per capita income per month and the share of production in GRP (formulas 1-3).

$$y_1 = 19.5 + 0.26*x_1 + 1.46*x_3, \quad (1)$$

$$y_2 = -198.3 + 2.8*x_1 + 5.6*x_2, \quad (2)$$

$$y_3 = -325 + 14.3*x_1 + 7.9*x_2 + 83.9*x_3, \quad (3)$$

where y_1 – Per capita income per month; y_2 – Investments per capita, y_3 - GRP per capita, x_1 – Share of extraction in GRP; x_2 – Share of Mineral extraction tax in the tax revenues, x_3 - Employment in the extraction industry.

Positive dependence of socio-economic variables on resource specialization variables with a high level of significance was revealed for all tested regressions. However, despite this positive correlation a significant proportion of Russian resource regions can be classified as enclaves. Isolation of the regions occurs mainly due to the localization of mineral resources in hard-to-reach areas with severe natural, economic, geological conditions.

Empirical analysis showed that high indicators of socio-economic development, especially per capita investment, are characteristic primarily of Arctic enclave resource regions. This can be explained by the low level of population and high incomes as a result of Russia's social policy to increase the incomes of the population in the regions Far North and equated to them. In addition, these regions often specialize mainly in the production of hydrocarbons - an industry that requires a high level of investment [14].

At the same time, most of Russia's resource regions have low levels of innovative production, which at the present stage of the world economy development becomes the most important factor of the country's economic growths.

Thus, the "resource curse" of the resource regions associated with relatively high living standards of the population due to the extraction of resources and low indicators of innovative development, processing, creation of industries with high added value is revealed.

4. Conclusion

Empirical analysis showed that relatively high indicators of socio-economic development are characteristic of resource regions, primarily enclaves, located in the Far North of Russia. Most of the enclave regions, which specialize mainly in the resource-extracting sectors, have socioeconomic indicators above the average for Russia. At the same time, low levels of innovation development indicators are observed for all enclave resource regions. Only for six out of twenty-seven resource regions the share of innovation production in GRP was higher than the average for Russia. In the absence of a developed intraregional economy and an innovative type of specialization, this can lead to a decline in production efficiency and economic growth. The directions of innovation development within the framework of smart specialization based on knowledge and innovations, resource-efficient, environmentally friendly and competitive economy should include the processing sector and related industries such as petrochemicals, engineering, construction, services, financial sector and others.

Acknowledgements
The research was supported by the grant No. 17-78-20218 of the Russian Science Foundation.

References

[1] Eder L V, Filimonova I V, Provornaya I V and Nemov V Y 2017 *IOP Conference Series: Earth and Environmental Science* **84** 012012
[2] Kontorovich A E, Eder L V and Filimonova I V 2017 *IOP Conference Series: Earth and Environmental Science* **84** 012010

[3] Eder L V, Filimonova I V, Provornaya I V, Nemov V U and Nikitenko S M 2017 *International Multidisciplinary Scientific GeoConference Surveying Geology and Mining Ecology Management, SGEM* **17 (53)** 727–34
[4] Gylfason T and Zoega G 2006 *The World Economy* **29 (8)** 1091–115
[5] Sachs J D and Warner A 2001 *European Economic Review* **45 (4-6)** 827–38
[6] Gilberthorpe E and Papyrakis E 2015 *The Extractive Industries and Society* **2 (2)** 381–90
[7] Van der Ploeg F 2011 *Journal of Economic Literature* **49 (2)** 366–420
[8] Venables A J 2016 *Journal of Economic Perspectives* **30 (1)** 161–84
[9] Eder L V, Filimonova I V, Provornaya I V, Komarova A V and Nikitenko S M 2017 *17th International Multidisciplinary Scientific GeoConference SGEM* **17(15)** 365–72
[10] Piirainen K A, Tanner A N and Alkærsig L 2017 *Technological Forecasting & Social Change* **115** 289–300
[11] Eder L V, Kontorovich A E, Filimonova I V, Provornaya I V and Goosen E V 2017 *International Multidisciplinary Scientific GeoConference Surveying Geology and Mining Ecology Management, SGEM* **17 (53)** 719–26
[12] Gulyaev P V *Indian Journal of Science and Technology* **9 (11)** 89423
[13] Islankina E and Thurner T W 2018 *Entrepreneurship and Regional Development* **March** 1–24
[14] Morgan K 2013 *Ekonomiaz* **83** 103–26

IOP Conf. Series: Earth and Environmental Science **206** (2018) 012021 doi:10.1088/1755-1315/206/1/012021

Public-private partnership as a form of development of resource regions of Russia

L V Eder[1,2], I V Provornaya[1,2], I V Filimonova[1,2], A V Komarova[1] and M V Mishenin[1]

[1]Trofimuk Institute of Petroleum Geology and Geophysics, Siberian Branch of Russian Academy of Sciences, 3 Koptyug ave, Novosibirsk, 630090, Russia
[2] Federal Research Center of Coal and Coal Chemistry, Siberian Branch of the Russian Academy of Sciences, 18 Sovetskiy ave, Kemerovo, 650000, Russia

E-mail: ederlv@yandex.ru

Abstract. The article shows that for Russia the innovative development of resource regions, based on the principles of smart specialization, should be accompanied by a significant amount of investment. A significant part of the Russian regions are enclaves, which is a characteristic feature of the economic and infrastructure isolation of the region from macro-regional and all-Russian economic relations. Such isolation is primarily due to the localization of mineral natural resources in remote areas with severe natural, economic and mining-geological characteristics. Such regions are unable to attract private investment to a large extent. In these circumstances, the actual task is to use the principles of public-private partnership (PPP).

1. Introduction

Public–private partnerships (PPPs) relate to a range of collaborative efforts between public and private organizations [1]. In its narrowest form, PPP refers to a particular type of contractual arrangement between public and private organizations [1-3] that is formed to fund, build, administer and sustain infrastructure (COM 2004 3271). Nevertheless, different types of public arrangements from traditional contracting to more innovative forms [4] are suggested to advantage from the characteristics of partnership relationships, for example, shared goals, reciprocity and trust. Therefore, the PPP concept is applied here to understand generally collaborative relationships between public and private organizations [5-6] and PPP is defined as an institutional arrangement between public and private organizations [7] that includes collaboration to reach shared goals of delivering public services. If properly managed, PPPs expand resource exchange, diminish transaction costs, increase risk sharing, clarify contract specifications and encourage stronger inter-organizational collaboration and thereby improve the quality and reliability of public services [8].

The paper aims to identify the possibility of using PPP practices to implement sustainable innovative development of Russia's resource regions.

To achieve this goal, the following tasks were formulated:

(1) distribute the resource regions of Russia in clusters, depending on the degree of readiness for smart specialization,

(2) to rank the resource regions according to the degree of readiness for smart specialization,

(3) to determine the relationship between the degree of readiness of resource regions for PPP and smart specialization,

IOP Conf. Series: Earth and Environmental Science **206** (2018) 012021 doi:10.1088/1755-1315/206/1/012021

(4) to research the directions and approaches for using PPP in Russia.

2. Methods of research

Based on the comparative analysis, the relationship between the degree of readiness of Russian resource regions for the application of smart specialization and PPP was determined. The division of regions into groups was carried out by the average value.

2.1. Cluster analysis for determining resource regions prepared for smart specialization

All 27 resource regions of Russia were initially divided into prepared, partially prepared, and not prepared for smart specialization (enclave regions). The study uses a hierarchical agglomerative method of cluster analysis. Clustering was performed by ward's method with Euclidean metric (formula 1)

$$d(X_i, X_j) = \left(\sum_{k=1}^{n} (x_{ik} - x_{jk})^2\right)^{1/2} \quad (1)$$

where x_{ik}, x_{jk} – values of characteristic k for the object i and the object j.

In the work, the cluster analysis was carried out using a specialized statistical data processing package Stata 13.

2.2. Model of the ranking of resource regions by the degree of readiness for smart specialization

Simultaneously with the distribution of resource regions for clusters, these regions were ranked by the degree of readiness for smart specialization for further determination of the relationship between the degree of readiness of Russia's resource regions for the application of smart specialization and public-private partnerships.

The ranking of regions can be carried out using 5 methods: grading, qualimetric, percentiles, interval partitioning, assignment of scores. In this work, a qualimetric method allowing to take into account the scatter of values is used:

$$K = \frac{X_i - X_{min}}{X_{max} - X_{min}} \quad (2)$$

where X_i is the current value for the selected indicator for the i region, X_{min} is the minimum value for the selected indicator, X_{max} is the maximum value for the selected indicator.

3. Results and discussion

Based on clustering methodological approaches, this study has revealed following groups: 1) economically developed resource regions of Russia, prepared for smart specialization by attracting private sector investment; 2) resource regions prepared for smart specialization using only public-private partnership, 3) enclave resource regions that initially require significant government support to enhance their investment attractiveness for private sector participation.

Because of the weak institutional framework, most enclave regions in practice are not provided enough to apply public-private partnerships. As such, the use of public-private partnerships and other forms of state support is a priority task at the initial stage, while regions with weak economic structure should receive support from the state until they become attractive for active private capital investing.

As a result of applying cluster analysis for resource-type regions in order to identify readiness for smart specialization, 10 regions were included in the first cluster, 11 for the second cluster, and 6 for the third cluster (table 1).

Table 1. Resource regions readiness for smart specialization.

Number of regions	Regions	Type of region	Description of type of regions
10	Belgorod region, Astrakhan region, Republic Of Bashkortostan, Republic Of Tatarstan, Udmurtia, Perm Krai, Orenburg region, Samara region, Krasnoyarsk Krai, Kemerovo region	Cluster No. 1 – Regions are ready to smart specialization	Resource regions, which are characterized by high population density, good infrastructure, well-developed manufacturing sector, relatively high rate of innovation development
11	Republic Of Karelia, Republic Of Komi, Arkhangelsk region, Murmansk region, Tyumen region, Republic Of Khakassia, Zabaykalsky Krai, Irkutsk region, Tomsk region, Republic Of Sakha, Amur region	Cluster No. 2 – Regions partially ready for smart specialization	Resource regions, which are characterized by low standard of living of the population, low value of innovative goods and services, low volume of investments in fixed capital
6	Nenets AO, Khanty-Mansi AO, Yamalo-Nenets AO, Sakhalin region, Chukotka AO, Magadan region	Cluster No. 3 – Resource enclave regions (not ready)	Resource regions, which are characterized by low population density, high standard of living of the population, relatively underdeveloped transport and processing infrastructure, high value of accumulated foreign investment

Taking into account the rating of the resource regions of Russia for the level of PPP development, as well as compiled rating of the resource regions, the regression relationship between these two indicators is built in terms of the level of readiness for smart specialization.

Three clusters were singled out in different colors in terms of the degree of readiness for PPPs, which was obtained by clustering at a previous stage. The first group consists of enclave regions with a low level of PPP development, the second group consists of regions that are partially ready for smart specialization and ready for the practice of PPP, the third group included regions ready for smart specialization and ready for the practice of PPP; the fourth group have a high level of PPP development, but at the same time only partially ready for smart specialization.

The second group included the Kemerovo region, which belongs to the regions ready for smart specialization, but at the same time has a low level of readiness for PPP, which is associated with poor quality of regulatory and legal acts. The Khanty-Mansiysk District, which is an enclave region, entered group IV, but at the same time it is characterized by a high level of development of the PPP sphere, which is connected with the improvement of the legislative base for ensuring access of private organizations to the provision of public services (figure 1).

The main part of PPP projects in the fuel and energy complex is carried out in regions with a long history of development of the oil and gas industry, where the transport and processing infrastructure are well developed [9-12].

In the oil and gas sector, the objects of PPP can include mineral deposits, production facilities (energy, railway transport, road facilities, ports, airports, utilities, etc.) and social infrastructure (housing, public buildings, education and health facilities, etc.) in the territories where extraction and processing of natural resources is conducted.

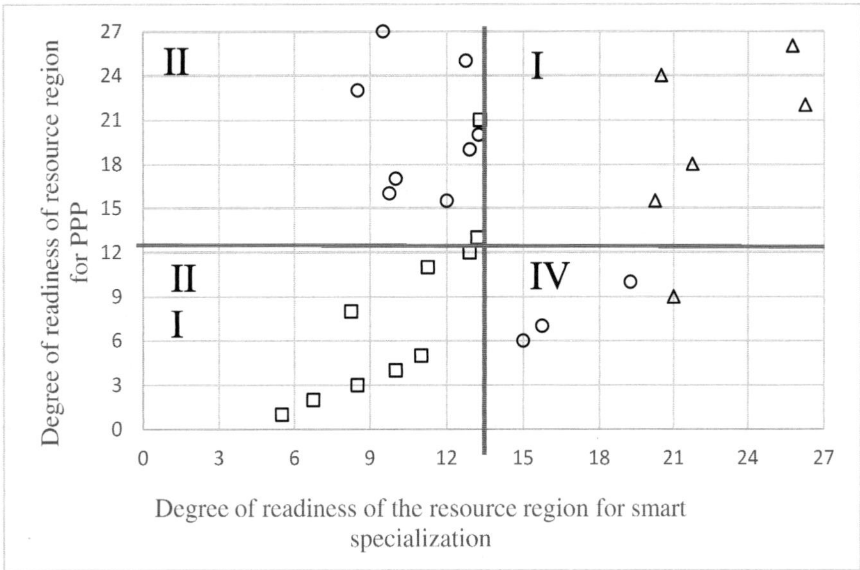

Figure 1. Dependence between the ranking of the regions of Russia depending on the level of PPP and on the level of readiness for smart specialization.

4. Conclusion

The paper shows that in the case of Russia the innovative development of resource regions based on the principles of smart specialization should be accompanied by a significant amount of investment.

The majority of the resource regions of Russia are characterized by a relatively low level of: the development of the institutional environment; legislative base for the PPP practice; experience in the implementation of PPP projects. As a result, most resource regions are not ready to apply the practice of PPP due to weak development of regulatory support, insufficient transparency and lack of investment attractiveness.

Thereby, nowadays an active task is to prepare the regions for the practice of PPP with the subsequent increase of their economic, infrastructural, innovative, personnel and qualification levels to that extent where it is possible to organize integrated exploitation with the necessary volume of private investment for sustainable development in the long-term perspective.

Acknowledgements

The authors are grateful for financial support to the Russian Science Foundation No. 16-18-10182.

References

[1] Jamali D 2004 *International Journal of Public Sector Management* **17 (5)** 414–30
[2] Chou J and Pramudawardhani D 2015 *International Journal of Project Management* **33 (5)** 1136–50
[3] Li B, Akintoye A, Edwards P J and Hardcastle C 2005 *Construction Management and Economics* **23 (5)** 459–71
[4] Gidman P, Blore I, Lorentzen J and Schuttenbelt P 1995 *Working Paper Series No. 4, UNDP/Habitat/World Bank, Nairobi* 1–11
[5] Erridge A and Greer J 2002 *Public Administration* **80 (3)** 503–22
[6] Parker D and Hartley K 2003 *Journal of Purchasing and Supply Management* **9 (3)** 97–108
[7] Hodge G A and Greve C 2007 *Public Administration Review* **67 (3)** 545–58
[8] Kwak Y H, Chih Y and Ibbs C W 2009 *California Management Review* **51 (2)** 51–78
[9] Eder L V, Filimonova I V, Provornaya I V, Nemov V U and Nikitenko S M 2017 *International Multidisciplinary Scientific GeoConference Surveying Geology and Mining Ecology*

Management, SGEM **17 (53)** 727–34

[10] Eder L V, Kontorovich A E, Filimonova I V, Provornaya I V and Goosen E V 2017 *International Multidisciplinary Scientific GeoConference Surveying Geology and Mining Ecology Management, SGEM* **17 (53)** 719–26

[11] Eder L V, Filimonova I V, Provornaya I V and Nemov V Y 2017 *IOP Conference Series: Earth and Environmental Science* **84** 012012

[12] Kontorovich A E, Eder L V and Filimonova I V 2017 *IOP Conference Series: Earth and Environmental Science* **84** 012010

IOP Conf. Series: Earth and Environmental Science **206** (2018) 012022 doi:10.1088/1755-1315/206/1/012022

The methodology for working out the techniques for geophysical monitoring of the process of directional hydraulic fracturing of a hard roof and the process of local hydrotreating of a coal seam face working space, its implementation within the framework of RSF project

A V Shadrin and V I Klishin

Federal Research Centre of Coal and Coal Chemistry at the Siberian Branch of the Russian Academy of Sciences, Leningradskiy ave 10, Kemerovo, 650065, Russia

E-mail: avsh-357@mail.ru

Abstract. The methodology of working out the techniques for geophysical monitoring of the process for directional hydraulic fracturing of a hard roof and the process of local hydrotreating of a coal seam face working space is introduced in the paper. The research is done in two basic directions: the first is predicting and managing geo-mechanic condition of a gas-saturated tight rock massif containing coal; the second one is monitoring and managing the processes of hard roof directional hydraulic fracturing and protective hydrotreating of a coal seam against rock bumps and sudden outbursts. To monitor the stress condition of a face working space a complex method that includes monitoring of stress condition by spectral-acoustic method (the analysis of a noise spectrum of an operating equipment) and gas factor control by gas monitoring equipment (controlling methane concentration near the fore-breast is justified. Some widely known options of applying spectral-acoustic method for controlling stress condition are analyzed and the application of the best-attested one is justified. To control the process of a jump-like growth of a fracture in a roof and in a coal seam the method of acoustic emission if justified.

1. Introduction

To decrease the danger of gas-dynamic phenomenon manifestation during the mining special measures are applied. They are: hydrotreating of the face working space, in particular, directional hydraulic fracturing of a hard roof and preventive hydrotreating of a coal seam [1]. Due to all these the development of the ways to monitor these measures during the operating process and estimating them after the operation are being worked out. Moreover, the ways of monitoring must be based on geophysical methods as all other methods applied for fulfilling these purposes are rather time and labour-consuming. The researches in the frame work of RSF grant No 17-17-01143 “Forecasting and controlling geo-mechanic condition of a rock mass within the period of formation and manifestation of dynamic squeezing of the main roof and its preventive hydro-treating with the purpose to prevent dynamic and gas-dynamic phenomena” are directed on researching these methods. This paper introduces the methodology of these researches and some freshly received results. The notion of methodology here is understood as an excepted and logically explained consequence of some practical tasks and the set of methods used for solving them.

2. Dynamic phenomena in the development of coal seams with hard roofs and the hydrotreating methods for combating them

The intensification of mining activity and deepening the development of coal deposits are complicated by dynamic phenomena. Starting with the depth of 500 m in stoping faces with hard roofs some intensive and dangerous sudden outsqueezing of coal with intensive gas emission take place. They can result in failing of face equipment causing serious danger for stope miners [1]. Moreover, poor caving of roof causes the concentration of rock pressure on coal massif in a stoping face zone and in junctions with mine workings provoking rock bumps. In some case local rock bumps can cause sudden outburst of coal and gas. Due to the influence of poor caving of roof with a significant area left out pillars and guarded development workings are subjected to the influence of great bearing pressure. It causes the destruction of mine workings and, correspondently, the failure of operating equipment and ventilation of faces [1].

It is also known that one of the reasons for methane explosion is collapsing of a roof on a large area during the movement of a longwall face that causes methane and coal dust emission into active mine workings. To prevent this it is important to provide artificial timely caving of the main roof. For this purpose the method of directional hydraulic fracturing is widely-applied. It allows excluding areal hanging of the coal and its sudden dynamic influence on mechanized complexes together with the safety of reused mine workings in stoping zones. [2, 3].

Regional hydrotreating is applied for heightening coal seam gas permeability. For this purpose fluid inflated into the boreholes drilled along the coal in hydraulic fracturing mode causes the formation of fracture network maximizing gas permeability and gas recovery of a seam [4]. To provide the mode it is necessary, in a short period, to create maximum possible area of fracture edges surface without significant coal tempering.

For heightening local gas-permeability and lessening the burden on a face working space fluid injection into the boreholes in hydro-sloughing mode is applied [5]. In this mode, recommended for seams with low gas and hydro permeability, under the influence of injecting fluid the face working space is squeezed out into the working. Therewith, due to the development of fracture network in face working space in-situ gas pressure lessens, coal strength decreases and bearing pressure zone moves into the depth of the massif reducing the chance for dynamic phenomena manifestation.

Regional and local hydrotreating of coal is also applied with the purpose of its moistening, as it is widely known that sudden outbursts never take place if the coal humidity is higher than 6 percent. It can be explained by the fact that under this humidity level gas is blocked in micro-fractures and pours not transforming into free state and not heightening in-situ pressure; the coal becomes more plastic shifting the bearing pressure into the depth of the massif decreasing rock bumps and outburst danger.

For moistening coal fluid is injected into a seam in a low-pressure mode or in a hydro-ploughing mode. In low-pressure moistening mode, recommended for seams with good hydro- permeability, additional fractures are not formed and the fluid is distributed through natural fractures into the pours moistening coal equally. In hydro-ploughing mode, recommended for coals with lesser hydro-permeability, apart from the abovementioned mode, injected fluid forms additional network of fractures which moistening the coal.

All the above-mentioned methods of coal seam and massif hydrotreating are characterized by a certain intensity of fracture development. As the fractures develop in a "jump-like" mode the fracture development intensity is understood here as a number of "jumps" of the developing fractures per time unit. However, there is no mechanism for controlling fracture development as the technology of all known hydrotreating methods provide only registering pressure and velocity of fluid injection. Meanwhile the developing fracture is the source of acoustic emission and the registration of this emission can help in recording, firstly, the "jump-like" growth of a fracture and secondly, its intensity in time. Under more complex modification of the method the coordinates of a growing fracture area can be determined and the energy of acoustic emission can be estimated. The value of the acoustic emission energy can indicate the fracture area growth. This allows us to set the task for controlling the hydrotreating process of a coal massif by geophysical method (method of acoustic emission) with the fluid injected mode determined by the previously set purpose.

3. Research directions and tasks

The research is planned to be developed in two directions:

1) Predicting and controlling geomechanical condition of a gas-saturated and coal containing rock massif with tight rocks;
2) Monitoring and controlling the processes of directional hydraulic fracturing of a hard roof and protective hydrotreating of a coal seam against rock bumps and sudden outbursts.

The following tasks are set to develop the first direction:

- To work out a physical model for geo-mechanic condition of a coal massif containing the in-seam working developed along the coal seam with a hard roof and to introduce geophysical methods for controlling basic parameters that determine stress condition of a face working space.
- As stress condition of a face working space, in-situ gas pressure and coal strength influence on the danger of dynamic phenomena manifestation it is planned to work out (justify) a complex geophysical method for predicting dynamic phenomena based on registering of these factors. To control the stress condition it is planned to apply spectral-acoustic method and to control in-situ gas pressure it is planned to use gas monitoring equipment as the concentration of methane at the face working is functionally connected with in-situ free gas pressure.
- For the developed geophysical method of prediction it is planned to justify a criterion for danger of dynamic phenomena manifestation and the procedure of its definition.
- For defining the sensitive depth of spectral-acoustic method while controlling stress condition of a face working space ahead of the development working it is planned to work out a laboratory facility that models the sound propagation from the source of the sound that imitates the cutting unit of the operating mining machinery in a modeled face working space with a regulated stress condition.
- To work out and construct the prototype (experimental sample) of a portable device to control stress condition of a face working space of a rock massif by spectral-acoustic method with the algorithm of processing the acoustic signal emitting source using the whole spectrum of a signal.

As for the second direction, the following tasks are set:

- To work out a model of fracture development in a coal seam roof.
- For monitoring the process of jump-like development of a fracture under the influence of fluid injected into it out of the borehole it is planned to apply acoustic emission method.
- To determine a functional connection between acoustic emission parameters and hydrotreating which can be applied to most frequently used pumping equipment with a hard operating characteristic (practically constant injecting tempo and the injected pressure is determined by a filtrating coefficient of a developed massif).
- For certain mining condition of a longwall with hard roof development it is necessary to work out a method for estimating the direction and the area of a directional hydraulic fracture development based on the dependence of acoustic emission parameters, pressure and injecting tempo on time.
- To justify the express-method for defining optimal values of acoustic emission parameters for providing controlling parameters of a local hydrotreating method under real condition.

4. Basic obtained results of the project

According to the task of the first direction the model of geo-mechanical condition of a rock massif containing an in-seam working driven along the seam with a hard roof was developed. The result is embodied in a computer software application for calculating stress condition near the mine working driven along the coal seam with a hydraulic fracture in a roof depending on mining and geological conditions [6]. The obtained calculations allow us to determine the influence of a hanging roof on stress condition in a face working space. It also helped to determine priority growth area of fractures in a roof and in a seam under the influence of injected fluid under various parameters of rock massif and the working. [7, 8].

Apart from the methods for estimating the influence of the hanging roof on the stress condition based on the industrial bending theory which is well-described in literal sources, this model applies methods of deformable solids and strength criterion of rock formation [6].

Interconnected complex of geo-physical methods (spectral-acoustic and gas analytical) for continuous control of a face working area stress condition in the process of mining is justified in the framework of the first direction. A distinctive peculiarity of this complex geo-physical method of controlling stress condition of a gas-saturated coal massif is that a controlled parameter in the form of high and low frequency noises of the working mining equipment that influence on the face is functionally connected with the value of average stresses in coal seam selvages, coal strengths and in-situ gas pressure [9, 10].

The analysis of the options for applying a spectral-acoustic method for controlling stress condition of a rock massif showed that to improve the accuracy of the control it is important to solve several tasks [11, 12]. One of them is such modification of the method when the whole spectrum of an acoustic signal is used to estimate the stress condition. It is important as the existing modifications do not use the whole spectrum but only some areas of low and high frequencies portion of spectrum. So it may turn out that spectral components with the highest changes of amplitudes when approaching to the zone with high stresses can appear in uncontrolled portion of spectrum. To avoid this failure of all the existing modifications it is offered to control stress condition by median of the amplitude-frequency response of an acoustic signal.

The idea of the proposition is in the following. Supposing that after digitalization of geophone analogue signal we will obtain a spectrum includes N harmonics. The total sum *A(r)* of these harmonics can be written as following:

$$A(r) = \sum_{i=1}^{N} A_{i,0} F(r) \exp\left(-Cf_i \frac{\sigma_l}{\sigma_c} r\right), \tag{1}$$

where: $A_{i,0}$ – is amplitude of i-th harmonic at the source of the signal ("noise" produced by operating equipment); i – harmonic's number, variable; 0 – means that the distance between the sound source and the receiver is equal to zero; $F(r)$ is a function that considers the directivity graphs of the signal source (for example, $F(r) = 1$ – for a plane wave signal; $F(r) = 1/r$ – for a spherical wave; r is the distance between the source and the receiver of the signal, m, variable; f_i - is i-th harmonic frequency, Hz (cps); σ_l and σ_c are correspondingly average limit value (a maximal value for the controlled seam area that precedes to the crashing of the massif caused by a dynamic phenomenon) and a current value of stress in a massif, Pa; l – notation "limit"; c – notation "current". Parameter C is equal to:

$$C = \frac{\alpha_0 \beta}{f_m},\ \text{m}^{-1}\text{Hz}^{-1},$$

where α_0 is the attenuation at a middle frequency of working range f_m in stressless position (under stressless condition), m^{-1}; m – notation "middle"; index 0 means stress absence; β is nondimensional coefficient of proportionality, characterized by acoustic properties of the massif, variable.

As it can be seen from equation (1) together with the growth of average current stresses the exponent value per each *i*- harmonic will increase with the number of the harmonic. It will cause the growth of *A(r)* value. Thus, a median number of harmonic corresponding to the half of *A(r)* sum, will also increase. The value of the slip characterizes the relative growth of the stresses acting in a controlled zone of a massif.

Technical requirements to assembly a portable prototype for the algorithm of determining the coefficient of relative stresses was worked out.

Applying spectral-acoustic method for predicting dynamic phenomena it is important to have a danger criterion. At present there are no reliable methods for determining danger criterion of dynamic phenomena manifestation for geo-physical methods. However, there are such methods for instrumental methods of predicting sudden coal and gas outbursts and rock bumps. They are based on the results of drilling control boreholes [5, 13]. Their reliability is conditioned by statistical processing of enormous aggregation of data obtained by the results of the prediction. Our offer to apply them for determining the danger criterion for geo-physical method is in the following:

The danger of a certain dynamic criterion manifestation in a certain face is controlled by instrumental and geo-physical methods simultaneously. Current values of danger for instrumental method $R_{\text{c,in}}$ and for geo-physical method $R_{\text{c,g}}$ are determined according to the relevant methods. Critical (maximum allowable) value of danger criterion for instrumental method is known, we take it as $R_{\text{l,in}}$. Then the danger criterion of a certain dynamic phenomena in a controlled face working can be determined for both methods of prediction as follows: $R_{\text{c,in}} / R_{\text{l,in}} = R_{\text{c,g}} / R_{\text{l,g}}$. Here we have:

$$R_{\text{l,g}} = \frac{R_{\text{l},in}}{R_{\text{c,in}}} R_{c,g}, \quad (2)$$

where: in –notation “instrumental”; g – notation “geophysical”.

Performing the current prediction in the development working it is important to know the issue of how deep the result of the perdition propagates ahead of the face. When applying a spectral-acoustic method this issue turns to be of important character. The reason of it is that the prediction method is based on spectral analysis of noises produced by operating equipment that passed the control zone. However, geophone which receives the acoustic signal is installed into the side of the working at a certain distance of the face. Thus, the transmitter of acoustic signals (cutting organ of the mining equipment) and geophone appear to be behind the controlled zone of a face working space. Analytically it is proved that as the signal propagates according to the Huygens' principle the depth of acoustic signal propagation ahead of the face and the one that reaches the geophone is comparable to the length of the wave, the lesser value of which is about 1 m [11, 12]. The increasing of the distance is caused by disjunctive and plicative dislocations ahead of the face, if there any, as the signal reflected from the dislocation will also reach the geophone. However, according to the results of analytical or experimental researches there are no reliable data on sensitivity depth of this prediction method ahead of the development working. Thus, in the framework of the project technical requirements for construction of laboratory facility for determining the sensitivity depth of spectral-acoustic method while controlling the stress condition of a face working space was worked out. In this laboratory facility the working frequencies, imitating the work of mining machinery together with the mechanisms of creating a zone of a high stress in a regulated distance from the face were chosen according to the similarity principle. We suppose that, for the first time, it will give us a chance to estimate the influence of the value of the excessive rock pressure ahead of the development working and the distance of excessive pressure zone from the wall face on the attenuation of the acoustic vibrations when the source of the sound is on the surface of the wall face and the receiver is installed into the side of the modeled working.

Within a framework of the second research direction the models of a fracture development separately in a roof and in a coal seam were developed. The difference between these two models is conditioned by their application purposes. The purpose of the model for the development of a fracture in a roof is in defining the trajectory of the fracture movement. The fracture is drilled by a “fracture drilling device” in a working wellbore. The model calculates the trajectory of a jump-like growing fracture under the influence of the injected fluid in uneven field of stresses conditioned by the influence of the working depending on the angle of the fracture towards the horizon, the distance to the ground surface and some parameters which characterize the strength of the crashing rock [6-8].

The application purpose of the model of fracture development in coal is in determining the relation between the length of the growing fracture, the injecting parameters (pressure, injecting tempo, fluid viscosity) and filtration and collecting properties of a massif (porosity, specific density of a fracture).

The application of the acoustic emission method for controlling the process of a jump-like fracture development under the influence of the fluid injected into it out of the wellbore is justified. It is determined here that under continuous injecting tempo which is a characteristic of the pump-set with hard performing capabilities used for hydrotreating of a rock massif (continuous injecting tempo and the injecting pressure depends on the length of the filtering part of a wellbore and filtering properties of a massif) the length of the hydrotreating fracture L can be determined by the registered numbers of acoustic emission impulses J under its jump-like growth from the initial length L_0 made by a fracture drilling device according to the following formula:

$$L = L_0 e^{J/k}, \qquad (3)$$

where the coefficient k is taken according to the results of testing injections.

Supposing that the shape of the growing fracture is close to a circle with diameter L we can determine its area S as follows:

$$S \approx \frac{\pi}{4} L_0^2 exp(2J/k). \qquad (4)$$

It is determined that under hydro ploughing of a coal seam by a pump-set with hard operating characteristic the optimal value of an injecting tempo q_{opt}, when the fluid of volume Q injected into the seam before breaking through the growing fractures into the closest working can be determined by the formula:

$$q_{opt} = \frac{Q}{a(Q-V)^2}, \qquad (5)$$

where coefficients a and V can be easily determined by the results of the testing injections.

It is justified theoretically and proved in practice that the optimal value of acoustic emission activity ΔJ_{opt} (the number of acoustic emission impulses per unit time) that accompanies the jump-like growth of the fractures changes in time t as follows:

$$\Delta J_{opt} = \frac{\delta_1}{t - \delta_2 \sqrt{t}}, \qquad (6)$$

where parameters δ_1 and δ_2 are also determined by the results of testing injection.

The effectiveness of a roof and a coal seam hydrotreating can be also estimated according to the decreasing of a stress condition in a face working space. For this purpose, in the framework of the project, in some Kuzbass mines several experimental researches on applying spectral-acoustic control method involving hardware-software complex that consists of acoustic data recorder (RIPAS) and "Geoscan-RIVAS" software has been started [18].

5. Conclusion

As a result of the researches it is justified that for automated monitoring and effective estimation of the process for directional hydraulic fracturing of a hard roof and the process of local hydrotreating of a face working space it is sufficient to apply the complex of interconnected methods of control: spectral-acoustic method, acoustic emission and gas-analytical methods.

Acknowledgments
The research was carried out due to the grant of Russian Science Foundation (project #17-17-01143).

References

[1] Klishin V I, Zvorygin L V, Lebedev A V and Savchenko A V 2011 *Safety Issues and New Technologies for Underground Development of Coal Deposits* (Novosibirsk: Novosibirskiy pisatel) p 524
[2] Klishin V I, Nikolskey A M, Opruk G Yu, Neverov A A and Neverov S A 2008 *Ugol* **11** 12–6
[3] Klishin V I, Opruk G Yu, Sentyurev A V and Nikolayev A V 2015 *Ugol* **11** 12–6
[4] Nozhkin N V 1979 *Preliminary Degassing of Coal Deposits* (Moskow: Nedra) p 271
[5] *Instructions for the safe conduct of mining operations in formations hazardous due to sudden emissions of coal (rock) and gas (RD 05-350-00). Prevention of gas-dynamic phenomena in coal mines (Collection of documents)* 2000 (Moscow: State Enterprise Scientific and Technical Center for Safety in Industry of the Gosgortekhnadzor of Russia) pp 120–303
[6] Cherdantsev N V 2017 *Occupational Safety in Industry* **12** 5–10
[7] Cherdanstev N V and Shadrin A V 2017 *Vestnik Nauchnogo tsentra po bezopasnosti rabot v ugolnoy promyshlennosti* **4** 18–26
[8] Cherdantsev N V and Shadrin A V 2018 *Calculating the Trajectory of a Fracture Movement in a Rock Mass Under Influence of Inside Pressure Caused by Fluid* certificate of registration for a computer software application No. 2018610989
[9] Shadrin A V 2016 *Journal of Mining Science* vol 52 issue 670–82
[10] Shadrin A V 2016 *Phisico-technicheskiye Problemy Razrabotki Poleznykh Iskopayemykh* **4** 48–62
[11] Shadrin A V and Kontrimas A A 2017 *Naukoyemkiye Technologii Razrabotki I Ispolzovaniya Mineralnykh Resursov: Scientific Journal* **3** 408–13
[12] Shadrin A V and Kontrimas A A 2017 *IOP Conf. Series: Earth and Environmental Science* **84** 012040
[13] *Federal norms and rules on industrial security "The instruction on forecasting dynamic phenomena and monitoring of a rock mass in developing coal deposits"* (approved by the order of Federal Ecological, Technological and Nuclear Supervision Service dated 15.08.2016 No. 339) p 129
[14] Shadrin A V and Teleguz A S 2017 *Vestnik Nauchnogo Tsentra Po Bezopasnosti Rabot V Ugolnoy Promyshlennosti* **3** 10–4
[15] Shadrin A V and Klishin V I 2017 *Vestnik Nauchnogo Tsentra VostNII Po Promyshlennosti I Ecologicheskoy Bezopasnosti* **3** 31–5
[16] Shadrin A V and Klishin V I 2018 *Vestnik Nauchnogo Tsentra Po Bezopasnosty Rabot v Ugolnoy Promyshlennosti* **1** 77–87
[17] Shadrin A V and Klishin V I 2018 *IOP Con. Series: Earth and Environmental Science* **134** 012057
[18] *The Procedure of Coal Massif Acoustic Sounding Applying Hardware-software Solution* 2016 (Moscow: *MNTL RIVAS) p* 36

IOP Conf. Series: Earth and Environmental Science **206** (2018) 012023 doi:10.1088/1755-1315/206/1/012023

Spectral acoustic and seismological monitoring of enclosing rock condition in the face area during roof softening by means of directional hydraulic fracturing (DHF)

V I Klishin[1], G Yu Opruk[1], A S Teleguz[1], A V Galkin[2]

[1]Federal Research Center of Coal and Coal Chemistry, Siberian Branch of the Russian Academy of Sciences, 10 Leningradsky ave, Kemerovo, 650065, Russia
[2]JSC "TopProm", 9a Ermakov ave, Novokuznetsk, 654007, Russia

E-mail: opruk@yandex.ru

Abstract. The article reveals information about the results of directional hydraulic fracturing practical application which implied the reduction of actual main roof rock caving steps. The authors calculated initial main roof caving steps taking the implementation of hydraulic fracturing into account. They demonstrate the scheme with boreholes location and present their estimated parameters. The authors controlled the implementation of hydraulic fracturing by means of acoustic sounding. They give the results of seismic activity control of enclosing rock in the extraction column profile at the time of the mechanized complex leaving the installation chamber. Considering the results of the conducted research, the authors have made a conclusion in relation to the effectiveness of the hydraulic fracturing implementation scheme under these conditions.

1. Introduction

At the moment industrial experimental works aimed at further introduction of directional hydraulic fracturing practice, when working face is functioning under tough roof, are taking place in Kuzbass [1, 2]. Roof rock softening was performed under mining and geological conditions of coal deposit extraction in longwall face 16-19 of seam 16 in a mine section of the JSC "Yubileynaya" Mine" when the mechanized complex was leaving the installation chamber.

The main roof of seam 16 is tough. Near installation chamber its thickness is from 20.0 to 27.0 meters. It consists of siltstones of various coarseness, from fine grain to coarse grain, as well as fine-grain sandstone. Rock hardness of siltstones varies from 3 to 9 according to Protodyakonov scale. As for sandstone, it is within the range between 6.0 and 12.0 (figure 1). The thickness of the immediate roof represented by fine and coarse siltstone is generally from 2.0 to 3.0 meters. The depth of installation chamber location in relation to the surface is 345 meters. Seam 16 is prone to sudden coal and gas outbursts, coal displacement, and dynamic destruction of bedrock starting from 268 meter deep point. It is hazardous in relation to rock bumps in the depth lower than 400 meters.

Thus, complex mining and geological conditions of coal extraction indicated that there is a possibility of considerable roof hanging at the time of the mechanized complex leaving the installation chamber and, as a result, increase of stress on the mechanized supports of the complex. That can lead to the possibility of gas-dynamic notions which are related to the rise of stress in the seam periphery [3].

2. Main part

In order to reduce the initial caving step the members of the Institute of Coal of the Federal Research Center for Coal and Coal Chemistry, Siberian Branch of the Russian Academy of Sciences, in collaboration with technical department of the JSC "Yubileynaya" Mine" made a decision to perform work aimed at main roof preliminary softening in seam 16 by means of directional hydraulic fracturing at the time when the mechanized complex was leaving the installation chamber. During installation work the authors were developing the roof softening scheme. Cut-off boreholes were drilled at an angle to the strata from the installation chamber and development workings (figures 1-4) [4].

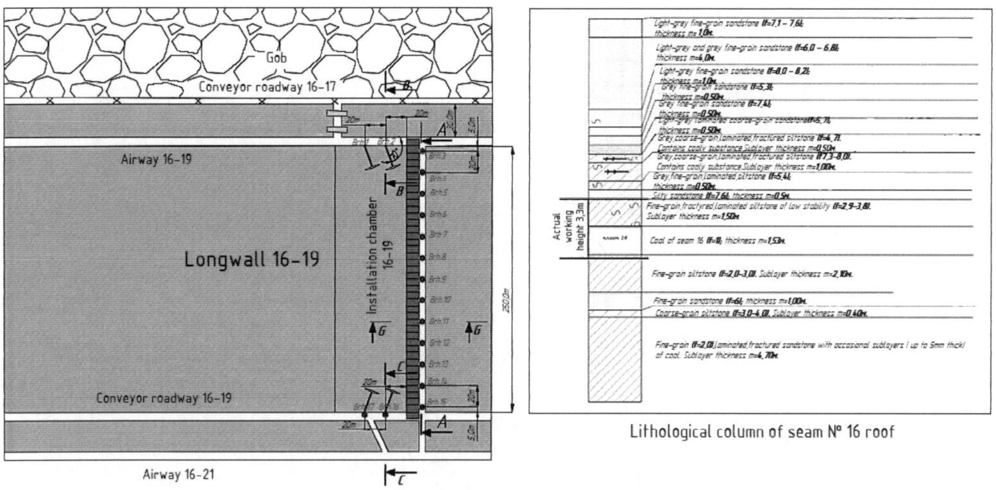

Figure 1. Process flow scheme with borehole location for implementation of directional hydraulic fracturing (DHF) method at the time of the mechanized complex leaving the installation chamber 16-19.

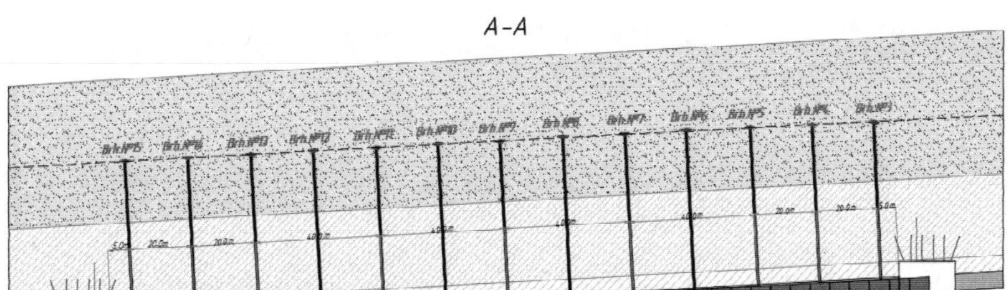

Figure 2. Vertical scheme of borehole location for implementation of directional hydraulic fracturing (DHF) from installation chamber 16-19 (cross section A-A, G-G).

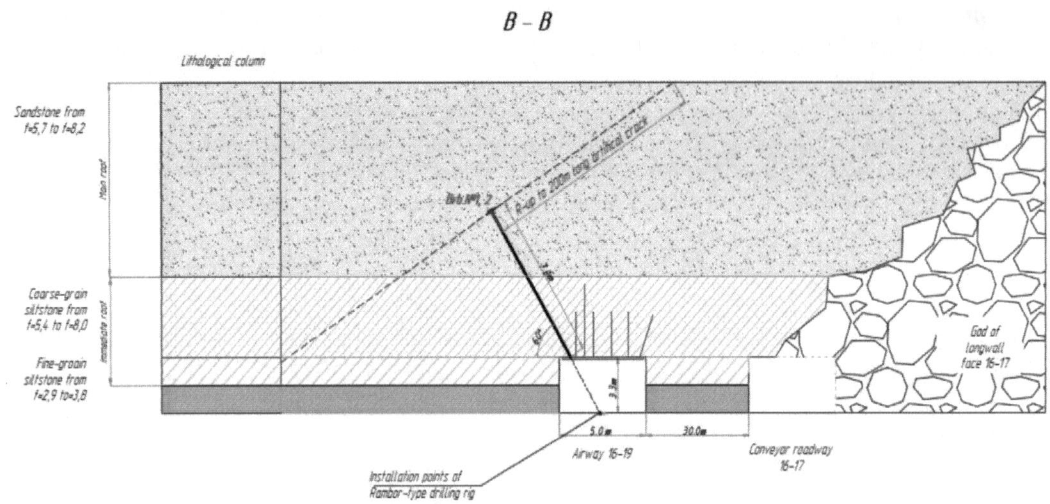

Figure 3. Vertical scheme of borehole location for implementation of DHF from airway 16-19 (cross section B-B).

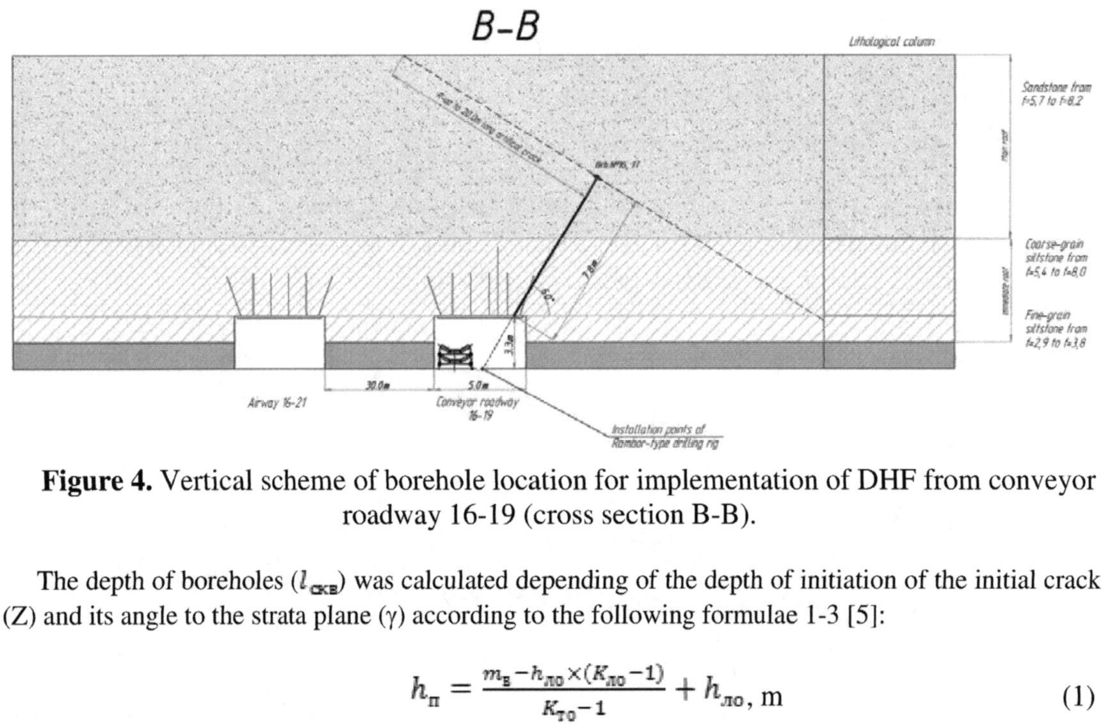

Figure 4. Vertical scheme of borehole location for implementation of DHF from conveyor roadway 16-19 (cross section B-B).

The depth of boreholes ($l_{скв}$) was calculated depending of the depth of initiation of the initial crack (Z) and its angle to the strata plane (γ) according to the following formulae 1-3 [5]:

$$h_{п} = \frac{m_{в} - h_{ло} \times (K_{ло} - 1)}{K_{то} - 1} + h_{ло}, \text{ m} \quad (1)$$

where: $h_{п}$ - roof rock caving height at which there is a cleavage of tough rock mass;
$m_{в}$– extracted stratum thickness;
$h_{ло}$ – thickness of the free-caving roof;
$K_{ло}$, $K_{то}$ – free-caving and tough rock fragmentation index.

$$Z = (h_{п} + h_{ло})/2, \text{ m} \quad (2)$$

$$l_{скв} = Z / \cos \gamma, \text{ м} \tag{3}$$

Estimated length of boreholes was 7.8 m.

Implementation of directional hydraulic fracturing was performed using both standard equipment for general use and special devices. The process consisted of the following stages:

- drilling of the inclined hole using rock crowns with 46 mm diameter and cutting of initiating crack in its bottom using a crack initiation device SHCHM – 45/1 (figure 5);
- promotion of sealer to the bottom of the borehole using high-pressure pipe set by means of their connection with the high-pressure system of the mechanized complex through flexible high-pressure hoses (figure 6);
- borehole sealing and pumping of the fracturing fluid in the initiating crack area.

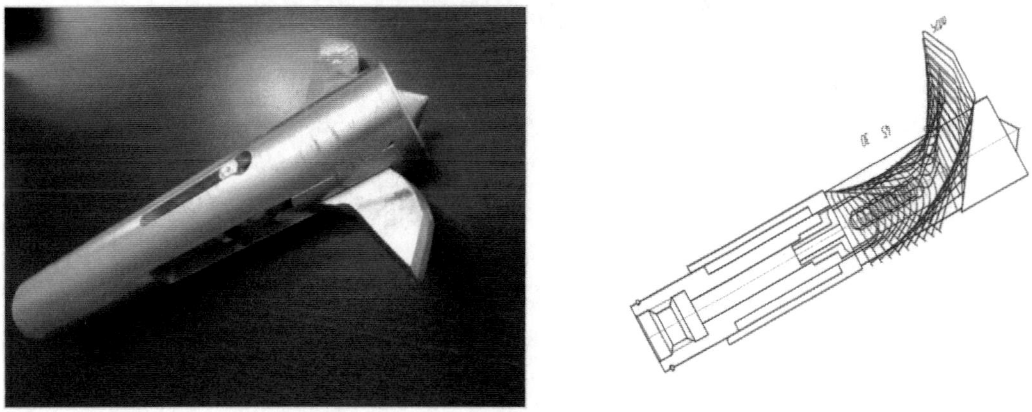

Figure 5. Crack initiation device SHCHM – 45/1.

Figure 6. Sealer "Taurus".

Roof rock softening was made after installation of the mechanized complex. Distance between boreholes was 20 m. During DHF implementation the authors had been monitoring the effectiveness of the method using spectral acoustic method. They applied software complex on the basis of RIPAS device (figure 7).

Acoustic sounding was carried out according to the "Procedure of rock mass acoustic sounding…" [7]. As a result of acoustic impulses processing, the authors calculated the coal seam relative stress index before and after the DHF. Figure 8 presents the results of the measurements taken in the airway 16-19 at the time of the mechanized complex removal from the installation chamber.

Dashed lines in figure 8 denote the location of boreholes for DHF in reference to the working face (in meters). X-axis shows the distance from the working face to the locations where geophon was installed, Y-axis shows the relative stress index.

Figure 7. Acoustic signal recording unit "RIPAS".

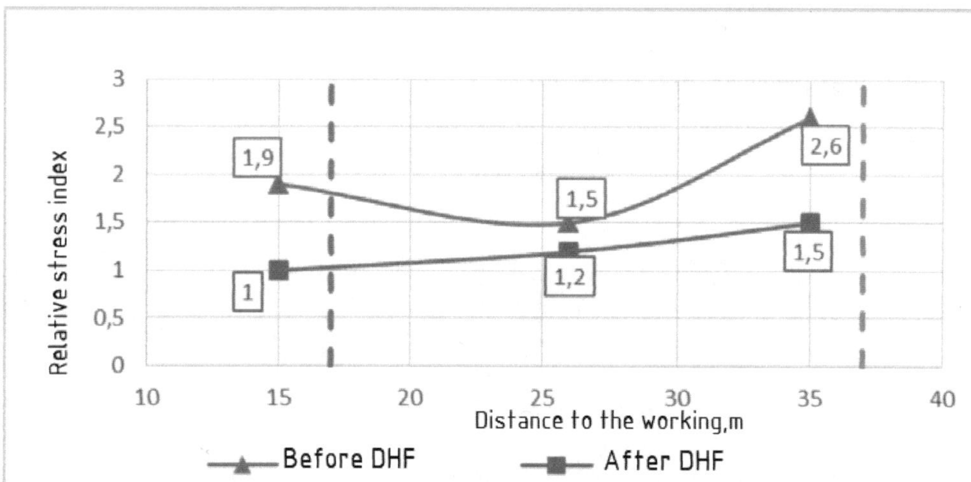

Figure 8. Results of relative stress index measurements.

Starting from the moment when the authors began proving scientific support for activities in longwall 16-19 located under tough roof, the scientific research institute VNIMI organized a system of continuous instrumental monitoring including seismic equipment recording. Besides, local network of seismic stations installed on the territory of the JSC "Yubileynaya" Mine" mining allotment was used. During seismic monitoring, at the time of the longwall face advance from the installation chamber at a distance of 35 m, the authors registered 18 seismic events almost on the entire territory of the examined area. These events were connected with roof caving which manifested itself even on the earth surface as perceptible ground vibrations preceded by the DHF. In figure 9 there is a graph showing registration of seismic events during seismic monitoring (figure 9-a) checked against the dates when DHF was made (figure 9-b) [8, 9].

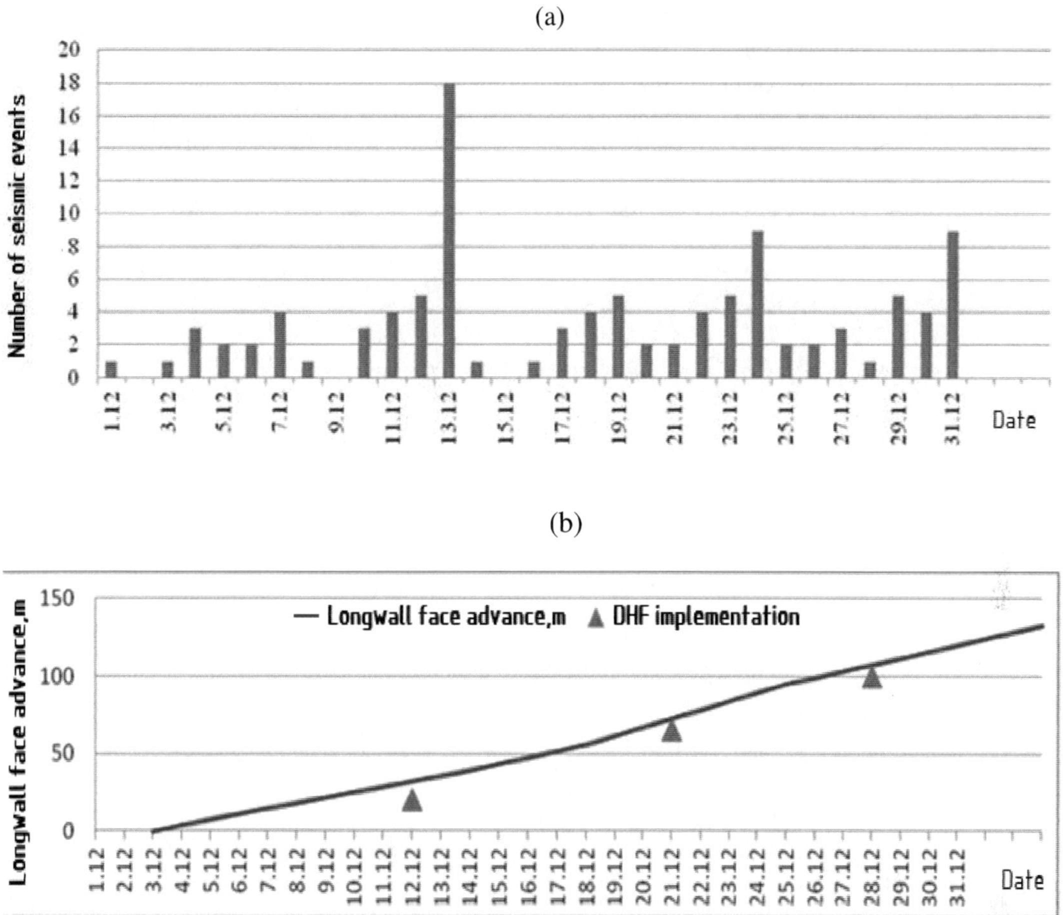

Figure 9. Seismic event records graph.

3. Conclusion

Thus, the authors determined that roof softening by means of DHF, when the mechanized complex was removed from the installation chamber, made it possible to implement safe initial roof caving. Besides, the authors recorded no enhanced gas emission which is usually due to sudden "instantaneous" gas displacement form the gob that happens at simultaneous caving of main roof rock that had been hanging in vast areas. The authors recorded the decrease of the relative stress index in the edges of the coal seam after DHF. That shows that there was a redistribution of stress in the coal and rock mass. The results of seismic monitoring confirm the cyclic character of the seismic vibrations reduction after roof softening and its increase in the following days.

Acknowledgements
The research was funded from the Russian Science Foundation grant (Project No.17-17-01143).

References

[1] Klishin V I, Opruk G Yu, Sentyurev A V and Nikolaev A V 2015 *Ugol* **11** 12–16
[2] Klishin V I, Zvorygin L V et al 2011 *Safety Issues and New Technologies of Coal Deposits Underground Mining* (Novosibirsk: Novosibirsky Pisatel) p 524

[3] *Instruction for Safely Mining Operations in Mines Extracting Coal from Coal Seams Prone to Rock Outbursts (RD 05-328-99)* 2000 (Moscow: Industrial Safety) 119
[4] Klishin V I, Opruk G Yu, Teleguz A S, Chernousov P A and Nikolaev A V 2017 *Knowledge Intensive Technologies for Mineral Resources Development and Usage* (Novokuznetsk: Siberian State Industrial University) **3** 177–181
[5] *Instruction on Selection of Methods and Parameters for Roof Softening in Extraction Areas* 1991 (Leningrad: VNIMI) 102
[6] Kurlenya M V, Klishin V I and Kokoulin D I *Crack Initiation Device* 2013 Patent RF No 129148 Bul 17
[7] *Procedure of Rock Mass Acoustic Sounding Using a Hardware and Software System* 2016 (Moscow: RIVAS Ltd.) p 36
[8] Kharkevich A S 2012 *Collection of Research Papers, VNIMI* 78–96
[9] Panin S F and Polyakov A I 2012 *Collection of Research Papers, VNIMI* 50–61

Assessment of elastic seismoacoustic vibration propagation through coal and rock mass within the extraction column during directional hydraulic fracturing (DHF) implementation

V I Klishin[1], O V Taylakov[1], G Yu Opruk[1], S V Sokolov[1] and A V Nikolaev[2]

[1]Federal Research Center of Coal and Coal Chemistry, Siberian Branch of the Russian Academy of Sciences, 10 Leningradsky ave, Kemerovo, 650065, Russia
[2]LLC "Mine Esaulskaya", LLC "RUK", 33 Kurako ave, Novokuznetsk, 652831, Russia

E-mail: opruk@yandex.ru

Abstract. The article reveals information about implementation of roof rock directional hydraulic fracturing technology when face was advancing under tough roof. The authors justify application of directional hydraulic fracturing. They developed the scheme for roof softening considering mining and geological conditions of "Esaulskaya" mine. The authors describe technological process taking place during implementation of directional hydraulic fracturing and give the list of the applied general and specialized equipment. Considering the results of the performed operations the authors made a conclusion in relation to the effectiveness of the conducted work. They assessed the qualitative characteristics of the performed activities aimed at roof softening. The authors determined physical and mechanical characteristics of the active roof on the basis of seismic survey data analysis.

1. Introduction

At present moment long pile mining which implies providing high-performance powered mining complex in working faces is the most wide spread coal field underground development system. The sphere of its application has grown during recent years due to mining of seams with tough roofs which makes longwall work significantly more complicated. Dynamic roof caving threatens mining safety, destroys equipment, machinery and workings. Besides, poor roof caving leads to concentration of rock pressure on unworked coal in the working face area and in places where the working face crosses the workings. That results in destruction of mine workings and, accordingly, disrupts standard operating conditions in the development mining sections and destabilizes face ventilation modes.

It is common knowledge that poor caving of longwall face roof rock where it is linked with airway and conveyor roadway promotes formation of vast cavities under the hanging roof where methane accumulates. Besides, a part of fresh air current that should ventilate the working face goes to gob. As a result, working face is not supplied with the estimated amount of air, which in its turn leads to emergency shut down due to exceeding threshold limit value of methane, and as a consequence, that brings about lower productivity. Increase of horizontal stress due to poor main roof caving after extraction of the previous panel leads to destruction of coal pillars and their fall in the workings [1, 2] (figure 1).

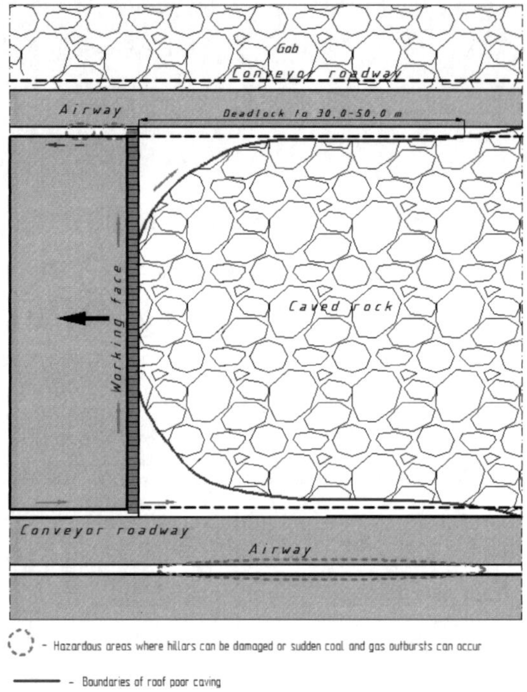

Figure 1. Roof rock impending in gob end sections.

2. Main Part

Industrial experimental works aimed at adaptation of directional hydraulic fracturing to different process charts have been taking place in Kuzbass.

Longwall face 26-32 of "Esaulskaya" mine has been mined under tough roof conditions. Mining-and-geological and mining-engineering characteristics of the seam are given in table 1.

Table 1. Mining-and-geological and mining-engineering parameters of the working area.

No.	Indicator name	Indicator characteristics	Value
1	Working area		26-32
2	Seam		26a
3	Coal grade		Zh (fat coal)
4	Operating ash content	%	30.7
5	Mining depth	m	430-450
6	Extracted seam thickness	m	2.15
7	Angle of seam inclination, from - to:	degree	
	along longwall face		0-2
	along the pillar		1-9
8	Length of the working area	m	876
9	Length of the mining face	m	210
10	Weight of loose coal in the 1st block	ton	566
11	Weight of loose coal at one meter face advance	ton	708

According to mining-and-geological forecast, the immediate roof of seam 26a with average thickness from 8 to 19 m is composed of fine and coarse siltstones. In some parts lower part of the immediate roof which is 0.10-0.32 meter thick acts as a false roof as it falls after coal extraction. Considering stability, immediate roof acts as semi-stable one. However, the main roof is composed of coarse siltstones and sand rock along the whole extraction area and is characterized as a tough roof. Impending of tough roof in vast areas in places where it is linked with development workings leads to higher pressure in the verges of coal and rock mass. To reduce negative effect of high rock pressure on the stability of mine workings with tough roof, the authors decided to use directional hydraulic fracturing for roof softening [7].

According to "Instruction on the choice of method and parameters of roof softening at the excavation sites" the staff members of the Institute of Coal of the Federal Research Center for Coal and Coal Chemistry (Siberian Branch of the Russian Academy of Sciences) in collaboration with "Esaulskaya" mine technical department members worked out a directional hydraulic fracturing implementation process flow to guarantee timely and stable roof caving after working face advance (figure 2) [3].

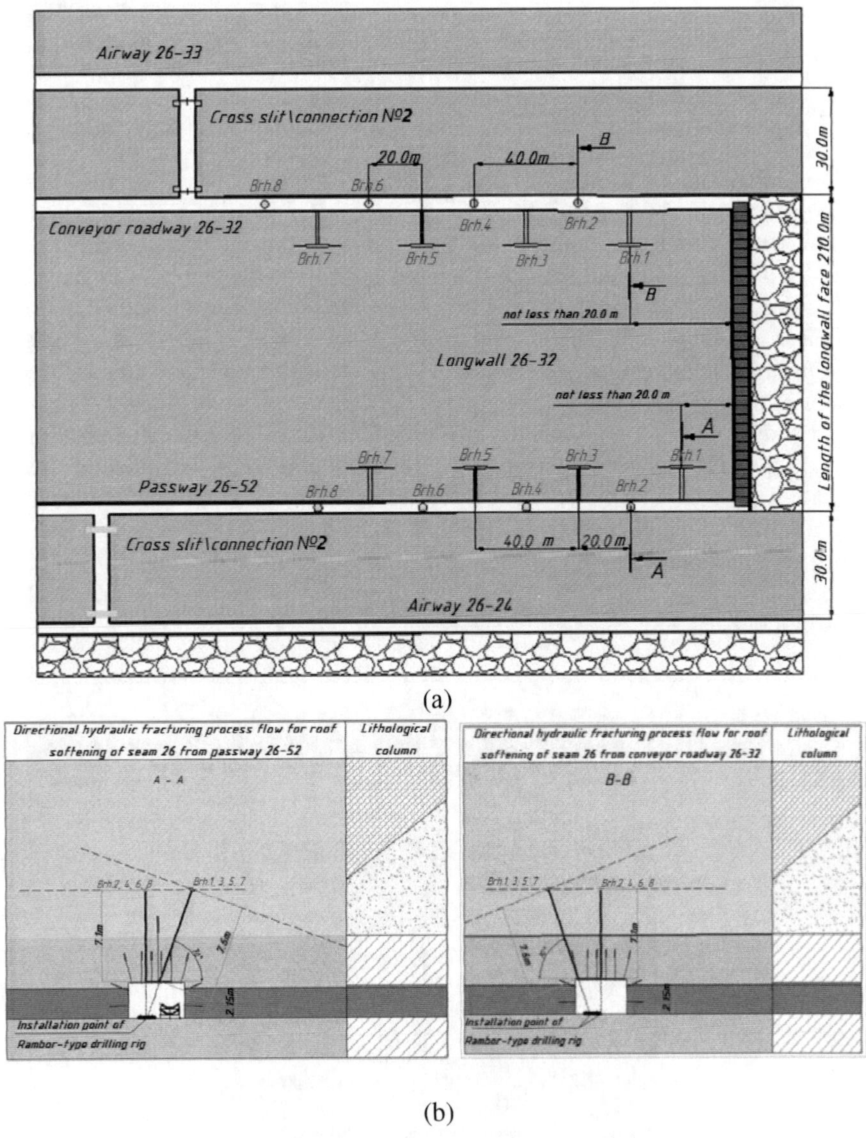

Figure 2. Directional hydraulic fracturing implementation process flow: a) location of wells for hydraulic fracturing; b) well drilling direction.

Directional hydraulic fracturing was implemented using standard and special mining equipment. Well drilling and initiating crack cutting was performed using pneumatic drilling roof bolting machine. For well drilling rock crowns with 46 mm diameter were used. Initiating cracks were cut using powered crack initiation devices (figure 3), which were set on the steel stems of the drilling machine instead of the crown [6].

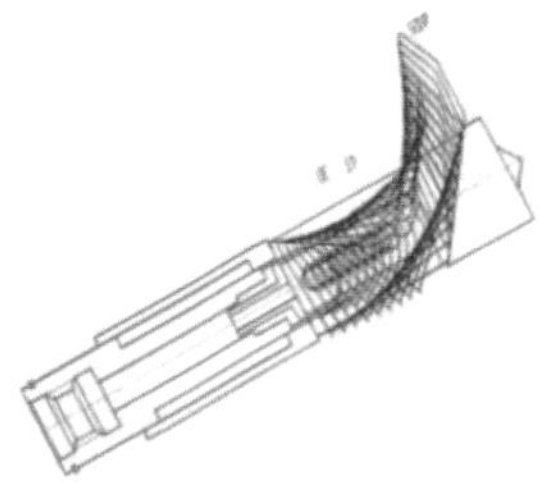

(a)

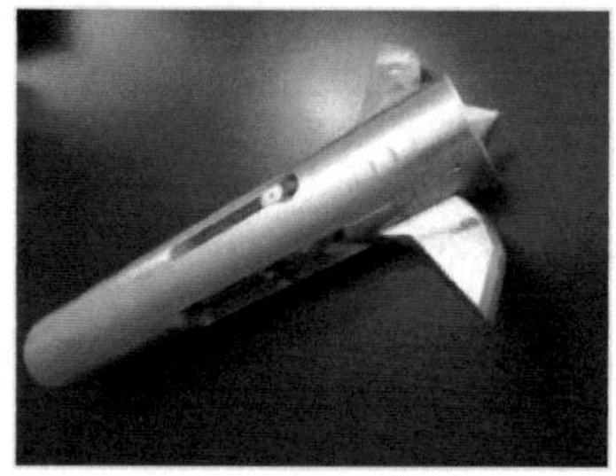

(b)

Figure 3. Crack initiation device SHCHM – 45/1: a) – motion direction; b) sample prototype.

Sealing of the initiation crack area was made using hydraulic sealer ("Taurus" type) (figure 4). Sealer was directed to the bottom of the well by means of special set of high-pressure pipes. Compression unit was connected with the end of the pipe projecting from the borehole through special adapter (collector) by means of high-pressure flexible hoses. Liquid pumping in the initiating crack area was made using high-pressure unit of the power-operated complex. Effectiveness of drilling and the process of directional hydraulic fracturing implementation were controlled using video endoscope. That helped ensure that the boreholes are drilled up to the designed depth, and the initiating crack was cut in the bottom of each well (figure 5).

First of all, hydraulic fracturing was made in the vertical wells where the artificial cracks expanded parallel to the strata. After that it was performed in the slant boreholes, and the crack expanded at an angle to the strata. In this case artificial crack extended up to the junction of the immediate and main roofs and spread further along the contact area (figure 4). When liquid was supplied into the borehole, the process of roof hydraulic fracturing started in 5-10 seconds after the emulsion had been pumped into the system. In addition to that, there was sudden decrease of pressure on a manometer in the area of the initiating crack [8, 9].

(a)

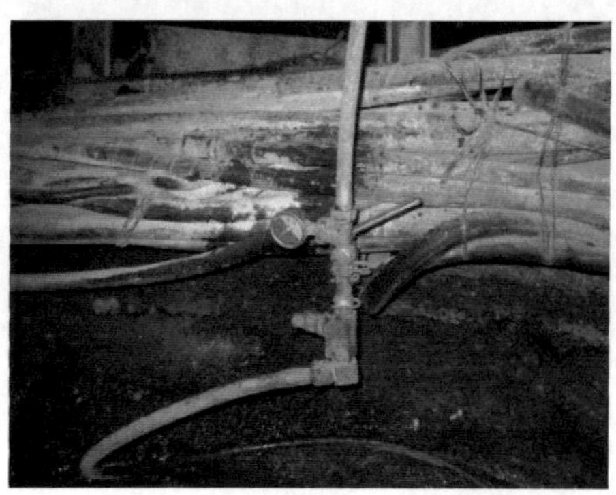

(b)

Figure 4. Sealer "Taurus": a) external view; b) placement in the mine opening.

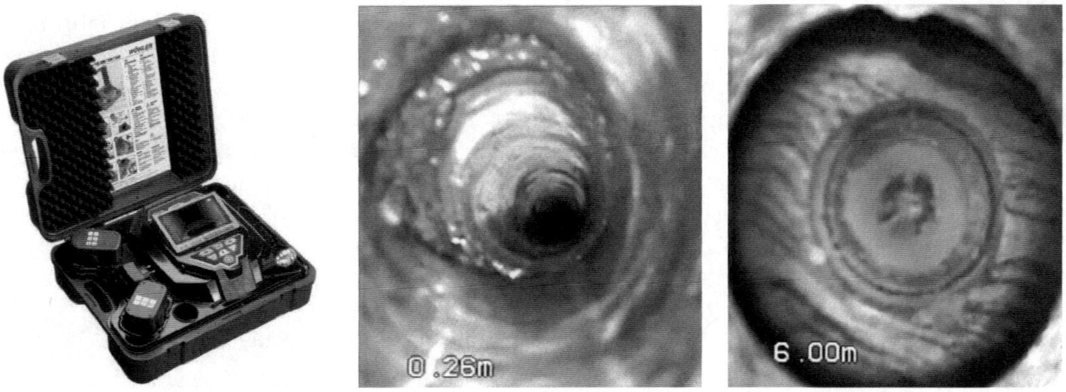

(a) (b) (c)

Figure 5. Borehole with initiating crack before hydraulic fracturing: a) video endoscope; b) borehole mouth; c) borehole bottom.

For assessment of the qualitative characteristics of the performed hydraulic fracturing the changes in coal and rock mass structure were monitored by means of a seismic survey [10-11], which in contrast to, for example, registration of the immediate changes in gas release from coal seams at hydraulic fracturing implementation [12-14], makes it possible to determine the spatial boundaries of hydrodynamic impact on the enclosing rocks.

To assess effectiveness of the taken measures aimed at roof softening the area near the extraction column was studied in the interval more than 150 m using seismic inspection method by means of transmitted waves. Moreover, more than 100 shots were made. They were performed in two stages of seismic vibrations impulse registration. First, registration of seismic data was made on the examined part of the extraction column which remained intact after hydraulic fracturing. Then seismic data was recorded on the same area when physical and mechanical state of the roof had changed as a result of hydraulic fracturing. The process of seismic shooting of the extraction column area included taking measurements in mine (coordinate reference system adjustment; placement of seismic recording equipment; signal initiation and registering) and office work (basic data preparation; data processing; data analysis and interpretation).

During taking measurements in mine, the authors used a system of seismic sensors. Seismic receivers (geophons) were placed at certain intervals along the sides of the working. They were connected with an autonomous seismic recording system. To perform geophysical study, the authors placed geophysical pickets for receiving and excitation of seismic vibrations along the sides of the parallel mine wordings at 10 meter intervals. Adjustment of coordinate reference system to the real conditions of the extraction column area was made based on the map of mine workings and taking the real placement of mine workings and other mine facilities into account. In compliance with safety requirements for initiation of acoustic waves in mines, the authors used mechanical impact vibration sources which had a time break system. Recording units and time break system were synchronized with each other just before taking measurements. Mistiming of the synchronized devices in this case was not more than 1 msec per day. The data with 1 msec discretization interval were continuously registered using autonomous seismic recording stations with filtration on the cut-off frequency equal to 500 Hz and recorded in the files containing up to 3 600 000 readings.

Velocity parameters of seismic signal were taken as the main source of data for assessment of physical and chemical characteristics of the active seam roof. Lines of correlation were defined on the seismic records. They characterized the compressional wave going through the active roof area. According to the results of picketing, the authors determined signal velocity parameters on different measurement points, which were later converted into the time of seismic wave travel and introduced into the software for tomography data processing. The surface area of the extraction column was analyzed using a range of tools for tomographic survey of the simulated environment that is located between two mine workings [16]. The authors determined instantaneous velocity values in different parts of the extraction column. The obtained results were saved in the system as an environment velocity model. After that velocity models were further used to design a two-dimensional representation of the examined part of the extraction column that indicated physical and mechanical properties of roof before and after hydraulic fracturing (figure 6). The authors detected a higher velocity background which is typical for sufficiently tough rock which is in a stress condition taking into account the results of the analysis considering the initial velocities of seismic waves travelling through roof which was not affected by hydraulic fracturing. The highest velocity values were compared with the maximum stresses registered in the area noted using symbol "+" in Fig. 6 within observation points (OP) OP 90-130 (Working No 1) and OP 40-90 (Working No 2). In mine Working No 1 within OP 20-60 there was an area with minimum seismic wave velocities (indicated as "-") which probably has more fractures and is possibly damp.

General speed background on the tomographic profile of the examined part of the roof, that was altered because of hydraulic fracturing, is lower. Maximum velocity values for the given state of the roof were observed in the area "+" (OP 90-120) in Working No 1. Minimum values – in areas "-": 1 – OP 20-80 within Working No 2; 2 – OP 10-60 within Working No 1. Thus, apart from the lower general velocity background, the analysis of velocities of seismic waves travelling through the roof before and after hydraulic fracturing showed the following results:

- Shrinking of the distribution area and redistribution of high stress of area "+";
- Increase of the area affected by the section with lower velocities "-", located within Working No 1;
- Formation of an area with high concentration of fractures, which is detected due to the minimum seismic wave velocities.

Moreover, to determine the effect the hydraulic fracturing had on coal and rock mass, the authors assessed amplitude characteristics in relation to lines of correlation of the recorded propagation of waves (figure 7) [15]. Signal amplitudes were calculated within different intervals of the digital representation of the extraction column, and on the basis of the analysis of their spatial distribution complex tomographic profiles were modelled which showed the changes in the following characteristics:

$\beta_V = V_1/V_2$, – ratio of seismic wave propagation velocities through intact roof of the examined part of the extraction column to the velocities of wave propagation through the roof of the examined section which had been altered due to hydraulic fracturing (figure 6, c).

$\beta_A = A_1/A_2$ – dependence of seismic waves relative amplitude values in intact roof of the examined part of the extraction column in the longwall face on the relative values of the wave amplitudes in the roof of the examined section which had been altered due to hydraulic fracturing (Fig. 8-a).

$\beta_{V,A} = \beta_V + \beta_A$ – relative energy characteristics factoring fluctuation of velocities and amplitudes of seismic waves (figure 8-b).

Based on the analysis of tomographic profile β_V, the authors assessed coal mass relief degree considering records of changes in seismic wave propagation velocities through the examined part of roof which were induced by hydraulic fracturing. Tomographic profile was modelled using average values of the recorded characteristics. That made it possible to detect areas of their sharp changes excluding objects irrelevant for tomographic profile. Area "P" most relieved after hydraulic fracturing was detected within the interval OP 0-100 where coal and rock mass shifted from Working No 1 towards Working No 2.

After that the authors analyzed the tomographic profile β_A. Taking into consideration different nature of the compared values, the behavior and direction of the area most relieved as a result of hydraulic fracturing resemble the area registered for β_V. The highest values of the examined characteristics were in two areas – "P": OP 20-110 in Working No 2 and 10-60 in Working No 1.

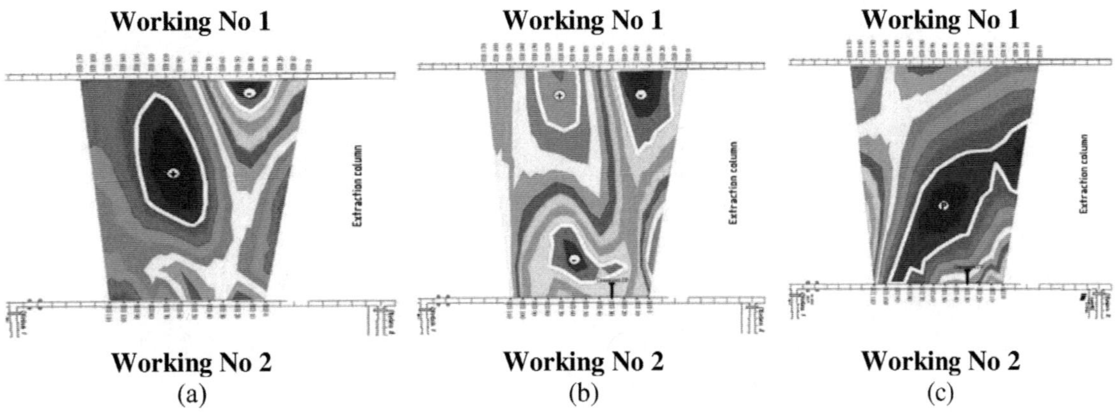

Figure 6. Visual representation of changes in coal and rock mass as a result of stress relief (based on measurements of seismic wave propagation velocities): a) distribution of velocity properties before hydraulic fracturing; b) distribution of velocity properties after hydraulic fracturing; c) ratio of velocities before hydraulic fracturing to velocities of wave propagation after hydraulic fracturing - $\boldsymbol{\beta_V}$.

To determine the location of the area most relieved after DHF more precisely, the authors analyzed tomographic profile $\beta_{V,A}$. That complex characteristic was used because earlier the authors identified that directions of the represented area for β_V and β_A are similar. Assessment of inversions of the relative energy value which relies on changes in different signal parameters (velocities and seismic wave amplitudes) made it possible to exclude possible alterations of each value which could be caused by their physical properties. The area most relieved as a result of hydraulic fracturing considering characteristic $\beta_{V,A}$ was determined within Working No 2 - OP 40-80.

On the basis of analysis and interpretation of the obtained data the following was established:

- Decrease of general velocity background after hydraulic fracturing implementation;
- Decrease of distribution area and redistribution of high stress in area "+";
- Increase of the affected area of the lower velocity section "-", located within Working No 1;
- Emergence of an area with high number of fractures characterized by minimum seismic wave velocities;
- Presence of an area in the roof most relieved as a result of DHF for β_V – "P" within the interval OP 0-100 with change of its position within the bulk of the face directed from Working No 1 to Working No 2;
- Presence of an area in the roof most relieved as a result of DHF for β_A with the highest values of the assessed characteristics on two sites – "P": OP 20-110 in Working No 2 and 10-60 in Working No 1;
- Presence of an area in the roof most relieved as a result of DHF for $\beta_{V,A}$ within working 2 OP 40-80.

To assess roof softening as a result of DHF the authors chose the criterion of high crack concentration in the examined rock mass. It implies the ratio of crack volume and cavities in the examined rock mass to the volume of rock mass. In the process of calculation the authors considered dependence of compressional wave velocity on the porosity factor, which implies ratio of pore volume to the volume of rock mass. [17, 18]:

$$\frac{1}{V_p} = \frac{k}{V_\Phi} + \frac{1-k}{V_\Pi}, \tag{1}$$

where k – porosity factor, which considers presence of cracks, pores and cavities in the examined rock; V_p – compressional wave velocity in the examined mass; $V_Ф$ – compressional wave velocity in fluid; $V_п$ – compressional wave velocity in the rock mass.

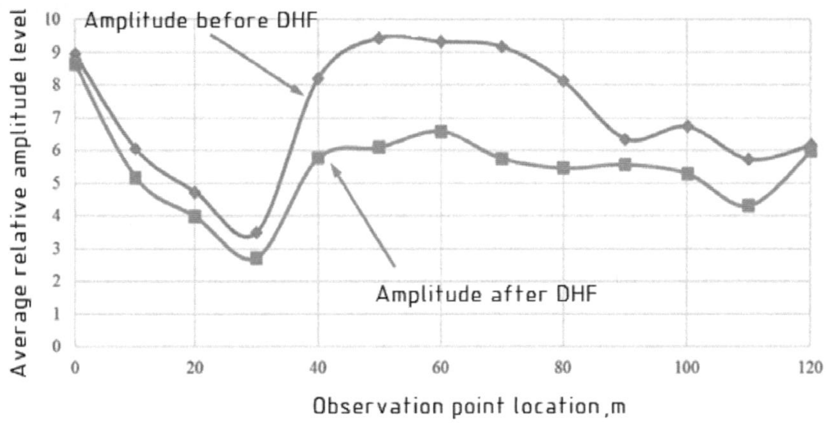

Figure 7. Dependence of average seismic wave amplitude level on the location of the observation point in Working No 2 in case when coal and rock mass was intact and affected by DHF.

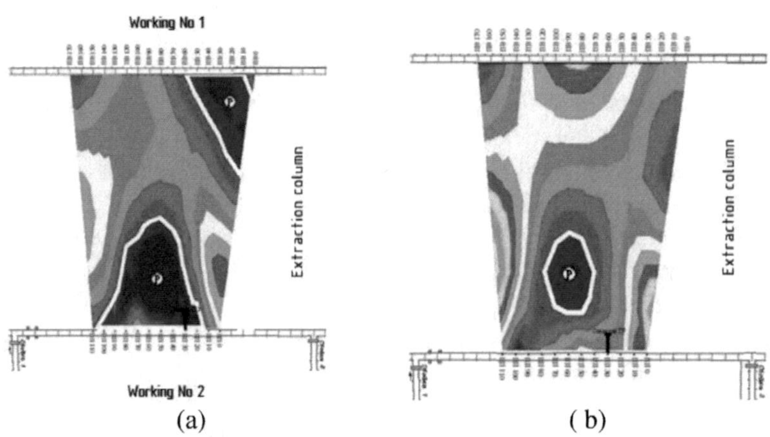

(a) (b)

Figure 8. a) Visual representation of changes of seismic wave relative amplitude values in the examined area affected by DHF - β_A; b) Visual representation of change in the value of the complex characteristic $\beta_{V,A}$ in the examined area affected by DHF.

The authors assumed that in the given individual case of compressional wave reception and initiation the whole volume of cracks having different directions formed as a result of roof softening had an impact on the signal parameters due to the difference in the level of parallel workings and alteration of vibration excitation points.

To assess changes in roof condition induced by DHF, the authors used fracture density k' instead of k. As $V_Ф$ the authors considered compressional wave velocity in methane-air medium; $V_п$ – wave velocity in the mass before DHF V_1; V_p - wave velocity in the mass after DHF. In this case transformed formula is the following (1):

$$\frac{1}{V_2} = \frac{k'}{V_\Phi} + \frac{1-k'}{V_1}. \quad (2)$$

Besides, initial fracture density of the coal and rock mass which was not affected by DHF was determined taking geological survey data into account. In case there were no such data, the value was calculated as a ratio of cracks and cavities volume in the examined coal and rock mass to the volume of rock mass. from formula (1) where V_p was replaced with V_1:

$$\frac{1}{V_1} = \frac{k}{V_\Phi} + \frac{1-k}{V_\Pi}. \quad (3)$$

Using formulae (2) and (3) the authors calculated k' and k respectively taking determined velocity values. Thus, they calculated fracture density characteristic after implementation of DHFk^*:

$$k^* = k \cdot k'. \quad (4)$$

On the basis of the assessment the authors determined that according to formula (2) the coefficient that characterizes fracture density in the examined coal and rock mass varies from 0 to 0.1. The distribution of its values within extraction column corresponds to distribution β_V of the ratio of velocities before DHF to wave propagation velocities after DHF.

3. Conclusions

As a result of industrial experiment works the authors performed roof softening using DHF. They determined that application of DHF secures timely roof caving after working face advance. It excludes the possibility of roof hanging in places adjacent to development workings and promotes stable ventilation in the extraction area. To assess the effectiveness of roof softening the authors took roof seismic measurements. As a result, no abnormal structural notions were detected in the examined area of coal and rock mass. DHF did not have any negative effect that could make mining more complex and difficult. Besides, the authors determined decrease of stress in the examined part of the extraction column, which appeared as decrease in energy characteristics of seismic signal. They determined the area of maximum relief which located within OP 40-80 of Working No 2. Calculated variations of the coefficient that characterizes roof fracture density lie within the range from 0 to 0.1.

References

[1] *Instruction* on *Safe Mining Operations in Coal Mines Developing Coal Seams Prone to Rock Bumps (RD 05-328-99). In the Corpus of Documents: Prevention of gas-dynamic Notions in Coal Mines (Corpus of Documents)* 2000 (Moscow: Scientific Technical Center "Industrial Safety") p 119

[2] Oganesyan S A 2004 *Ugol* **6** 25–28

[3] Tsivka Yu V and Petrov A N 2005 *Ugol* **7** 49–50

[4] *Instruction* on selection of methods and parameters for roof softening in Extraction Areas. VNIMI 1991 (Leningrad) p 102

[5] Artemiev V B et al 2009 *Protection of Development Workings by Means of Pillars in Coal Mines: Monograph* (Saint-Petersburg: Nauka) p 231

[6] Kurlenya M V et al 2013 *Crack Initiation Device patent RF* No 129148 Bul 17

[7] Klishin V I 2002 *Adjustment of Powered Roof Supports to Dynamic Loading Conditions* (Novosibirsk: Nauka) p 200

[8] Klishin V I et al 2015 *Ugol 11* 12–16

[9] Klishin V I et al 2017 *Proc. of the Int. Sci. Conf. on Knowledge Intensive Technologies for Mineral Resources Development and Usage* (Novokuznetsk: SibSIU) vol 3 pp 177–181

[10] Azarov N Ya and Yakovlev D V 1988 *Seismoacoustic Method for Potential Assessment of Mining and Geological Conditions of Coal Deposits Development* (Moscow: Nedra) p 199
[11] Taylakov O V et al 2017 *Mining Industry* **6(136)** 72–75
[12] Nozhkin N V 1979 *Coal Deposits Advance Degassing* (Moscow: Nedra) p 271
[13] Taylakov O V et al 2014 *Vestnik of Kuzbass State Technical University* **6** 13–16
[14] Taylakov O V et al 2017 *Ugol* **11** 24–29
[15] Taylakov O V et al 2012 *Proc. of the Int. Sci. Conf. on Knowledge Intensive Technologies for Mineral Resources Development and Usage* (Novokuznetsk: SibSIU) pp 266–267
[16] Salnikov A S et al 2012 *Seismic Survey Technologies* **2** 74–88
[17] Makeev M P et al 2013 *Gornyatskaya Smena. Collected Papers of the Russian Scientific Conference for Students, Post-Graduate Students and Young Scientists with Scientific School Elements* vol 3 pp 286–287
[18] Averbukh A G 1982 *Study of Rock Composition and Properties at Seismic Surveying* (Moscow: Nedra) p 232

IOP Conf. Series: Earth and Environmental Science **206** (2018) 012025 doi:10.1088/1755-1315/206/1/012025

Remounting of a mechanized complex using Petitto Mule equipment in the conditions of mines in the south of Kuzbass

A M Nikitina[1], S V Rib[1], D M Borzykh[1] and R A Dadynsky[2]

[1]Siberian State Industrial University, 42 Kirova street, Novokuznetsk, 654007, Russia
[2]LLC “Department of Mounting, Remounting and Maintenance of Mining Equipment”, 3 Shakhtovaya street, Osinniki, 652807, Russia

E-mail: nik.am_78@mail.ru

Abstract. It is proposed to reduce the duration of mounting and dismantling works before starting to develop the next extraction panel in mines that operate according to the scheme ‘mine – longwall’, which will improve the safety and eliminate emergencies during mining operations.

1. Introduction

In the conditions of modern market requirements for the stable shipment and supply of coal, the mines operating according to the scheme ‘mine – longwall’ have problems with downtimes during coal mining due to remounting the mechanized complex.

In the ‘mines – longwalls’ there is no reserve extraction front. As practice shows, the duration of the remounting of the mechanized complex can increase by 1.5-2 times or more, compared to the corresponding figure calculated without considering the time spent on eliminating the consequences of roof collapses and the enclosing rocks inrush into the working space. Duration of mounting and dismantling works significantly exceeds the duration of a stop necessary from the technical point of view. So, in the mines of the South Kuzbass, the actual duration of faces downtime during the mechanized complex remounting is 20-150 days or more. At the same time, the economic damage associated with the irrecoverable losses of production reaches 3-7 million rubles per day [1].

2. Production experience

The carried out analysis of the mechanized complex remounting in the flat seams showed that the time of remounting works is mainly affected by the time of dismantling, since the mounting of the mechanized complex takes place together with the dismantling and on average is approximately two months. In this regard, the dismantling of the mechanized complex requires detailed consideration and solution of problems in this area. With an increase in the depth of development, due to a more active manifestation of rock pressure, the process of mechanized complex dismantling has become more complex and the possibility of a high rate of injuries would arises [2].

Based on the experience of dismantling works, the scheme with the use of polymer dismantling overlap and pilot sections of mechanized roof supports (figure 1) has got widespread use.

For many decades and today, to extract the mechanized roof supports in Kuzbass mines a standard set of equipment and mechanisms (winches LPK-10B, LMT-150.01, LShM, pump station SNT-32) is used. This technology of dismantling is distinguished by a high number of units of energy consumption and starting equipment, rather low indicators of performance, level of safety, costs for the acquisition and maintenance of a fleet of dismantling equipment. In case of insufficient tractive effort, pulley systems are applied using several assembly blocks to increase the tractive effort.

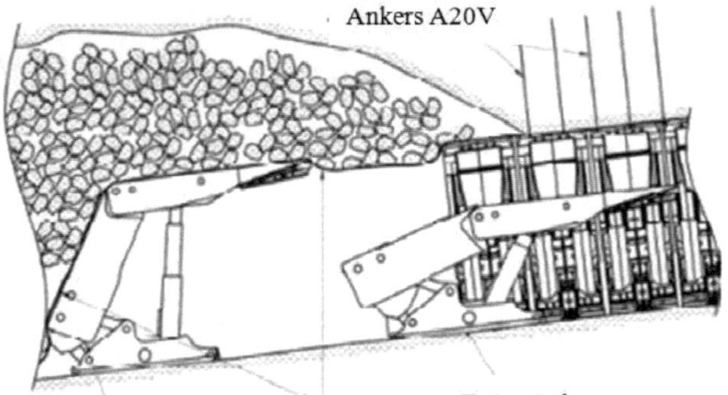

Figure 1. Scheme with the joint application of polymer mesh and pilot sections.

The first machines for extracting heavy support sections Petitto Mule appeared in the United States about 30 years ago, and now there are already several hundred of such machines working in many coal mines around the world, which confirms the reliability and quality of these machines. All machines have an option – they can be disassembled into four parts for easy descending into the mine. There are also cars in diesel version, explosion-proof, designed to extract the support sections weighing 50, 60, and 70 tonnes (figure 2).

Figure 2. Machine Petitto Mule model 1550 is designed to extract support sections with a mass up to 36 tonnes.

Advantages and disadvantages of Petitto Mule equipment:
- high cost of dismantling equipment (about 100 million rubles);
- increase in the width of the dismantling path;
- complicated organization of the place of its reloading on the diesel locomotive in the mounting chamber (poor compatibility with the monorail transportation technology);
- need to strengthen the soil (or concreting) for the movement of Petitto Mule (on weak soils);
+ high performance of dismantling works;
+ a small number of units of electrical equipment;
+ high level of dismantling safety, the absence of extensive hazardous areas.

So, for example, the mounting and dismantling works at the mine "Alardinskaya" are carried out by equipment that was used to remounting the mechanized complexes, the weight of the support sections of which was within 20 tonnes (KM-142, KM-130, FAZOS): the winches LPK-10B, LMT-

150.01, LShM, the pumping station SNT-32. At the moment, Glinnik 22/47Poz cleaning complex is used at the mine, the weight of the support section of which is 32 tonnes. When using the above mentioned equipment for mounting and dismantling works, a number of problems arise [3]:

- equipment failure due to overload;
- labour-intensive and time-consuming preparatory and face end operations due to the use of a pulley system;
- daily equipment remounting, switching of starting equipment;
- delivery of sections from the dismantling chamber (by traction) takes a lot of time due to the limited speed of the rope;
- maintenance and repair of starting equipment and mechanisms;
- high rate of injuries.

Let us analyze the time spent on each process separately while dismantling support sections of a mechanized complex according to the schedule of works (figures 3 and 4).

№ n/n	The name of the operation	Duration, min	Number of employees per shift	1 shift							
				8	9	10	11	12	13	14	15
1	Shift reception, transfer, inspection of production, equipment	10	5								
2	Preparation for dismantling (hook-up, hydraulics switching, shortening)	15	4								
3	Dismantling the section with PM	5	4								
4	Issuance of sections to the place of loading on the vent. drift (200m)	10	2								
5	Loading on a diesel locomotive (mine)	15	2								
6	Shunting operations of the RM on the vent line	10	2								
7	Loading of fastening materials	5	2								
8	Moving the RM to the place of dismantling	10	2								
9	Mounting the dismantling space	35	2								
10	Moving pilot sections with PM	15	4								
11	Maintenance of oil stations	460	1								
Total		70, min/sec	5								

Figure 3. Schedule of dismantling works using winches.

The dismantling and turning of the support section takes 90 minutes of working time. This is due to a large number of preparatory and face end operations, (stretching the rope, hanging blocks) that are repeated 4-5 times per cycle, disconnection of the section to be dismounted from the general hydraulic system of the mechanized complex. The movement of the pilot sections also consists of a variety of manipulations with rope and blocks. This leads not only to a great waste of time, but this process is also very labour intensive. Also, people who dismantle the support sections are in close proximity to the dangerous zones of the rope action.

When moving a support section from a line a number of the roof rocks falls into the working space of the dismantling chamber, which subsequently prevents the pushing-in of the support pilot sections. At the mine "Alardinskaya", the roof rocks of the seams are represented by sandstones with a large collapse step, due to which the rock inrushes are quite large blocks of sandstone. The clearing of the workplace is carried out manually using small-scale mechanization (a jackhammer) and shipment of the rock mass into the transportation container, which is also time consuming.

№ n/n	The name of the operation	Duration, min	Number of employees per shift	1 shift							
				8	9	10	11	12	13	14	15
1	Reception - transfer of shift, inspection of production, equipment	10	8								
2	Preparation for dismantling (hook-up of ropes, hinge of blocks, hydraulics	30	4								
3	Reassignment of mounting blocks	15	4								
4	Turning the sections on the dismantling track	15	4								
5	Delivery of fastening materials from the unloading place	10	4								
6	Mounting the dismantling chamber	25	2								
7	Cutting of the monorail road, delivery to the place of loading	25	2								
8	Moving pilot sections (2 pcs)	30	4								
9	Issuance of sections for a monorail road	15	3								
10	Loading and moving sections	110	3								
11	Moving the demounting winches (3s / day = 1 hour / cm)	60	6								
12	Maintenance of oil stations	460	1								
Total		125 min/sec	8								

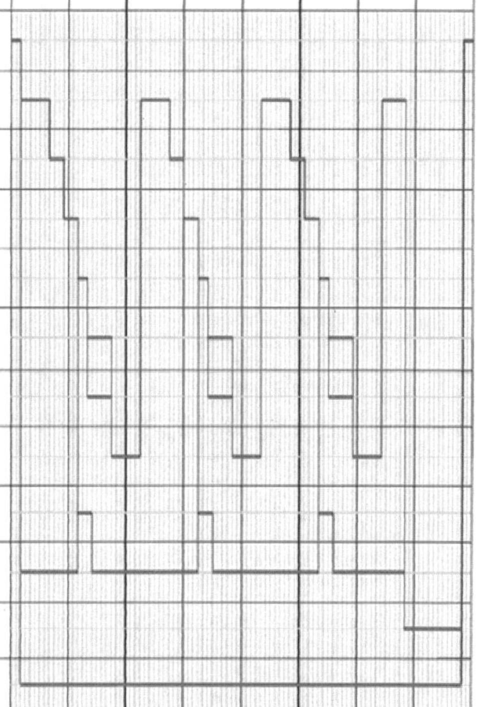

Figure 4. Schedule of dismantling works using Petitto Mule.

The return of the section in the dismantling chamber is carried out by diesel locomotive DLZ-110 FERRIT or by LSHM winches dragging over the soil of the dismantling chamber using the system of pulley blocks (in the case of a decrease in the spacing of the hydraulic props of the complex as a result of the effect of rock pressure on the mechanized support section). The average time of return of the mechanized support section in the dismantling chamber is 55 minutes.

The process of lining the place of dismantling is one of the most labor-intensive processes when dismantling the support sections of the mechanized complex. Lining is carried out by a pitprop with a diameter of 200-240 mm. At the same time, after every 3-5 dismantled support section of the mechanized complex in the goaf side of the dismantling chamber, a pigsty crib 2500x2500 is lined from pitprops. All operations are performed manually, which also leads to a great waste of time.

The repair shift almost all the time is preparing for dismantling works (shifting, mounting of dismantling winches, switching electrical equipment, moving of fire-extinguishing means). With such schedule of works performance, the time for dismantling sections of the mechanized complex will be 32-42 days.

Based on the experience of dismantling operations in the conditions of the mine “Alardinskaya”, it is proposed to conduct dismantling operations under the protection of the roof overlap with a net from polymeric materials. The calculation of the sections dismantling time is given in table 1.

Table 1. Calculation of the sections dismantling period.

Parameter	Units amend.	Index	
		Winches	Petitto Mule
Number of sections	PC	130	
Time for dismantling of the 1st section	minutes	125	70
Removable capacity technological scheme (1 shoulder)	sec / cm	2.4	4.9
The daily productivity of the technological scheme	sec / day	9.6	19.7
Duration of dismantling of sections (1 shoulder)	day	13.5	6.6
Duration of dismantling of sections (2 arms)	day	6.8	3.3

3. Recommendations and results of the use of Petitto Mule equipment

It is recommended to use the equipment for mechanized clearing the dismantling place and the dismantling chamber by the ripper loader Deilmann-Haniel L513. Dismantling of sections should be produced by the machine Petitto Mule. The delivery of mechanized support sections should be carried out by the hydraulic winch MW-250. The lining of the place of dismantling should be performed with a system of wooden pigsty cribs LINK-N-LOCK. It is also recommended to use the equipment delivery from the dismantling chamber to the mounting chamber on inclined workings to accumulating turnouts located on the ventilation drift of the excavation column being prepared by the outboard diesel locomotives DLZ-110F. From the accumulating turnout to the mounting chamber, the delivery should be performed by a multifunctional LHD with the trailer CHT-50.

Reduction of time for the main operations of the mechanized complex dismantling:

- dismantling of the mechanized support section from 90 minutes to 15 minutes;
- lining the place of dismantling from 40 minutes to 30 minutes;
- shifting of the pilot sections from 50 minutes to 20 minutes.

These measures will allow the number of cycles per day to be increased from three to ten, the period of remounting to be shortened by 20 days, which in turn will save up to 140 million rubles.

4. Conclusion

The dismantling with the help of the self-propelled dismantling machine Petitto Mule allows the shortest possible time for re-installing cleaning equipment to be achieved, as well as the risk of injury to personnel during dismantling to be reduced.

References

[1] Kozemaslov V A and Nikitina A M 2013 *Proceedings of the All-Russian Scientific Conference of Students, Postgraduates and Young Scientists* vol 17 part 2 (Novokuznetsk: SibSIU) pp 53–54

[2] Domrachev A N, Rib S V and Nikitina A M 2016 *Izvestiya Tula State University. Earth Sciences* **4** 81–90

[3] Mikunov V V, Nikitina A M and Rib S V 2017 *Proceedings of the All-Russian Scientific Conference of Students, Postgraduates and Young Scientists* vol 21 part 2 (Novokuznetsk: SibSIU) pp 78–84

Technical requirements to laboratory facility for stressed state rock massive spectral-acoustic control method sensitivity depth research if front of a development heading

A V Shadrin[1], V I Klishin[1], M S Nikitenko[1,2] and D S Trukhmanov[2]

[1]Federal Research Center of Coal and Coal Chemistry, Siberian Branch of the Russian Academy of Sciences, 18 Sovetskiy ave, Kemerovo, 650000, Russia
[2]T.F. Gorbachev Kuzbass State Technical University, 28 Vesennyaya street, Kemerovo, 650026, Russia

E-mail: ashadr1951@mail.ru

Annotation. The article presents the basic requirements for a laboratory facility creation that allows one to investigate the spectral-acoustic monitoring method sensitivity depth when testing the face area stressed state in front of a development heading. These data are necessary for establishing the scope of the spectral-acoustic method use for forecasting the dynamic phenomena in the process of various technological operations performance when mining: forecasting the outburst danger, controlling the anti-outburst measures effectiveness, assessing the anchor support quality and the hard-to-collapse roof directed hydraulic fracturing efficiency and so on. The reported study was funded by Russian Scientific Foundation grant (project No. 17-17-01143).

1. Introduction

Mining operations intensification and the coal deposits development depth increasing are complicated by the dynamic and gas dynamic phenomena manifestation. Beginning from 500 m depth, intensive and dangerous sudden extrusions of coal with increased gas emission and other types of dynamic and gas dynamic phenomena (hereinafter – DPh) occur at the extraction faces with thick hard-to-collapse roof, as a result of which face equipment is damaged and serious danger for mining workers occurs [1,2]. In addition, the hanging roof causes the of rock pressure concentration on the coal massif in the area of the extraction face and on its junction with the mine workings, which provokes a rock shock. In some cases, a local rock shock can develop into a sudden coal and gas outburst. Due to the significant area hanging roof influence, the abandoned pillars and protected development workings are subject to the action of high support pressure. This leads to the mine workings' destruction and, accordingly, to the mining transport normal operating mode and face ventilation failure.

2. Research objective

In Kuzbass gas dynamic phenomena (sudden outbursts of coal and gas and similar phenomena - in fact, underdeveloped outbursts) mainly occur in development headings [3, 4]. To exclude them at depths below the critical by the factor of outburst danger, it is necessary to carry out the current forecast of the indicated phenomena [5].

The stressed condition control at the walls of the opening is also necessary for other technological operations of mining: for assessing the state of the anchor support [6], at directed hydraulic fracturing of a hard-to-collapse roof [7], for performing anti-outburst and anti-shock hydraulic treatment of coal seams [5] and in a number of other technological operations [8].

Since for these purposes the spectral-acoustic method is now increasingly used, it was decided to justify the field of its application, since the method has a feature that raises the question of the radius of its sensitivity. This feature is due to the fact that when using this method, the face area is probed by acoustic vibrations generated by the cutting device of the combine (or other operating mechanisms). In this case, the receiver of acoustic vibrations (geophone) is installed in the wall of the development heading at a certain distance from the face behind the controlled area. It has been experimentally established that, in spite of that, this method allows to fix an approach to a dangerous DPh manifestation area ahead of the face [9, 10]. And the revealed danger, as shown by theoretical and experimental studies [11], when using this method is due primarily to the growth of stresses. At the same time, currently there is no scientific basis characterizing the method sensitivity depth ahead of the development heading. Therefore, the authors set the task to develop and to make a laboratory facility for modeling the stressed condition of the face area ahead of the development heading, probing it with acoustic oscillations and receiving them with the sensors installed behind the face.

The list of basic requirements to laboratory facility was compiled based on the laws of wave acoustic processes, on the similarity theory elements when scaling the design of a mining opening for the creation of a laboratory facility and proceeding from the spectral-acoustic control method.

3. Requirements for design, parameters and functionality

- Dimensions of the simulated and the real development heading width have the ratio 1:20.
- Small-grained sand is used as a material for coal seam modeling.
- The stress field in the face area in front of the simulated development heading should be smoothly changed within 0-5 MPa with the possibility of adjusting the ratio between the maximum and minimum stress values within the range of 1-1.5.
- The installation should ensure that the position of the maximum created stress zone moves relatively to the opening face front at a distance of 0.2 to 1.4 times of the opening width discretely with a step equal to 0.2 of the opening width. Thus, the total number of bands for the stressed state creation is seven. They are located perpendicular to the opening axis and parallel to the floor and the roof and are arranged sequentially from the face front to the depth of the massif.
- The width of the modeled part of the rock mass at the walls of the opening to the beginning of the acoustic traps should be two widths of the opening model (0.4 m).
- The width of acoustic traps at the walls of the opening, in front of the modeled face area and behind the massif at the walls of the opening is 0.5 m.
- The thickness of the modeled part of the floor and roof above the walls of the opening is equal to the opening width (0.2 m for floor and roof); to provide access to equipment at the face of the opening, the space above the opening is not filled with anything.
- Measurement of the stress field in front of the opening must be carried out by sensors located in the middle (to the width of the opening) of at least nine sectional planes parallel to the face in their lower part and separated from each other by a distance equal to 0.2 part of the opening width, when the first plane is removed from the face at a distance equal to the model metal wall thickness.
- The model should provide continuous emission of acoustic vibrations and their registration with sensors installed on the wall of the opening model. The range of simulated frequencies should correspond to the operating range of the spectral-acoustic method equipment for monitoring the stressed state of a rock massif lying in the range of 20-3500 Hz. The simulated frequencies should be evenly distributed in this range. Their number should be around 10. The approximate values of the simulated frequencies are 50, 400, 750, 1100, 1500, 1850, 2300, 2650, 3000 and 3500 Hz.

- If it is not possible to create continuous radiation at these frequencies, pulsed radiation can be applied, followed by the determination of the signal spectrum by means of a Fourier fast transform.
- The acoustic vibrations emitter should not touch the body of the model, but be isolated from it by a sound-absorbing gasket. It must be installed in one of three possible locations on a straight line running along the middle of the height of the development heading, parallel to the floor, and the centers of the two of them being removed from the walls of the opening at a distance equal to 0.1 part of its width, and the third – to be in the middle.
- The model should measure the amplitude of the acoustic signal by three-coordinate receiving sensors installed at the wall of the modeled opening at distances equal to one, two, three and four values of the opening width (i.e., the centers of the receiving sensors are removed from the face front at distances of 0.2, 0.4, 0.6, and 0.8 m).
- In order to suppress the reflected acoustic oscillations from the model walls, it must have acoustic plugs on the two side, one front and one back surfaces of the model.
- Operating conditions of the laboratory facility – average room conditions (temperature: 20°-25°C, relative humidity: 40-50%).
- Laboratory facility storage conditions – heated and ventilated rooms located in any macroclimatic regions.
- Conditions of laboratory equipment transportation:
  - The laboratory facility can be transported by a covered motor vehicle in accordance with the rules and conditions for cargoes loading and fastening that act on motor vehicles;
  - Transportation of the laboratory facility is carried out in wooden boxes specially made to the sizes, determined by the installation dimensions.

4. Requirements for the emitter and receiver of acoustic oscillations and to the measuring equipment for monitoring the stresses and the acoustic signal amplitude-frequency characteristics

4.1. Requirements for the emitter (or set of emitters) of acoustic oscillations

- Frequency range: 20-20000 Hz.
- Dimensions of the emitter surface – no more than 20 mm.

4.2. Requirements for the receiver of acoustic oscillations (converter of acoustic oscillations into electrical ones)

- Type – three-coordinate (three-component) – for recording longitudinal and transverse acoustic waves.
- Frequency range: 20-20000 Hz.
- Uneven frequency response – up to 10%.
- Dimensions: any linear dimension – no more than 20 mm.

4.3. Pressure sensor requirements

- Measured pressure range 0-5 MPa.
- Overall dimensions:
  length, width – no more than 30 mm; thickness – no more than 15 mm.

5. Control requirements for laboratory facility

Control of the creation of a laboratory facility is carried out in two stages:

- At the first stage, the mechanical strength of the installation is monitored by creating a limiting planned stressed state at 5 MPa throughout the whole simulation area of the stressed condition zone.

- At the second stage, the control consists in checking its operability. For this, the following is done:
  - An initial stressed condition is created in the modeled face area. When the stresses increase, they are convinced of the pressure sensors operability.
  - The source of acoustic vibrations is connected to one of the transitional devices, through which sound is introduced into the simulated coal-rock environment.
  - The outputs of the acoustic receivers via the appropriate terminal block are connected to the recorder (oscilloscope or voltmeter).
  - By switching on/off the source of acoustic oscillations, on the appearance of signals from the outputs of piezoelectric accelerometers they are confirmed of the laboratory facility operability.

Subsequent work on the laboratory facility to investigate the sensitivity depth of the spectral-acoustic method for monitoring the stressed state of the face area in front of the development heading is carried out following the research procedure methodology.

6. Conclusion

The specified list of requirements in the opinion of the authors should provide an opportunity of stressed state rock massive spectral-acoustic control method sensitivity depth research if front of a development heading.

Acknowledgements

The reported study was partially funded by Russian Science Foundation according to the research project No. 17-17-01143 "Forecasting and controlling geo-mechanic condition of a rock mass within the period of formation and manifestation of dynamic squeezing of the main roof and its preventive hydro-treating with the purpose to prevent dynamic and gas-dynamic phenomena".

References

[1] Klishin V I, Zvorygin L V, Lebedev A V and Savchenko A V 2011 *Safety Problems and Coal Deposits Underground Development New Technologies* (Institute of Mining SB RAS, Novosibirsk: Novosibirsky pisatel) p 254

[2] Klishin V I et al 2008 *Coal* **11** 12–16

[3] Polevshchikov G Ya 2003 *Dynamic Gas Phenomena in the Process of Development Opening Heading in Coal Mines* (Kemerovo: Institute of Coal and Coal Chemistry, SB RAS) p 317

[4] Zykov V S 2010 *Sudden Outbursts of Coal and Gas and Other Gas Dynamic Phenomena in Mines* (Kemerovo: Institute of Coal and Coal Chemistry, SB RAS) p 333

[5] Federal rules and regulations in the field of industrial safety *"Instruction for the dynamic phenomena forecast and rock massif monitoring during mining of coal deposits"* (approved by the Order of the Federal Service for Environmental, Technological and Nuclear Supervision on August 15, 2016 No. 339) pp 129

[6] Voznesenski Ye A 2012 *Substantiation and development of a spectral acoustic method for monitoring the anchor support to ensure the mine opening stability* (Extended abstract of candidate's thesis. Moscow: MGGU) pp 24

[7] Klishin et al 2015 *Coal* **11** 12–16

[8] Federal rules and regulations in the field of industrial safety *"Safety Rules in Coal Mines"* 2014 (approved by the Federal Service for Environmental, Technological and Nuclear Supervision on 19.11.2013) (Moscow: ZAO NTC PB) pp 197

[9] Mirer S V, Khmara O I and Shadrin A V 1999 *Spectral-acoustic Forecast of Coal Seam Outburst Danger* (Moscow, Kemerovo: Kuzbassvuzizdat) pp 92

[10] Korol V I and Skobenko A V 2013 *Acoustic Method of Gas-dynamic Phenomena Forecasting in Coal Mines* (Dnepropetrovsk: National Mining Unversity) p 181

[11] Shadrin A V 2016 *Journal of Mining Science* vol 52 issue 4 pp 670–682

The method for determining the rock resistance to brittle fracture

L T Dvornikov and V A Korneyev

Siberian State Industrial University, 42 Kirova street, 654007, Novokuznetsk, Russia

E-mail: korneev_va@list.ru

Abstract. In the paper the most common and rapid methods of determining the strength properties of rocks in domestic and foreign practice are considered. The authors proposed a fundamentally new way of assessing the rock resistance to brittle fracture. The developed method can be successfully used to determine the properties of rocks both on samples and outcrop of the massif, as well as in the wells drilled in the massif.

1. Introduction

Strength characteristics of rocks are one of the main factors that determine the rationale for making certain decisions in the mining industry and machine building. The most important among them are the coefficient of rock strength on the scale of Professor M.M. Protodyakonov and the tensile strength in uniaxial compression. The first indicator is used mainly in domestic practice and in the CIS countries, the second – abroad.

Rapid methods are of particular relevance when determining the strength properties of rocks, ensuring the prompt receipt of information in the conditions of mining. This article is devoted to the analysis of standard rapid methods for determining the mechanical properties of rocks used in domestic and foreign practice. In the paper the authors also propose a fundamentally new way of assessing the rock resistance to brittle fracture.

2. Standard rapid methods for determining the strength properties of rocks

Coefficient of rock strength and its tensile strength in uniaxial compression are used for calculation of the parameters of mine workings support, performance of drilling and blasting operations, and also for determining the required technological measures in mines that develop coal seams prone to rock bumps and sudden coal outbursts.

The considerable labor content and low efficiency of laboratory methods used for determining the physical and mechanical properties of rocks contributed to the development and application of rapid methods that allow the research to be performed in the conditions of in-situ rock.

The main features of such methods are:

- the possibility of carrying out studies on samples of irregular shape;
- rapidness of analysis;
- simple implementation in the conditions of mining;
- obtaining of acceptable result with a small number of measurements.

In foreign practice, the rapid methods for determining the strength properties are used: method for determination of the Schmidt Hammer rebound hardness [1,2], method for determination of the Shore scleroscope hardness [2,3], method for the needle penetration test [4], method for determination of the point load strength index of rock [5]. These methods are generally recognized and recommended by the International Society for Rock Mechanics (ISRM).

These methods are based on the determination of indirect indices (Schmidt hardness, Shore hardness, needle penetration index, rock strength index) characterizing the strength properties of the rock, with further switching to standard strength characteristics or technological parameters of the mining equipment operation [6].

The method of measurement by the Schmidt hammer (figure 1) is based on the determination of the shock pulse that appears after the impact by the striker on the rock with a given force. The strength properties of the rock under study are determined by the height of the striker rebound according to the established calibration curve.

The Shore scleroscope (figure 2) is based on a similar principle of action with the only difference that the impact of the striker is carried out by free fall from a constant height. The value of hardness is estimated according to the height of the striker rebound.

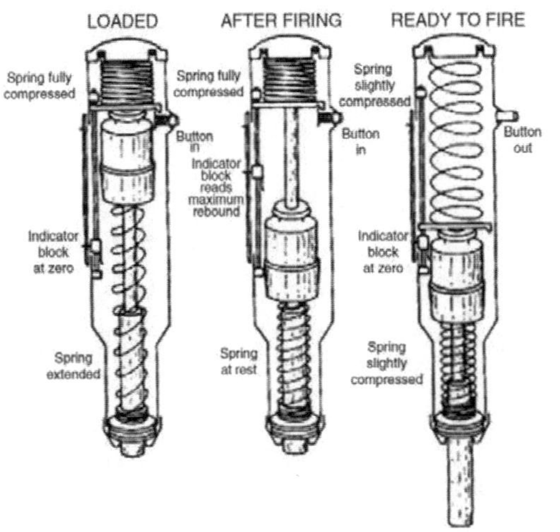

Figure 1. Working principle of a Schmidt hammer [1].

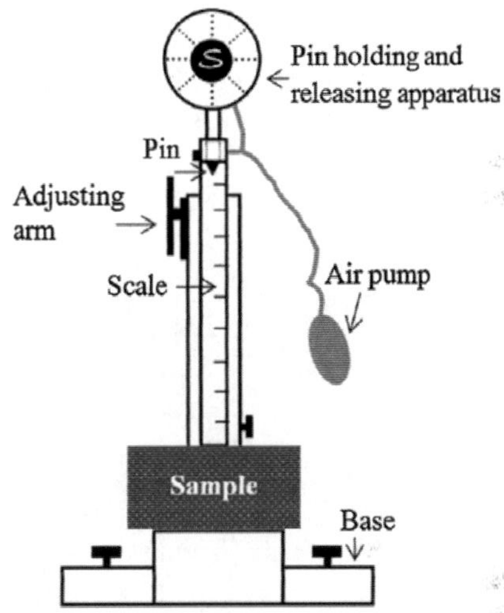

Figure 2. Schematic view of a Shore scleroscope instrument [6].

The properties of the rock using the penetrometer (figure 3) are determined according to the needle indentation index, which is defined by the impact of a special needle on the rock and is calculated depending on the its penetration depth and the load required for it. The needle is made of hardened steel and is a rod 0.84 mm in diameter, ending with a conical tip (figure 4). This sewing needle, designated as JIS S 3008 (No. 2)

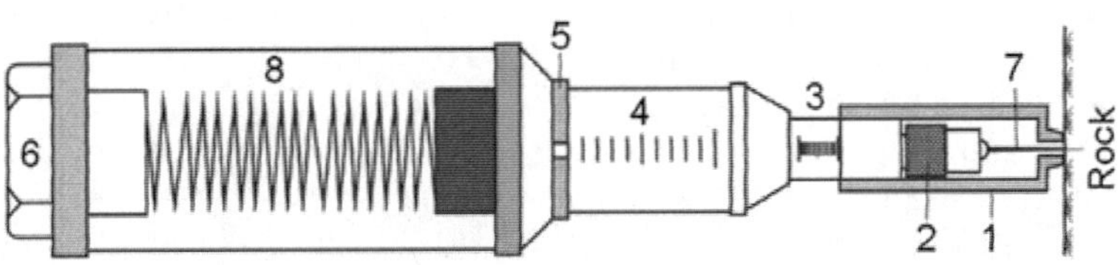

Figure 3. Needle penetrometer and its parts: 1 – presser, 2 – chuck, 3 – penetration scale, 4 – load scale, 5 – load indicating ring, 6 – cap, 7 – penetration needle and 8 – spring [4].

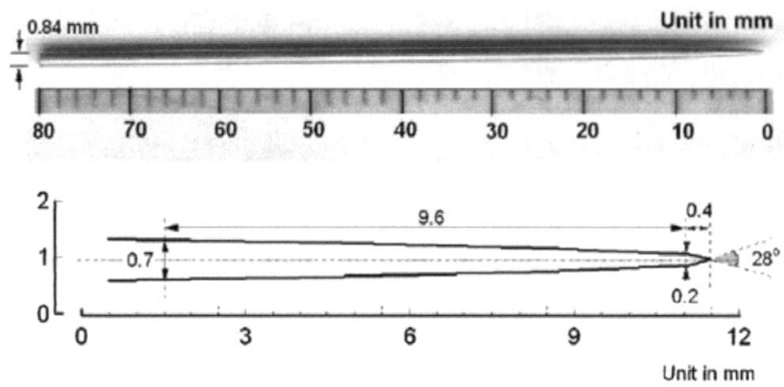

Figure 4. Sewing needle designated as JIS S 3008 (No. 2) and its geometry [4].

The point load method is realized by crushing the samples between conical punches (figure 5) with the fixation of the force necessary for their destruction. In accordance with the recommendations of the International Society of Rock Mechanics, the method can be used for core tests with diametrical or axial load applications, as well as for testing block or lumpy samples. The result of the tests is the rock strength index, on the basis of which it becomes possible to calculate the ultimate strength for uniaxial compression according to the correlation dependence.

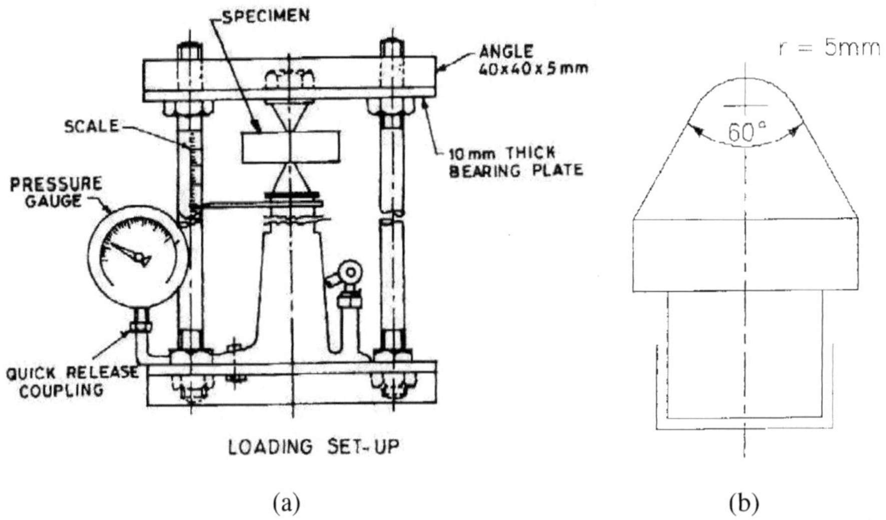

Figure 5. Schematic view of a Point Load Testing Machine (a), platen shape (b) [5].

Such rapid methods as the method of determining the hardness coefficient on Protodyakonov scale Russian National Standard (GOST) 21153.1-75 and the method for determining the contact strength (GOST R 50834-95) are used among numerous standard methods in Russia.

Determination of the hardness coefficient in accordance with GOST 21153.1-75 is carried out on samples of irregular shape by dropping the load with the help of a special device POK-1 (figure 6). The device POK-1 consists of a cup – 1, with inserted in it a tubular impactor – 2, balance weight – 3 of 2.4 kg with a handle – 4, to which the twine is tied. The tubular impactor has in the upper part a limiter of the weights lifting – 5 to a height of 0.6 m. The device kit includes a volumeter, consisting of a cup – 6 and a plunger – 7 with a measuring scale, and a sieve with a hole size of 0.5 mm for screening the crushed pieces of rock samples. Determination of the strength is carried out on the basis of the fraction volume of the crushed rock after screening.

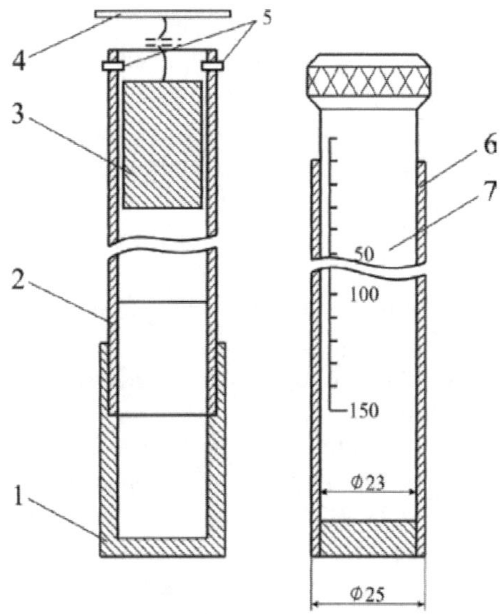

Figure 6. Scheme of the device POK-1 for determining the rock strength by the method of grinding in accordance with GOST 21153.1-75.

Determination of contact strength in accordance with GOST R 50834-95 is carried out by pressing indenters from a hard alloy with flat round bases $d_1 = 1.0 \pm 0.1$ mm and $d_2 = 3.0 \pm 0.1$ in diameters, having the form of a truncated cone with an angle at the apex 60° (figure 7). The contact strength is determined based on the values of the destructive force obtained by pressing indenters with smaller and larger bases. The advantage of this method is the ability to perform measurements by pressing the indenter into the borehole wall.

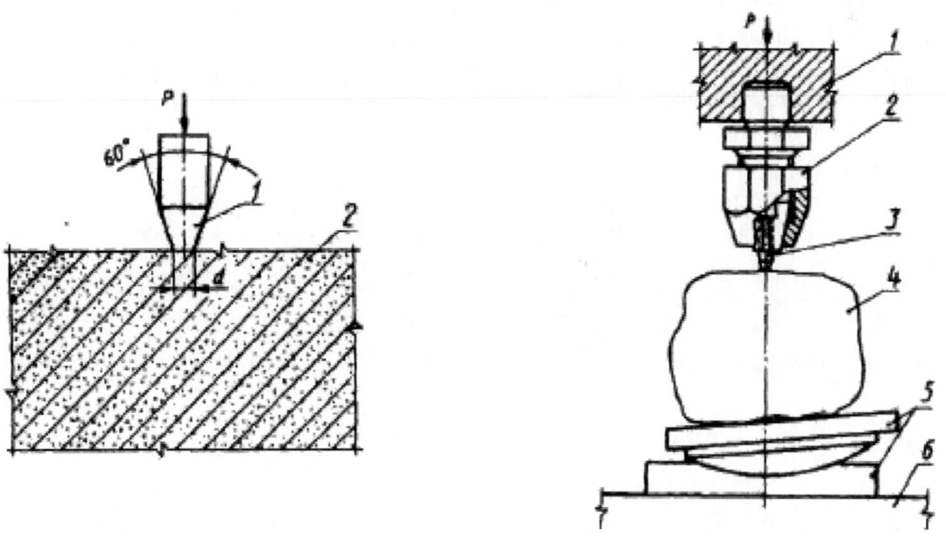

(a) (b)

Figure 7. Determination of the contact strength according to GOST R 50834-95 (a) the rock loading scheme (1 – indenter, 2 – rock, *d* – indentor diameter, *P* – applied force), (b) load device diagram (1 – upper press plate , 2 – clamping chuck, 3 – indenter, 4 – sample, 5 – spherical joint, 6 – lower press plate, *P* – applied force).

3. Results and discussion

Existing standard express methods, with the exception of the method for determining the contact strength according to GOST R 50834-95, can be used only in the study of rock samples and outcrops of the massif. However, mining practice has shown that for the implementation of various technological tasks, the information on the properties of rocks in different sites of the massif is often required.

Performance of cores examination by the above given methods is rather labor-consuming and takes long time, as in this case the core drilling is required. The use of the method for determining the contact strength in the well according to the recommendations of GOST R 50834-95 does not provide with the objective data for a number of reasons.

First, in accordance with this method pressing of indenters in the form of a truncated cone into the cylindrical surface of the borehole wall leads to the formation of a stress concentrator and the destruction of the rock with less effort, which distorts the measurement results.

Second, the small contact surface of the indenters used in the method does not allow reliable information to be obtained about the rock resistance to brittle fracture.

Third, the method lacks the analytically justified mechanism for interpretation of the massif properties by the fixed indicators.

In addition, in accordance with the method the use of two indenters with different diameters of the contact surface complicates the research and makes it last longer in time.

Taking this into account, the authors have developed a method to determine the rock resistance to brittle fracture [7], which can be successfully used for studying the properties of rocks both in samples and massif outcrops and in wells drilled in the massif. In accordance with this method, the strength properties of rocks are determined by the magnitude of the effort required for brittle fracture of the rock by the indenter of a special shape.

The indenter used in the method is shown in figure 8, where: 1 – the body of the indenter made in the form of a straight circular cylinder with diameter D, equal to its length L; 2 – ends of the indenter having roundings with radius R equal to the radius of the straight circular cylinder forming the body 1 of the indenter, 3 – the rock under test, P – the force applied to the indenter. The action of the indenter on the rock 3 to be destroyed is carried out along the generatrix of the quadric cylinder forming the body 1 of the indenter. When measuring in wells, the indenter is positioned in such a way that the generatrix of the cylinder constituting its body is directed parallel to the generatrix of the cylindrical surface of the well.

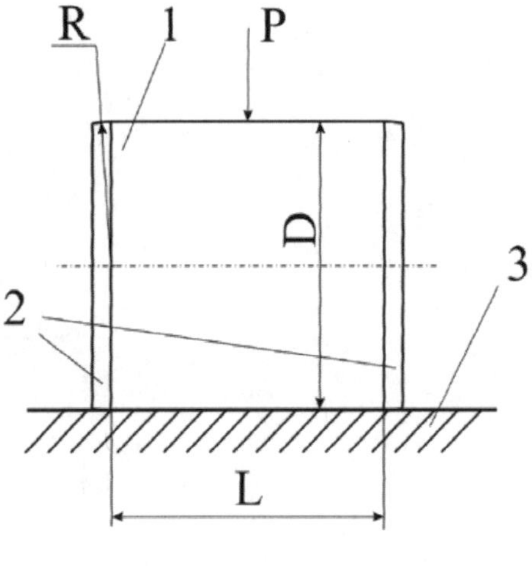

Figure 8. The indenter used in the proposed method [7].

Taken into account the ideas of K.V. Ruppeneyt considered in [8] for the simulation of rotary drilling, the minimum volume of rock that retains its properties can be represented in the form of a cube, on each facet of which there must be at least 30 rock-forming grains. In this case, the length τ of the edge of such cube can be related to the average grain diameter d by the dependence:

$$\tau = 6.7d \tag{1}$$

Taking real dimensions of the grains of homogeneous rock within 0.1 ... 1.0 mm, it can be determined that the length of the edge of the cube of minimum volume should be not less than 6.7 mm. Thus, the length L of the indenter body in the developed method is assumed to be not less than this value.

4. Conclusion

The feature of the proposed method is the use of the indenter with rounded ends, which makes it possible to exclude the presence of a stress concentrator. As a result, it becomes possible to obtain acceptable measurement results without thoroughly ensuring the parallelism of the cylinder generatrix and the rock.

An advantage of the method is also the possibility of achieving the indenter penetration into the rock under the study with a slight effort. This effect is due to the fact that at the initial stage of the measurement the interaction of the indenter with the rock takes place only along the thin line of contact, which is the generatrix of a quadric cylinder forming the indenter body. This feature, on the one hand, ensures the coverage of the required amount of rock grains to achieve an acceptable measurement accuracy (in comparison with the cylindrical and conical indenters), and, on the other hand, the possibility for conducting studies in wells where the use of indenters with a flat contact surface does not allow reliable results to be obtained.

Determination of the rocks properties of samples in accordance with the developed method is supposed to be carried out on a manual hydraulic press. To perform the research in the well conditions, the authors proposed a special device Strength Tester PSSh-1 ("PSSh-1") [9].

Determination of rocks strength properties by the developed method is supposed to be carried out with the help of empirical dependencies involving the known analytical equations of the theory of contact problems. A detailed description of the theory of contact problems can be found in the works of I.Ya. Shtayerman, A.I. Lurye, S.D. Ponomarev et al. [10-12].

Acknowledgments

The paper was prepared with the financial support of the grant of the President of the Russian Federation for young Russian scientists – candidates of sciences No. MK-6689.2018.8.

References

[1] Aydin A 2009 *International Journal of Rock Mechanics & Mining Sciences* **46** 627–34
[2] 1978 *Int. J. Rock Mech. Min. Sci. & Geomech. Abstr.* Vol. **15** 89-97
[3] Rabia H and Brook N 1979 *Int. J. Rock Mech. Min. Sci. & Geomech. Abstr.* vol **16** 335–6
[4] Ulusay R et al 2014 *Journal Rock Mechanics and Rock Engineering* **47** 1073–8
[5] 1985 *Int. J. Rock Mech. Min. Sci. & Geomech. Abstr.* **22** 51-60
[6] Tumac D 2014 *Rock Mech. Rock Eng.* **47** 703–15
[7] Dvornikov L T and Korneyev V A 2018 *Patent Application* 2018114343 RU
[8] Alimov O D and Dvornikov L T 1976 *Drilling Machines* (Moscow: Mechanical engineering) p 295
[9] Pygachev E V, Korneyev V A and Korneyev P A 2015 *Bulletin of the Siberian State Industrial University* **2** 42–5
[10] Shtaerman I Ya 1949 *The Contact Problem of the Theory of Elasticity* (Moscow:

IOP Conf. Series: Earth and Environmental Science **206** (2018) 012027 doi:10.1088/1755-1315/206/1/012027

Gostekhizdat) p 271

[11] Lyrie A I 1955 *Spatial Problems in the Theory of Elasticity* (Moscow: State Publishing House of Technical and Theoretical Literature) p 493

[12] Ponomarev S D et all 1958 *Calculations for Strength in Mechanical Engineering* vol II (Moscow: State Scientific and Technical Publishing House of Machine-building Literature) p 974

Mechanical characteristics of the doubly-fed machine and its use in the hoist electric drive

V Yu Ostrovlyanchik, I Yu Popolzin and D A Marshev

Siberian State Industrial University, 42 Kirova street, Novokuznetsk, 654007, Russia

E-mail: eidoline@yandex.ru

Abstract. Currently, a large number of existing mine hoisting units use the electric drive based on asynchronous motor with a rotary station. Despite the fact that such drives have a large overload capacity and linear (within the working areas) mechanical characteristics, they have drawbacks, first of all, significant energy losses. The promising way of their modernization, as well as design of new drives based on a phase-wound rotor motor, is the use of the scheme of a doubly-fed electric machine. The paper outlines the requirements for the electric drive of a mine hoisting unit, in particular, for its mechanical characteristics; the analysis of the mechanical characteristics of the doubly-fed machine in asynchronous and synchronous modes, as well as changes in the active resistance of the rotor was carried out; the working sections of characteristics are indicated; the operating modes of the machine are compared to the sections of the speed diagram of the hoisting unit; the rationale for the combined control of the machine with the change in operating modes directly during operation is shown.

1. Introduction

Most hoisting systems are equipped with electric drives with asynchronous motors with a phase rotor and a rotor station. The main disadvantages of these electric drives are significant losses of electrical energy in the rotor station, as well as dynamic overloads in the mechanical part of the drive that arise during the switching of the station stages. Therefore, the work related to the elimination of these shortcomings in existing hoisting systems is relevant.

When upgrading mine hoisting systems (MHS), it is promising to use the mode of a doubly-fed machine (DFM) for phase-wound rotor motors, in which an additional voltage from the frequency converter is added to the rotor circuit of the motor. In this case, frequency converters are used in the rotor circuit, which are able to consume the slip energy, to recuperate it into the network and to feed the engine, that is, to work both in the forward and reverse directions. The advantage of this scheme, in particular, is the possibility of recuperation of slip energy into the electricity network.

2. Modes of operation of the mine hoisting unit and the doubly-fed machine

For the MHS, five main characteristic modes (sections of the speed diagram) should be identified [1].

- Torque control on the looked motor. It is necessary to exclude the reverse stroke of the machine drum when removing the mechanical brake and starting;
- Movement at speed when exiting curves or in the mode of "Revision";
- Acceleration to the set speed;
- Movement at a given speed;
- Slowing down.

The main criterion determining the working areas of mechanical characteristics is constant stiffness $\beta = \frac{d\omega}{dM} = const$ (where ω – the angular velocity, M – the moment of the electric drive) and the required overload capacity $\lambda = \frac{M_{cr}}{M}$ (where M_{cr} – the critical moment on the specified characteristic).
Let us consider the possible modes of DFM operation and compare them with the speed diagram of the hoisting unit. The mode of DFM operation depends on the relation between the frequency of the internal rotor EMF and the frequency of the additional voltage introduced into the rotor circuit.

If the frequency of the auxiliary voltage is equal to the frequency of the eigen EMF, then this mode of operation is called asynchronous [2-4], if the frequency of the additional voltage does not depend on the load and rotor speed – the mode is called synchronous [2, 3, 5]. In asynchronous mode, the frequency of the auxiliary voltage is a parameter dependent on the speed, and in the synchronous mode – the independent parameter determining the speed [2], [4].

3. DFM mechanical characteristics in asynchronous mode

The equation of DFM mechanical characteristics for an asynchronous mode was obtained in [2] and [4]. It is written as:

$$M = \frac{3}{2} z_p \frac{k_S^2 U_S^2}{\omega_0^2 \sigma L_R} \frac{s_{cn} s}{s_{cn}^2 + s^2} \left[1 - \frac{U_R^*}{s} \left(\cos\delta + \frac{s}{s_{cn}} \sin\delta \right) \right], \tag{1}$$

where k_S – the stator coupling coefficient; L_R – rotor inductance; z_P – the number of the machine pole pairs; M – electromagnetic moment of the machine; ω_0 – speed of the ideal idle motion of the machine; $\sigma = 1 - \frac{L_\mu^2}{L_S L_R}$ – the scattering coefficient (L_μ – the mutual inductance of the stator and the rotor, L_S – the inductance of the stator); k_S – the coupling factor of the stator; $s_{kn} = \frac{R_R}{\omega_0 \sigma L_R}$ – rated critical slip of the machine on the natural mechanical characteristic (R_{Rn} – rated rotor active resistance; X_{Kn} – rated X-reactance of the machine short-circuit); $s = \frac{\omega_0 - \omega}{\omega_0}$ – current slip; ω – current rotational speed of the machine rotor; $U_R^* = \frac{U_R}{k_S U_S}$ – relative value of the voltage amplitude on the rotor (U_R, U_S – the effective values of the voltages on the rotor and stator, respectively); δ –phase shift between the additional voltage on the rotor and the voltage on the stator.

The expression for the mechanical characteristic (1) can be divided into two components:

a) $M_e = \frac{3}{2} z_p \frac{k_S^2 U_S^2}{\omega_0^2 \sigma L_R} \frac{s_{cn} s}{s_{cn}^2 + s^2}$ – equation of the natural mechanical characteristic of the phase-wound rotor motor (in the absence of additional voltage source in the rotor circuit);

b) $k_y = 1 - \frac{U_R^*}{s} \left(\cos\delta + \frac{s}{s_{cn}} \sin\delta \right)$ – component (coefficient) describing the change in the natural mechanical characteristic of the machine when an additional voltage is introduced into the rotor circuit.

This component depends on the amplitude of additional voltage and its phase. In [7] the condition for maximizing this component with respect to the phase shift of the additional voltage was obtained. According to [7], to ensure the maximum value of the moment at a given speed, it is necessary to control the phase of the additional voltage on the rotor according to the law

$$\delta = arctg\left(\frac{s}{s_{cn}}\right). \tag{2}$$

Under condition (2), the MDP mechanical characteristics have the longest sections with negative stiffness (working areas).

The speed control in the asynchronous mode is performed by changing the relative amplitude of the additional voltage on the rotor U_R^*. As follows from (1), an increase in the amplitude of the additional voltage on the rotor reduces the speed of the machine at a constant load moment. This also reduces the speed of ideal idle motion on the artificial characteristics of the machine, the critical slip and the critical moment due to the fact that when U_R^* increases k_y decreases.

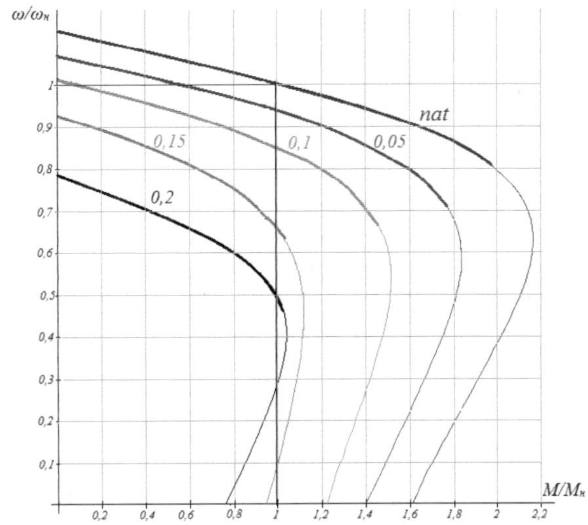

Figure 1. Family of DFM mechanical characteristics in the asynchronous mode depending on the relative amplitude of the additional voltage on the rotor (the working sections of the characteristics are marked with bold lines).

The family of DFM mechanical characteristics with the maximization of the moment by condition (2) is shown in figure 1.

As it can be seen from the family of mechanical characteristics shown in figure 1 that when the amplitude of the additional voltage on the rotor increases, the speed of the machine decreases, but at the same time the overload capacity of the machine drops and the working section of mechanical characteristics decreases (a section with negative stiffness). Also the speed regulation in the DFM asynchronous mode with constant load torque is inappropriate below the speed of $0.6\omega_0$, since the overload capacity of the machine becomes unsatisfactory. The given family of characteristics is constructed for the experimental DFM based on the crane engine, the working section of the natural mechanical characteristic of which has relatively low stiffness. Since the MHS uses powerful asynchronous motors with low nominal slip (and, therefore, with a greater stiffness of the working areas), the lower limit of the speed control of the DFM-based hoisting system can be estimated as $0.5\omega_0$.

Thus, the DFM in asynchronous mode at the nominal load moment can operate only in the speed range $0.5\omega_0<\omega\leq\omega_0$, and the speed control range in this mode does not exceed 2:1 down from the rated speed of the machine. Consequently, the asynchronous mode is suitable only for MHS running at a constant maximum speed or for acceleration/deceleration in the speed range $0.5\omega_0<\omega<\omega_0$.

4. DFM Mechanical characteristics in synchronous mode

The synchronous operating mode takes place when the frequency of the auxiliary voltage is independent of the frequency of the eigen EMF of the rotor [2, 3, 5]. The rotor speed in this mode is constant and is determined by the expression:

$$\omega = \omega_0 \frac{f_S - f_R}{f_S} = \omega_0 s, \tag{3}$$

where f_S, f_R – the stress frequencies on the stator and the additional voltage on the rotor, respectively. The parameter s in this case has the meaning of the relative voltage frequency on the rotor.

An expression for the mechanical characteristic of a DFM operating in synchronous mode was obtained in [5] and [7]. It has the following form:

$$M = \left(\frac{L_\mu}{L_S}\right)^2 \frac{3sR_R'U_S^2}{\omega_0\left(R_R'^2 + X_K^2 s^2\right)} + \frac{L_\mu}{L_S} \frac{3U_S U_R}{\omega\sqrt{R_R'^2 + X_K^2 s^2}} \sin\gamma, \tag{4}$$

where R_R' – the value of the rotor's active resistance adjusted to the stator; $\gamma = \theta + arctg\frac{sX_{sc}}{R_R'}$ – phase shift between the stator field and the rotor field, θ – load angle, defined similarly to the load angle of the synchronous machine, X_{sc} – reactive short circuit resistance of the machine.

It follows from expression (2) that the DFM mechanical characteristic in synchronous mode has two components:

a) $M_e = \left(\frac{L_\mu}{L_S}\right)^2 \frac{3sR_R'U_S^2}{\omega_0\left(R_R'^2 + X_{sc}^2 s^2\right)}$ – the component equivalent to the natural mechanical characteristic of the phase-wound rotor motor without a source of additional voltage in the rotor (asynchronous component);

b) $M_c = \frac{L_\mu}{L_S} \frac{3U_S U_R}{\omega\sqrt{R_R'^2 + X_{sc}^2 s^2}} \sin\gamma$ – the component describing the electromechanical transformation of energy in the machine as a result of the stator field interaction and the winding of the rotor, over which the current flows due to the presence of an additional voltage source in the rotor circuit.

Since the relative frequency s in this mode depends only on f_R, it is an independent parameter in (4) and does not depend on the moment on the machine shaft, its parameters and the amplitude of the stresses on the stator and rotor. In this case, the asynchronous component (4) is constant for a given value of the relative frequency, and the synchronous component depends only on angle γ conditioned by the internal load angle of the machine θ.

In practice, due to the limited speed of the frequency converter in the rotor circuit, the non-zero time of the transient processes in the rotor circuit when the frequency of the additional voltage is varied and the assumptions made when deriving expression (2), the machine speed in this mode depends on the load due to fluctuations in the additional voltage frequency, and the mechanical characteristics have a greater but finite stiffness. Nevertheless, for the purposes of this study, it can be assumed that the stiffness of mechanical characteristics is infinite, and the rotor rotation speed depends only on f_R and γ.

Since the speed of rotor rotation does not depend on the load and is set forcibly, the load change leads to a change in the angles θ and γ. Thus, increasing the load on the shaft of the machine, the angle γ and the synchronous component of the moment Mc increase. The synchronous component of the moment is maximal at $\gamma=\pi/2$. When $\gamma>\pi/2$, the synchronous component of the moment begins to fall with increasing load, the machine falls out of synchronism and stops. Thus, the overload capacity of the machine in synchronous mode is limited by the maximum possible value of the synchronous torque component. In this case, the synchronous component, as follows from (4), can be increased if to increase the amplitude of the additional voltage on the rotor.

The family of DFM mechanical characteristics corresponding to expressions (3) and (4) is shown in figure 2. The characteristics were obtained at the amplitude of additional rotor voltage *0.15US*. The dashed lines show the areas of mechanical characteristics that correspond to the unstable operation of the machine after falling out of step.

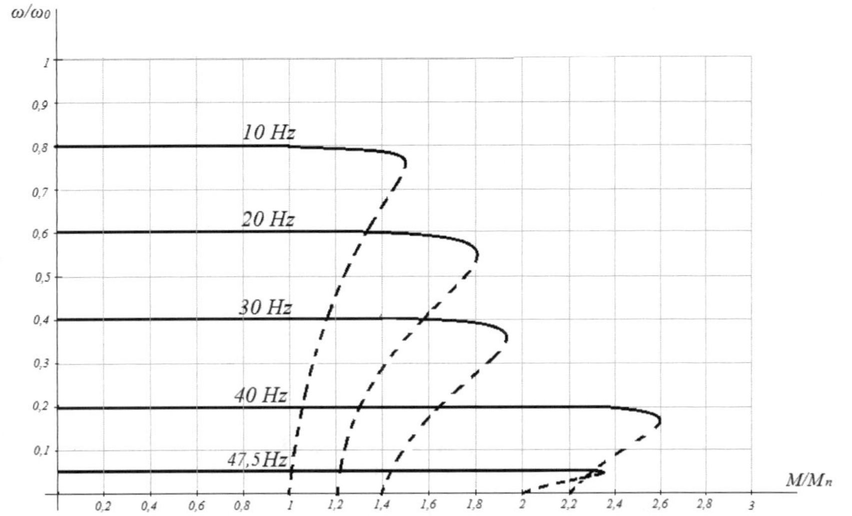

Figure 2. The family of DFM mechanical characteristics in synchronous mode.

According to the characteristics shown in figure 2, it can be seen that the working sections of the DFM mechanical characteristics in the synchronous mode are stiff, and when the load moment exceeds the critical moment the machine "capsizes", these parts of the mechanical characteristics have a positive stiffness, and the machine operation in this mode is unstable.

When the frequency of the auxiliary voltage on the machine rotor increases, the rotor speed decreases in accordance with (2). Taking into account the error in regulating the frequency at the output of the converter in the rotor circuit, it is advisable to take the maximum frequency of the additional voltage within 47.5-49 Hz, which corresponds to the relative rotation speed of the rotor 0.02-0.05.

In general, the DFM mechanical characteristics in synchronous mode are similar to the mechanical characteristics of a synchronous motor. However, the DFM overload capacity depends on the speed (figure 2). The greatest overload capacity is observed in the region of low velocities (below $0.5\omega_0$). Besides, as the frequency of the additional voltage in the rotor decreases, the losses in the machine magnetic system increase (similar to a decrease in the stator voltage frequency [8]). Also, when the voltage frequency decreases, switching of the power valves of the frequency converter becomes difficult. Therefore, the frequency of the additional voltage in the DFM synchronous mode is reasonable to vary in the range 25 Hz – 47.5 ÷ 49 Hz, which corresponds to the machine operation in the speed range 0.02 ÷ 0.05 $\omega_0 \leq \omega \leq 0{,}5\omega_0$. Thus, the synchronous mode is suitable for the MHS operation in the specified speed range, including acceleration and deceleration.

5. DFM mechanical characteristics with a change in the rotor active resistance

As it follows from the above carried analysis of asynchronous and synchronous operation modes, none of these modes is suitable for MHS operation with a locked motor. In synchronous mode, the machine does not develop torque with the locked motor, and in asynchronous mode the overload capacity is too low to realize the stable machine operation in the speed range. Therefore, to implement the MHS work, the mode of changing the rotor active resistance should be used. In this case, the equation of the DFM mechanical characteristic in this mode is analogous to one for the asynchronous motor with a phase rotor [9], and it can be written as:

$$M = \frac{3U_S^2 k^2 R_{R\Sigma}}{\omega_0 s\left[\left(R_S + k^2 \frac{R_{R\Sigma}}{s}\right)^2 + X_{sc}^2\right]}, \tag{5}$$

where $R_{R\Sigma}$ – the total active resistance of the rotor chain; k – the transformation ratio of the machine. Since the moment is realized only by the active component of the rotor current, we can assume that in expression (5) $X_K = 0$ and write (5) in the form:

$$M = \frac{3U_S^2 k^2 R_{R\Sigma}}{\omega_0 s\left[\left(R_S + k^2 \frac{R_{R\Sigma}}{s}\right)^2\right]}. \tag{6}$$

The analysis of expression (5) shows that with an increase in the active resistance of the DFM rotor circuit, the critical moment remains unchanged, and the speed at the unchanged moment decreases, that is, the stiffness of mechanical characteristics in this mode is inversely proportional to the total active resistance of the rotor chain. The family of DFM mechanical characteristics in this mode is shown in figure 3.

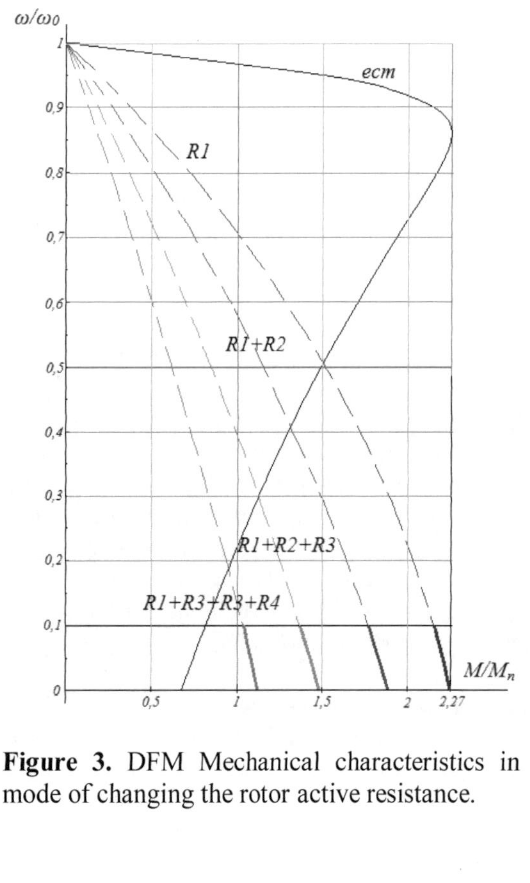

Figure 3. DFM Mechanical characteristics in the mode of changing the rotor active resistance.

The bold lines indicate the areas of performance that ensure the machine operation at zero speed, in the braking off mode and in the speed range $0 \le \omega \le 0{,}1\omega_0$. If we simulate the change in the active resistance smoothly (which is equivalent to the increment in the rotor active resistance), then it is possible to achieve the constancy of the moment developed by the machine.

The working zone of the DFM mechanical characteristics in this mode is limited by the speed determined from the parameters of the machine rotor chain and the value of the additional resistance in the rotor circuit. In practice, it is expedient to choose this speed as $0.05 \div 0.1\omega_0$ so that after the braking off the machine could exit the mode of acceleration or low speed ($0 \le \omega \le 0{,}1\omega_0$) and change to a synchronous mode. Thus, it is recommended to use the controlled current converter in the rotor circuit for regulating the speed in the range $0 \le \omega \le 0{,}1\omega_0$, while

DFM in its characteristics is similar to the AM PR characteristics with the active resistance in the rotor circuit.

6. Conclusion

Thus, we can draw the following conclusions:

- The main criterion for determining the working areas of the mechanical characteristics of the MHS electric drive is constant negative stiffness and overload capacity;
- The asynchronous mode, the synchronous mode and the active resistance change mode should be combined to achieve a large speed range in the MHS drive.
- For the machine operation in the range, it is advisable to use a controlled current converter in the rotor circuit with an imitation of the change in the active resistance of the rotor circuit, since it provides good overload capacity at a low speed and the possibility of smooth acceleration.
- Synchronous operation mode is recommended to use for acceleration, deceleration and operation at a low speed in the speed range $0.02 \div 0.05\omega0 \leq \omega \leq 0.5\omega0$.
- It is recommended to use asynchronous mode for MHS operation at the maximum hoisting speed, as well as during acceleration and braking of the machine in the speed range $0.5\omega0 < \omega \leq \omega0$.
- The specified combination of modes allows the speed control of the machine in the range of at least 30:1 to be achieved while maintaining the acceptable overload capacity of DFM in the entire speed control range.

References

[1] Ostrovlyanchik V Yu 2004 *Automatic Electric Drive of Direct Current of Mining and Metallurgical Production* (Novokuznetsk) p 382
[2] Onischenko G B and Lokteva I L 1979 *Asynchronous Valve Cascades and Dual-feed Motors* (Moscow: Energia) p 200
[3] Firago B I and Pavlyachik L B 2006 *Regulated Electric Drives of Alternating Curr*ent (Minsk: Technoprospectiva) p 363
[4] Ostrovlyanchik V Yu et al 2016 H*igh Technology of Development and Use of Mineral Resources* **3** 303–309
[5] Klyuchev V I 2001 *Theory of Electric Drive* (Moscow: Energoatomizdat) p 704
[6] Ostrovlyanchik V Yu et al 2017 *IOP Conference Series: Earth and Environmental Science* **84** 012030
[7] Ostrovlyanchik V Yu 2016 *Proc. of the VII All-Russia Sci. Pract. Conf. on Automated Electric Drive and Industrial Electronic*s pp 66–78
[8] Ostrovlyanchik V Y and Popolzin I Yu 2016 *Bulletin of KuzSTU* **2(114)** 75–84
[9] Chilikin M G et al 1974 *Fundamentals of Automated Electric Drive* (Moscow: Energia) p 568

Physical and technical fundamentals of technology used to increase the wear resistance of working surfaces of large volume excavator buckets

V E Gromov, S V Raykov, V E Kormyshev, D A Kosinov, O A Kondratova and D A Romanov

Siberian State Industrial University, 42 Kirova street, 654007, Novokuznetsk, Russia

E-mail: gromov@physics.sibsiu.ru

Abstract. Structural-phase states and tribological properties of the wear-resistant coating formed by electric arc surfacing with flux-cored wire on the low-carbon steel Hardox 450, used to protect large volume excavator buckets, are established.

1. Introduction

Intensification of mining technological processes demands higher standards of the mechanical properties of working surfaces of mining mechanisms and equipment.

To protect metals and alloys from various types of external influences – wear, corrosion, statistical and dynamic loads, electric arc surfacing is used, which ensures high performance properties of products.

A careful analysis of the relation "wear parameters – hardness – microstructure" is necessary for the scientifically substantiated practical use of various types of surfacing in products of critical use. Only in this case it is possible to obtain products with high performance parameters.

This makes it possible to solve one of the most important problems of ensuring an optimal correlation between the properties of the surface and the material volume. In this case, there is no need to use volumetric-alloyed materials and it becomes possible to solve to a certain extent the important problem of increasing the reliability and durability of parts in the operating conditions [1-3].

The purpose of this work was to analyze the research results of mechanical and tribological properties of the layer formed on Steel Hardox 450 by the electric arc surfacing of the *Fe-C-Ni-B* flux-cored wire.

2. Material and methods of research

As the base material the sheets of Hardox 450 steel were used (wt%): 0.19 *C*, 0.70 *Si*, 1.6 *Mn*, 0.025 *P*, 0.010 *S*, 0.25 *Cr*, 0.25 *Ni*, 0.25 *Mo*, 0.004 *B*, the rest – *Fe*). Because of the low content of alloying elements, it is well welded and processed.

Sheets of Hardox 450 steel are lining plates protecting the working surfaces of the excavator bucket Harnischfeger 2800 with a capacity 33 m^3. Depending on the operating conditions, the service life of such lining is 12-18 months.

Surfacing of the reinforcing layer was carried out using the MIG/MAG method (Metal Inert Gas / Metal Active Gas – arc welding by the melting metal electrode with an automatic feeding of the filler wire) in the gas environment (Ar = 98%, CO_2 = 2%) with a welding current 250-300 A and voltage

30-35 V. The flux-cored wire with the following chemical composition (wt.%) was used as the welded electrode: 0.7 *C*; 2.0 *Mn*; 1.0 *Si*; 2.0 *Ni*; 4.5 *B*; the rest is *Fe* (figure 1).

Figure 1. Armoring of lining plates made of Hardox 400 steel by semi-automatic surfacing of flux-cored wire: a, b – semiautomatic surfacing of the grid; c, d – grid view; d – the bottom of the bucket; e – overall view of armoring the front bucket wall.

Tribological tests were carried out on a tribometer "CSEM Tribometer High Temperature S/N 07-142", CSEM Instruments; counterbody – a ball with a diameter of 2 mm made of a hard alloy VK6, the wear rate was estimated by the cross-sectional area of the wear track using the 3D-profilometer MICRO MEASURE 3D station by STIL. The structure of the volume of the modified layer was analyzed by the cross-section method, for which the samples were cut into two parts perpendicular to the surface of the modification.

The defective structure of the material was studied by optical (microscope "Microvisor metallographic μVizo - MET-221"), scanning (scanning electron microscope "SEM-515 Philips") and transmitting diffraction (EM-125 FET Tecnai 2062 TWIN) electron microscopy [4-7]. The elemental composition of the surface layer was determined by methods of X-ray microanalysis (microanalyzer EDAX ECON IV, which is an add-on device to electronic scanning microscope SEM-515 "Philips"). The phase composition of the surface layer was analyzed by X-ray diffraction methods (diffractometer XRD-7000s, Shimadzu, Japan) [8-9].

3. Results and discussion

The main difference of Hardox 450 steel from the widespread wear-resistant steels is a low content of alloying elements. Due to this fact, the steel is well welded and processed. High hardness of Hardox steels is achieved due to a special sheet hardening system, which allows a fine-grained structure to be obtained. The essence of hardening is in the rapid cooling of the rolled sheet without subsequent tempering.

The formation of a layer deposited on the steel surface by the electric arc method is accompanied by poorly controlled heating of the material. This results in a tempering of the hardened state. The steel tempering leads to the release of particles of the carbide phase (cementite) located in the volume of the plates and at their boundaries. The particles have an acicular shape, typical for cementite, formed during low-temperature tempering of hardened steel [10, 11].

Investigations of the mechanical and tribological properties of the deposited layer showed that the microhardness at one pass varies within 10.5÷12.5 GPa and ≈15 GPa at a double pass, which is 2 and 3 times, respectively, higher than microhardness of the base (figure 2) .

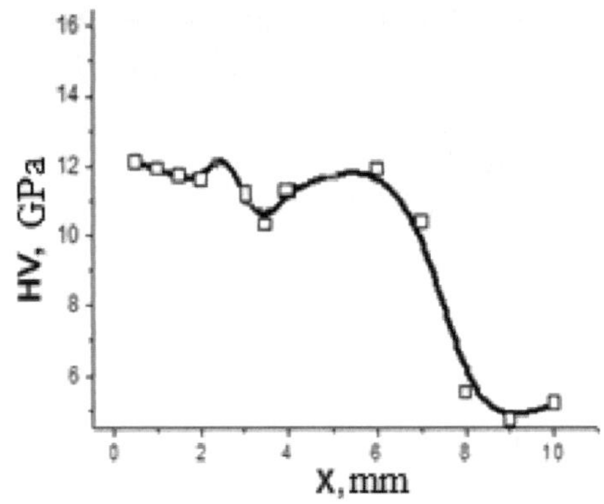

Figure 2. The profile of system microhardness "welded layer/steel".

At one pass, the wear resistance is 2 times higher, and the coefficient of friction is ≈2.2 times lower than the corresponding values of steel in the initial state.

The difference in the hardness of the deposited layers is due to the difference in the phase composition of a single and double layers. After one pass in the deposited layer, an iron boride of composition Fe_2B is formed, after a double pass – *FeB*, the hardness of which differs by 1.5 times.

In addition to that, the increased mechanical and tribological properties of the deposited layer are due to the formation of a multiphase submicro- and nanoscale structure caused by the hardening effect (the formation of an ultra-small martensitic structure of α phase) and the presence of inclusions of iron borides of composition Fe_2B and *FeB* of submicron sizes forming a plate-type eutectic. Iron borides do not contain a dislocation substructure, which is apparently due to their high hardness (figure 3). The scalar dislocation density reaches 10^{11} cm^{-2} in α phase.

The presence of flexural extinction contours indicating the formation of internal stress fields at the phase interfaces $Fe_2B/\alpha\text{-}Fe$ and Fe_2B/FeB was noted. In the contact layer on the side of Hardox 450 steel a substructure of the plate type with a scalar dislocation density ~$5\cdot10^{10}$ cm^{-2} were revealed. In the volume and at the boundaries of the plates, the cementite particles 20-30 nm in size and borides Fe of composition FeB were detected. On the side of the surfacing in the contact zone, the structure of the plate eutectic *α-Fe*+ borides *Fe* of composition Fe_2B was formed.

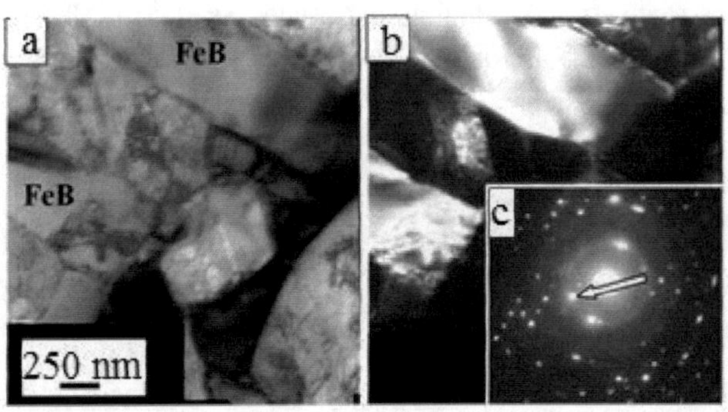

Figure 3. Electron microscopic image of the structure of the deposited layer with iron borides: a – bright field; b – dark field obtained in reflexes [103]FeB + [110]α-Fe; c – microelectronogram (the arrow indicates the reflexes in which the dark field is obtained).

The application of the surfacing technology by flux-cored wire on the lining plates made of Hardox 450 steel makes it possible:

1. To reduce the amount of money spent on protection of the bucket with 33 m^3 capacity from 5 million rubles with a service life 12-18 months to 4.3 million rubles for 21 months.
2. To reduce the time between repair of the excavator, including overhaul and rearmouring by 6 days.
3. To reduce the dependence of enterprises-customers on the supply of imported components.

4. Conclusion

The study of the structure, phase composition, defective substructure, mechanical and tribological properties of coatings deposited on a low-carbon steel by electric arc method in different modes in one and two passes by a powdered boron-containing wire was performed by the methods used in the modern physical material science.

The surfacing created by *Fe-C-Ni-B* wire forms a high-strength layer of thickness ≈7 mm with a microhardness 10.5-12.5 GPa at a single pass and thickness 10 mm with a microhardness ≈15 GPa at a double pass, that is 2 and 3 times, respectively, higher than the microhardness of the base metal. The wear resistance of the deposited layer is two times higher than the wear resistance of the original steel, and the coefficient of friction is 2.2 times lower.

When surfacing in one pass by the wire, the eutectic of a plate-type containing submicron borides of composition Fe_2B is formed, and for the double deposited layer – *FeB*. The formation of the hardened ultra-fine martensitic structure of α-phase, a high scalar dislocation density ($\sim10^{11}$ cm^{-2}), and the presence of a large number of flexural extinction contours contribute to further increase in mechanical and tribological properties.

The results of the work have been implemented at mining enterprises to protect buckets of excavators from abrasive wear.

Acknowledgements

The study was carried out with the financial support of the Ministry of Education and Science of the Russian Federation within the framework of the state task No. 3.1283.2017/4.6 and the Russian Foundation for Basic Research No. 18-48-420001 p_a.

References

[1] Kapralov E V 2014 *Structure and Properties of Composite Wear-Resistant Weld Deposit on Steel* (Novokuznetsk: SibSIU) p 109

[2] Kapralov E V, Raikov S V, Budovskikh E A, Gromov V E, Vaschuk E S and Ivanov Yu F 2014 *Bulletin of the Russian Academy of Sciences. Physics* vol 78 **10** 1015–1021

[3] Gromov V E, Kapralov E V, Raikov S I, Ivanov Yu F and Budovskikh E A 2014 *Progress in the Physics of Metals* **15** 211–232

[4] Utevsky L M 1973 *Diffraction Electron Microscopy in Metallurgy* (Moscow: Metallurgy) 584

[5] Andrews K, Dyson D and Kioun S 1971 *Electronograms and Their Interpretation* (Moscow: Mir) p 256

[6] *Practical Methods in Electron Microscopy ed M Audrey Gloucher* 1980 (Leningrad: Mechanical Engineering) p 375

[7] Smirnova A V, Kokorin G A and Polonskaya S M et al 1985 *Electron Microscopy in Metal Science: Reference* (Moscow: Metallurgy) p 192

[8] Brandon D and Kaplan U 2006 *Microstructure of Materials. Methods of Research and Control* (Moscow: Technosphere) p 384

[9] Krishtal M M, Yasnikov I S, Polunin V I et al 2009 *Scanning Electron Microscopy and X-ray Spectroscopic Microanalysis in Examples of Practical Applications* (Moscow: Technosphere) p 208

[10] Kurdyumov V G, Utevskii L M and Entin R I 1977 *Transformations in Iron and Steel* (Moscow: Nauka) p 236

[11] Ivanov Yu F, Kornet E V, Kozlov E V and Gromov V E 2010 *Tempered Structural Steel: Structure and Mechanisms of Hardening* (Novokuznetsk: SibSIU) p 174

Electroexplosive electrical erosion resistant coatings of the Ag-W system used for electrical contacts of power mine equipment

D A Romanov, S V Moskovsky, V E Gromov and K V Sosnin

Siberian State Industrial University, 42 Kirova street, 654007, Novokuznetsk, Russia

E-mail: romanov_da@physics.sibsiu.ru

Abstract. In this paper the coatings of the Ag-W system obtained by electroexplosive coating on copper electrical contacts of power mine equipment were studied. The modes of deposition and sample weights of silver and tungsten powders were chosen, metallographic research by the methods of light and atomic force microscopy, microhardness tests were carried out. The results of the tests show that the copper electrical contacts of KPV-604 controller have increased values of microhardness and are characterized by a submicrocrystalline structure based on tungsten and silver. The optimum mode of electroexplosive deposition is established at a voltage value 2.5 kV. In this mode, the coating layer has the greatest average microhardness value, in comparison with other investigated regimes. This value is 457.5 ± 55.2 HV, which is 3.8 times higher than the average value of microhardness in the copper substrate. The analysis of the thin section roughness showed that the average roughness value of the coating is greater than the average roughness values of the substrate and the layer with the altered state by 50.015 nm and 22.849 nm, respectively. The coating of the Ag-W system on the copper contacts improves their mechanical and physical properties. The studied treatment modes make it possible to increase significantly the microhardness of the surface contact layer, as well as to increase their electroerosion resistance due to the presence of tungsten particles in the coating, and to maintain the necessary electrical conductivity, facilitated by the presence of silver.

1. Introduction

Many metal contacts are exposed to electrical erosion, that is, the destruction of a conductive material under the influence of electrical discharges. It is well known that the magnitude of erosion failure decreases with the increase in the melting temperature of the material [1]. The application of coatings from refractory materials on contacts makes it possible to protect them from electrical erosion alongside with the preservation of electrical conductivity, thereby extend the products life cycle [2-4].

At present, scientists use different methods of coating by concentrated energy flows: deposition of galvanic coatings, gas-thermal method, method of electron-beam evaporation–condensation in vacuum, method of applying protective coatings by plasma spraying, ion-plasma coating methods, etc.

Electroexplosive deposition plays an important role in the modern development of methods for the formation of protective coatings. Electroexplosive deposition is a method of applying hardening coatings from electric explosion products of foil and powder samples to the materials surface. This method makes it possible to obtain a modified material with increased strength, durometric and tribological properties on a metallic substrate, as it was noted in [5-10]. Hardening is achieved due to the formation of coatings with the formation of finely dispersed phases in a viscous metal matrix [5].

Thus, the aim of this paper is to study coatings of the Ag-W system obtained by means of electrospray deposition on copper contacts.

2. Materials and methods of research

In this study, composite coatings of the Ag-W system were applied to a copper substrate (KPV-604 contacts) by the method of electroexplosive deposition on the electric explosive device EVU 60/10 (Siberian State Industrial University, Novokuznetsk). The electric explosion unit has the following plasma parameters formed during the electrical explosion of a silver foil with a sample weight of tungsten powder: the plasma action time is ~ 100 μs, the absorbed power density at the jet axis is ~ 8.2 GW/m^2, the pressure in the shock-compressed layer near the surface is ~ 18.8 MPa.

An end scheme of explosion was applied [5], in order to increase the intensity of the thermal action on the material surface prior to its melting and to provide the conditions for the deposition. A silver foil with weight 0.2463 g was clamped between coaxial electrodes, to which a regulated voltage was applied via a vacuum discharger. The silver foil was filled with a tungsten powder with weight of 0.434 g. The coatings are applied in the conditions of thermal action, which causes the heating of the substrate surface up to the melting point under in three energy modes, different in the values of the absorbing power density at the coaxial electrodes: U1 = 2.4 kV, U2 = 2.5 kV, U3 = 2.6 kV. When the capacitive accumulator is discharged, the peripheral region of the foil adjacent to the external electrode-nozzle becomes the source of the condensed phase of the explosion products, and the foil section above the central electrode, where the tungsten powder is filled, acts as a source of ionized vapour [10].

After the electric-explosive spraying, the samples obtained under different conditions were examined for microhardness by the Vickers method (microhardness meter HVS-1000A). The load was constant for all modes and was 0.05 HV.

The study of the structure of straight thin sections was carried out on the metallographic microscope Olympus GX-51. The study of the coating porosity and the zone of thermal effect of the samples was performed on the atomic force microscope NT-MDT Solver “NEXT”.

With the help of Image Analysis 3.5 software built onto the software interface of the atomic force microscope, the analysis of the coating roughness, the layer with the changed state and the substrate material (copper) was performed, the distribution of the surface relief height of the straight thin section depending on the distance was studied.

3. Results and discussion

When analyzing the structure on the metallographic cross-sectional microscope of all three samples, the formation of a multilayer structure is revealed, which consists of a low-porosity coating, slightly varying thickness, a liquid-phase alloying layer, and a heat-effected layer (figure 1). The coating thickness equals to 49.04 ± 0.7 μm for mode 1, 68.5 ± 0.9 μm for mode 2 and 61.26 ± 0.6 μm for mode 3. Coating thickness is measured using the vertical secant method.

Coatings in modes 2 and 3 were more uniform in width than in mode 1, which can be explained by the higher temperature of the jet and, therefore, the rate of coating diffusion to cooling was higher.

It can be seen from figure 1 that in the first treatment mode, pores of 3 to 30 μm in size are present in the coating. In treatment mode 2, the pore size decreases compared to mode 1. The average pore size in the second mode is 16 μm. In the third treatment mode, the average pore size is 8 μm. Thus, as the absorbing power density increases, the average pore size in the coating of the Ag-W system decreases.

The average value of the layer thickness with the changed state between the coating and the substrate is 14 μm for mode 1, 18.5 μm for mode 2 and 20 μm for mode 3. The width of the layer of the changed state between the substrate and the coating increases as the values of the absorbing power density on the coaxial electrodes increases as well.

The analysis of the transition layer between the coating and the substrate (figure 2) showed that the boundary is not even. A zone of mutual mixing of the coating with the substrate is formed.

Analyzing the data of the comparative histogram (figure 3), it can be concluded that the processing mode 2 has the maximum average microhardness of the coating layer, in comparison with other investigated modes. It is 457.5 ± 55.2. In the substrate the microhardness values are smaller than in the coating layer and are 119.4 ± 2.5 HV and 122.0 ± 3.3 HV at a distance of 5 and 40 μm from the coating, respectively.

It is also important to note that the average microhardness at a distance of 5 μm from the coating for all treatment modes is less than the average microhardness at 40 μm from the coating. It is possible to make an assumption that the reason for this is the heat effect, which is realized when the coating of the Ag-W system is applied by the electroexplosive deposition method [11].

The final stage of the complex study of the regularities in the formation of electroexplosive electroerosion-resistant coatings of the Ag-W system was the atomic force microscopy of the coating obtained in the optimal exposure mode (mode 2). The topography of the sample direct thin section, treated in accordance with mode 2, is shown in figure 4.

Figure 1. The structure of the cross-sections of the electroexplosive coatings of the Ag-W-Cu system obtained in various treatment modes (a – mode 1, b – mode 2, c – mode 3).

The atomic-force image made it possible to establish that the application of coatings of the Ag-W system on the copper contact by the method of electroexplosive deposition in the optimal treatment mode leads to the formation of a structure consisting of a coating, a layer of changed state and a substrate material. The average thickness value of the layer with the changed state located between the coating and the substrate is 16 μm, which correlates with the data obtained in the metallographic analysis.

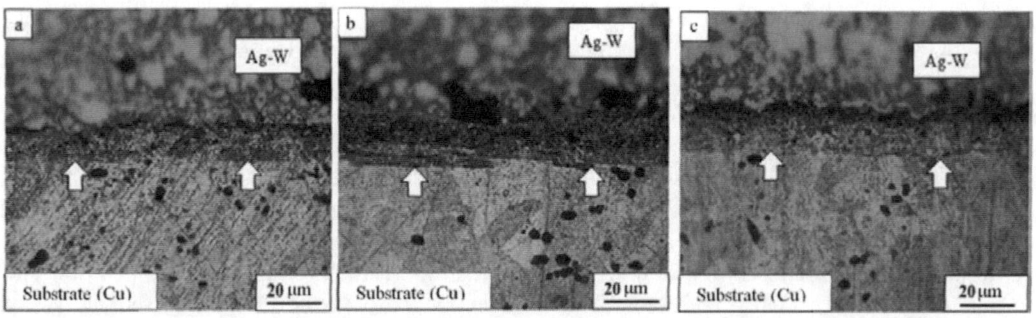

Figure 2. The structure of electroexplosive coatings cross-sections of the Ag-W-Cu system at the coating boundary with the substrate, obtained in different treatment modes (a – mode 1, b –mode 2, c – mode 3). The arrows indicate the areas of mutual mixing of the coating with the substrate.

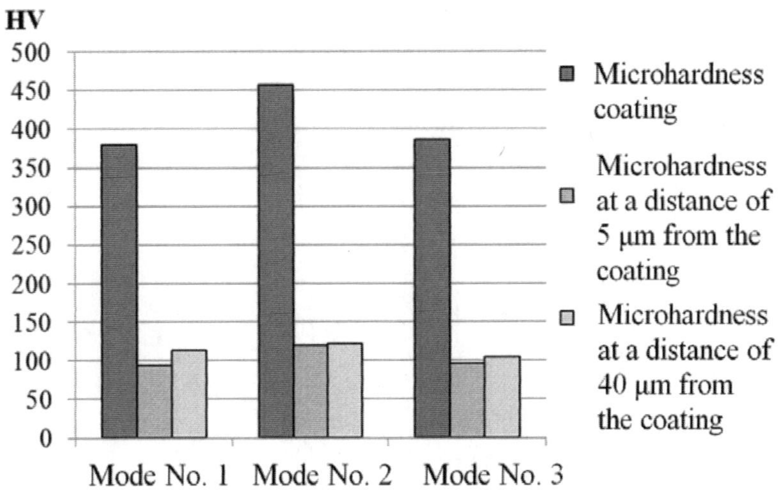

Figure 3. The comparative histogram of coating microhardness of the Ag-W system obtained in different modes of KPV-604 copper contacts treatment.

The analysis of the presented in figure 5 distribution graph of the surface height Z from the distance X (the original image is given in figure 4) shows that the height of the coating profile is higher than the height of the substrate, as well as the layer with the changed state. The greatest value of relief height is achieved at 40 microns (figure 4). The average height at this point is 661.5 nm – this point corresponds to the boundary of transition from the layer with the changed state to the coating layer.

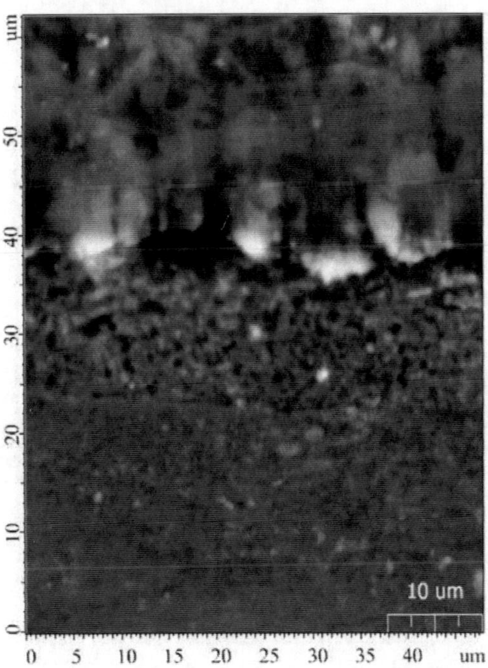

Figure 4. Atomic force microscopy of the surface profile of the Ag-W coating system deposited on the copper contact (mode 2).

The study of the coating roughness, the layer with the changed state, and the substrate material (copper) in mode 2 revealed the following roughness values:

1. The average substrate roughness is 25.397 nm.
2. The average roughness value of the layer with the changed state is 52.563 nm.
3. The average coating roughness is 75.412 nm.

Thus, the analysis of the atomic force image showed that the average roughness value of the coating is larger than the average roughness values of the substrate and the layer with the changed state by 50.015 and 22.849 nm, respectively.

The study of the distribution of the surface relief height depending on the distance is shown in figure 5.

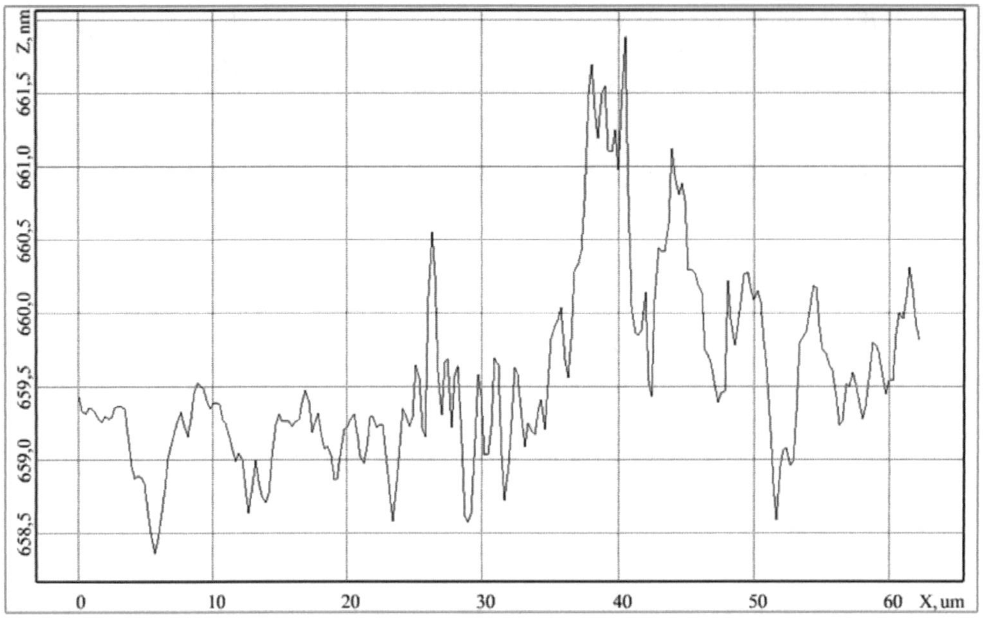

Figure 5. Distribution of the surface relief height depending on the distance (mode 2).

4. Conclusions

In the present research devoted to the investigation of electroexplosive coatings of the Ag-W system formed on copper contacts KPV-604, the modes of deposition and the sample weights of Ag-W powder were chosen, coatings were applied by electroexplosive deposition on copper contacts in various modes, metallographic studies for microhardness were carried out and atomic force microscopy of the coatings obtained.

The results of microindentation made it possible to reveal the optimum deposition mode (U2 = 2.5 kV), in which the coating layer has the greatest average microhardness value in comparison with other investigated modes. This value is 457.5 ± 55.2 HV, which is 3.8 times higher than the average value of microhardness in the copper substrate.

The analysis of the thin section roughness in Mode 2 showed that the average roughness value of the coating is greater than the average roughness values of the substrate and the layer with the changed state by 50.015 nm and 22.849 nm, respectively.

Thus, coating of the Ag-W system on copper contacts improves their mechanical and physical properties. The treatment modes studied can significantly increase the microhardness of the surface contact layer, as well as increase their electroerosion resistance due to the presence of tungsten particles in the coating, and also maintain the necessary electrical conductivity due to the presence of silver particles.

The results of the study show that coatings of the Ag-W system obtained on copper contacts can be recommended for use in design developments aimed at reduction of contacts electroerosion and extension of their service life.

Acknowledgements

The research was carried out with the financial support of the Russian Foundation for Basic Research in the framework of scientific projects No. 16-32-60032 mol_a_dk, 18-48-420001 r_a, 18-32-00075 mol_a, Grant of the President of the Russian Federation for state support of young Russian scientists – candidates of sciences MK-1118.2017.2 and the state task 3.1283.2017/4.6.

References

[1] Urvantsov L A 1966 *Erosion and Protection of Metals* (Moscow: Mechanical Engineering)
[2] Lobanov N F and Kamensky M N 2014 *Proceedings of Tula State University. Technical Sciences* **2** 62–67
[3] Safonov A L and Safonov L I 2007 *Technologies in the Electronic Industry* **6(18)** 54–58
[4] Safonov A L and Safonov L I 2009 *Technologies in the Electronic Industry* **3(31)** 48–54
[5] Vashchuk E S, Romanov D A, Budovskikh E A and Ivanov Y F 2011 *Steel in Translation* **6** 469–474
[6] Romanov D A, Budovskikh E A and Gromov V E 2011 *Journal of Surface Investigation: X-ray, Synchrotron and Neutron Ttechniques* **6** 1112–17
[7] Bagautdinov A Ya 2007 *Physical Fundamentals of Electrospray Alloying of Metals and Alloys* (Novokuznetsk: SibGIU)
[8] Vashchuk E S 2011 *Steel in Translation* **41(6)** 469–474
[9] Ivanov Y F et al 2012 *Journal of Surface Investigation: X-ray, Synchrotron and Neutron Techniques* **6** 499–501
[10] Granovskii A Y, Sarychev V D and Gromov V E 2013 *Tech. Phys.* **58(10)** 1544–47

IOP Conf. Series: Earth and Environmental Science **206** (2018) 012031 doi:10.1088/1755-1315/206/1/012031

Method of researching the influence of isothermic annealing modes on welding of hauling track rails used in mine workings

R A Shevchenko, N A Kozyrev, A I Kutsenko, A A Usoltsev and A A Kutsenko

Siberian State Industrial University, 42 Kirova street, Novokuznetsk, 654006, Russia

E-mail: kozyrev_na@mtsp.sibsiu.ru

Abstract. Modern problems of butt welding of hauling tracks in mine workings are considered. The article describes the method for studying the influence of isothermal annealing modes of a welded joint of samples from rail steel. The cutoff layout of samples used for experiments and the equipment are provided. The developed method allows the temperature to be measured in the weld metal and the heat-affected zone in the process of welding the samples from rail steel. Using the developed method it is proposed to study the effect of thermal cycles of butt welding on the structure of rail steel. The result of using this method is shown as an example.

1. Introduction

When constructing and repairing continuous welded railroad tracks, electrocontact and aluminothermic welding of rails is used. In Russia, the greater part of rails is welded by electrocontact method [1, 2]. Welding by the method of continuous flash welding with preheating was changed by the method of pulsed flash welding.

This method of contact welding is the most economical and technological in comparison with continuous flashing. During the contact welding of rails, as well as using other types of welding, heating and continuous cooling of metal in the heat-affected zone (HAZ) occurs. Depending on the chemical composition of the steel, the welding process is selected using the existing methods of flushing – continuous or pulsed that determine the linear magnitude and temperature fields in the welded joint HAZ [3, 4]. The choice of the thermal mode is based on the exclusion of the hardening structures formation (martensite and bainite), causing additional stresses and cracks, which lead to the destruction of rails [5]. Thus [6, 7], the development of modes for rails of high-speed railways made of chromium steel is of particular importance.

It is known that the increase in the chromium content in steel causes a shift to the right of the C-shaped curves of the austenite decay beginning and end on the diagram of austenite isothermal decay, which in turn leads to a decrease in the critical cooling rate at which the austenite is converted to martensite [8, 9]. A high-strength layer with a martensite structure is formed during the rapid heating of the welded joint by the method of pulsed flashing and subsequent intensive cooling of HAZ in microvolumes with a high content of chromium, nickel and carbon. Martensite sections [5] play the role of stress concentrators and lead to the formation of defects in welded joints (development of fatigue cracks in the head, web and base of a rail with a brittle fracture).

This problem in the process of production of a seamless welded rail is solved by the mandatory heat treatment of the welded joint. Heat treatment is carried out using induction plants, which in turn increases costs. This disadvantage in practice is suggested to be corrected by combining continuous and pulsed flash methods, changing the heating intensity during welding and controlling the cooling rate [6]. At the same time, partially using the method of continuous flashing, there might be defects of welding nature inherent to continuous reflow.

The alternative way of solving the problem is proposed, which allows any of the considered flashing methods to be used. After upsetting and cooling the welded joint at the time of reaching the required temperature, it is proposed to perform isothermal annealing by passing pulses of alternating electric current through the welded joint [10].

2. Methods of research

This article describes the method used for researching the effect produced by modes of isothermal annealing of samples welded joints from rail steel by passing pulses of alternating electric current through the welded joint. To avoid the formation of hardening structures, the aging temperature is selected based on obtaining a finer dispersed structure of the weld metal. The exposure time is determined by the incubation period of the formation of the required structure and is controlled by the number of current pulses.

3. Results and discussion

Samples with dimensions 90 mm × 30 mm × 10 mm are cut from the rail steel for the test (figure 1), further the contact butt welding is performed in different isothermal tempering modes.

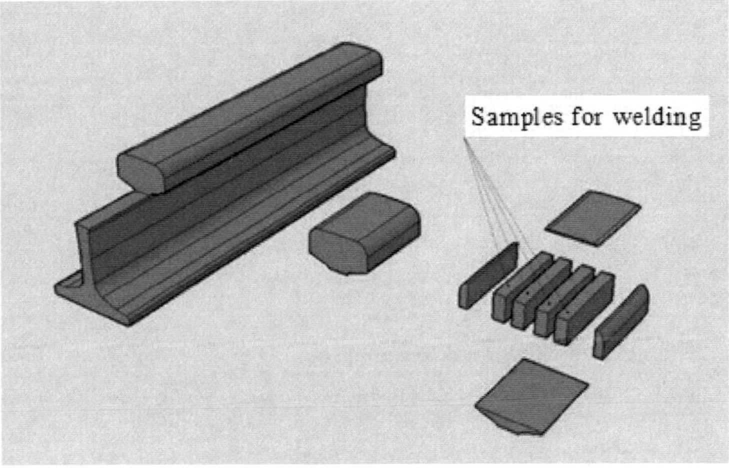

Figure 1. Layout of cutting samples with sizes 90 mm × 30 mm × 10 mm.

Welding of samples is carried out on the modernized machine for contact butt welding MS-2008M (figure 3). The welding modes are controlled by a personal computer.

To obtain information on the thermal effect on the metal structure, a method for measuring the temperatures in the heat-affected zone during welding has been developed. In this case, the welding current is measured for further adjustment of modes. The scheme of equipment interaction is shown in figure 2.

The machine for contact butt welding by flashing operates in the following way:

Before the welding begins, the welded parts are installed and clamped in the movable 2 and fixed 3 clamps, using the clamping devices of the welded parts. If the parts are installed correctly, the operator on the control panel switches on welding. The contactor of the welding current 11 comes into action, which in turn switches on the welding transformer 7. At the same time, the electric motor 12 of the flashing and tempering 12 is turned on. The flashing process begins. As the ends of the welded parts draw together, at the end of flashing and at the beginning of the upsetting, the limit switch is activated used to give a signal to the controller 13, installed on the block of limit switches 10, about the beginning of the heat treatment, the contactor of the welding current turning on switches off. The upsetting ends without current. At the end of the upsetting, the limit switch is activated, which is responsible for the operation of the electric motor 12 of flashing and upsetting driver, and it stops.

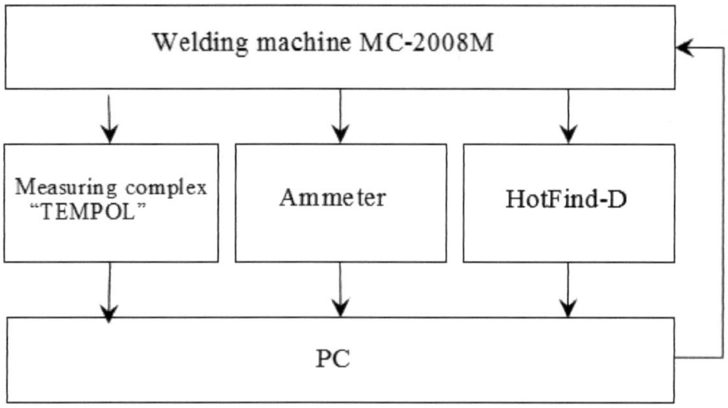

Figure 2. The scheme of equipment interaction.

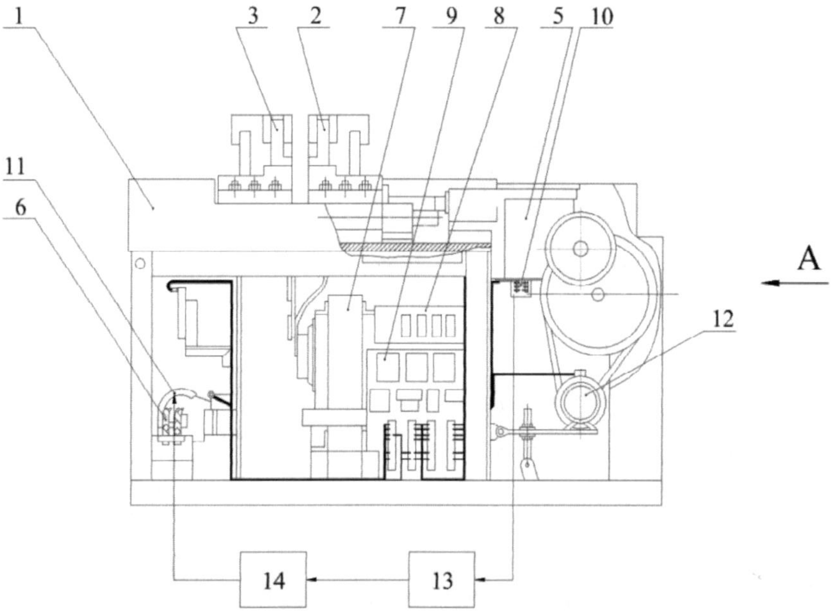

Figure 3. General view of the machine for contact butt welding MS-2008M.

In this case, the controller 13 supplies a voltage in the programmed time interval to the symistor 14 to activate the contact relay, which in its turn switches on the contactor of welding current turning on 11, thus, switching on the welding transformer 7. Through the clamped part in clamps 1 and 2 from the welding transformer 7 flows electric current, as a result of which the welded joint is heated to a certain temperature. At the end of the programmed time, the controller 13 stops sending a signal to the symistor 14, disconnecting the welding transformer 7. The welded joint is cooled to a certain temperature. The switching on of the welding transformer 7 is repeated the required number of times and at the necessary intervals according to the program in the controller 13. Thus, the process of welded parts heat treatment the takes place. At the end of the process, the operator removes the welded product for further cooling in the specially designated area.

IOP Conf. Series: Earth and Environmental Science **206** (2018) 012031 doi:10.1088/1755-1315/206/1/012031

The temperature is measured at points T1, T2, T3, T4. The proposed zone of thermal influence varies in different modes from 15 to 25 mm from the seam center, which corresponds to the maximum distance of the thermocouple placement at point T4. Since it is impossible to measure the temperature in the center of the seam, the temperature is measured by the HotFind-D thermal imager at point T1. The location of the points in which the temperature is measured is shown in figure 4.

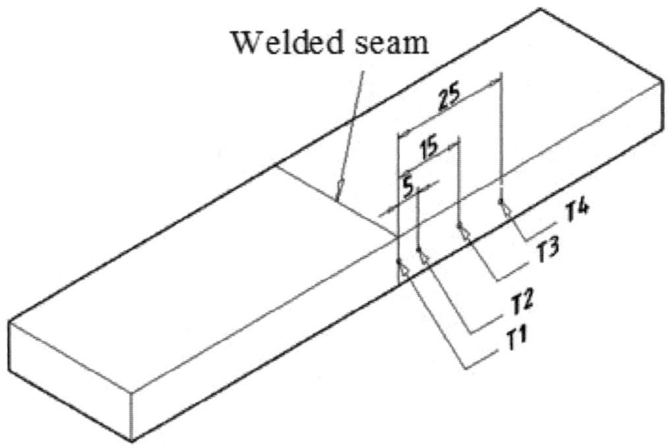

T1 – is measured by the thermal imager, T2, T3, T4 – are measured by thermocouples.

Figure 4. The location of the points in which the temperature is measured.

Thermocouples XA (chromel-alumel) are used for measuring the metal temperature. The standard calibration of thermocouples XA is provided in accordance with the state standards GOST 3044-84 (ST SEV 1059-85). The thermocouples are made of a thermoelectrode wire with diameter 0.1-0.5 mm by brazing (using a protective quartz cap) or by stranding (when measured by a bare junction). To isolate thermoelectrodes from each other quartz straw (diameter 0.5-1.0 mm), two-channel corundum and porcelain tubes (outside diameter 3.5 mm) are used.

The collection and processing of data obtained from thermocouples is carried out with the help of the measuring complex "Tempol". The microprocessor modules used in the complex allow the measurement of the analog signal in the range from 15 mV to 2.5 V to be performed. A PC controls the complex using the developed software operating in MS DOS and WINDOWS 95/98/2000/XP environments. The complex allows the collection, registration and storage of analog signals on the hard disk of the PC to be performed with the possibility of scaling up to 2048 channels. The software of the measuring complex includes two programs written in the Basic programming language for MS DOS and MS Windows operating systems and Microsoft Excel spreadsheet. The software allows the link with the I-7018 module to be set up, which is connected to the "SOM1" serial port of the computer through I-7520 converter, and the measurements from 8 sensors to be record sequentially into the table of a specified format. At the end of the measurement process, the resulting data array is converted into a graphic image depending on the research tasks.

The HotFind-D thermal imager allows temperature measurements up to 1500 ° C to be performed. The thermal imager is equipped with a non-coolable micro-bolometric matrix in the focal plane of the lens with a resolution 160 × 120 pixels. The transfer of the video image of thermograms to the PC is carried out using the video capture card of an analog signal in the NTSC format with a frequency 60 Hz.

Figure 5 shows an example of how to use the method for determining the temperature change in the welding process.

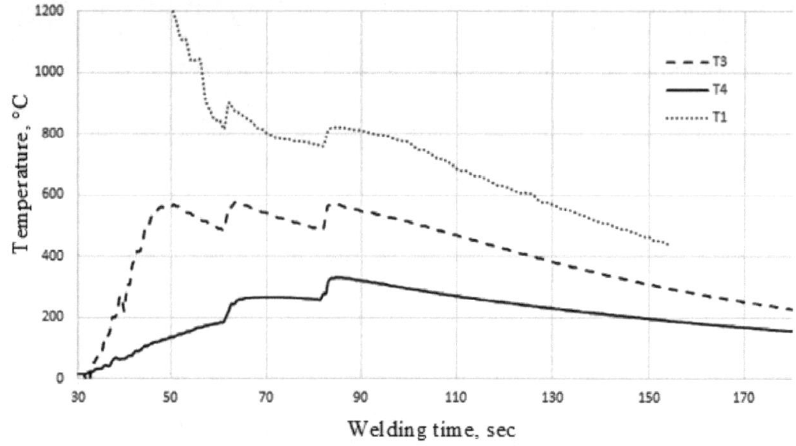

Figure 5. The example of the measured thermal cycle of welding samples.

4. Conclusion

The proposed method allows the thermal cycles of contact butt welding of rail steel samples to be measured on the machine MS-2008M, and the cooling rate of weld metal and the zone of thermal influence to be calculated, also the resulting structure of steel with a certain chemical composition to be predicted. The measurement of the welding current makes it possible to correct the heat input during welding.

References

[1] Kozyrev N A et al 2017 *IOP Conference Series: Materials Science and Engineering* **253** 012002
[2] Kuchuk-Yatsenko S I et al 2008 *Welding International* vol 22 issue 5 338–341
[3] Mitsuru F et al 2015 *JFE Technical Report* **20** 159–163
[4] Saita K et al 2013 *Nippon steel & Sumitomo Metal Technical Report* **105** 84–92
[5] Shur E A 2006 *Effect of Structure on the Operational Stability of Rails* (Ekaterinburg: UIM) pp 37–63
[6] Shur E A and Rezanov V A 2012 *Bulletin of VNIIZhT* **3** 20–22
[7] Gavrilov V V 2012 *Metro and Tunnels* **1** 42–43
[8] Gulyaev A P and Gulyaev A A 2012 *Metal Science* (Moscow: Alliance) p 644
[9] Shevchenko R E 2016 *Proc. of the All-Russian Sci. Conf. of Students, Graduate Students and Young Scientists in Natural and Technical Sciences* (Novokuznetsk: SibSIU) pp 196–198
[10] Shevchenko R A et al 2017 *Proc. of XX Int. Sci. and Pract. Conf. in Metallurgy: Technologies, Innovations, Quality* (Novokuznetsk: SibSIU) pp 332–336

Use of silicomanganese slag and ladle electric steelmaking slag in manufacturing the welding fluxes for surfacing the mining equipment

N A Kozyrev, A R Mikhno, A A Usoltsev, R E Kryukov and A A Umanskiy

Siberian State Industrial University, 42 Kirova street, Novokuznetsk, 654007, Russia

E-mail: kozyrev_na@mtsp.sibsiu.ru

Abstract. The possibility of manufacturing a welding flux based on silicomanganese slag and ladle electric steelmaking slag is discussed. Silicomanganese slag and ladle electric steelmaking slag were used in the experiments. Flux additive was introduced in a ratio of 5, 10, 20, 30, 50% to the silicomanganese slag. The possibility in principal of using technogenic wastes of metallurgical production is shown: slag from silicomanganese production and flux additives based on ladle electric steelmaking slag as surfacing flux. It is determined that in connection with the low concentration of iron oxide and increased basicity, it is possible to use this material for alloyed metal deposition. The analysis of the contamination level with nonmetallic inclusions and an increase in the sulfur content in the welded metal showed that without quality deterioration it is possible to introduce a flux-additive based on the ladle slag of steelmaking production in the amount not exceeding 20%. Increase in the flux additive in the amount not more than 20% worsens the welding quality, the number of non-metallic inclusions increases, as well as the sulfur content. When increasing the amount of the flux additive over 30%, the appearance of the welded layer deteriorates.

1. Introduction

The important and promising direction in welding production is the development and use of new welding fluxes based on the waste of metallurgical production. When surfacing the fluxes with an increased content of basic oxides and a low FeO content are used [1-10].

In recent years, a number of studies using technogenic wastes has been carried out. A number of studies are devoted to improving the composition of welding fluxes for mining equipment using new flux additives based on technogenic metallurgical wastes (slags and sludges) [11-17]. In particular, new welding fluxes produced using slag from silicomanganese production have been proposed [18, 19]. The use of technogenic metallurgical wastes as initial components for the production of welding fluxes can significantly reduce the cost of production.

2. Methods of research

The chemical composition of the considered welded samples was determined according to state standards GOST 10543-98 by the X-ray fluorescence method on XRF-1800 spectrometer and by the atomic-emission method using DFS-71 spectrometer. Investigation of longitudinal samples of the deposited layer for the presence of nonmetallic inclusions was performed in accordance with GOST 1778-70 using the optical microscope OLYMPUS GX-51 in a bright field in the range of magnifications × 100.

IOP Conf. Series: Earth and Environmental Science **206** (2018) 012032 doi:10.1088/1755-1315/206/1/012032

3. Results and discussion

In this paper, we considered the possibility of using silicomanganese slag produced by the West Siberian Metallurgical Plant with a chemical composition, wt%: Al_2O_3 6.91-9.62%, CaO 22.85-31.70 %, SiO_2 46.46-48.16 %, FeO 0.27-0.81 %, MgO 6.48-7.92 %, MnO 8.01-8.43%, F 0.28-0.76%, Na_2O 0.26-0.36% K_2O до 0.6 2 %, S 0.15-0.17 %, P 0.01 %. With ladle slag from electric steelmaking production of rail steel by JSC "EVRAZ ZSMK" with the chemical composition, wt%: 1.31% FeO, 0.22% MnO, 36.19% CaO, 36.26% SiO_2, 6.17% Al_2O_3, 11.30% MgO, 0.28% Na_2O, 0% K_2O, 3.34% F, <0.12% C, 1.26% S, 0.02% P.

Slag from silicomanganese production of fraction 0.45-2.5 mm was mixed with a flux-additive from ladle electric steel-smelting slag and a mixture with liquid glass.

The production of flux additives was carried out in the following way: a ladle electric steelmaking slag with fraction less than 0.2 mm was mixed with a liquid glass in a ratio of 62% and 38%, respectively. After this, the mixture was kept at room temperature for 24 hours, followed by drying in the oven at 350 °C, then the mixture was cooled, crushed and sieved with a fraction 0.45-2.5 mm.

In the experiments a flux additive with a chemical composition was used, wt. %: FeO – 0.84; MnO – 0.06; CaO – 35.47; Al_2O_3 – 4.71; MgO – 5.01; Na_2O – 1.92; K_2O – 0.11; S – 0.98; P – 0.013; Cr_2O_3 – 0.04; TiO – 0.33. The flux-additive was mixed with silicomanganese slag in various ratios (table 1).

The study of the possibility of using different compositions of welding fluxes was carried out in laboratory conditions on the welding tractor ASAW-1250 in the surfacing mode: I = 680 A; U = 28 V; V = 28 m/h. Submerged-arc welding was carried out on samples 300 × 150 mm of 20 mm thickness from sheet steel grade 09G2S using wire Sv-08GA with diameter 4 mm.

Table 1. Ratio of silicomanganese slag and flux-additive, %.

Flux marking	Silicomanganese slag (SiMn), wt %	Flux-additive, wt %
90	100	0
91	95	5
92	90	10
93	80	20
94	70	30
95	50	50

The appearance of the welded beads is satisfactory for surfacing with a flux additive content up to 30%. The chemical compositions of fluxes, slag crusts, and weld metal are given in tables 2-4, respectively. The metallographic study was carried out on micro-sections without etching by OLYMPUS GX-51 optical microscope with a magnification × 100. The type of nonmetallic inclusions in the welded bead zone is shown in figure 1, the evaluation of nonmetallic inclusions, carried out in accordance with GOST 1778-70, is given in table 5.

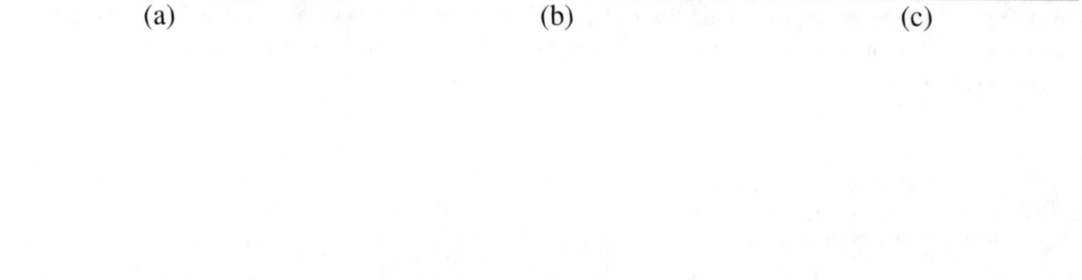

(a) (b) (c)

(d) (e) (f)

Samples: a) 90; b) 91; c) 92; d) 93; e) 94; f) 95

Figure 1. Non-metallic inclusions in the zone of welded samples.

Table 2. Chemical composition of the flux.

Flux	Mass fraction of elements, %						
	FeO	MnO	CaO	SiO_2	Al_2O_3	MgO	Na_2O
90	0.50	7.97	31.34	46.09	6.61	5.74	0.40
91	0.38	7.59	32.71	45.04	6.60	4.34	1.07
92	0.42	7.28	33.12	45.10	6.33	4.52	1.40
93	0.42	6.90	32.06	46.20	6.85	4.03	1.40
94	0.57	4.62	34.03	43.80	5.09	4.58	3.28
95	0.74	2.58	35.64	43.33	5.19	4.92	3.99

Continuation of **table 2.**

Flux	Mass fraction of elements, %						
	K_2O	S	P	ZnO	Cr_2O_3	F	TiO_2
90	0.01	0.33	0.011	0.004	0.05	0.45	0.07
91	0.081	0.28	0.012	0.003	0.031	0.77	0.10
92	0.082	0.32	0.011	0.003	0.029	0.87	0.11
93	0.082	0.34	0.011	0.003	0.024	0.89	0.11
94	0.14	0.54	0.012	0.004	0.036	1.63	0.17
95	0.17	0.73	0.013	0.005	0.041	1.99	0.21

Table 3. The chemical composition of slag crusts.

Flux	Mass fraction of elements, %						
	FeO	MnO	CaO	SiO_2	Al_2O_3	MgO	Na_2O
90	1.69	7.78	32.35	42.50	6.59	5.55	0.30
91	1.54	7.61	32.42	44.28	7.87	4.28	0.59
92	1.62	7.04	32.47	43.39	6.63	4.24	0.89
93	1.78	6.36	33.10	43.13	7.23	4.38	1.19
94	1.69	5.34	33.28	44.55	6.05	4.64	2.05
95	1.66	3.90	34.48	44.61	5.59	4.87	2.78

Continuation of **table 3.**

Flux	Mass fraction of elements, %						
	K_2O	S	P	ZnO	Cr_2O_3	F	TiO_2
90	0.01	0.21	0.011	0.012	0.04	0.37	0.07
91	0.001	0.16	0.011	0.003	0.033	0.53	0.094
92	0.081	0.20	0.011	0.004	0.033	0.68	0.11
93	0.088	0.23	0.012	0.004	0.034	0.83	0.12
94	0.11	0.33	0.011	0.005	0.032	1.24	0.15
95	0.14	0.46	0.012	0.005	0.036	1.66	0.17

IOP Conf. Series: Earth and Environmental Science **206** (2018) 012032 doi:10.1088/1755-1315/206/1/012032

Table 4. The chemical composition of the welded beads.

Flux	Mass fraction of elements, %						
	C	Si	Mn	Cr	Ni	Cu	Ti
90	0.07	0.43	1.16	0.05	0.11	0.14	-
91	0.04	0.44	1.30	0.04	0.08	0.13	0.001
92	0.04	0.42	1.30	0.04	0.09	0.13	0.001
93	0.04	0.41	1.22	0.04	0.09	0.13	-
94	0.04	0.38	1.25	0.04	0.08	0.13	-
95	0.04	0.35	1.22	0.04	0.09	0.12	-

Continuation of **table 4.**

Flux	Mass fraction of elements, %					
	V	Mo	Al	Nb	S	P
90	0.007	0.021	-	0.003	0.019	0.012
91	0.002	0.01	0.002	0.010	0.019	0.012
92	0.003	0.01	-	0.010	0.024	0.011
93	0.001	0.01	-	0.11	0.031	0.010
94	0.002	0.01	-	0.011	0.033	0.009
95	-	0.01	-	0.011	0.036	0.008

Table 5. Non-metallic inclusions in the zone of welded seams.

Sample	Non-metallic inclusions, point	
	nondeformable silicates	spot oxides
90	1b.2b	1a
91	2b.1b	2a.1a
92	1b.2b	1a.2a
93	1b. rare 2b	1a
94	1b. rare 2b	1a
95	2b.3b. rare 4b	1a.2a

Thus, due to the low concentration of iron oxide and increased basicity it is possible to use this material for depositing alloyed metal. The analysis of the contamination level with nonmetallic inclusions and an increase in the sulfur content in the welded metal showed that it is possible to introduce a flux additive based on the ladle slag of steelmaking production without quality deterioration in the amount not exceeding 20%.

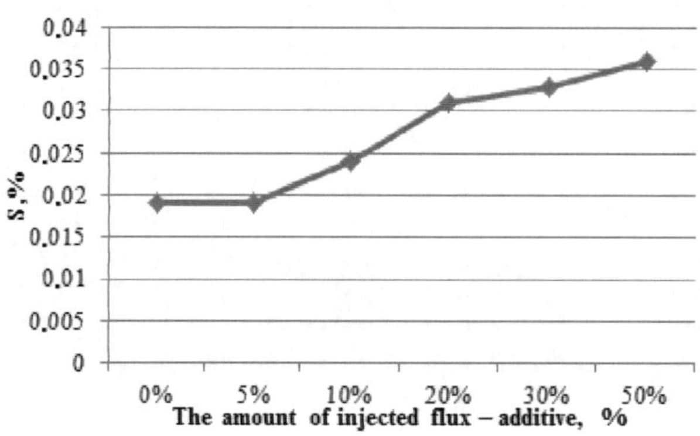

Figure 2. The effect of the additive introduced into the deposited layer on the sulfur concentration.

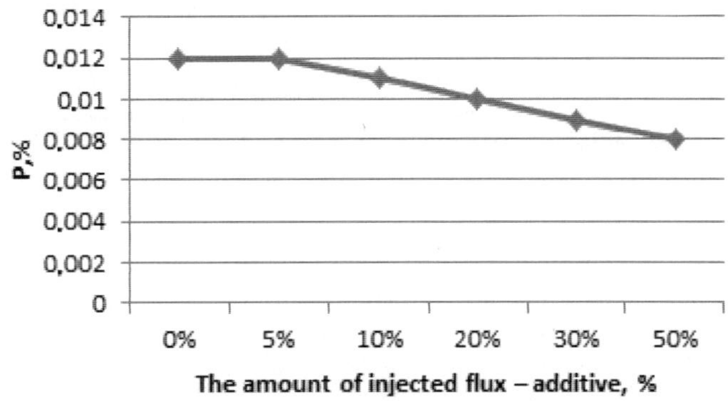

Figure 3. The effect of the additive introduced into the deposited layer on the phosphorus concentration.

4. Conclusions

1) The possibility in principal of using technogenic wastes from metallurgical production is shown: silicomanganese slag and flux additives based on ladle electric steelmaking slag as surfacing flux for the mining industry.

2) When exceeding the amount of the introduced flux additive over 30% the appearance of the welded layer deteriorates.

3) The introduction of the flux additive increases the sulfur content and reduces the phosphorus content in the deposited layer.

4) When the amount of the flux-additive exceeds more than 20%, the quality of the weld metal deteriorates, the number of non-metallic inclusions increases, as well as the sulfur content.

References

[1] Amado Cruz Crespo et al 2009 *Welding International* vol 23 **2** 120–131
[2] Golovko V V and Potapov N N 2011 *Welding International* vol 25 **11** 889–893
[3] Rafael Quintana Pucho et al 2009 *Welding International* vol 23 **2** 132–140
[4] Crespo A C et al 2007 *Welding International* vol 21 **7** 502–511
[5] Volobuev Yu S, Volobuev O S, Parkhomenko A G, Dobrozhela E I and Klimenchuk O S 2012 *Welding International* vol 26 **8** 649–653
[6] Volobuev Yu S, Surkov A V, Volobuev O S, Kipiani P N, Shestov D V, Pavlov N V and Savchenko A I 2010 *Welding International* vol 24 **4** 298–300
[7] Potapov N N and Kurlanov S A 1987 *Welding International* vol 1 **10** 951–954
[8] P L Babushkin P L and Persits V Yu 1991 *Welding International* vol 5 **9** 741–742
[9] Podgaetsky V V and Rabkin D M 1954 *Fluxes for Automatic and Semiautomatic Welding* (Kiev: Academy of Sciences of the Ukrainian SSR) p 56
[10] Potapov N N and Konishchev S A 1989 *Welding Materials for Arc Welding: A Reference book* vol 1 (Moscow: Mechanical Engineering) p 544
[11] Pavlov P N 1988 *Fundamentals of Engineering Calculation of Machine Elements for Fatigue and Durability* (Leningrad: Mashinostroyeniye) p 209
[12] Tolstov I A and Korotkov V A 1990 *Handbook on Surfacing* (Chelyabinsk: Metallurgiya) p 384
[13] Tyilkin M A 1971 *Improving the Durability of Parts of Metallurgical Equipment* (Moscow: Metallurgiya) p 610
[14] Shekhter S Ya and Shvartser S Ya 1981 *Surfacing of Metallurgical Equipment Parts: A Reference Book* (Moscow: Metallurgiya) p 160
[15] Hazin M L 2006 *Theory and Methods of Restoration of Mining Pachines Parts* (Ekaterinburg: UGGU) p 319
[16] Kobernik N V and Orlik A G 2010 *Proceedings of MSTU n.a. N.E. Bauman* pp 49–52

[17] Chernyshov G G, Kobernik N V, Orlik A G and Chernyshova T A 2011 *Physics and Chemistry of Materials Processing* **5** 44–50
[18] Kryukov R E et al 2017 *IOP Conf. Series: Materials Science and Engineering* **253**
[19] Kozyrev N A et al 2017 *Ferrous Metallurgy. Bulletin of Scientific, Technical and Economic Information* vol 5(1409) 85–89

IOP Conf. Series: Earth and Environmental Science **206** (2018) 012033 doi:10.1088/1755-1315/206/1/012033

Use of barium – strontium modifier in manufacturing welding flux based on silicomanganese slag for welding and surfacing mining equipment

N A Kozyrev, A R Mikhno, R E Kryukov, N F Yakushevich and A A Provodova

Siberian State Industrial University, 42 Kirova street, Novokuznetsk, 654007, Russia

E-mail: kozyrev_na@mtsp.sibsiu.ru

Abstract. The paper discusses the possibility in principle of barium-strontium modifier application as a refining and gas-protective additive for welding fluxes based on silicomanganese slag for surfacing mining equipment. In the experiments the barium-strontium flux additive was used, prepared in two ways: barium-strontium modifier mixed with liquid glass and the dust of barium-strontium modifier with a fraction less than 0.2 mm. The additives were mixed at a ratio of 2-10% from weight of silicomanganese slag. It is shown that the use of a mixture of barium-strontium modifier with liquid glass as an additive is more preferable to the use of additive as a dust. It was established that the use of not more than 8% of a barium-strontium additive is optimal from the point of view of metal contamination with nonmetallic inclusions.

1. Introduction

The development of materials and the use of recovery technology that significantly increase the wear resistance of the working surfaces of machines and mechanisms of mining equipment exposed to abrasive and impact wear during operation is an important task. The most promising methods of recovery include the use of submerged-arc welding of the parts wear surfaces.

To restore the mining equipment, welding fluxes such as ANF-6-1, AN-348, AN-60, AN-20, AN-28 are used. To reduce the costs for restoration of mining equipment, the application of a new welding flux based on wastes from metallurgical production such as silicomanganese slag and flux – additives containing barium and strontium modifier is suggested [1-18]. The studies on the introduction of a barium-strontium modifier additive into welding fluxes based on silicomanganese slag, as well as the quality of the weld beads produced, were studied previously [19-20], the further studies are presented in this paper.

2. Methods of research

The chemical composition of the welded samples was determined according to the state standards GOST 10543-98 by the X-ray fluorescence method on XRF-1800 spectrometer and by the atomic-emission method using DFS-71 spectrometer. Investigation of longitudinal samples of the deposited layer for the presence of nonmetallic inclusions was performed in accordance with GOST 1778-70 using the optical microscope OLYMPUS GX-51 in a bright field within the range of magnifications × 100.

3. Results and discussion

To produce the flux, slag from silicomanganese production with the chemical composition given in table 1 was used. The preparation flux additive containing barium-strontium modifier with the chemical composition (table 2) was carried out by mixing the barium-strontium modifier with liquid glass in a ratio of 75% and 25 %, respectively. After that, the mixture was kept at a room temperature for 24 hours, dried in the oven at 300 °C, further cooled, crushed and sieved to obtain fraction of 0.45-2.5 mm. Further, the flux and flux-additives were mixed in various proportions. The investigated flux and flux-additive ratios are given in table 3. The silicomanganese slag with the fraction 0.45-2.5 mm was applied in the experiments. In the first option (with liquid glass), the additive was introduced in an amount of 2, 4, 6, 8, 10% (corresponding samples M2, M4, M6, M8, M10). According to the second option, the additive (dust) was introduced in an amount of 2, 6, 8, 10% (corresponding samples M21, M61, M81, M101). The flux made of silicomanganese slag without additives (M) was used for comparison.

The surfacing was performed on samples with sizes 300 × 150 mm and 20 mm in thickness from sheet steel grade 09G2S. The process was carried out using S-08GA welding wire by ASAW-1250 welding tractor in the following modes: I = 680 A; U = 28 V; V = 28 m/h.

The samples were cut from the welded plates, and the X-ray spectral analysis of the deposited layer composition and metallographic studies of the deposited layer were carried out. The chemical composition of the welding flux is provided in table 4. The chemical composition of the slag crust is given in table 5, the chemical composition of the weld metal is given in table 6.

Table 1. Chemical composition of silicomanganese slag.

Content, %										
Al_2O_3	CaO	SiO_2	FeO	MgO	MnO	F	Na_2O	K_2O	S	P
6.91-9.62	22.85-31.70	46.46-48.16	0.27-0.81	6.48-7.92	8.01-8.43	0.28-0.76	0.26-0.36	до 0.62	0.15-0.17	0.01

Table 2. Chemical composition of barium – strontium modifier.

Content, %											
BaO	SrO	CaO	SiO_2	MgO	K_2O	Na_2O	Fe_2O_3	MnO	Al_2O_3	TiO_2	CO_2
13.0-19.0	3.5-7.5	17.5-25.5	19.8-29.8	0.7-1.1	2.5-3.5	1.0-2.0	1.5-6.5	до 0.4	1.9-3.9	0.7-1.1	16.0-20.0

Table 3. Investigated ratios of flux – flux-additives, %.

Flux marking	The amount of barium-strontium additive, wt%.		The amount of silicomanganese slag, wt%
	With liquid glass	Dust	
M	-	-	100
M2	2	-	98
M4	4	-	96
M6	6	-	94
M8	8	-	92
M10	10	-	90
M21	-	2	98
M61	-	6	94
M81	-	8	92
M101	-	10	90

As it can be seen from figures 1-4, when the additive flux is introduced, barium, strontium are reduced, and the concentration of sulfur and phosphorus get lower. Investigation of longitudinal samples of the deposited layer for the presence of nonmetallic inclusions was carried out on thin sections without etching by OLYMPUS GX-51 optical microscope with a magnification × 100 comparing with the reference scales according to GOST 1778-70.

The presence of non-deformable silicates, spot and sulfide oxides was established in the weld metal (figure 5) after studying the nature of its contamination with nonmetallic inclusions. The characteristics of nonmetallic inclusions of the deposited layer are given in table 5.

It was shown that when using a mixture of barium-strontium modifier in the amount 2-8% with a liquid glass, the degree of weld metal contamination with nonmetallic inclusions is less, in comparison with the absence of liquid glass in the additive, as well as 10% of the liquid glass in it.

Table 4. Chemical composition of the flux.

(a)

Flux	Mass fraction of elements, %									
	FeO	MnO	CaO	SiO_2	Al_2O_3	MgO	Na_2O	K_2O	S	P
M	0.50	7.97	31.34	46.09	6.61	5.74	0.40	0.01	0.33	0.011
M2	1.39	7.68	31.27	46.31	6.48	5.40	0.52	0.04	0.20	0.022
M4	0.77	7.69	30.56	46.11	6.42	5.42	0.49	0.03	0.38	0.022
M6	1.11	7.12	29.27	45.52	6.87	4.75	0.75	0.20	0.17	0.037
M8	1.04	7.14	31.65	43.93	6.22	3.71	0.79	0.27	0.13	0.044
M10	2.08	7.09	30.16	43.90	7.17	3.58	0.62	0.20	0.12	0.027
M21	1.32	8.14	32.09	45.19	6.20	5.35	0.36	0.01	0.43	0.014
M61	0.33	8.02	30.97	46.38	6.71	6.01	0.29	0.01	0.18	0.0.13
M81	1.89	7.90	30.79	40.07	5.71	3.03	0.48	0.57	0.13	0.090
M101	1.01	7.22	31.62	44.99	6.71	3.29	0.40	0.32	0.14	0.061

(b)

Flux	Mass fraction of elements, %					
	ZnO	Cr_2O_3	F	BaO	SrO	TiO_2
M	0.004	0.05	0.45	отс.	отс.	0.07
M2	0.008	0.03	0.48.	0.35	0.18	0.09
M4	0.0014	0.04	0.36	0.35	0.21	0.08
M6	0.0012	0.04	0.45	0.79	0.41	0.14
M8	0.003	0.033	0.41	0.95	0.44	0.16
M10	0.011	0.039	0.34	1.45	0.60	0.13
M21	0.009	0.05	0.40	0.15	0.087	0.07
M61	0.009	0.03	0.42	0.16	0.092	0.07
M81	0.005	0.028	0.52	1.57	0.99	0.22
M101	0.003	0.021	0.40	2.97	1.11	0.15

Table 5. The chemical composition of slag crusts.

(a)

Flux	Mass fraction of elements, %									
	FeO	MnO	CaO	SiO_2	Al_2O_3	MgO	Na_2O	K_2O	S	P
M	1.69	7.78	32.35	42.50	6.59	5.55	0.30	0.01	0.21	0.011
M2	2.07	7.54	31.91	43.63	6.52	5.92	0.31	0.01	0.16	0.012
M4	2.11	7.15	31.45	45.31	6.38	5.42	0.43	0.03	0.18	0.017
M6	1.93	7.20	31.37	44.30	7.46	5.26	0.39	0.02	0.23	0.017
M8	2.00	7.83	31.19	44.55	7.35	3.68	0.53	0.13	0.11	0.023
M10	1.12	7.16	30.60	44.90	6.48	3.10	0.84	0.32	0.14	0.047
M21	2.34	7.42	31.97	43.30	6.59	5.56	0.31	0.3	0.21	0.014
M61	2.21	6.95	30.26	45.55	7.06	4.99	0.33	0.13	0.17	0.017

M81	1.84	7.78	31.94	45.21	6.87	3.94	0.34	0.077	0.12	0.019
M101	2.04	7.36	31.25	44.43	7.50	3.75	0.38	0.20	0.12	0.030

(b)

	Mass fraction of elements, %					
Flux	ZnO	Cr_2O_3	F	BaO	SrO	TiO_2
M	0.012	0.04	0.37	-	-	0.07
M2	0.006	0.10	0.45	0.19.	0.098	0.08
M4	0.008	0.07	0.41	0.38	0.20	0.09
M6	0.008	0.05	0.46	0.34	0.19	0.09
M8	0.005	0.037	0.42	0.64	0.39	0.12
M10	0.013	0.026	0.56	1.52	0.58	0.16
M21	0.004	0.07	0.38	0.34	0.20	0.07
M61	0.011	0.04	0.38	0.64	0.39	0.11
M81	0.004	0.039	0.38	0.83	0.41	0.11
M101	0.003	0.035	0.48	1.03	0.54	0.14

Table 6. The chemical composition of the welded beads.

(a)

	Mass fraction of elements, %							
Flux	C	Si	Mn	Cr	Ni	Cu	V	Mo
M	0.07	0.43	1.16	0.05	0.11	0.14	0.007	0.021
M2	0.08	0.26	0.83	0.05	0.13	0.15	0.003	0.020
M4	0.07	0.29	0.84	0.05	0.13	0.15	0.005	0.020
M6	0.09	0.26	0.77	0.04	0.10	0.15	0.004	0.020
M8	0.04	0.41	1.28	0.05	0.08	0.13	0.001	0.01
M10	0.04	0.36	1.20	0.05	0.08	0.12	0.001	0.01
M21	0.09	0.31	0.76	0.05	0.13	0.15	0.004	0.019
M61	0.08	0.23	0.77	0.05	0.13	0.14	0.004	0.017
M81	0.05	0.41	1.25	0.04	0.08	0.13	0.003	0.01
M101	0.04	0.41	1.26	0.04	0.08	0.13	0.001	0.01

(b)

	Mass fraction of elements, %				
Flux	Nb	S	P	Ba	Sr
M	0.003	0.029	0.018	отс.	отс.
M2	0.004	0.026	0.015	0.0041	0.0004
M4	0.003	0.025	0.014	0.0054	0.0006
M6	0.003	0.021	0.011	0.0065	0.0008
M8	0.012	0.013	0.011	0.0096	0.0010
M10	0.010	0.014	0.012	0.0119	0.0011
M21	0.002	0.028	0.016	0.0049	0.0001
M61	0.003	0.026	0.014	0.0058	0.0003
M81	0.011	0.013	0.011	0.0086	0.0005
M101	0.011	0.012	0.011	0.0112	0.0006

The use of a strontium-barium modifier dust in the amount of 2-8% also reduces the level of contamination of the welded layer with non-metallic inclusions compared with the use of a larger amount of additive (10%).

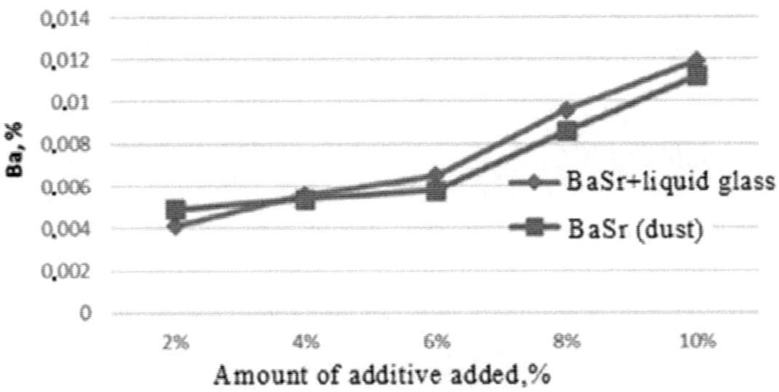

Figure 1. The effect of the additive introduced into the deposited layer on the barium concentration.

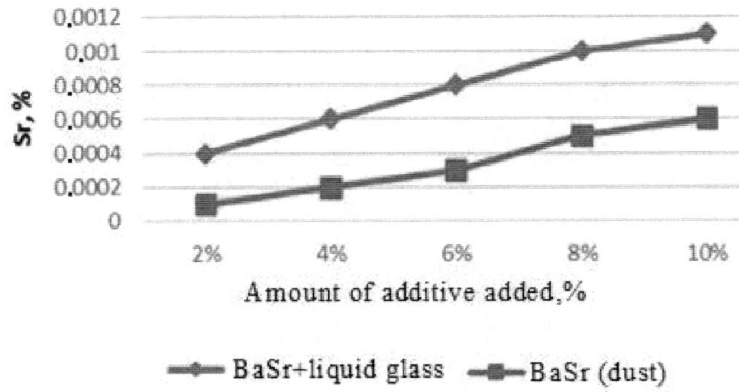

Figure 2. The effect of the additive introduced into the deposited layer on the strontium concentration.

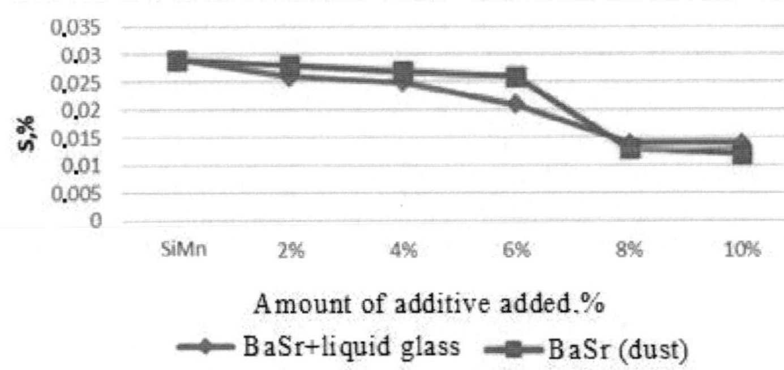

Figure 3. The effect of the additive introduced into the deposited layer on the sulfur concentration.

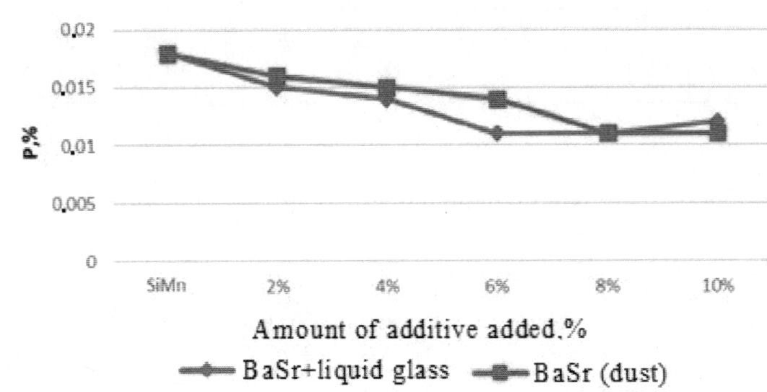

Figure 4. The effect of the additive introduced into the deposited layer on the phosphorus concentration.

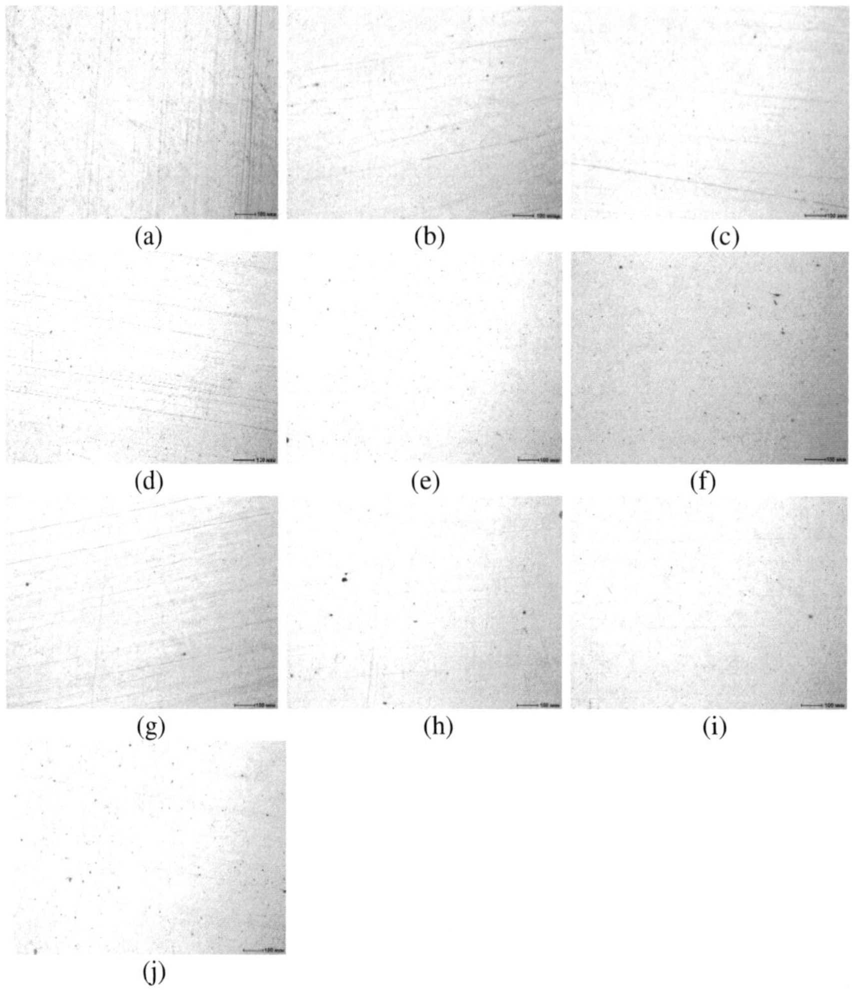

Samples: a) M; b) M2; c) M4; d) M6; e) M8; f) M10; g) M21; h) M61; i) M81; j) M101.

Figure 5. Nonmetallic inclusions in the deposited layer.

Table 7. Characteristics of nonmetallic inclusions in the weld metal.

Sample	Non-metallic inclusions, point		
	nondeformable silicates	spot oxides	sulfides
M	1b,2b, rare 3b	1a	1a
M2	1b,2b	1a	-
M4	1b,2b	1a	-
M6	1b,2b	1a	-
M8	1b,2b	1a	-
M10	1b,2b, rare 3b	1a	1a
M21	1b,2b	1a	-
M61	1b,2b	1a	-

M81	1b,2b	1a	-
M101	1b,2b, rare 4b	1a	2b

4. Conclusions

1) The possibility in principle of applying barium-strontium carbonatite as a refining and gas-protective additive for welding fluxes is shown. The use of a mixture of barium-strontium carbonatite with a liquid glass as an additive is more preferable with respect to the use of a dust additive.

2) It is determined that the use of barium-strontium additive in the amount not more than 8% is optimal from the point of view of the contamination degree of the weld metal with nonmetallic inclusions. This flux is recommended for welding and surfacing of mining equipment.

References

[1] Ivakin V L, Chernyak S S and Pimnev D Yu 2003 *A New Technology for Improving the Quality of Metals and Alloys with Barium-cesium Carbonate* (Irkutsk) p123

[2] Rozhikhina I D, Nokhrina O I, Dmitrienko V I and Platonov M A 2015 *Izvestiya VUZov. Ferrous Metallurgy* vol 58 **15** 871–875

[3] Crespo A C, Puchol R Q and González L P 2009 *Welding International* vol 23 **2** 120–131

[4] Golovko V V and Potapov N N 2009 *Welding International* vol 25 **11** 889–893

[5] Volobuev Y S, Volobuev O S, Parkhomenko A G, Dobrozhela E I and Klimenchuk O S 2012 *Welding International* vol 26 **8** 649–653

[6] Volobuev Y S, Surkov A V, Volobuev O S, Kipiani P N, Shestov D V, Pavlov N V and Savchenko A I 2010 *Welding International* vol 24 **4** 298–300

[7] Potapov N N and Kurlanov S A 1987 *Welding International* vol 1 **10** 951–954

[8] Babushkin P L and Persits V Y 1991 *Welding International* vol 5 **9** 741–742

[9] Pavlov I V and Oleinichenko K A 1995 *Welding International* vol 9 **4** 329–332

[10] Chigarev V V and Kosenko A A 1994 *Welding International* vol 8 **10** 808–809

[11] Kurlanov S A, Potapov N N and Natapov O B 1993 *Welding International* vol 7 **1** 65–68

[12] Bublik O V and Chamov S V 2010 *Welding International* vol 24 **9** 730–733

[13] Crespo A C, Puchol R Q, González L P 2009 *Welding International* vol 23 **2** 120–131

[14] Golovko V V and Potapov N N 2011 *Welding International* vol 25 **11** 889 893

[15] Crespo A C, Puchol R Q, Goncalez L P, Sanchez L G, Gomez Perez C R and Cedre E D 2007 *Welding International* vol 21 **7** 502–511

[16] Volobuev Y S, Volobuev O S, Parkhomenko A G, Dobrozhela E I and Klimenchuk O S 2012 *Welding International* vol 26 **8** 649–653

[17] Volobuev Y S, Surkov A V, Volobuev O S, Kipiani P N, Shestov D V, Pavlov N V and Savchenko A I 2010 *Welding International* vol 24 **4** 298–300

[18] Potapov N N and Kurlanov S A 1987 *Welding International* vol 1 **10** 951–954

[19] Kozyrev N A, Kryukov R E, Burnakov M A, Mikhno A P and Fedotov E E 2017 *Proc. of the XX Int. Sci. and Pract. Conf. on Metallurgy: Technologies, Innovations, Quality* (SibGIU-Novokuznetsk) part 1 pp 296–299

[20] Kozyrev N A, Kryukov R E, Kryukov N E, Kovalsky I N and Kozyreva O E 2017 *Ferrous Metallurgy. Bulletin of Scientific, Technical and Economic Information* **6(1410)** pp 95–98

Development of a flux-cored wire for surfacing mining equipment operating in the conditions of shock-abrasive wear

A I Gusev, N A Kozyrev, A A Usoltsev, R E Kryukov and A R Mikhno

Siberian State Industrial University, Novokuznetsk, 42 Kirova street, 654007, Russia

E-mail: kozyrev_na@mtsp.sibsiu.ru

Abstract. The effect of introducing various elements into the charge of a powder surfacing wire was studied. The dependences of the deposited layer hardness and its wear resistance on the mass fraction of the elements that make up the powder wires were determined by the method of multifactor correlation analysis. A new composition of the flux-cored wire used for surfacing mining equipment operating in the conditions of abrasive wear was developed.

1. Introduction

One of the most important reserves for increasing the production volume and extension of the equipment lifetime is the increase in the efficiency of the actuation component of shearer-loaders and tunneling combines operating in the conditions of shock abrasive wear. The development of new materials that significantly increase the wear resistance of such parts and the use of technology for their recovery is a very important and urgent task [1-10]. Welding wires of Fe-C-Si-Mn-Cr-Ni-Mo systems of type A and B, according to the classification of MIS [11], have become widespread for surfacing the wearing parts.

2. Methods of research

The laboratory studies were carried out according to the scheme described in [12-13]. Multilayer surfacing of the samples was carried out with preheating of the plates to 3500C and subsequent (after surfacing) slowed cooling. Surfacing on the plates was carried out by the welding tractor ASAW-1250 using the produced flux-cored wire. The following components were introduced into wire samples: nickel, cobalt, tungsten; amorphous carbon was replaced by carbon-fluorine-containing dust, with the following chemical composition, wt. %: Al_2O_3 = 21-46; F = 18-27; Na_2O = 8-15; K_2O = 0.4-6%; CaO = 0.7 - 2.3; SiO_2 = 0.5-2.5; Fe_2O_3 = 2.1-3.3; C_{tot} = 12.5-30.2; MnO = 0.07-0.9; MgO = 0.06-0.9; S = 0.09-0.19; P = 0.10-0.18. The wire was manufactured on a laboratory machine.

The diameter of the manufactured wire was 5 mm, the casing was made of St3 tape. Powder-like materials were used as a filler: powder of iron of grade PZhV1 according to state standards GOST 9849-86, ferrosilicon powder of grade FS 75 according to GOST1415-93, powder of high-carbon ferrochromium grade FK900A according to GOST 4757-91, powder of carbon ferromanganese FMn 78 (A) according to GOST 4755-91, nickel powder PNK-1L5 according to GOST 9722-97, ferromolybdenum powder of FMo60 grade according to GOST 4759-91, ferrovanadium powder of grade FV50U 0.6 according to GOST 27130-94, cobalt powder PC-1U according to GOST 9721-79, tungsten powder PVN TU 48-19-72-92.

The chemical composition of the weld metal was determined by the X-ray fluorescence method on XRF-1800 spectrometer and by the atomic-emission method using DFS-71 spectrometer (table 1). The hardness of the test samples was measured by the Rockwell method in accordance with the requirements of GOST 9013-59.

Metallographic examination of thin sections was carried out using an optical microscope OLYMPUSGX-51 in a bright field in the magnification range ×100-1000 after etching the samples surface in the 4% solution of nitric acid. The grain size was determined in accordance with GOST 5639-82 with an magnification × 100. The martensite dispersion was evaluated by comparing the structure with the standards of the corresponding scales and the sizes of martensite needles with the data of table No. 6 of GOST 8233-56. Determination of the martensite needles length was carried out using a package of applied software for metallographic studies SiamsPhotolab 700. The investigation of longitudinal samples of the welded layer for the presence of nonmetallic inclusions was carried out in accordance with GOST 1778-70. The polished surface was studied with an increase ×100 by LaboMet-1I metallographic microscope. The abrasion rate of the welded layer of the prototypes was determined by testing for wear on the machine 2070 SMT-1 according to the disk-welding die scheme.

3. Results and discussion

The chemical composition of the weld metal is given in table 1. Metallographic studies (table 2) shows that the microstructure of the deposited layer by the flux-cored wire of the Fe-C-Si-Mn-Cr-Mo-Ni-V-Co system consists of martensite formed within the boundaries of the former austenitic grains, residual austenite, present in a small amount in the form of separate islands, and thin δ-ferrite layers located along the boundaries of the former austenite grains, in the absence of chromium it has a ferrite-pearlitic structure.

It is believed that the most favorable bases for wear-resistant alloys used for restoration of equipment parts is martensitic and austenitic. In this case, depending on the general structure of the alloy and the presence of excess phases, in different cases one or another base may be preferable. The combination of these two bases (martensite with a certain amount of austenite) can provide the most favorable combination of the welded alloy properties.

The analysis of the obtained results shows that the degree of influence of various chemical elements on the hardness of the welded layer and the samples wear rate is different. Within the studied ranges carbon, chromium, molybdenum, nickel, manganese and, to an insignificant extent, vanadium simultaneously increase the hardness of the deposited layer and decrease the wear rate of the samples. When studying the effect of the use of tungsten and cobalt on the properties of the deposited layer, it has been established that an increase in the concentration of tungsten slightly increases the hardness of the weld metal, but at the same time the wear resistance decreases. This, apparently, is associated with a high-strength solid martensitic matrix, in which more solid carbides of tungsten are "built-in". The low viscosity of the matrix does not allow the tungsten carbides to be retained on the surface, as a result of which the wear is not carried out according to the scheme of uniform surface abrasion, but according to the scheme of high-strength carbide particles pitting from the matrix, as a result of which cracks are formed in the matrix, thus, contributing to additional wear of the matrix.

Table 1. Chemical composition of deposited layers.

No.	C	Si	Mn	Cr	Mo	Ni	Al	Co	Cu	W	Ti	V	S	P
G5	0.4	0.7	0.8	5.2	0.5	0.4	0.02	0.1	0.0	0.00	0.00	0.05	0.03	0.02
G6	0.4	0.8	0.7	4.9	0.5	0.5	0.02	0.11	0.0	0.00	0.00	0.04	0.04	0.02
G7	0.2	0.7	0.7	5.5	0.4	0.6	0.02	0.08	0.1	0.00	0.00	0.04	0.04	0.01
G8	0.3	0.6	0.8	4.9	0.4	0.8	0.02	0.09	0.0	0.00	0.02	0.04	0.03	0.02
G9	0.1	0.7	0.6	4.1	0.3	0.3	0.10	0.05	0.0	0.00	0.01	0.02	0.05	0.02
G10	0.1	0.6	0.6	4.0	0.3	0.3	0.06	0.05	0.0	0.00	0.02	0.03	0.05	0.01
G11	0.2	0.5	0.6	4.1	0.3	0.3	0.03	0.12	0.0	0.00	0.00	0.02	0.04	0.01
G12	0.2	0.6	0.6	4.0	0.3	0.3	0.05	0.19	0.0	0.00	0.02	0.03	0.05	0.02
G13	0.2	0.5	0.5	0.0	0.3	0.3	0.01	0.05	0.1	7.74	0.00	0.00	0.07	0.01
G14	0.2	0.5	0.4	0.0	0.3	0.2	0.02	0.07	0.0	7.42	0.00	0.01	0.04	0.01
G15	0.2	0.5	0.5	0.0	0.3	0.2	0.05	0.07	0.0	7.55	0.00	0.01	0.03	0.01

G16	0.2	0.5	0.5	0.0	0.3	0.2	0.05	0.06	0.0	7.65	0.00	0.02	0.03	0.01
G131	0.2	0.7	1.4	7.1	0.3	0.3	0.08	0.00	0.0	0.00	0.01	0.02	0.03	0.00
G141	0.2	0.7	1.3	5.9	0.3	0.2	0.09	0.	0.0	0.00	0.03	0.03	0.02	0.01
G151	0.2	0.7	1.2	6.3	0.3	0.3	0.08	0.00	0.0	0.00	0.00	0.02	0.03	0.01
G161	0.2	0.7	1.1	6.0	0.3	0.3	0.07	0.	0.0	0.00	0.02	0.04	0.03	0.01
G17	0.1	0.5	0.9	3.9	0.2	0.2	0.02	0.00	0.0	0.03	0.00	0.00	0.03	0.01
G18	0.1	0.6	1.2	6	0.3	0.3	0.01	0.00	0.1	0.02	0.00	0.00	0.03	0.01
G19	0.1	0.5	1.1	5.9	0.3	0.3	0.00	0.00	0.0	1.64	0.00	0.00	0.03	0.01
G20	0.1	0.4	0.9	4.1	0.2	0.2	0.00	0.00	0.0	0.02	0.00	0.03	0.03	0.01
G21	0.1	0.5	1.1	6.2	0.3	0.4	0.00	0.00	0.0	0.02	0.00	0.04	0.03	0.01
G22	0.2	0.6	0.9	4.1	0.4	0.2	0.03	0.01	0.0	0.04	0.00	0.00	0.02	0.01
G23	0.2	0.6	0.9	3.5	0.3	0.2	0.02	0.02	0.0	4.66	0.00	0.00	0.03	0.01
G24	0.2	0.7	1.0	4.1	0.3	0.2	0.03	0.19	0.0	0.08	0.00	0.00	0.03	0.01

Table 2. Characteristics of the deposited metal.

Sample No.	Structure	Austenite grain size	HRC	Abrasion rate, G/rev
G5	Martensit	7	49	0.0000140
G6	Martensit	6 and 7	52	0.0000056
G7	Martensit	6	50	0.0000710
G8	Martensit	6 and 7	52	0.0000140
G9	Martensit	6 and 7	44.5	0.0000710
G10	Martensit	6 and 7	42.5	0.0000390
G11	Martensit	6 and 7	42.5	0.0000440
G12	Martensit	6 and 7	37	0.0000730
G13	Ferrito-perlite	5 and 6	22.5	0.0002060
G14	Ferrito-perlite	6	25.5	0.0000480
G15	Ferrito-perlite	5	22	0.0000360
G16	Ferrito-perlite	6	26	0.0000390
G131	Martensit	7	55	0.0000280
G141	Martensit	7	41	0.0000550

Continuation of **Table 2.**

Sample No.	Structure	Austenite grain size	HRC	Abrasion rate, G/rev
G151	Martensit	7	45	0.0000074
G161	Martensit	7	45	0.0000340
G17	Martensit	6-7;	40	0.0000280
G18	Martensit	6	45	0.0000540
G19	Martensit	6	49	0.0000430
G20	Martensit	6-7;	38	0.0001420
G21	Martensit	6	48	0.0000550
G22	Martensit	6	43	0.0000330
G23	Martensit	6-7;	49	0.0000650
G24	Martensit	6-7;	42	0.0000300

The introduction of cobalt into the composition of the charge does not significantly affect the hardness and abrasive wear of the deposited layer, which is associated with the production of a more viscous but less rigid matrix. In the absence of solid carbide particles embedded in the matrix, the effect of cobalt introduction is negative. It should be also noted that the influence of tungsten on the deposited layer hardness and the samples wear rate is primarily related to the structure of the metal matrix, and the replacement of the martensitic structure in the deposited layer with ferrite-pearlitic one also has little positive effect on wear resistance.

IOP Conf. Series: Earth and Environmental Science **206** (2018) 012034 doi:10.1088/1755-1315/206/1/012034

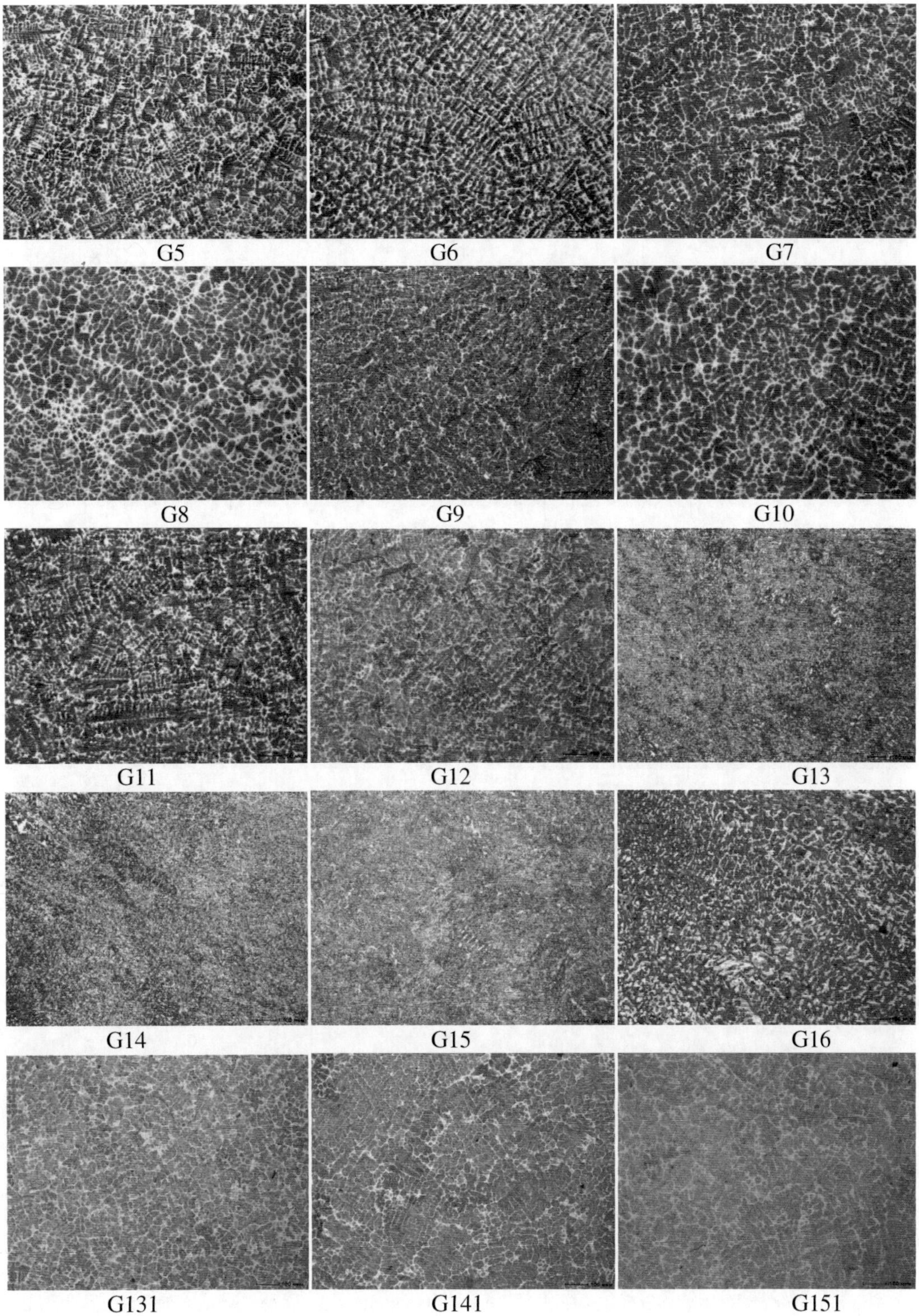

IOP Conf. Series: Earth and Environmental Science **206** (2018) 012034 doi:10.1088/1755-1315/206/1/012034

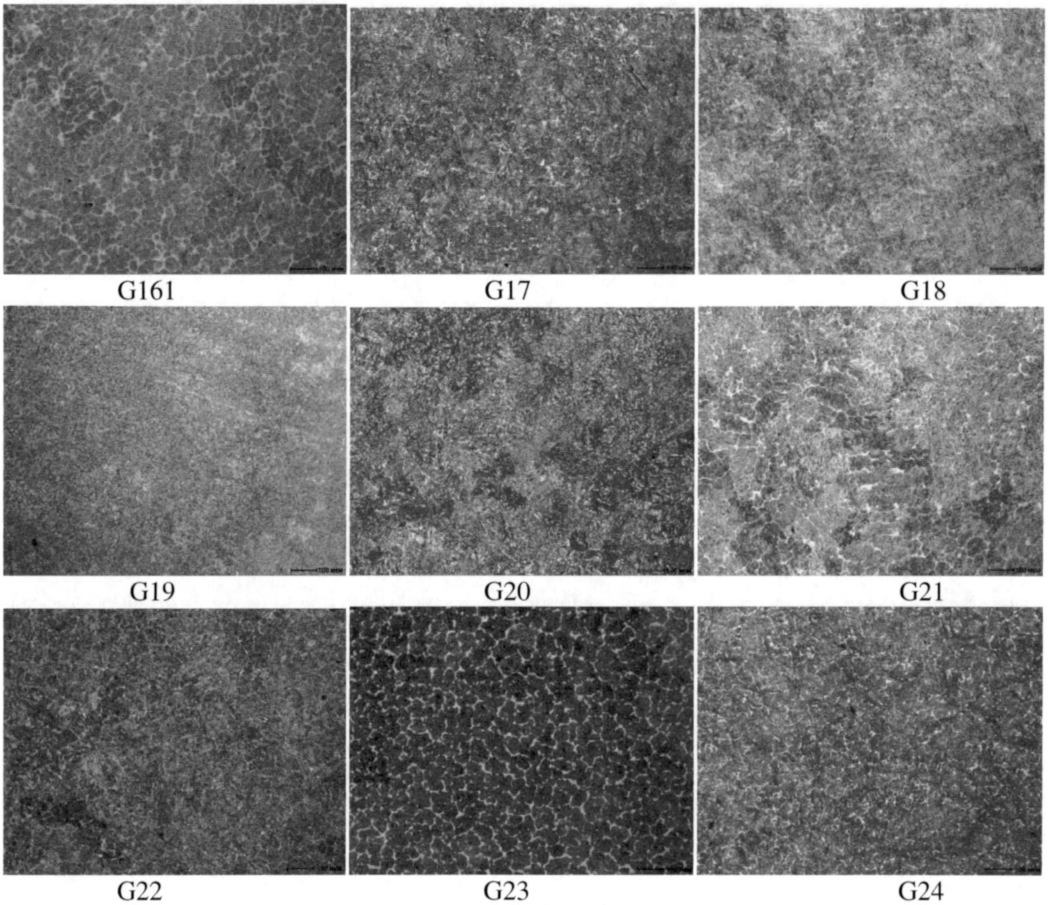

Figure 1. Microstructure of the deposited layers ×100.

The evaluation of the influence of powder wires chemical composition on the wear rate and hardness of the welded layer was carried out by mathematical and statistical methods that allow the patterns of variation of the resulting index depending on various factors to be studied using the methods described in [14, 15].

The dependences of the deposited layer hardness and its wear resistance on the mass fraction of elements were obtained.

Hardness of the welded layer:

$HRC=35.884+22.840\cdot C+11.079\cdot Si-15.903\cdot Mn+4.785\cdot Cr+10.118\cdot Mo+1.399\cdot Ni-38.964\cdot Al-24.126\cdot Co-65.841\cdot Cu-0.057\cdot W+17.887\cdot Ti-76.116\cdot V-76.351\cdot S-239.206\cdot P$ (approximation error is 4.2%);

Rate of samples wear (W):

$W=-0.00025-0.00034\cdot C+0.000171\cdot Si+0.000144\cdot Mn-0.000008\cdot Cr-0.00014\cdot Mo+0.0000308\cdot Ni-0.00076\cdot Al-0.000088\cdot Co+0.000227\cdot Cu+0.0000074\cdot W+0.00017\cdot Ti+0.000633\cdot V+0.002836\cdot S+0.005644\cdot P$ (approximation error is 8.14%).

Testing of the new flux-cored wire was carried out on the plates installed on drums 2 m in diameter of the shearer-loader JOY 4LS20 (figure 2). The results of the flux-cored wire research are protected by the patent of the Russian Federation [16].

Figure 2. Protective plates on the drums of the shearer-loader JOY 4LS20.

4. Conclusions

1) Based on the research results and the multifactor correlation analysis, the dependences of the welded layer hardness and its wear resistance on the mass fraction of the elements that make up the powder wires of the Fe-C-Si-Mn-Cr-Mo-Ni-V-Co system are determined.

2) The obtained dependences are used to predict the hardness of the welded layer and its wear resistance when the chemical composition of the weld metal changes. The results of the flux-cored wire investigations were patented.

References

[1] Kirchgaßner M, Badisch E and Franek F 2008 *Wear Journal* **265** 772–779
[2] Teplyashin M V and Komkov V 2013 *Scientific Notes of TOGU* vol 4 **4** 1554–1561
[3] Teplyashin M V, Komkov V G and Starienko V A 2013 *Scientific Notes of TOGU* vol 4 **4** 1543–1549
[4] Cheylyakh Ya A and Chigarev V V 2011 *Bulletin of Priazovsky State Technical University. Engineering Sciences* **22** 103–108
[5] Metlitskii V A 2008 *Welding International* vol 22 796–800
[6] Kejžar R and Grum J 2005 *Welding International* vol 20 961–976
[7] Liu D S, Liu R P and Wei Y H 2013 *Materials Science and Technology* vol 30 316–322
[8] Lim S C, Gupta M, Goh Y S and Seow K 1997 *Surface Engineering* vol 13 247–250
[9] Zhuk Yu 1999 *Materials Technology* **14** 126–129
[10] Hardell J, Yousfi A, Lund M, Pelcastre L and Prakash B 2014 *Tribology – Materials, Surfaces & Interfaces* **8** 90–97
[11] Paton B E 1974 *The Technology of Electrical Welding of Metals and Alloys by Flashing* (Moscow: Metallurgiya) p 768
[12] Gusev A I, Kibko N V, Popova M V, Kozyrev N A and Osetkovsky I V 2017 *Izvestiya Vuzov. Ferrous Metallurgy* vol 60 4 318–323
[13] Gusev A I, Kibko N V, Kozyrev N A, Popova M V and Osetkovsky I V 2016 *IOP Conf. Series: Materials Science and Engineering* **150** 012033
[14] Ventzel Ye S 2003 *Probability Theory and Its Engineering Applications* (Moscow: Akademiya) p 464
[15] Gmurman V E 2003 *Probability Theory and Mathematical Statistics* (Moscow: Vyisshaya shkola) p 479
[16] Kozyrev N A, Gusev A I, Galevskii G V, Kryukov R E, Osetkovskii I V, Usol'tsev A A and Kozyreva O A Pat. 26641590 IPC8B23 K35 / 36V 23 K35 / No. 2016125085/02 (039292)

Development of a new wear-resistant flux cored wire for welding armor on the buckets of mining equipment

N A Kozyrev, R E Kryukov, A A Usoltsev, N V Kibko and A S Nepomnyashchikh

Siberian State Industrial University, 42 Kirova street, Novokuznetsk, 654007, Russia

E-mail: kozyrev_na@mtsp.sibsiu.ru

Abstract. In the paper, the possibility of producing a flux-cored wire on the basis of the dust from gas purification of silicomanganese production and dust from a gas purification of aluminum production at a different ratio of components was studied. The possibility in principal of producing a flux-cored wire for surfacing is discussed, the qualitative indices are studied and the manganese recovery coefficients with various component ratios were calculated. Statistical processing of the obtained results was carried out, statistical dependences of the influence of the component composition on the deposited layer properties were constructed.

1. Introduction

In the process of mining equipment operation, a number of problems arise that lead to emergency stops. First and foremost, this is due to the operation of the main parts of mining equipment in the conditions of heavy wear, the working surfaces get destroyed, so, it requires the application of technological measures increasing their wear resistance. For this reason, the task of improving the technology of manufacturing new powder wires for mining equipment with high wear resistance becomes urgent. [1-12].

2. Methods of research

In the wear test, a method of mass loss of the sample was used, which is based on a change in the mass of the sample when performing a test disc- welding die. The wear of the samples was carried out on the machine 2070 CMT-1. The chemical composition of the weld samples was determined according to the state standards GOST 10543-98 by the X-ray fluorescence method on the XRF-1800 spectrometer and by the atomic-emission method using the DFS-71 spectrometer. The hardness of the welded layers was measured by hardness tester MET-DU. The evaluation of nonmetallic inclusions was carried out according to GOST 1778-70 on thin sections without etching by OLYMPUS GX-51 optical microscope with a magnification $\times 100$.

3. Results and discussion

Wire manufacturing was carried out on a laboratory unit for the production of powder wires with passing it through drawing-blocks. The diameter of the produced wire was 6 mm, its cladding was made from St3 tape. The filler consisted of: dust from gas-purification system of aluminum production, wt. %: Al_2O_3 = 21-46.23; F = 18-27; Na_2O = 8-15; K_2O = 0.4-6; CaO = 0.7-2.3; Si_2O = 0.5-2.48; Fe_2O_3 = 2.1-3.27; C = 12.5-30.2; MnO = 0.07-0.9; MgO = 0.06-0.9; S = 0.09-0.19; P = 0.1-0.18 and dust from silicone manganese gas-purification system, wt. %: Al_2O_3 = 2.43; Na_2O = 1.32;

K_2O = 5.56; CaO = 6.4; SiO_2 = 29.19; BaO = 0.137; MgO = 7.54; S = 0.23; P = 0.04; Fe = 1.067; Mn = 27.69; Zn = 2.687; Pb = 3.833.

Surfacing was carried out under a flux made of silicomanganese slag with chemical composition: AL2O3 = 6.91-9.62%, CaO = 22.85-31.70%, SiO2 = 46.46-48.16%, FeO = 0, 27-0.81%, MgO = 6.48-7.92%, MnO = 8.01-8.43%, F = 0.28-0.76%, Na_2 O = 0.26-0.36 % K_2O = 0.6 2%, S = 0.15-0.17%, P = 0.01%, using the ASAW-1250 welding tractor on steel plates 14-16 mm thick with a size of 100x500 mm. The parameters of the flux-cored wire are given in table 1. The chemical composition of the welded metal layer and the chemical composition of slag crusts is given in tables 2, 3, hardness and the results of the durability test in table 4.

Table 1. Parameters of the flux cored wire.

No	dust from silicomanganese gas-purification system		dust from gas-purification system at aluminum production		Total weight, g	Cladding weight, g	Powder weight, g	Coefficient of filling,%	Manganese recovery, %
	g	%	g	%					
81	100	89.89	11.25	10.11	19.759	18.333	1.426	7.217	73
82	100	81.63	22.5	18.37	17.816	16.523	1.293	7.258	82
83	100	74.77	33.75	25.23	18.975	17.735	1.240	6.535	88
84	100	68.96	45	31.04	19.510	18.501	1.009	5.172	126
85	100	59.70	67.5	40.3	19.089	17.983	1.106	5.794	124
86	100	50	100	50	19.428	18.082	1.346	6.928	124

Table 2. Chemical composition of the weld metal layer.

Marking	Mass fraction of elements, %											
	C	Si	Mn	Cr	Ni	Cu	Ti	V	Mo	Al	Nb	S	P
81	0.08	0.38	1.31	0.03	0.08	0.11	0.001	0.003	0.02	0.00	0.006	0.079	0.017
82	0.08	0.57	1.35	0.05	0.06	0.19	-	0.004	-	0.05	0.012	0.074	0.014
83	0.08	0.40	1.19	0.04	0.07	0.010	-	-	-	0.01	0.009	0.063	0.015
84	0.09	0.49	1.25	0.03	0.08	0.09	0.001	0.003	0.01	0.01	0.006	0.076	0.017
85	0.16	0.46	1.19	0.02	0.07	0.07	0.001	0.003	0.01	0.01	0.006	0.073	0.016
86	0.22	0.50	1.19	0.03	0.07	0.07	0.001	0.003	0.01	0.00	0.006	0.083	0.014

Table 3. Chemical composition of slag crusts.

Marking	Mass fraction of elements, %													
	FeO	MnO	CaO	SiO_2	Al_2O_3	MgO	Na_2O	K_2O	S	P	ZnO	Cr_2O_2	F	TiO_2
81	1.80	8.37	29.95	43.65	7.27	5.30	0.39	0.16	0.17	0.010	0.040	0.048	0.63	0.07
82	2.25	8.16	30.05	44.45	7.70	5.47	0.39	0.13	0.14	0.012	0.038	0.059	0.62	0.06
83	1.77	8.13	30.17	43.84	7.51	5.17	0.37	0.07	0.17	0.011	0.014	0.050	0.62	0.08
84	2.75	7.52	31.62	43.12	7.56	5.33	0.30	0.09	0.15	0.011	0.012	0.053	0.47	0.08
85	2.11	7.93	30.57	42.95	8.41	5.58	0.46	0.08	0.18	0.011	0.016	0.043	0.77	0.07
86	2.11	7.75	30.30	42.47	8.47	5.42	0.46	0.05	0.17	0.011	0.011	0.036	0.83	0.07

Table 4. Hardness and wear of samples.

No	Initial weight, g	Weight after abrasion, g	Rotation, n	Loss of sample weight, g	Wear, g/rot.	Hardness, HB
81	82.587	82.139	3070	0.448	0.000146	126
82	92.395	91.960	3300	0.435	0.000132	136
83	89.288	88.780	4990	0.508	0.000102	132
84	84.935	84.495	3500	0.44	0.000126	133
85	113.299	112.974	3410	0.325	0.000095	143
86	115.758	115.338	3300	0.42	0.000127	167

The error of the speed meter of the lower sample shaft is ± 3%, the limit for the permissible value of the random component mean square deviation of the friction torque meter reduced error in the static loading mode is – 1%. The tests were carried out in the following conditions: the load was 30 mA (78.4H), the frequency was 20 rpm, the samples loading was performed with a help of a spring mechanism, the rotation speed was measured by the tachogenerator on the motor shaft, and the number of rotations – by the non-contact sensor. During the test, the sample interacted with a block made of steel R18.

The studies shows that the manganese recovery factor (the ratio of the manganese concentration obtained in the weld metal to the total amount of manganese introduced) is associated not only with the reduction of manganese oxide from the manganese containing flux, due to the carbon contained in the flux cored wire, but also with a significant excess of carbon in flux cored wire made from manganese flux, with manganese recovery being more than 100% (table 1). The manganese recovery depends on the filling ratio of the flux-cored wire (figure 1), as well as on the concentration of carbonaceous material in the charge composition (dust from gas purification system of aluminum production) and, respectively, the carbon concentration in the weld metal (figures 2, 3).

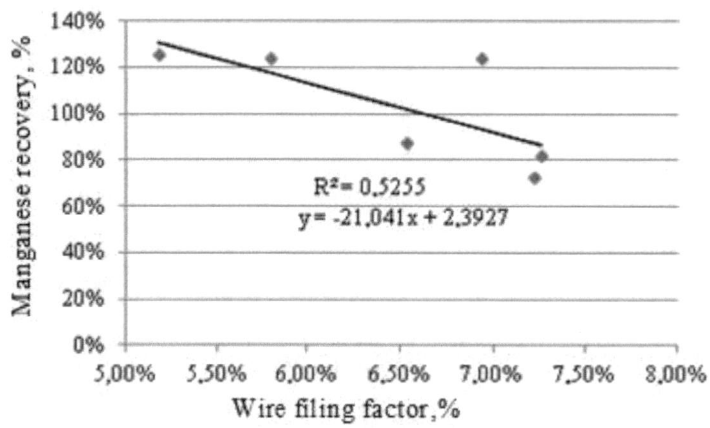

Figure 1. Dependence of manganese recovery on the filling ratio of the flux cored wire.

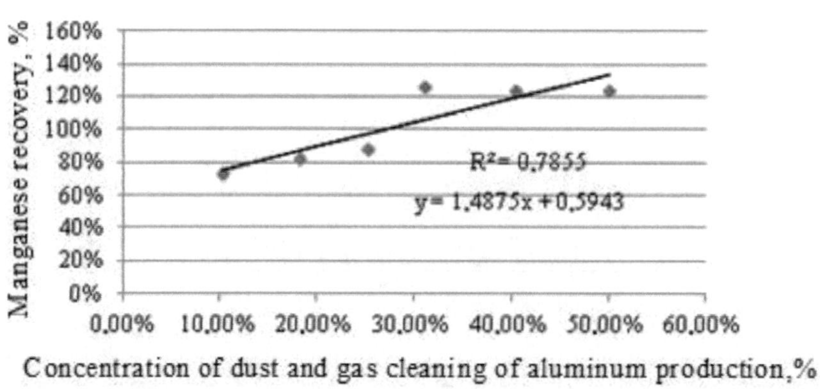

Figure 2. Dependence of manganese recovery on concentration of dust from gas purification system of aluminum production.

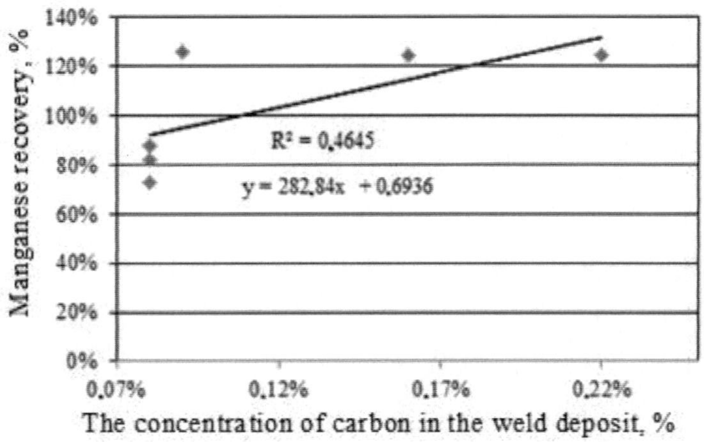

Figure 3. Dependence of manganese recovery on the carbon concentration in the deposited metal.

Figure 4 shows non-metallic inclusions in the welded bead zone; the evaluation of nonmetallic inclusions, carried out in accordance with GOST 1778-70, is presented in table 5.

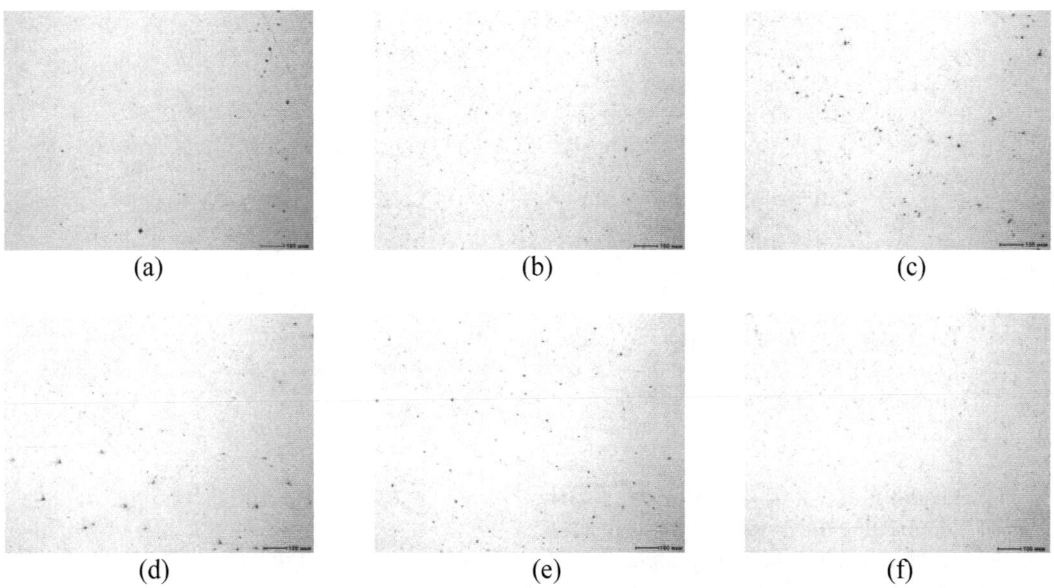

Samples: a) 81; b) 82; c) 83; d) 84; e) 85; f) 86

Figure 4. Non-metallic inclusions in the welded zone of samples.

Table 5. Non-metallic inclusions in the zone of welded seams.

Sample	Non-metallic inclusions, point	
	nondeformable silicates	spot oxides
81	1b,2b	1a, 2a, 3a
82	1b	1a, 2a, 3a
83	2b, rare 4b	1a,2a
84	2b, 1b	1a

85	2b, 1b	1a
86	1b, rare 4b	1a

4. Conclusions

1) The possibility in principal of producing a flux-cored wire for wear-resistant surfacing, using dust from gas purification system of silicomanganese and dust from gas purification system of aluminum production with a different components ratio is shown.

2) The optimal deposition modes were determined, qualitative indices were studied, and manganese recovery for various component ratios was calculated. Statistical processing of research results was carried out, statistical dependences of the influence of the component composition on the deposited layer properties were constructed.

3) This flux-cored wire is recommended for use.

References

[1] Metlitskii V A 2008 *Welding International* vol **22** 796–800
[2] Filippov M A et al 2015 *Welding International* vol **29** 819–822
[3] Liu D S, Liu R P and Wei Y H 2013 *Materials Science and Technolog*y **30** 316–322
[4] Kejžar R and Grum J 2005 *Welding International* **20** 961–976
[5] Li R He et al 2014 2014 *Surface Engineering* **30** 784–790
[6] Ma H R et al 2016 *Surface Engineering* **46** 1–7
[7] Lim S C et al 1997 *Surface Engineering* **13** 247–250
[8] Zhuk Yu 1999 *Materials Technology* **14** 126–129
[9] Hardell J et al 2014 *Tribology – Materials, Surfaces & Interfaces* 8 90–97
[10] Deng X T, Fu T L, Wang Z D, Misra R D and Wang G D 2016 *Materials Science and Technology* **32** 320–327
[11] Kirchgaßner M, Badisch E and Franek F 2008 *Wear Journal* **26**5 772–779
[12] Mitsuru F et al 2015 *JFE Technical Report* **20** 159 – 163
[13] Saita K et al 2013 *Nippon steel & Sumitomo Metal Technical Report* **105** 84–92
[14] Kozyrev N A et al 2014 *Proc. of the 16th Int. Sci. and Pract. Conf. on Technologies for Hardening, Coating and Repair: Theory and Practice* (St. Petersburg: SPPU) vol 1 pp 69–71
[15] Tyilkin M A 1971 *Increase in the Durability of Parts of Metallurgical Equipment* (Moscow: Metallurgy) p 210
[16] Teplyashin M V, Komkov V G and Stariyenko V A 2013 *Scientific Notes of TOGU* vol 4 **4** 1543–49
[17] Cheylyakh Ya A and Chigarev V V 2011 *Vestnik of Priazovsky State Technical University. Engineering Sciences* **22** 103–108
[18] Povolotsky D Ya 2000 *Fundamentals of Steel Production Technology* (Chelyabinsk: SUSU) p 189
[19] Kozyrev N A et al 2018 *Ferrous Metallurgy. Bulletin of Scientific, Technical and Economic Information* **1(1417)** 77–86
[20] Osetkovsky I V, Kozyrev N A, Kryukov R E, Usoltsev A A and Gusev A I 2017 *IOP Conference Series: Earth and Environmental Science* **84** 012017

IOP Conf. Series: Earth and Environmental Science **206** (2018) 012036 doi:10.1088/1755-1315/206/1/012036

Models of functioning and resources allocation of road and transport infrastructure

V A Buyvis[1], A V Novichikhin[1] and V O Kaledin[2]

[1]Siberian State Industrial University, 42 Kirova street, Novokuznetsk, 654007, Russia
[2]Novokuznetsk Institute (Branch) of Kemerovo State University, 23 Tsiolkovskogo street, 654041, Russia

E-mail: Buyvis_va@mail.ru

Abstract. Specific features of functioning of a modern road and transport infrastructure on the example of Kemerovo region are determined. As a tool for justifying management decisions and the formation of projects for short-term periods, a set of indicators of functioning and resources allocation of the road and transport infrastructure was proposed. Taking into account the situation in Kemerovo region, three types of scenarios for the development of projects in the road and transport infrastructure are considered, depending on the investment potential: optimistic, mild and pessimistic. To eliminate uncertainties in the planning of functioning and resources allocation of road and transport infrastructure at the strategic level, the model for the problem of functioning and resources allocation in the road and transport infrastructure within the fuzzy-set theory was formulated. The modeling of scenarios based on the developed fuzzy model of functioning and resources allocation in the road and transport infrastructure at a qualitative level was performed. The proposed models for the functioning and resources allocation in the road and transport infrastructure made it possible to obtain quantitative and qualitative assessments of the system condition for assessing the effectiveness of functioning in operational and strategic planning in the conditions of uncertainty.

1. Introduction

The existing condition of the road and transport infrastructure does not fully meet the modern requirements of both the economy of Kemerovo region and the economy of the entire Russian Federation [1], the socio-economic development is largely constrained by the condition and level of highways development. At present, with an excess of the normative load, almost 27% of the road network is operated, while the local road network is not developed enough [2]. In Kemerovo region 26.73% of public roads do not meet the regulatory requirements for transport performance indicators, including 66.36% of regional or inter-municipal roads and 11.32% of local roads. The accelerated increase in the country automobile fleet has not yet led to a corresponding increase in the volume of construction, reconstruction and repair of the road network, and repair of roads in recent years has even somewhat decreased. For example, with the increase in the length of public roads by 15% over the past 10 years, the fleet of vehicles has increased by more than 1.7 times. The axle load of 11.5 tonnes was observed on sections of federal roads with a total length of 2613 km (5.2% of the total length of the network) with a demand in 22000 km, a 10-tonne load – on 11900 km (23.8% of the total network length) with a demand in 26 thousand km, respectively [2].

2. Tools for justifying the management decisions and the formation of projects for short-term periods

One of the reasons for the existing situation in the road and transport infrastructure is that the system of resources allocation functions in the conditions of uncertainty [3, 4].

As a tool for justifying management decisions and project formation for short-term periods, the following set of indicators for the functioning and resources allocation in the road and transport infrastructure is proposed [5]:

- infrastructure indicator Ind_1, characterizing the length of the sections of transportation lines, on which there are limitations of traffic and carrying capacity due to non-compliance with regulatory requirements;
- indicator of transport work Ind_2, characterizes the volume of transport using the reserve routes due to the non-compliance of highways, in which the main routes are laid, with the regulatory requirements and transport performance indicators;
- operational indicator Ind_3, characterizing the value of consignments delivered by motor transport within the time limits exceeding the standard (contractual) period;
- social indicator Ind_4, characterizes the value of additional time spent by the population on the road due to the non-compliance of roads to regulatory requirements;
- economic indicator Ind_5, characterizes the efficiency of investments directed to the road and transport infrastructure, it is proposed to use the net present value as an indicator (NPV).

To develop projects in the road and transport infrastructure, a scenario approach is used, which is determined by the level of possible financing. In this connection, three types of scenarios (optimistic, mild and pessimistic) are considered for the conditions of Kemerovo region depending on the investment potential. The length of sections of transportation lines with restrictions on traffic and carrying capacity due to non-compliance with regulatory requirements for the beginning of scenarios is 5070 km [6].

The optimistic scenario presupposes a high level of investment, compliance of the speed of construction works and current maintenance of road facilities, with the level of motorization and planned volumes of transportation, the violation of delivery deadlines depends only on the extent of the reserve routes, which will be used only during the construction works. In the mild scenario, the level of investment is 50% of the required level, the speed of construction works and current maintenance of road facilities are comparable or slightly ahead of their wear rate, the violation of delivery deadlines is due not only to the length of reserve routes, which are supposed to be used during the construction works, but also the presence of sections on the transportation lines, on which there are limitations for traffic and carrying capacity because of the non-compliance with regulatory requirements.

According to the pessimistic scenario, the level of investment is 25% of the required level, the speed of construction works and current maintenance of the road facilities are comparable or slightly lower than the rate of their deterioration, the duration of delivery delays is due not only to the length of reserve routes, which will be used during the construction works, but also to the presence of sections on the transportation lines with limitations on traffic and carrying capacity because of the non-compliance with regulatory requirements.

The simulation results are presented in figure 1, the period is 12 years (from 2019 to 2030) in accordance with the Transport Strategy of the Russian Federation until 2030 [2].

3. Problem model of functioning and resources allocation of road and transport infrastructure within the fuzzy-set theory

To eliminate uncertainties in the planning of functioning and resources allocation of road and transport infrastructure at the strategic level, it is proposed to use the fuzzy-sets method [7] to develop appropriate mathematical models.

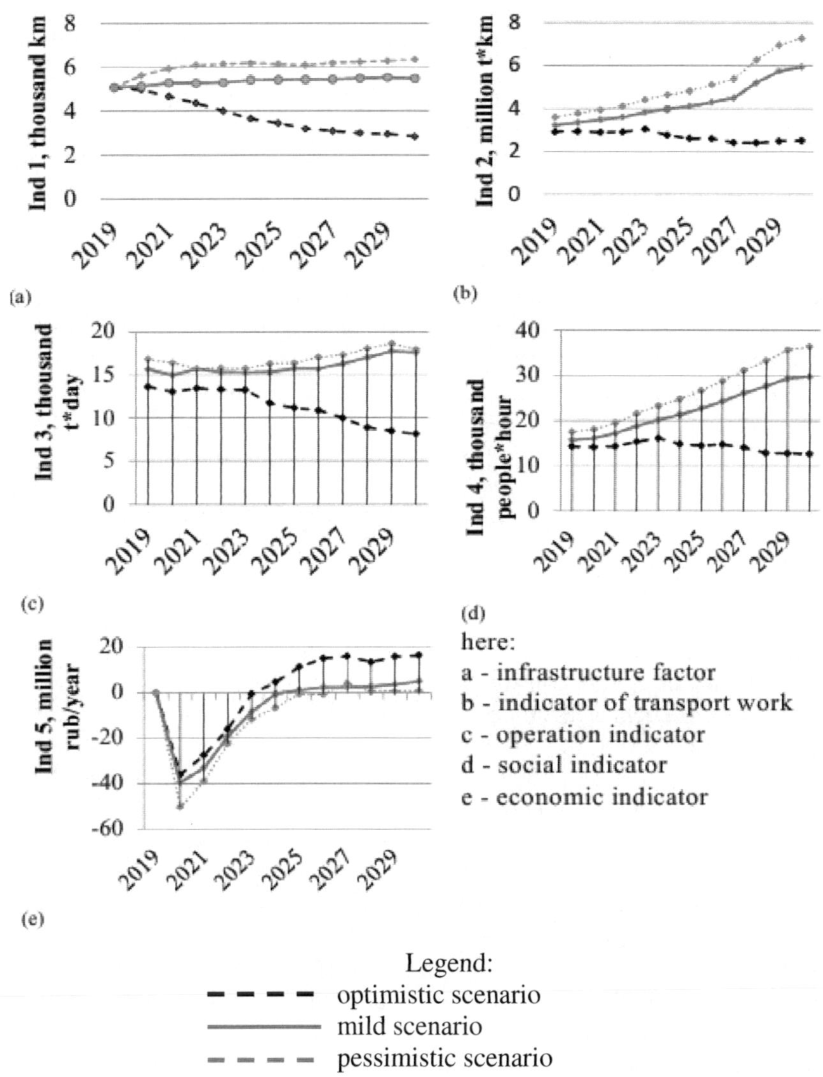

Figure 1. Forecast estimates of indicators of road and transport infrastructure functioning and resources allocation in it – Ind_1, Ind_2, Ind_3, Ind_4, Ind_5 (at the quantitative level).

Thus, the application of fuzzy models of road and transport infrastructure functioning and resources allocation in it will help to solve the following tasks:

1. Definition of a set of criteria, constraints, factors that detail the assessment of the system condition.
2. Formation of judgments about the interrelations and interdependencies of factors and constraints.
3. Selection of environmental factors (combinations of factors) that are of greatest importance for modeling the planning system of road and transport infrastructure functioning and resources allocation in it.

Let us formulate the problem model of functioning and resources allocation of road and transport infrastructure within the fuzzy-set theory:

$$\mathrm{SCRA} = < T, R|M, \mathrm{A}, P, E, I, O, F(x), B, X_{\mathrm{d}}, \mathrm{X_c}, \mathrm{K} > \quad (1)$$

where SCRA is the system condition of resources allocation in the road and transport infrastructure;

T – the period of time for which the assessment of the system condition of resources allocation in the road and transport infrastructure is performed;

R – the resources necessary for functioning of the system of road and transport infrastructure;

$M = \{M_1, M_2, \dots, M_i\}$ – the set of formalized descriptions of material resources, which are an important factor for functioning of the system of road and transport infrastructure;

$A = \{A_1, A_2, \dots, A_i\}$ – the set of conditions (desired, planned, required) of the planning system of the road and transport infrastructure functioning and resources allocation in it;

$P = \{P_1, P_2, \dots, P_k\}$ – the set of formalized descriptions of human resources that are an important factor for functioning of the road and transport infrastructure;

$E = \{E_1, E_2, \dots, E_j\}$ – the set of formalized descriptions of economic resources that are an important factor for the functioning of road and transport infrastructure;

$I = \{I_1, I_2, \dots, I_n\}$ – the set of formalized descriptions of infrastructure resources, which are an important factor for the functioning of road and transport infrastructure;

$O = \{O_1, O_2, \dots, O_R\}$ – the set of formalized descriptions of organizational and managerial resources, which are an important factor for the functioning of the system of the road and transport infrastructure;

$F(x)$ – the function that characterizes the importance of a combination of resources for achievement of the planned condition of resources allocation in the road and transport infrastructure;

$B = \{B_1, B_2, \dots, B_x\}$ – the set of constraints that must be taken into account when developing alternative trajectories for the achievement of the planned condition of resources allocation in the road and transport infrastructure;

$XC = \{X_{C1}, X_{C2}, \dots, X_{Cd}\}$ – the set of actual estimates of the significance of analyzed factors (their impact on the functioning of the road and transport infrastructure);

$Xd = \{X_{d1}, X_{d2}, \dots, X_{dc}\}$ – the set of importance estimates of environmental factors combinations in terms of the importance and appropriateness of taking into account their influence on the possibility of achieving the planned condition of the system of resources allocation in the road and transport infrastructure;

K – the criterion for choosing a set of preferred factors and their combinations for taking into account the development of alternative trajectories of achieving the planned condition of the system for resources allocation in the road and transport infrastructure;

The results of modeling the scenarios proposed above on the basis of the developed fuzzy model of the road and transport infrastructure functioning and resources allocation it are presented in figure 2.

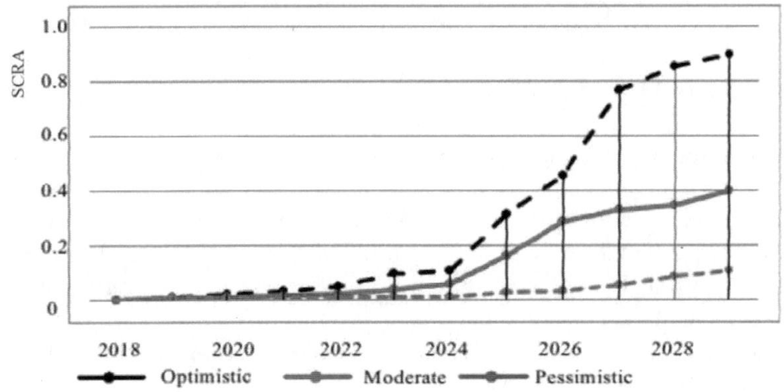

Figure 2. Forecast estimates of the system condition of resources allocation in road and transport infrastructure (at a qualitative level).

IOP Conf. Series: Earth and Environmental Science **206** (2018) 012036 doi:10.1088/1755-1315/206/1/012036

Forecast estimates allow to correct the development strategy of the road and transport infrastructure taking into account the formalized presentation of qualitative estimates of its indicators.

4. Conclusion

The proposed models of the road and transport infrastructure functioning and resources allocation make it possible to obtain quantitative and qualitative estimates of the system condition for assessing the functioning efficiency in operational and strategic planning in the conditions of uncertainty.

References

[1] *Road Economy of Russia in Figures and Facts 2003* (Moscow: Rosavtodor) p 325

[2] *Road guidance document No. 2018-00-2005* (Moscow: Rosavtodor) p 9

[3] *Decree of the Government of the Russian Federation "On Approval of the Transport Strategy of the Russian Federation for the Period till 2030" No. 1734-r from November 22, 2008* http://docs.cntd.ru/document/902132678

[4] Kuzmin E A 2012 *Uncertainty and Certainty in the Management of Organizational and Economic Systems* (Ekaterinburg: Institute of Economics, UrB RAS) p 184

[5] Borch K H 1968 *The Economics of Uncertainty* (Princeton University Press) p 227

[6] Buyvis V A, Novichikhin A V and Temlyantsev M V 2017 *IOP Conference Series: Earth and Environmental Science* **84** 012026

[7] *Decree of the Administration Board of Kemerovo region "On Approval of the Regional Program "Development of the Network of Public Roads of Kemerovo Region" for 2015-2022" No. 212 from June 30, 2015* http://docs.cntd.ru/document/428628801

[8] Zadeh L A 1978 *Fuzzy Sets and Systems* **1** pp 3–28

Geomechanical justification based on the results of numerical modeling of combined technology parameters of working areas development with the use of robotized means of underground mining

V N Fryanov[1], **L D Pavlova**[1] **and A A Cherepov**[2]

[1]Siberian State Industrial University, 42 Kirova street, Novokuznetsk, 654007, Russia
[2]Raspadskaya Coal Company, LLC, 33 Kurako ave, Novokuznetsk, 654027, Russia

E-mail: ld_pavlova@mail.ru

Abstract. Based on the results of numerical modeling the assessment of the options for selective and sequential development of the working areas and coal mine blocks by fully-mechanized longwalls with a complete collapse of the roof and short faces with the roof controlled by rib pillars. As the criteria for assessing the effectiveness of geotechnological options, the coefficients of vertical stress concentrations in the selvedge of the coal seam in the vicinity of extraction galleries and the height of the unloading zone of the roof rocks over the gob are adopted. Factors that limit the use of longwall faces are the presence of disjunctive disturbances within the mine field, which leads to coal losses and irrational use of subsoil. It was established that successive development of excavation sites with long and short faces results in a more even distribution of the vertical stress concentration coefficients, reduces the probability of zones with increased rock pressure and gas dynamic phenomena, and also provides an opportunity to increase the coal extraction factor.

1. Introduction

The modern technology of underground mining of flat coal seams is characterized by a daily load of 5-30 thousand tonnes for the mining face, which is achieved by selective development of coal seams, as a rule, sections of a rectangular shape with a length of an extraction pillar up to 3 km and a face up to 400 m. Factors that complicate the use of high-performance complex mechanized longwalls (CML) are geological disturbances of discontinuous type, zones of increased rock pressure and threatened by gas-dynamic phenomena, areas of coal seams with increased fracturing and the tendency for roof fall, variable thickness and seam inclination, high gas content of the coal-methane seam within the extraction pillar.

These factors, taking into account the complex shape of the boundaries of the mining allotment, the variability of mining and geological parameters and the need to reduce production costs due to the high competition of coal products on the market, lead to the selective mining of coal seams. At the same time, separate seam development is carried out at the first stage by high-performance CML, and at the second stage the residual reserves are mined with special technologies. Accordingly, all the balance reserves of coal within the mine field fall into two categories: technological for the use of long mechanized faces and residual, temporarily conserved and designed to be developed in future periods by new technologies.

In this paper we propose an option of simultaneous combined development of rectangular working areas by longwall mining (LM) and oblique areas with limited coal reserves by shortwall mining (SM).

Based on the analysis results of mining layouts, preparation and development of the coal layers of the mines "Antonovskaya", "Polosukhinskaya", "Yubileynaya", "Alardinskaya" and others in Kuzbass, the most typical mining situation was reconstructed with the arrangement of high-performance long mechanized faces (extraction pillars 4 and 7 in figure 1) and blocks of limited sizes or in the form of oblique figures (blocks 1-3,5,6 in figure 1). According to the tectonics of Kuzbass, disjunctive disturbances are revealed within the mining allotment of the mine (in figure 1 are shown by square hatching), which regulate the spatial position of extraction pillars 4 and 7 for their development by LM and blocks with coal extraction in short faces.

It is obvious that after selective mining of coal reserves in extraction pillars 4, 7 by long complex mechanized faces, the following options for the development of mining operations are possible (figure 1):

- cleaning-up of residual coal reserves in blocks 1-3, 5, 6 by peripheral technologies, for example, short faces;
- conservation of mining operations in the seam.

When cleaning-up the residual coal reserves, the following complicating factors are possible:

- limited access to reserves, since when developing the seam with long mechanized faces in sections LM 4, LM 7, the ventilation roads 4, 7 that square them up and the belt roads 4, 7 will be gobbed, so there will be a need for new workings;
- lack of progressive technological solutions for mining seam sections of limited dimensions or in the form of oblique figures;
- location of areas with residual coal reserves in the zones of bearing pressure, which occurs under the influence of the suspended underworked roof rocks in extraction pillars 4, 7, which can lead to the emergence of concentrators of elastic energy and bumps;
- the need to ensure the level of technical and economic indicators, not lower than those achieved at the initial stage of seam mining by long mechanized faces.

Thus, the separate sequential development of a seam at the first stage by high-capacity CMLs, and the development of residual reserves by special technology at the second stage are associated with both an increased risk of hazardous industrial situations and a decrease in technical and economic indicators at the second stage.

However, the combined option of simultaneous mining of coal seams reserves in complex mining and geological conditions is also possible. The essence of the method consists in the simultaneous preparation and mining the rectangular-shaped sites by long mechanized faces and oblique sites with limited coal reserves by short faces.

Advantages of such organization of mining operations in accordance with the combined technology are as follows:

- increase in the extraction factor of balance reserves in the coal seam due to partial excavation of coal in sites with an oblique form and limited coal reserves;
- reduction in the specific volume of workings due to their use in development of technological and non-technological coal reserves for long mechanized faces;
- compensation for reduction in the mine production level during the re-installation of the purification equipment due to a periodic increase in the production simultaneously in several oblique sites with limited coal reserves;
- reduction of risks of gas dynamic phenomena occurrence through forecasting the parameters of geomechanical processes, rational spatial and temporal location of long and short faces and the order of mining operations within the mine field.

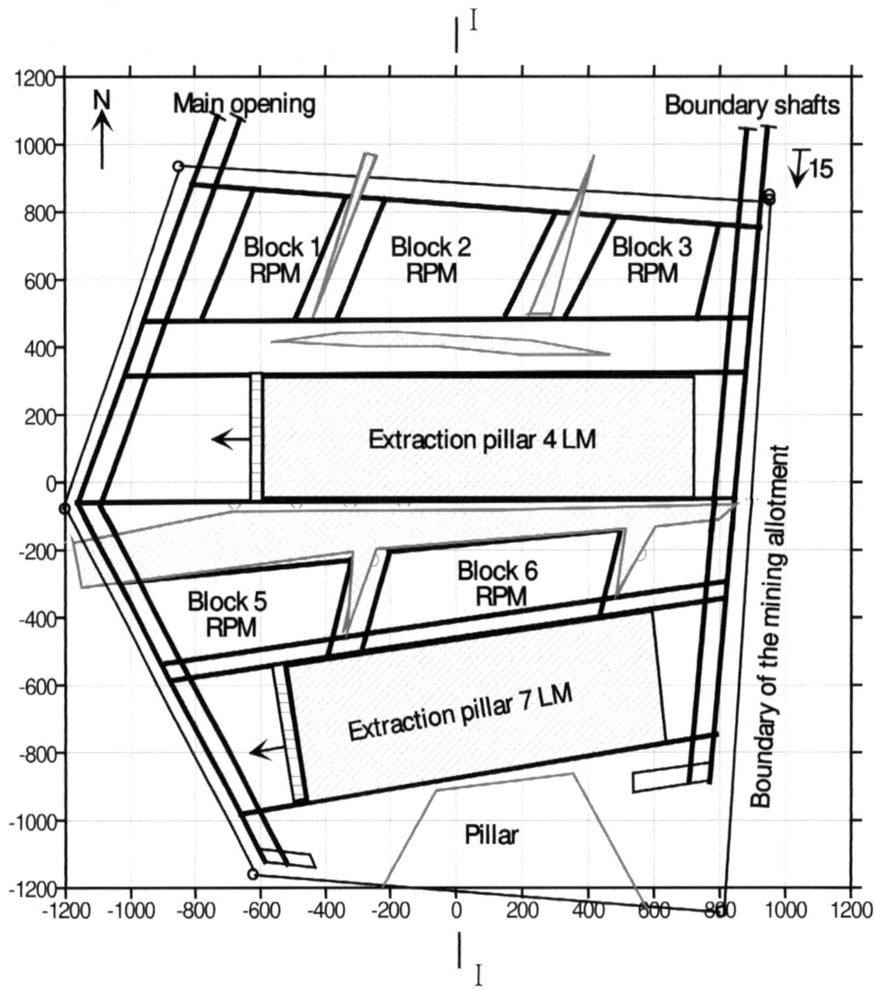

Figure 1. Schemes of opening, preparation and mining the coal seam in difficult mountain-geological conditions.

Having regard to the above, scientific research providing rational spatio-temporal location of mining excavations with a simultaneous combined mining of rectangular-shaped sites by long mechanized faces and oblique areas with limited coal reserves by short faces is currently relevant. As an alternative geotechnology for mining the oblique sites with limited coal reserves, special methods can be used, with elements of automation and robotization of technological processes [8 - 11].

2. Methods of research

For the geomechanical substantiation of the parameters and the choice of the rational spatial and temporal location of the mine workings with the simultaneous combined mining of sites within the mine field, the following methods were used: analysis and generalization of the experience of mining the sections of coal seams in accordance with a sequential scheme, numerical modeling on the basis of the finite element method for quantitative prediction of geomechanical parameters under the mutual influence of long and short cleaning faces, scientific generalization of the research results.

3. Results and discussion

To solve the task, the following research program was developed and implemented: the formation of a geomechanical model of the rock massif with different options of mine workings location, numerical modeling of geomechanical processes; identification of regularities in the distribution of geomechanical parameters of coal bed development systems, the choice of a rational option of a simultaneous combined scheme for mining rectangular shaped sections by long mechanized faces and oblique blocks with limited coal reserves by short faces.

The geomechanical model of the rock massif with different options of location of mine workings included vertical sections along the line of incidence of the seam (I-I in figure 1).

The rock massif in the model is assumed to be layered with an isotropic distribution of layers in each layer, the number of layers is 100. The thicknesses of layers and the physical and mechanical properties of the rocks are accepted for the conditions of Yerunakovskiy district in Kuzbass. The field of investigation is limited from above by the earth's surface, from below by the boundary of the zone of overwork, adopted according to [1, 2]. At the lateral boundaries of the model vertical displacements were determined in the gravitational stress field according to the Dinnik theory [3]. On the bottom of the model, the vertical displacements are assumed to be zero. The dip angle of the seam is equal to 0°, the extractable seam thickness is 3.00 m. The model dimensions for seam dip, according to figure 1, are equal to 2400 m, the maximum depth of the bed is 400 m. In the zones of disjunctive disturbances, the increased fracture and fall zone was taken into account by the coefficient of structural weakening according to [2].

A quantitative forecast of geomechanical processes was carried out by the finite element method in a plane-deformed two-dimensional formulation of the problem using the author's package of problem-oriented programs [4-8].

Based on the results of numerical modeling, vertical, horizontal and principal stresses, as well as vertical and horizontal displacements and deformations of rocks within the entire model were determined. The calculations were carried out in stages by solving the following tasks:

1) elastic without taking into account the influence of mine workings;

2) elastic, taking into account the influence of mine workings and disjunctive disturbances;

3) elastic-plastic by changing the modulus of elasticity in the "stress-strain" diagram by the secant deformation module with a preliminary estimate of the out-of-limit state of the rocks according to the Mohr-Coulomb strength and strength energy theory.

In the process of numerical simulation several options of the location of workings and mined sites are considered (table 1). Considering the large amount of information for comparing the geomechanical parameters of the rock massif, one integral indicator is taken – the coefficient of of vertical stresses concentration.

Table 1. Options for location of the exhausted working areas.

Option No.	Layout of workings	Maximum coefficient of vertical stresses concentration	
		at the short face	at the long face
1	Worked extraction pillars LM 4, LM 7	2.48	-
2	Worked blocks RPM 1-3	-	1.77
3	Worked blocks RPM 1-3, pillar RPM 4	2.82	2.44
4	Worked blocks RPM 1-3, pillar RPM 4, blocks RPM 5-6	3.02	2.48
5	Worked blocks RPM 1-3, pillar RPM 4, blocks RPM 5-6, pillar RPM 7	3.61	2.62

Figure 2 shows the isolines of the distribution of the vertical stress concentration factor in the rock massif after selective excavation by long faces (figure 2a) and sequential mining by short and long faces (figure 2b).

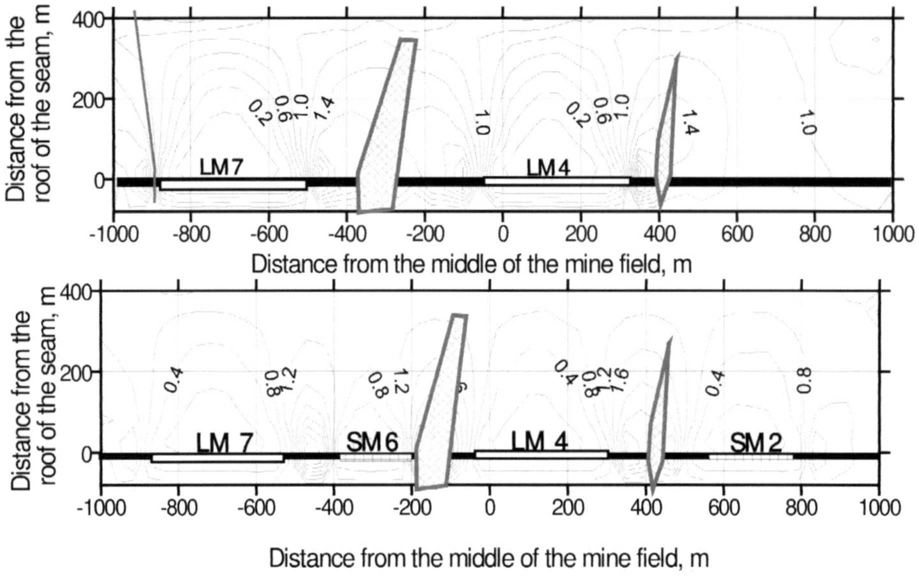

Figure 2. Distribution of the coefficient of of vertical stresses concentration in the rock massif: a – after selective mining of working areas RPM-4 and RPM-7 (option 1 in table 1); b – after subsequent mining of working areas SM 1-3, RPM-4, SM 5-6 and RPM-7 (option 5 in table 1).

Based on the comparison results of five options for placement of mining blocks and pillars, the following dependencies are established:

- the coefficient of vertical stresses concentration during mining of a single block by short faces is 1.4 less than the corresponding coefficient for mining the seam by long faces;
- the height of the zone of tensile vertical stresses (unloading zones) over the gob is 100-120 m, that is, approximately 30% of the length of the gob;
- sequential mining of blocks by short faces and pillars by long faces leads to redistribution of vertical stresses with a gradual increase in the concentration coefficient over the coal pillars between the gobs up to 3.61;
- disjunctive disturbances are concentrates of increased stresses and are the centers of probable occurrence of gasdynamic phenomena.

Features of the distribution coefficient of vertical stresses concentration in the vicinity of blocks mined by short faces are shown in figure 3. The parameters of the zone of bearing rock pressure at the marginal sections of the seam taking into consideration the influence of the disjunctive disturbance are compared. It is established that in the zone of influence of the disjunctive (to the left from block RPM 1-3 in figure 3) the width of the zone of bearing rock pressure reaches 200 m, and in the absence of violation – 130 m. Therefore, when choosing the spatial position of the development workings and the width of the pillars, disjunctive violation as a source of occurrence of gas-dynamic processes should be taken into account.

The distribution isolines of the coefficient of vertical stresses concentration with the mutual influence of the gobs after mining the seam by short and long faces in the presence of disjunctive disturbances are shown in figure 4.

According to figure 4 the distribution of the vertical stress concentration coefficient in the roofs mined by long as well as short faces is identical. The most dangerous by geomechanical criteria is the coal pillar between the pillar LM 4 and blocks 1-3, since the contouring preparatory workings will be located in the dangerous zone in the vicinity of disjunctive disturbances. Especially it should be noted that in this zone there is a high probability of occurrence of gas dynamic phenomena in the form of mountain bumps or sudden outbursts.

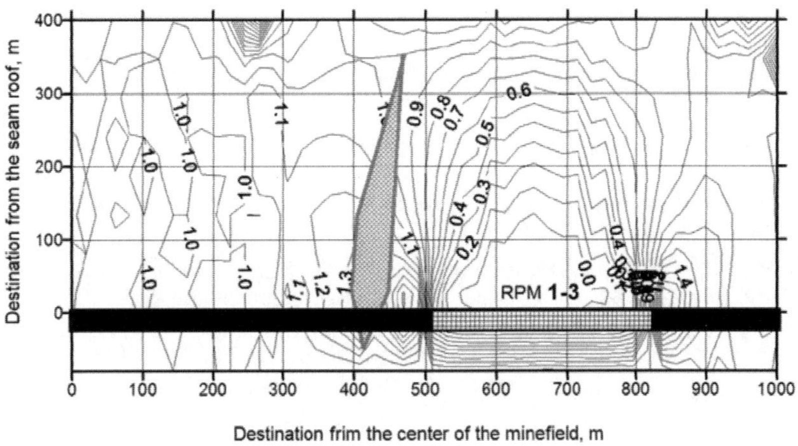

Figure 3. Distribution of the coefficient of concentration of vertical stresses in the rock massif after mining the extraction blocks RPM 1-3 (option 2 in table 1).

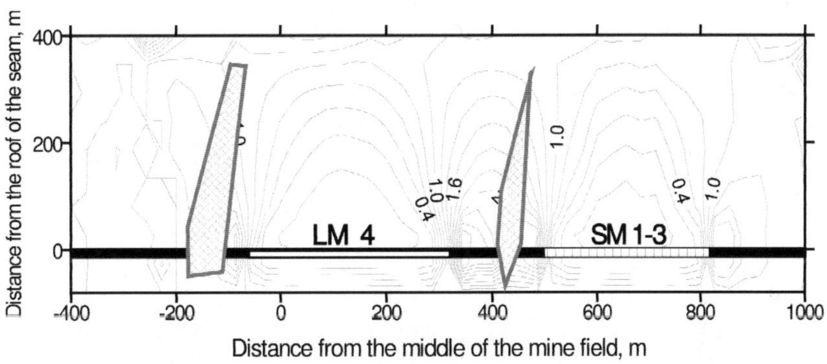

Figure 4. Distribution of the coefficient of vertical stresses concentration in the rock massif after mining the extraction blocks RPM 1-3 and LM 4 (option 3 in table 1).

On the basis of the forecast of geomechanical situation, it follows that the technology of preparatory workings and the extraction of coal in the faces must include robotic elements ensuring the performance of basic operations without the presence of people. To reduce the dangerous consequences of the occurrence of dynamic phenomena in the form of a rock outbursts or a sudden release of coal, rock and gas in the vicinity of blocks mined by short faces, it is also necessary to use the means of robot stoping of the coal seam.

4. Conclusion

Thus, the sequential mining of a coal seam within the mine field by combined technology is possible by geomechanical conditions and can be recommended for use in combination with the application of robotic means of coal extraction and transportation of rock mass.

The proposed approach will ensure an increase in the coal mining factor, reduce the probability of accidents and incidents and increase production efficiency by reducing the specific capital costs and the repeated use of development workings for combined mining by short and long faces.

IOP Conf. Series: Earth and Environmental Science **206** (2018) 012037 doi:10.1088/1755-1315/206/1/012037

Acknowledgements
The article was prepared with the financial support of the Russian Foundation for Basic Research within the framework of a research project 18-45-420003 p_a and under the agreement No. 17 dated 24-07-2018.

References

Reva V N, Melnikov O I and Rayskye V V 1995 *Maintenance of Mine Workings* (Moscow: Nedra) p 270
Guidance on the Rational Location, Protection and Maintenance of Mine Workings in the USSR Coal Mines 1986 (Leningrad: VNIMI) p 222
Dinnik A N, Morgaevsky A B and Savin G N 1938 *Proceedings of the Meeting on Mining Pressure* (Moscow: Academy of the USSR, Department of Technical Sciences) pp 7–55
Fryanov V N and L D Pavlova 2016 *IOP Conference Series: Earth and Environmental Science* **45** 012002
Fryanov V N, Pavlova L D and Temlyantsev M V 2017 *IOP Conference Series: Earth and Environmental Science* **84** 012001
Tsvetkov A B, Pavlova L D and Fryanov V N 2017 *IOP Conference Series: Earth and Environmental Science* **84** 012005
Cherepov A A and Pavlova L D 2017 *The Bulletin of KuzSTU* **6** 61–8
Kornev E S, Pavlova L D and Fryanov V N 2014 *Numerical Modeling of Geomechanical Processes in Mining of Seams by Short Faces: Monograph* (Prague: Vědecko Vydavatelské Centrum “Sociosféra-CZ”) p 206
Pavlova L D and V N Fryanov 2018 *Mountain Journal* **2** 48–52
Nikolaev P I and Zinoviev V V 2016 *Bulletin of Kuzbass State Technical University* **4** 26–33
Fryanov V N and Pavlova L D 2016 *Fundamental and Applied Questions of Mining Sciences* vol 2 **3** 202–10

IOP Conf. Series: Earth and Environmental Science **206** (2018) 012038 doi:10.1088/1755-1315/206/1/012038

Development of an automated process control system for the coal preparation plant

L P Myshlyaev[1], G V Makarov[1], V V Grachev[1], M V Lyakhovets[1], K G Wenger[2] and I A Leontyev[2]

[1]"Scientific-research Center for Control Systems" LLC, 42 Kirova street, Novokuznetsk, 654007, Russia
[2]"Stroyservis" CJSC, 121 Kuznetskiy ave, Kemerovo, 650055, Russia

E-mail: mail@nicsu.ru

Abstract. The experience of modernization of the automated process control system for the coal preparation plant in the conditions of technical re-equipment of the coal preparation plant "Barzasskoye Tovarischchestvo" LLC (Beryozovsky) is considered. The modernized automated process control system (APCS), the modernization goals and ways of their achievement are presented, the main APCS subsystems and the integrated functional and technical structure of the upgraded system are given. The procedure of commissioning of the upgraded system is described.

1. Introduction

The efficiency of mining enterprises directly depends on the composition and quality of the minerals mined, which cannot be affected. It is possible to increase the quality of the mined mineral, and therefore, the economic efficiency of mining in the process of enrichment, that is why coal preparation plants have been long existing side by side with mines. In addition, modern market realities are such that in the presence of constant demand it is necessary to increase the production capacity of the enterprise. Thus, any modern mining enterprise is a "living" developing system, constantly changing the technological structure and automated process control system, accordingly. So, the coal preparation plant "Barzasskoye Tovarischchestvo" LLC (Beryozovsky) started as a reloading station with the simplest sorting functions and currently it is a complex integrated enterprise that includes technological processes in the following branches

- coal reception and coal preparation;
- coal enrichment in the main building, including dense medium separation, thickening and dehydration, etc.;
- flotation-filtering department;
- drying department;
- storage and loading.

2. Brief characteristics of the control object

Within the current stage of technical re-equipment, two departments have been re-equipment: flotation-filtering department and drying department, also the existing technological complex of the concentrate storage was modernized [1].

The operating coal preparation plant is a complex facility with various technological equipment, continuous handling lines with their territorial distribution. Technological processes are automated, and the task of modernization includes, among other things, a targeted change in the existing automated system that allows the automation of newly introduced equipment and technological processes with minimal material and time costs to be implemented without disrupting the overall system performance.

The next stage of re-equipment is the modernization of the enrichment processes taking place in the main building and the completion of a new intermediate module between coal receiving and enrichment in the main building with the addition of a dense medium separation facility.

At present, the modernization of the technological complex is closely linked with the modernization of automation systems of the corresponding technological processes.

The integration of the APCS of the reconstructed loading complex and the projected APCS of the coal preparation plant under construction is ensured by

- creation of a single information space in the system of centralized collection, processing and storage of information in the automated system of operational dispatch control of the plant technological complex, the general discipline of data visualization and dispatching management;
- inclusion of the control algorithms of technological regimes coordination and mutual blocking of equipment at the connection points of the existing and projected technological complex in the APCS mathematical support;
- partial use of the APCS software and hardware of the reconstructed loading complex for the implementation of APCS functional systems of the coal preparation plant under construction.

The APCS, implemented within the framework of modernization, is intended for complex automation of control and control over production and technological processes, equipment of the re-equipped loading facility and the coal preparation plant, including the solution of the following additional tasks (taking into account the fact that the existing system fully performs the functions assigned to it and does not require the modernization of any component): centralized control, analysis and display of information about the state of production and technological processes and equipment of the coal preparation plant and the reconstructed loading complex.

3. The purpose of the system development

The purpose of modernization is to increase the efficiency of the enrichment process by integrating the following components into a single APCS: the systems of complex control at the flotation-filtering department, drying department, concentrate storage, as well as modernization of the department for coal receiving and loading and the main building. As a result, technical and economic indicators of its functioning improve: the unit costs for production decrease, the yield of product increases and the quality of mark products improves.

The set goal is achieved in the following areas.

1. Automation of information and control functions in solving problems of: immediate formation and analysis of information on changes in operation modes and condition of technological processes, units and equipment; immediate consistent correction of tasks according to the mode parameters of technological processes; immediate implementation of control decisions and regulation of technological parameters; control, recording and analysis of technological and production discipline violations, control effectiveness.

2. Improvement of reliability of the automated process control system, efficiency and quality of control due to: the use of modern technical means, methods and algorithms of automatic control, analysis, diagnostics and control of technological processes and equipment; complex and detailed display of information on the state of equipment and units, changes in technological parameters, actions of operational personnel in the system; use of modern microprocessor controllers with high reliability, long operation time between failures as a technical base on the lower level of the system, simple replacement of failed components, expansion of technical structure, modification of

mathematical support and software; minimization of the number of unreliable electromechanical devices in local systems of logical control and control over individual mechanisms and assemblies.

4. Functional structure of the system

The automated process control system at floating-filtering, drying departments and concentrate storage (APCS FFD) is developed as a component part of the integrated automated process control system at the coal preparation plant and the reconstructed technological facility of coal loading (APCS CPP) of "Barzasskoye Tovarischchestvo" LLC on the basis of the existing software and hardware of the APCS CPP operating in the technological complex of enrichment of the I and II stage and the loading facility.

The designed APCS are developed taking into account the maximum possible use of the resources of the already existing technical means of APCS, unification of the software and hardware of the projected and operating APCS. All previously commissioned and newly designed APCS of technological complexes are connected through a common interface, unified hardware and a uniform approach to the control of both local units and complexes as a whole.

The enlarged functional structure of the integrated APCS CCP is shown in figure 1, a. It consists of the following main systems:

1) the existing APCS CPP of the first and second stages, that includes

- an automatic system for the centralized collection, processing and storage of information of the coal preparation plant (CPP);
- engineering support system;
- automation control system of production (ACSP);
- automated dispatch control system (ADCS) of the CPP technological complex;
- local system of centralized control and loading control;
- systems for automatic control and control of technological complexes (ACS TK) of coal receiving facility, enrichment and loading of the I and II stages of the factory;

2) projected APCS CPP, that includes

- automatic system of the centralized collection, processing and storage of information of the floating-filtering and drying-furnace departments;
- automated dispatch control system of the technological complex of the floating-filtering and drying-furnace departments (ADCS FFD and DFD);
- system of automatic control and loading control of the newly installed equipment at the concentrate storage facility;
- system of automatic control and technological complex control of the floating-filtering department (SAC TCC FFD);
- system of automatic control and technological complex control of the drying-combustion department, supplied alongside with the technological equipment, including the local system of centralized control and control of drying units (SAC TCC DFD).

The enlarged functional structure of the automated process control system of the reconstructed technological complex for coal loading and the coal preparation plant being built is given in figure 1, b. It consists of the following main systems

1) the reconstructed APCS *for the loading complex, that includes*

- the system of centralized collection, processing and storage of information of the reconstructed loading complex;
- engineering support system;
- industrial process control system;
- automated dispatch control system (ADCS) of the reconstructed loading facility;
- local system of centralized control (enrichment module);
- systems for automatic control of the existing enrichment and coal receiving facilities;
- system of automatic control of the main building;

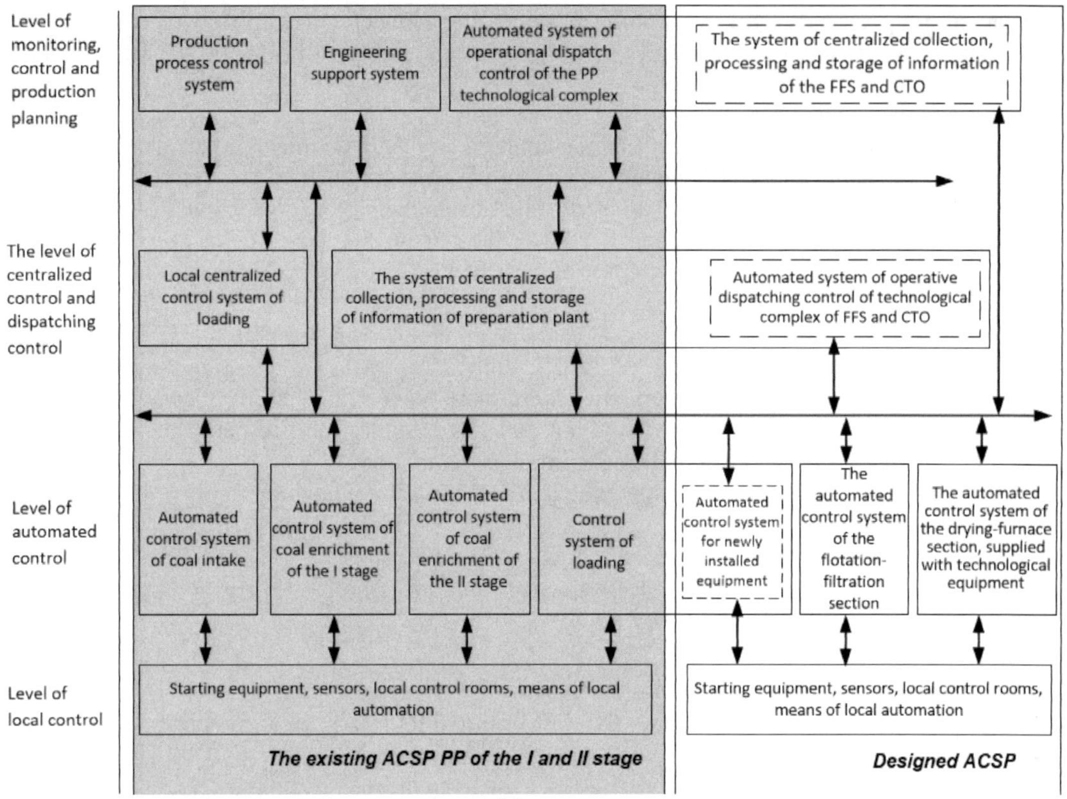

(a)

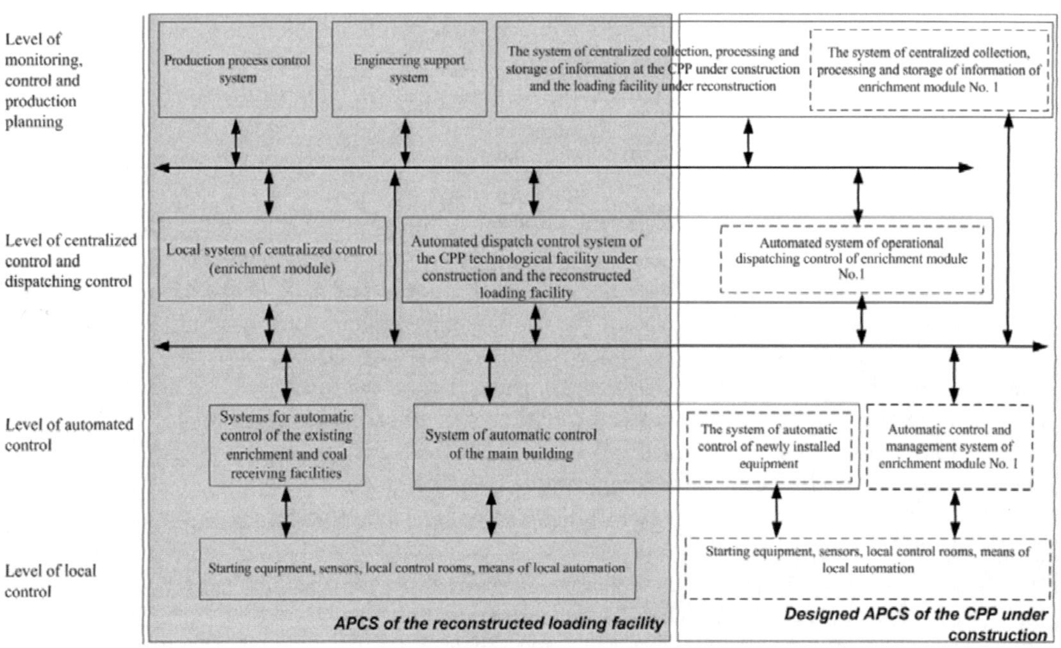

(b)

Figure 1. Schemes of the functional structures of integrated APCS CPP of "Barzasskoye Tovarischchestvo" LLC.

2) the projected APCS of the coal preparation plant under construction, that includes

- system of the centralized collection, processing and storage of information of the enrichment module No. 1;
- automated system of dispatch control of the enrichment module No. 1;
- system of automatic control of the newly installed equipment in the main building;
- system of automatic control of the enrichment module No. 1.

The automatic system of centralized collection, processing and storage of information on the technological complex of the floating-filtering department and the ADCS FFD is implemented on the basis of the software and hardware of the existing APCS CCP of the first and second stages by expanding its information support and software.

The projected APCS is a two-level system.

The upper level

- automated system of dispatch control and control of technological complexes of the floating-filtering and drying-furnace departments;
- automated system of dispatch control of the enrichment module No. 1.

The lower level

- automatic control systems for technological complexes at the floating-filtering department, drying-furnace department, loading facility (concentrate storage facility).
- system for automatic control of the enrichment module No. 1, the system for automatic control of the newly installed equipment in the main building.

The application of this kind of two-level automation systems has been tested at other coal industry enterprises and has proved its efficiency.

5. Technical structure of the system

The Scheme of the enlarged technical structure of the automated process control system of the technological complex at the coal loading facility and the coal preparation plant of the open pit mine "Barzasskiy" is presented in figure 2. The scheme of the enlarged technical structure of the automated process control system at the flotation-filtering and drying-furnace departments is given in [1]. The system is based on microprocessor programmable controllers, servers, personal computers and touchpads.

The hardware implementation of the upper-level systems, including existing systems of the integrated APCS CPP (control system for production processes, engineering support system, system for centralized control over the I and II loading stages; system of the centralized collection, processing and storage of information; automated dispatch control system and control of the technological complex of the plant), as well as the projected system of the APCS FFD (the automated control system for technological processes, flotation-filtering, drying-furnace departments and a concentrate storage facility) are executed on the basis of servers and personal computers of corporation Hewlett-Packard, DELL, HMI-terminals by Omron.

The existing software package Genesis32 from ICONICS v.9.21 is used as the basic software of the upper level of the newly introduced APCS FFD. The function characteristics of this software package is sufficient to perform all the automated functions, the purchase of additional software modules is not required.

The *lower-level* subsystem is implemented in microprocessor programmable logic controllers CJ2 by Omron and provides: reception and processing of discrete signals about the state of equipment and units, generation of discrete signals for units control, realizing the functions of local, remote and automatic control of unit equipment; reception and processing of signals of the measuring information from local systems of measurement and control of technological parameters; it is possible to receive standard analog and discrete signals; the exchange of data between controllers via a specialized controller network Controller Link; data exchange via the Ethernet information network with workstations, with servers and with HMI terminals of the upper-level system; the formation of

IOP Conf. Series: Earth and Environmental Science **206** (2018) 012038 doi:10.1088/1755-1315/206/1/012038

start/stop commands of the units automatically or by commands of the dispatcher in accordance with the current state of the process and units, the regulations requirements.

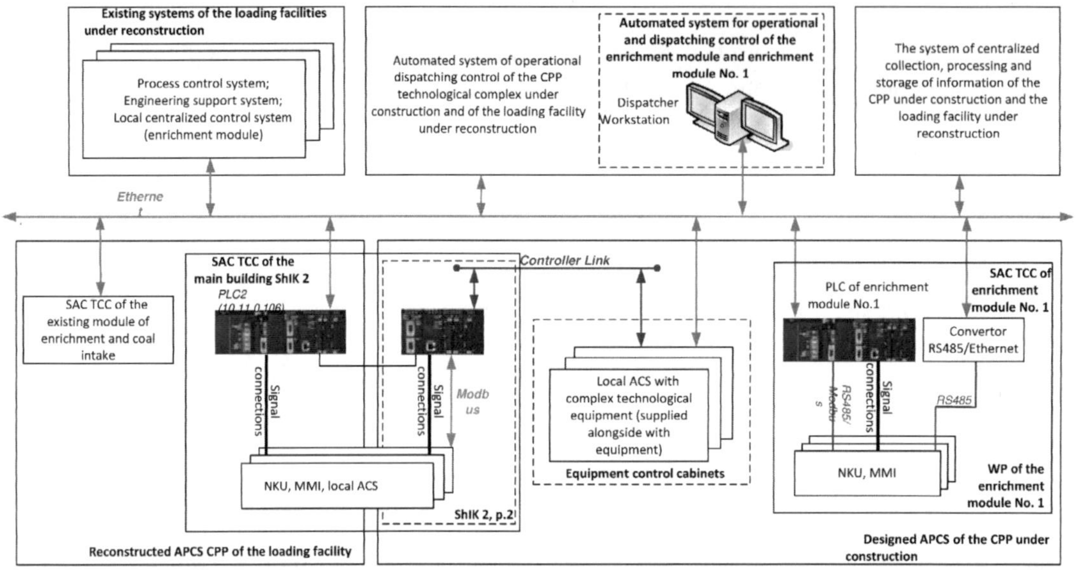

Figure 2. Scheme of the enlarged technical structure of the APCS CPP.

The selection of input-output modules, construction of a system for collecting, processing and transmitting information and control actions are based on the scheme of the enlarged technical structure of the system (figure 2), the composition of equipment and units of the plant technological complex.

The interconnection of the APCS controllers with the means for measuring and controlling the parameters of technological processes and the state of the equipment is provided by analog (4-20mA) and discrete (24VDC) signals. To transfer control actions to the control circuits of electric drives, “dry contacts” are used. The quick mounting system Omron G70A-ZOC16-3 is used to ensure galvanic isolation of discrete outputs of controllers.

6. System installation

The APCS was put into operation in the following way.

At the first stage, the upgraded automatic control and management system for the technological complex of the concentrate storage facility was put into operation.

At the second stage, the automatic control system of the technological complex at the floating-filtration department was put into operation.

At the third stage, it is planned to put into operation an automatic control system for the technological complex of the drying-furnace department, supplied alongside with technological equipment, including a local system for centralized control of drying units.

The further reconstruction includes the addition of new enrichment modules and modernization of the existing ones with the subsequent integration of all APCS CPP into a single system with a centralized control.

The task of modernization of the existing automation system is complicated by the fact that the CPP is not closed for reconstruction and cannot be stopped for a long time remaining an active production.

The automation system put into operation within the framework of the reconstruction is directly connected with the existing technological complex and its control system, for example, an automated system of dispatch control. Integration of the newly introduced system with the existing system must be performed on the stopped technological equipment.

However, in practice, the approved production plans allow the production equipment to be stopped only for the performance of preventive maintenance (PM), which is often not enough to perform integration and maintenance procedures. Thus, the commissioning of the updated automation systems is done piece by piece during the PP works, competing in time with the work carried out by the maintenance personnel of the production facility. This significantly delays the commissioning period [2].

Therefore, it is more rational to start the modernization of the system from the formation and installation of cabinets with new equipment, given the close integration of the installed equipment with the existing one. Then, step by step (either by units or by technological subsystems) to transfer the functions of the old system to the new one with the possibility of parallel operation of both systems. Dismantling of obsolete equipment is done in the end, only after the new system is put into operation.

When the control cabinets are ready and the signal lines from the control panels of the units to the PLC cabinets are completed the following is performed

- the alternate inclusion of temporary control schemes of process units that can be stopped for commissioning; and it is necessary to provide for the possibility of functioning of part of the units under the control of the old system, and part of the units – under the control of the new system;
- with the time scheme, the control of the upgraded units is transferred to the controller, however, for the purposes of units control, the signal lines of the old system are used; the overall control of the technological complex and other units remains unchanged, the control also remains on the existing mimic diagram; at this stage, the debugging of the software of the system being put into operation is performed;
- after debugging the software of the controllers of all units, a transfer to the final control scheme of the units;
- when the means of upper level control (SCADA-system) are ready, the dispatching control is transferred from the mimic diagram to the workstation;
- after debugging the local control system of each unit, the control of the whole technological complex is transferred with the help of the controllers of the updated automated control system.

7. Conclusion

The application of this kind of phased modernization, on the one hand, requires a great deal of time to put the updated system into operation, on the other hand, it allows all routine maintenance of the system modernization to be performed without long stops of technological equipment [3].

An important positive factor for the creation of the APCS CPP of “Barzasskoye Tovarischchestvo” LLC is the understanding by the customer of the need for the consistent development of a control system by the specialists of one organization providing uniform solutions for software, technical, algorithmic, and information support and maintenance of the previously adopted principles at all levels of control.

Even in the presence of a large number of local automation systems, it is necessary to ensure that the uniform designations and approaches adopted in the single space of the APCS CPP are used to give the operator an opportunity to concentrate on performing his basic functions, rather than studying heterogeneous interfaces of local control systems or individual complexes provided by different suppliers, that are not always can be integrated into a single environment. Analyzing the accumulated experience [4-6] it can be clearly seen that it is necessary to carry out the reconstruction of one stage with regard of the APCS further development and to make decisions taking into account the subsequent “seamless” integration of the created subsystems into a single modern control system at a complex multi-stage production such as “Barzasskoye Tovarischchestvo” LLC.

Acknowledgement
The work is carried out by the State Task of the Ministry of Education and Science of the Russian Federation No. 8.8611.2017/8.9.

Rerferences

[1] Liakhovets M V, Wenger K G, Myshlyaev L P, Shipunov M V, Grachev V V and Melkoserov M Yu 2017 *Proc. of the XI All-Russian Sci. and Pract. Conf. on Automation Systems in Education, Science and Production* (Novokuznetsk: SibSIU) pp 151–156
[2] Myshlyaev L P, Lyakhovets M V, Leontiev I A, Wenger K G and Salamatin A S 2017 *Proc. of the XI All-Russian Sci. and Pract. Conf. on Automation Systems in Education, Science and Production* (Novokuznetsk: SibSIU) pp 55–57
[3] Ivushkin A A, Grachev V V and Myshlyaev L P 2007 *Proc. of Int. Conf. on High Technologies in Development and Use of Mineral Resources* (Novokuznetsk: SibSIU) pp 247–252
[4] Mochalov S P, Ivushkin A A, Myshlyaev L P, Wenger K G, Murko V I, Kutsenko A I, Fedyaev V I and Karpenok V I 2012 *Coal* **10** 49–53
[5] Ivushkin K A, Shipunov M V, Grachev V V and Myshlyaev L P 2013 *Proc. of Int. Conf. on High Technologies in Development and Use of Mineral Resources* (Novokuznetsk: SibSIU) pp 185–189
[6] Grachev V V, Shipunov M V, Ivushkin K A, Myshlyaev L P, Tsiryapkina A V, Linkov A A and Makarov G V 2015 *Proc. of the X All-Russian Sci. and Pract. Conf. on Automation Systems in Education, Science and Production* (Novokuznetsk: SibSIU) pp 191–195

Using of the modified approach "hardware-in-the-loop" while developing an process control system for belt conveyor

S S Zhuravlev, S V Rudometov, V V Okolnishnikov and S R Shakirov

Institute of Computational Technologies, Siberian Branch of the Russian Academy of Sciences, Ak. Lavrentieva ave, 6, Novosibirsk, 630090, Russia

E-mail: okoln@mail.ru

Abstract. The method to test application software of a process control system for belt conveyor on the base of the modified approach "hardware-in-the-loop" is proposed. The description and example of using of specialized software-hardware complex for testing of application software of process control systems is given.

1. Introduction

Alterations of software and hardware of the process control systems during commissioning and start up slows down the process of implementation and can cause emergency conditions.

Such alterations can be required as the need arises to integrate the new or modified software and hardware into existing process control systems. In this connection not only adjustment and testing is required but checking of compatibility of the new system components with the components of previous versions working on the operating enterprise. It is possible to solve the problem with the help of adjustment and testing of application software for process control systems with the help of specialized software-hardware complex adapted to checking of compatibility of the tested system with previous versions.

2. Specialized software-hardware complex

The complexity of complex debugging and testing of process control systems resides on laboriousness of artificial shaping the complete set of agreed concurrent signals of the real technological equipment.

In order to develop, debug and test the application software of process control systems such facilities as software and hardware simulators of signals and interfaces are used. But specialized systems with the problem oriented models of automated technological processes are more effective. They ensure the agreed generation of sensor and control signals.

To solve the problem of debugging and testing of application software for process control systems there exist different specialized systems with the problem oriented models, for example [1, 2] and others. The specialized software-hardware complex for testing the application software for process control systems in mines is developed in the Institute of Computational Technologies of Siberian Branch of the Russian Academy of Sciences [3].

This specialized software-hardware complex tests the application software by substituting the actual signals from sensors of operating technological equipment and controller commands to virtual signals formed on the base of values of simulated parameters of the technological equipment.

The complex has two level structure containing software and hardware levels. The software level contains the SCADA system, the simulation environment MTSS [4], a communication dispatcher, input/output modules of process control system, simulation models of technological processes, simulation models of technological equipment, control programs for the upper and the lower levels of process control system.

The hardware level contains a workstation of the model operator, a workstation of the SCADA operator, data transmission environment, a unit for forming actual signals, and equipment of process control system (controllers, input and output equipment, sensors, etc.). The unit for forming actual signals generates corresponding analogous and discrete signals basing on model data.

The simulation model takes into account the peculiarities of the hardware, i.e. parameters of technological equipment (pumps, engines, etc.), the project data on the technological equipment configuration, and the parameters of the actual technological processes.

The complex is used in the Institute of Computational Technologies of Siberian Branch of the Russian Academy of Sciences while testing the application software for mine process control systems.

3. Process control system for coal mining

The Institute of Computational Technologies of Siberian Branch of the Russian Academy of Sciences is developing process control systems of dangerous industrial projects including technological processes of coal mining and transportation.

According to functions the process control systems which is develop in the Institute can be divided into two parts – underground (lower) level and surface (upper) level. Lower level provides control and data collection from different technological equipment of the mine. Upper level performs the functions of central calculating processor and dispatcher's workplace.

As the process control systems develop more and more complex algorithmic tasks and functional requirements were entrusted on them. Moreover different teams of developers participated while developing versions of the process control systems and there can be cases of interaction of different versions or implementations in the frame of integrated automation system. In this connection there exists the necessity to check the compatibility of components of the developed process control system with the previously developed ones.

4. Modified approach "hardware-in-the-loop"

In the frames of the existing software-hardware complex the task the checking the compatibility of components of the developed process control system with the previously developed versions or implementations was solved by modifying the approach "hardware-in-the-loop". The approach "hardware-in-the-loop" involves automatic testing of process control system (controllers) with signals which numerical value are formed with the usage of mathematical model and transformation of model signals into virtual is done by corresponding transformers.

The approach is modified in such a way that in order to test the controller together with the model the connected in parallel controller of previously version or implementation (the reference controller) is used. This allows carrying out the test and obtaining identical functioning of the test and the reference controllers on the identical sets of input signals.

Figure 1 depicts the diagram of implementation of the modified approach. The simulation model of the technological process generates virtual signals and the unit of generation of the actual signals transforms them into their electric analogues. The parallelization of discrete and analogous signals is carried out in the unit of duplication of actual signals. After that the signals go simultaneously to the test controller and the reference controller. The control commands of the upper level of the process control system pass to the controllers through the interface RS-485. The signals formed by the controllers go to the unit of generation of actual signals and after transforming into digital form go to the technological process model where in the specialized unit the comparison of signals from the test controller and the reference controller is performed. The model of technological process simulates control commands from the test controller.

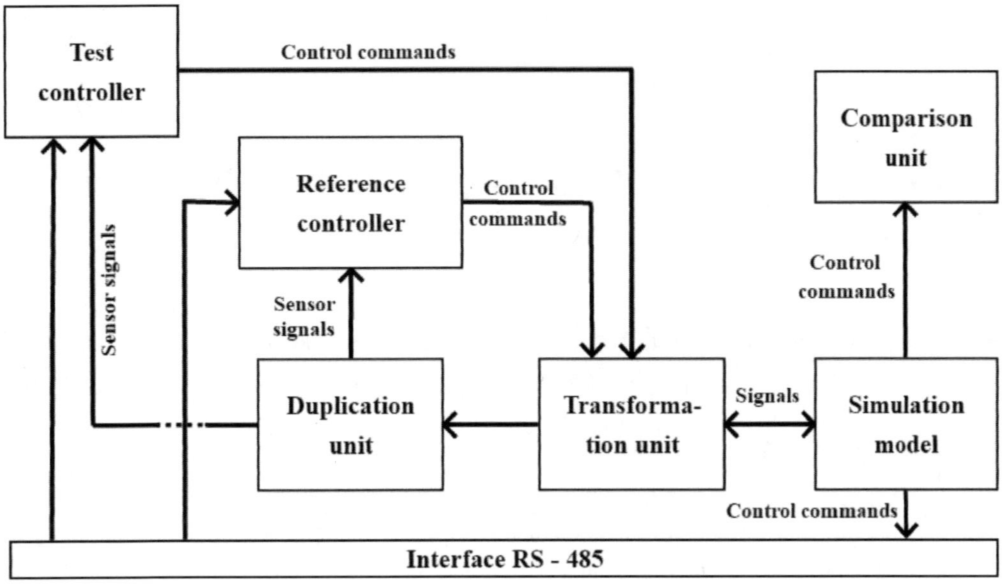

Figure 1. The diagram of interaction of main modules of software-hardware complex while using modified approach "hardware-in-the-loop".

5. Conclusion

The software-hardware complex with the modified structure was applied to develop one of process control system designed to automate the process of coal mining and transportation. The controllers tested on the software-hardware complex were successfully introduced on the mine Gramoteynskaya (Kuzbass) to control of belt conveyers.

The peculiarity of the software-hardware complex is the existence of the mode of operation when several controllers work concurrently on the identical sets of test signals. The simulation model works as the source of signals and analyses responses of control system. It allows to test the compatibility of input signals of the test controller and the reference controller, to accelerate the verification of the developed applied software.

The developed software-hardware complex can be applied to debugging and testing of the applied software for perspective control process systems for other industrial branches for example while developing the process control system for movement of roof supports when the coal mining is produced by longwall.

Acknowledgements

This research was partly financially supported by the Russian Foundation for Basic Research (project 16-07-01179).

References

[1] Ryssel U, Ploennigs J and Kabitzsch K 2012 *Science of Computer Programming* **77** 83–95

[2] Kelemenová T, Kelemen M, Miková L, Maxim V, Prada E, Lipták T and Menda F 2013 *American Journal of Mechanical Engineering* **1(7)** 276–281

[3] Okolnishnikov V V, Rudometov S V, Shakirov S R and Zhuravlev S S 2017 *International Journal of Control Systems and Robotics* **2** 61–6

[4] Rudometov S V and Okolnishnikov V V 2013 *Proc. of 7th IFAC Conf. on Manufacturing Modelling, Management, and Control* (Saint Petersburg: IFAC) pp 653–8

IOP Conf. Series: Earth and Environmental Science **206** (2018) 012040 doi:10.1088/1755-1315/206/1/012040

Operation algorithm of the hoist electric drive based on a doubly-fed machine with combined control

V Yu Ostrovlyanchik, I Yu Popolzin and V A Kubarev

Siberian State Industrial University, 42 Kirova street, Novokuznetsk, 654007, Russia

E-mail: eidoline@yandex.ru

Abstract. Currently, a large number of existing mine hoisting installations use the electric drive based on an induction motor with a rotary station. Despite the fact that such drives have a large overload capacity and linear (within the working areas) mechanical characteristics, they have drawbacks, first of all – significant energy losses. A promising way for modernization of such drives, as well as design of new ones based on a wound-rotor induction motor, is to use a doubly-fed machine. In the article the operation algorithm of the hoist electric drive on the basis of a doubly-fed machine is given. The algorithm realizes a combined method of controlling the machine in the speed function, combining asynchronous and synchronous modes of operation with the mode of changing the active rotor resistance. The control of the hoist electric drive in accordance with the proposed algorithm allows the required range of machine speed control to be reached while maintaining the overload capacity on the entire control range.

1. Introduction

The working cycle of a mine hoisting systems (MHS) contains several characteristic modes conditioned by the speed diagram and the diagram of the unit driving forces [1]. These modes include:

- The torque control on the locked motor. This mode is necessary for the realization of the machine retraction, that is to exclude the reverse stroke of the machine drum when the mechanical brake is removed and the movement starts. In this mode, the motor must develop the required torque with the stalled rotor;
- Movement at a low speed. In this mode, the vehicle exits the dumping curves, also the vehicle moves in the inspection mode;
- Speeding-up to the specified speed with a given acceleration;
- Movement at a given speed;
- Slowdown with a specified acceleration.

The presence of these operating modes imposes the following requirements for the MHS electric drive:

1) A large range of speed regulation due to the ratio of maximum and minimum stabilized speeds of vehicles. The required control range should be at least 30:1;

2) Constant stiffness $\beta = \frac{d\omega}{dM} = const$ (where ω is the angular speed, M is the moment of the electric drive) and the required overload capacity $\lambda = \frac{M_{cr}}{M}$ (where M_{cr} is the critical moment on the specified characteristic) of the drive mechanical characteristics over the entire speed control range.

The majority of existing hoisting installations use an electric drive based on a wound-rotor induction motor (WRIM) and a rotary station. The specified type of a drive satisfies the above requirements, but has the following disadvantages:

1) Useless energy losses due to the rotary station resistances, which can reach 30% of the power consumed by the drive motor;

2) Dynamic overloads in the mechanical part of the drive, which occur when switching stages of the rotary station.

2. The combined control of the double-feed machine in the hoist electric drive

Promising is the application of the scheme of a doubly-fed machine (DFM), which is realized by connecting the frequency converter to a WRIM rotor.

For the DPM, two operating modes are possible [2], [3], [4]:

1) Asynchronous mode. In this mode, the frequency of the additional rotor voltage is set equal to the current frequency of the eigen EMF of the rotor, and the speed is controlled by changing the amplitude and phase of this voltage. The DFM mechanical characteristics in this mode are similar in their shape to the mechanical characteristics of the induction motor. In [5] the condition for maximizing the machine torque over the phase of additional voltage is obtained; if it is observed or the phase shift is fixed at a selected constant value, it is possible to adjust the DFM speed only by changing the amplitude of the rotor voltage. In this mode, because of the drop in the machine overload capacity, its speed control is possible only in the range 0.5ω0 <ω≤ω0 (where ω_0 – the ideal idling speed of the WRIM), which makes this mode suitable for the MHS operation at the specified maximum speed and for speeding and slowing-down the vehicle in the specified speed range. In this mode, it is possible to recuperate the energy of slip into the network.

2) Synchronous mode. In this mode, the frequency of the auxiliary rotor voltage is set independently of the frequency of the eigen EMF of the rotor in accordance with the expression:

$$\omega = \omega_0 \frac{f_S - f_R}{f_S} = \omega_0 - \omega_R, \tag{1}$$

where f_S, f_R – frequencies of stresses on the stator and boost voltage on the rotor, respectively; $\omega_R = 2\pi f_R$.

The speed in this mode is controlled by changing the frequency of boost rotor, the amplitude of the voltage controls the machine overload capability, and the phase shift controls the reactive power consumed. The mechanical characteristics of the machine in this mode are similar to the mechanical characteristics of a induction motor – they have an infinite negative stiffness within the working area. If the overload capacity is exceeded, the DFM goes into synchronous mode on an unstable part of the mechanical characteristic. The speed regulation in this mode is expedient within the speed range 0.1ω0≤ω≤0.5ω0, since in the indicated speed range the DFM in synchronous mode has the maximum overload capacity.

From the analysis it follows that the above-described modes (asynchronous and synchronous) ensure the operation of the machine within the speed range 0.1ω0≤ω≤ω0 and are unsuitable in the retraction mode, during speed adjustment when exiting the curves and working in the inspection mode. In this case, the power consumed by the engine is about 10% of the rated power and it is possible to apply the regime with a controlled current converter in the rotor circuit. In this case, the initial equation for the analysis and construction of the algorithmic structure can be the equation of the machine mechanical characteristics:

$$M = \frac{3U_S^2 k^2 R_{R\Sigma}}{\omega_0 s \left[\left(R_S + k^2 \frac{R_{R\Sigma}}{s} \right)^2 \right]}. \tag{2}$$

where U_S –the effective value of the stator voltage; k – coefficient of machine transformation; s – slip; R_S– active stator resistance; $R_{R\Sigma}$ – total active resistance of the rotor chain.

Thus, the combination of the above-described modes of DFM operation makes it possible to realize all sections of the MHS speed diagram:

1) The control of the machine torque within the speed range $0 \leq \omega \leq 0{,}1\omega_0$ is realized by controlling the active component of the rotor current;

2) The work of the MHS and its acceleration/deceleration in the speed range $0.1\omega_0 \leq \omega \leq 0.5\omega_0$ are expedient in synchronous mode;

3) The MHS operation its acceleration/deceleration within the speed range $0.5\omega_0 < \omega \leq \omega_0$ are expedient in the asynchronous mode.

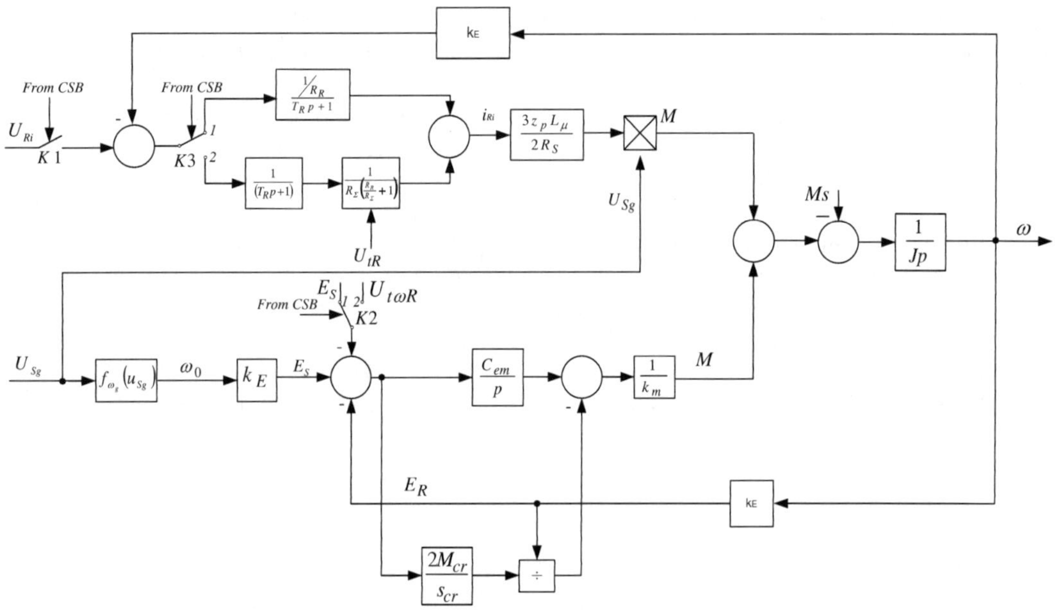

Figure 1. Structural diagram of a double-feed machine.

Thus, the electric drive of the double-feed MHS and controlled current converter in the rotor circuit satisfies the requirements for the drive of hoisting machines provided that combined control is performed combining these three operating modes of the machine.

The block diagram of the electric drive using the combination of all three operating modes will take the form shown in figure 1.

In figure 1 the following notations are used:

U_{Ri} – the active component of the boos rotor voltage;

R_R – active resistance of the rotor;

$R_\Sigma = R_R + R_{add}$ – total active resistance in the rotor circuit;

R_{ad} – additional active resistance in the rotor circuit;

T_R – electromagnetic time constant of the rotor circuit;

k_E – coefficient of internal feedback on the rotor EMF;

i_{Ri} – active component of the rotor current;

Z_P – the number of machine pole pairs;

L_μ – the mutual inductance of the stator and rotor;

R_S – active resistance of the rotor;

IOP Conf. Series: Earth and Environmental Science **206** (2018) 012040 doi:10.1088/1755-1315/206/1/012040

U_{Sg} – stator voltage;

$f_{\omega_g}(u_{Sg})$ – functional relation between the amplitude and frequency of the voltage at the stator;

E_S – stator EMF;

$U_{t\omega R}$ – voltage of the setting the boost voltage frequency on the rotor (speed in synchronous mode);

U_{tR} – voltage of the setting the value of additional active resistance in the rotor circuit;

C_{em} – electromagnetic stiffness;

M_{cr} – critical moment on the natural mechanical characteristic of the WRIM;

S_{cr} – critical slip on the natural mechanical characteristic of the WRIM;

K_m – motor transmission ratio;

M – electromagnetic moment;

M_s – moment of loading (static moment);

J – total moment of inertia of the MHS mechanical part reduced to the rotor of the hoisting motor;

ω – angular velocity of rotor rotation;

K1, K2, K3 – switches (logic functions) specifying the mode of DFM operation;

CSB – converter synchronization block (see figure 2).

For the total active resistance of the rotor circuit, the following condition is fulfilled:

$$0 < R_{\Sigma} \leq \infty \ . \tag{3}$$

The diagram shown in figure 1 works as follows. If it is necessary to operate the machine within the speed range $0 \leq \omega \leq 0{,}1\omega_0$, the switch K1 is open, K2 is in position 1, K3 is in position 2, and the machine is powered only by voltage U_{Sg}. In this case, the voltage U_{3R} controls the value of the additional active resistance in the rotor circuit so that the moment of the machine is equal to the specified value.

If it is necessary to provide synchronous operation of the machine, K1 is closed, K2 is in position 2, K3 is in position 1. In this case, a boost voltage in the rotor circuit is acting with an amplitude U_{Ri} and the specified frequency ω_R independent of the rotor speed of the machine ω. The rotor speed in this mode is set by the expression $\omega=\omega_0-\omega_R$.

If it is necessary to provide an asynchronous mode of operation, the switch K1 is closed, K2 is in position 1, K3 is in position 1. The boost voltage is applied to the rotor with an amplitude U_{Ri}, the frequency of which is equal to the frequency of the rotor EMF, and the value of the active rotor current i_{Ri}, determining the moment of the machine, is controlled by the change in the amplitude of the boost voltage.

The following system of logical expressions corresponds to the description provided above (we assume conditionally that the position of the switches K2 and K3 in position 1 corresponds to logical "0", and in the position 2 – to logical "1", the closed position K1 corresponds to "1" and the open "0"):

$$\begin{cases} (K1=0)\wedge(K2=0)\wedge(K3=1) - \text{mode of the controlledcurrent converter;} \\ (K1=1)\wedge(K2=1)\wedge(K3=0) - \text{synchronous mode;} \\ (K1=1)\wedge(K2=0)\wedge(K3=0) - \text{asynchronous mode} \end{cases} \tag{4}$$

Thus, when implementing a combined DFM control, the drive has a variable structure. The functional diagram of MHS electric drive based on DFM with a combined control, corresponding to the structural diagram in figure 1, is given in figure 2.

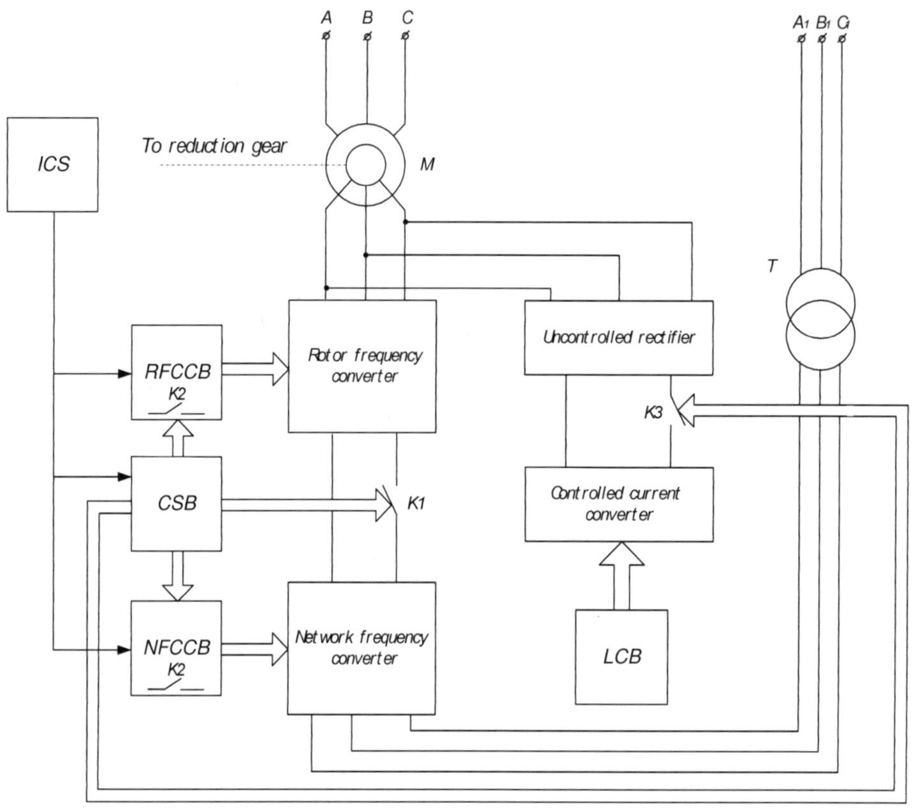

Figure 2. Functional diagram of MHS electric drive based on DFM with a combined control.

In figure 2, the following designations are used: M – wound rotor induction motor; T – power transformer; RFCCB – rotor frequency converter control block; NFCCB – network frequency converter control block; CSB – converters synchronization block; LCB – load control block; ICS – information and control system; K1, K2, K3 – switches, corresponding to the switches in figure 1.

The scheme shown in figure 2 works as follows:

1) With the stalled rotor, the switch K1 is open, K2 is in position 1, and no boost voltage is applied to the rotor. Switch K3 is in position 2, an uncontrolled rectifier is included in the rotor circuit, in the circuit of which a controlled current transducer (CCT) is included . Regulation of the motor torque occurs by changing the rectified rotor current in the CCT. Such regulation is equivalent to a change in the active resistance of the rotor chain.

2) After removing the mechanical brake and the start of movement, the current in the circuit of the CCT decreases and the machine begins to accelerate. When achieving the speed 0.1 ω_0 the regulation with the help of CCT becomes impractical, therefore it is necessary to switch the DFM to synchronous mode. For this, the frequency inverter FI is switched to the rectifying mode, the rotary FI – into inverted mode, the frequency of the voltage at the output of the rotary FI is set equal to the current frequency of the rotor EMF, the voltage amplitude is set according to the required overload capacity. Upon reaching the specified threshold speed, K3 switches to position 1, K1 closes, connecting the rotary FI to the rotor circuit, K2 switches to position 2.

3) The voltage frequency at the output of rotor FI gradually decreases while maintaining the given amplitude, the machine accelerates in synchronous mode. If it is necessary to accelerate to a full speed, the acceleration in synchronous mode is performed up to the speed $0.5\omega_n$. Upon reaching it, K2 switches to position 1, K1 is closed, K3 is in position 1. The voltage frequency of the rotor starts to be dependent on the current frequency of the eigen rotor EMF. The amplitude of the rotor voltage begins to gradually decrease until the machine reaches the specified speed. When reaching the specified speed, the machine operates in asynchronous mode.

4) When the deacceleration starts, the voltage amplitude on the rotor starts to increase, the machine starts to slow down. The phase shift is set according to the condition [6], the frequency of the boost voltage on the rotor is equal to the frequency of the eigen rotor EMF.

5) After the machine reaches the speed $0.5\omega_n$, it is switched to synchronous mode, the amplitude of the boost voltage is constant, the frequency of the boost voltage is set independently of the frequency of the eigen rotor EMF. When the frequency of the boost voltage decreases, the machine slows down. Deacceleration in synchronous mode is possible until the machine is completely stopped and the mechanical brake is used.

If there is a necessity for a speed reversal, it should be carried out by changing the alternation of the stator phases with the locked rotor by means of a reverser (it is not shown in figure 2).

3. The algorithm of combined drive control

The above description of the algorithm can be formalized:

1) Torque control at zero speed:

$$If \quad \omega = 0: \begin{cases} K1 = 0, \\ K2 = 0, \\ K3 = 1, \\ R_{add} = \text{var}. \end{cases} \tag{5}$$

2) Movement at a low speed:

$$If \quad 0 < \omega \le 0.1\omega_0: \begin{cases} K1 = 0, \\ K2 = 0, \\ K3 = 1 \\ R_{add} = const. \end{cases}$$

$$If \quad 0.1\omega_0 < \omega \le 0{,}5\omega_0: \begin{cases} K1 = 1, \\ K2 = 1, \\ K3 = 0, \\ U_R = const, \\ f_R = const. \end{cases} \tag{6}$$

3) Machine acceleration:

$$If \quad 0 < \omega \le 0{,}1\omega_0: \begin{cases} K1 = 0, \\ K2 = 0, \\ K3 = 1, \\ R_{add} \downarrow. \end{cases}$$

$$If \quad 0{,}1\omega_0 < \omega \le 0{,}5\omega_0 : \begin{cases} K1 = 1, \\ K2 = 1, \\ K3 = 0, \\ U_R = const, \\ f_R \downarrow . \end{cases}$$

$$If \quad 0{,}5\omega_0 < \omega \le \omega_0 : \begin{cases} K1 = 1, \\ K2 = 0, \\ K3 = 0, \\ U_R \downarrow, \\ f_R = \dfrac{\omega_0 - \omega}{\omega_0} f_S . \end{cases} \tag{7}$$

4) Machine operation at the specified maximum speed ω_m:

$$If \quad \omega = \omega_m : \begin{cases} K1 = 1, \\ K2 = 0, \\ K3 = 0, \\ U_R = 0, \\ f_R = \dfrac{\omega_0 - \omega}{\omega_0} f_S . \end{cases} \tag{8}$$

5) Machine deacceleration:

$$If \quad 0{,}5\omega_0 < \omega \le \omega_0 : \begin{cases} K1 = 1, \\ K2 = 0, \\ K3 = 0, \\ U_R \downarrow, \\ f_R = \dfrac{\omega_0 - \omega}{\omega_0} f_S . \end{cases}$$

$$If \quad 0 \le \omega \le 0{,}5\omega_0 : \begin{cases} K1 = 1, \\ K2 = 1, \\ K3 = 0, \\ U_R = const, \\ f_R \downarrow . \end{cases} \tag{9}$$

The operation of the circuit, shown in figure 2 according to the algorithm (5-9), makes it possible to implement a combined electric drive of the hoisting unit with a dual-feed machine that provides speed control in the range not less than 30:1 (according to the authors' estimate the range, depending on the type of speed diagram of particular unit, can reach 40-50:1).

4. Conclusion

All required modes can be realized on the basis of a combination of methods of electric drive control, which consists in the combination of control over frequency and the amplitude of boost voltage on the DFM rotor and the application of a controlled current converter in the rotor circuit.

The advantages of this method for controlling the electric drive are obvious, since the combined method makes it possible to take advantage of all known control methods, and, therefore, significantly improve the efficiency of control.

References

[1] Ostrovlyanchik V Yu 2004 *Automatic Electric Drive of Direct Current of Mining and Metallurgical Production* (Novokuznetsk) p 382
[2] Onischenko G B and Lokteva I L 1979 *Asynchronous Valve Cascades and Dual-feed Motors* (Moscow: Energia) p 200
[3] Fiargo B I and Pavlyuchik L B 2006 *Regulated Electric Drives of Alternating Current* (Minsk: Technoprospectiva) p 363
[4] Klyuchev V I 2001 *Theory of Electric Drive* (Moscow: Energoatomizdat) p 704
[5] Chilikin M G et al 1974 *Fundamentals of an Automated Electric Drive* (Moscow: Energia) p 568
[6] Ostrovlyanchik V Yu et al 2017 *IOP Conference Series: Earth and Environmental Science* **84** 012030

IOP Conf. Series: Earth and Environmental Science **206** (2018) 012041 doi:10.1088/1755-1315/206/1/012041

Design principles of modern control systems for main ventilation fan installations

V Yu Ostrovlyanchik, V A Kubarev, D E Modzelevsky, D A Marshev and N V Verigin

Siberian State Industrial University, 42 Kirova street, 654007, Novokuznetsk, Russia

E-mail: kubarev.sibsiu@list.ru

Abstract. In the paper, based on the analysis of the functional properties of control objects, the basic principles of automation of the fan installation are considered, the functional structure of the control object is given, the main functional connections of the controlling object with the distributed principle of controlling the objects of the fan installation are determined. The technique for developing a control system using the principle of modularity is proposed, which makes it possible to reduce the labor costs for the implementation and debugging of the system of technological automation and protection of the fan installation.

1. Introduction

Up to now, outdated automation and protection systems are being used at the main mine fans. The use of outdated automation equipment and control systems of the electric drive does not allow the fan installation to be integrated into the dispatching system, and, besides, it does not allow the required level of safety of the fan installation to be ensured.

The development of modern fan control systems will improve the quality of their operation and reduce the cost of repairs and maintenance.

Thus, the problems solution for designing modern control systems for fan installations, the definition of the basic design principles and the stages of designing control systems is important.

2. The proposed implementation

The purpose of this work is to develop the functional structure of the control systems for main ventilation fan installations based on the design principles of fan installation automation systems.

LLC "Scientific Research Institute of Automation and Electrical Engineering SibSIU" together with the Department of Electrical Engineering, Electric Drive and Industrial Electronics at SibSIU developed a fan control system. The system consists of separate autonomous modules, which allows, if necessary, the composition and functionality of the system to be changed [1]. The functional diagram of the fan installation is shown in figure 1.

The functional diagram (figure 1) presents:

- Controllig objects: operator (OP), operator workstation (OpW), FICS1 – fan installation control system No. 1, FICS2 – fan installation control system No. 2;
- Controlled objects for fan installation No. 1: synchronous motor (SM1), fan blades (WO1), guide vane (GV1), straightener vane (SV1), fan doors (FD1, FD2), with position sensors D_1 ... D_4, electromagnetic brake (EB1); for fan No. 2 SM2, WO2, GV2, SV2, FD3, FD4, EB2 accordingly.

IOP Conf. Series: Earth and Environmental Science **206** (2018) 012041 doi:10.1088/1755-1315/206/1/012041

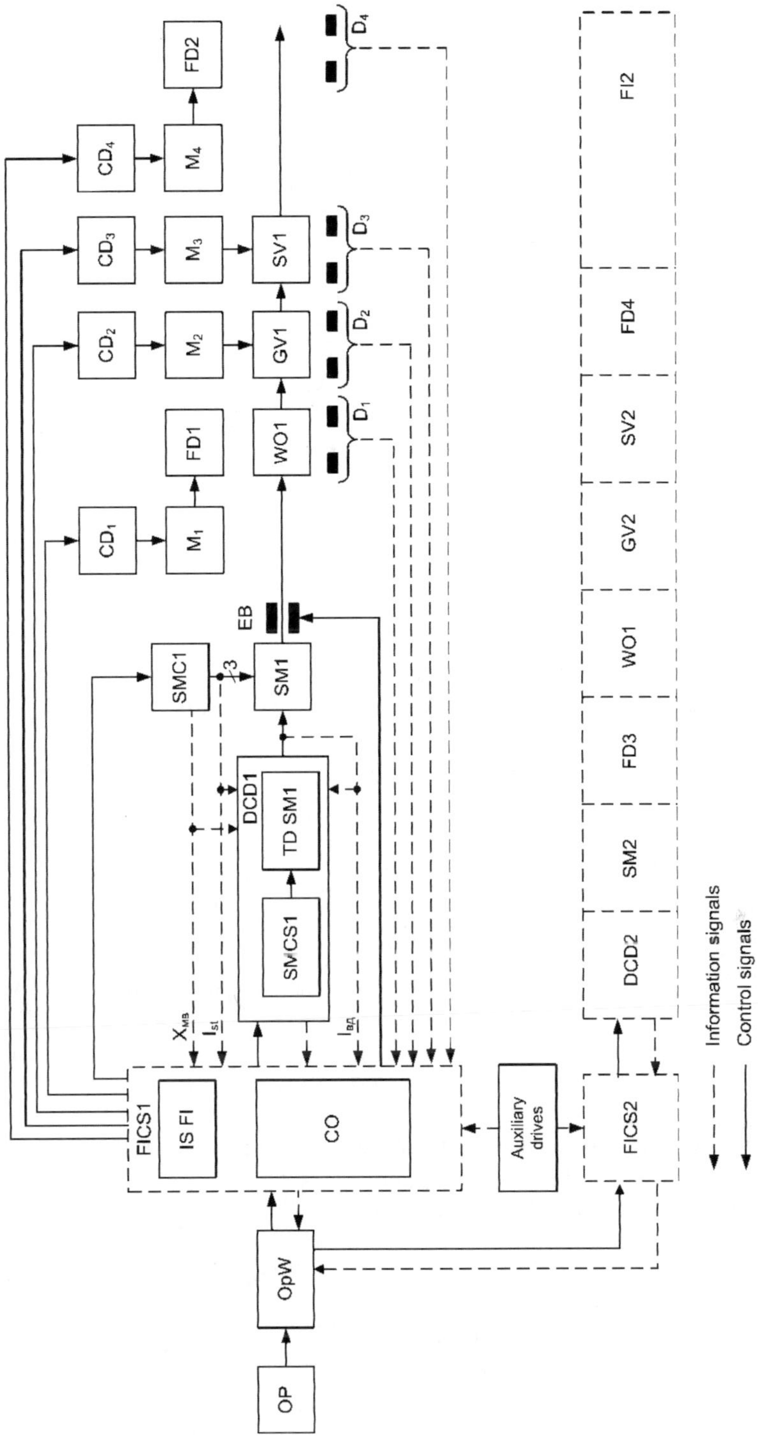

Figure 1. Functional diagram of the fan control system.

According to the functional features in the diagram in figure 1, the following types of control devices can be distinguished. Single-position devices are devices that have only two conditional states: "on' and "off". Such a device can include a synchronous motor, the starting mode of which is carried out by switching on by means of a high-voltage switch and automatic excitation control. This operation is performed automatically by the synchronous motor excitation system, all protections and interlocks are performed directly in the control system. Therefore, the synchronous motor as a whole can be considered as a controlled element [3], controlled switching on, that is, by signaling: "on", "off", "reverse" (figure 2).

The controlled values: angular speed and parameters of the synchronous motor (ω – speed, I_{rot} - rotor current, I_{st} – stator current).

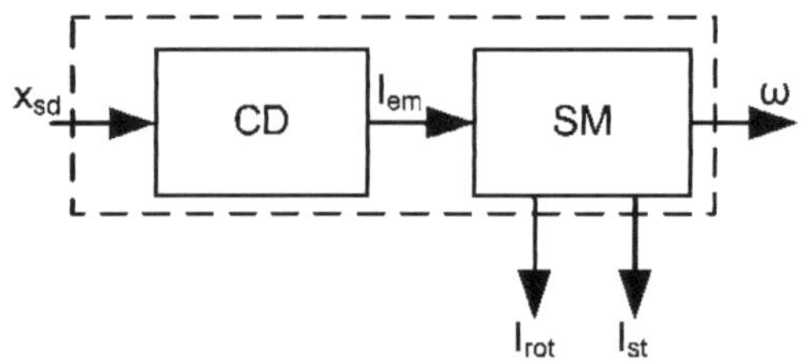

Figure 2. Functional diagram of the synchronous motor as a single-position device.

To the second type of devices – two-position devices, one can refer devices with two output states, for example, doors (figure 3).

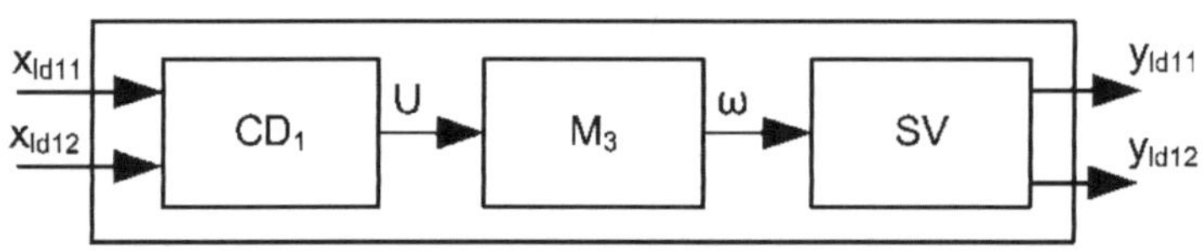

Figure 3. Functional diagram of the two-position device.

For this type of device, two states of the mechanism can be distiguished, for example: "open" $y_{ld11} = 1$, $y_{ld12} = 0$ and "closed" $y_{ld11} = 0$, $y_{ld12} = 1$. Accordingly, the control signals "to open" $x_{ld11} = 1$, $x_{ld12} = 0$ and "to close" $x_{ld11} = 0$, $x_{ld12} = 1$.

To the third type belongs multi-positioning devices: guiding and directing devices (figure 4).

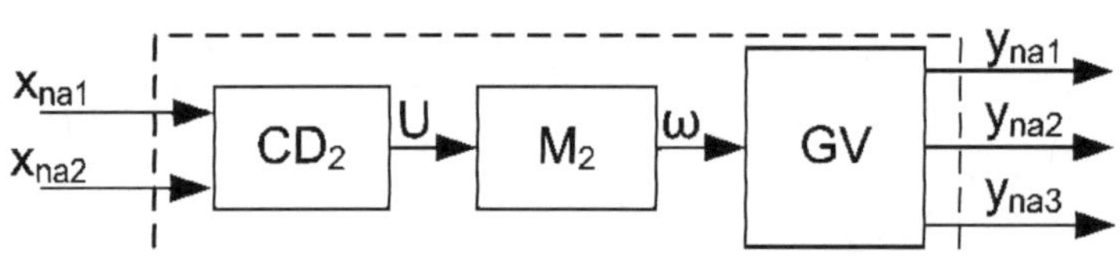

Figure 4. Functional diagram of the multi-position device.

For this type of device, there are also one or more intermediate states of the actuator between the two end positions, for example between the positions y_{na1}, and y_{na3}, the intermediate position y_{na2}.

The functional diagram of the control object of the fan installation is given in figure 5, information and control signals of the fan control system are presented in table 1.

Table 1. Controlling and controlled signals.

No.	Mechanism	Functional features	Control signals		Controlled signals and output values	
			Designation	Type	Designation	Type
1	Doors No.1…No.4	Two positions 1. Closed 2. Open	x_{ld11} — No.1 to close x_{ld12} — No.1 to open x_{ld21} — No.2 to close x_{ld22} — No.2 to open x_{ld31} — No.3 to close x_{ld32} — No.3 to open x_{ld41} — No.4 to close x_{ld42} — No.4 to open	Discrete	y_{ld11} — No.1 closed y_{ld12} — No.1 open y_{ld21} — No.2 closed y_{ld22} — No.2 open y_{ld31} — No.3 closed y_{ld32} — No.3 open y_{ld41} — No.4 closed y_{ld42} — No.4 open	Discrete
2	Synchronous motor	Three positions 1. Forward 2. Stopped 3. Reverse	x_{sd1} — MB work x_{sd2} — MB reverse	Discrete	y_{sd1} — MB in work y_{sd2} — MB in reverse I_{rot} – rotor current I_{st} – stator current ω – speed	Discrete Analog
3	Directing vanes	Two positions 1. Open 2. Closed	x_{sa1} — to close x_{sa2} — to open	Discrete	y_{sa1} — closed y_{sa2} — open	Discrete
4	Guide vanes	Three positions 1. Work 2. Closed 3. Reverse	x_{na1} — to close x_{na2} — to open	Discrete	y_{na1} — closed y_{na2} — open y_{na3} — reverse	Discrete
5	High-pressure pump station	Two positions 1. Switched on 2. Switched off	x_{ms1} — to switch off x_{ms2} — to switch on	Discrete	y_{ms1} — switched off y_{ms2} — switched on	Discrete
6	Brake unit	Two positions 1. Braked 2. Unbraked	x_{et1} — to unbrake	Discrete	y_{et1} — braked y_{et2} — unbraked	Discrete

The control device includes the following modules:

1. OP – operator;
2. OpW – the console (workstation) of the operator;
3. RRU – the unit of reset and availability control unit. Checks the condition and availability of all mechanisms;
4. RMRRU – the unit of reset and availability control unit in the reverse mode. Cheks the condition and availability of all mechanisms for the reverse mode;
5. RPU – block for reverse preparation. Prepares the reverse of the fan installation;
6. SM – synchronous motor;
7. SMCD – synchronous motor control device;
8. IS FI - information system of the fan installation.

The basic design principle of a functional diagram is based on the simulation of the process of manual control. The process of starting the fan installation can be represented in the following sequence:

1. Selection of the fan installation to be put into operation.

IOP Conf. Series: Earth and Environmental Science **206** (2018) 012041 doi:10.1088/1755-1315/206/1/012041

2. Selection the type of control “manual”/”automatic” and the operating mode for the fan installation: “work”/”reverse”.
3. Check the condition and the required initial position of all mechanisms.
4. Bringing the mechanisms in the required starting position.
5. Fixing the initial state of the mechanisms (“readiness”) and start-up.

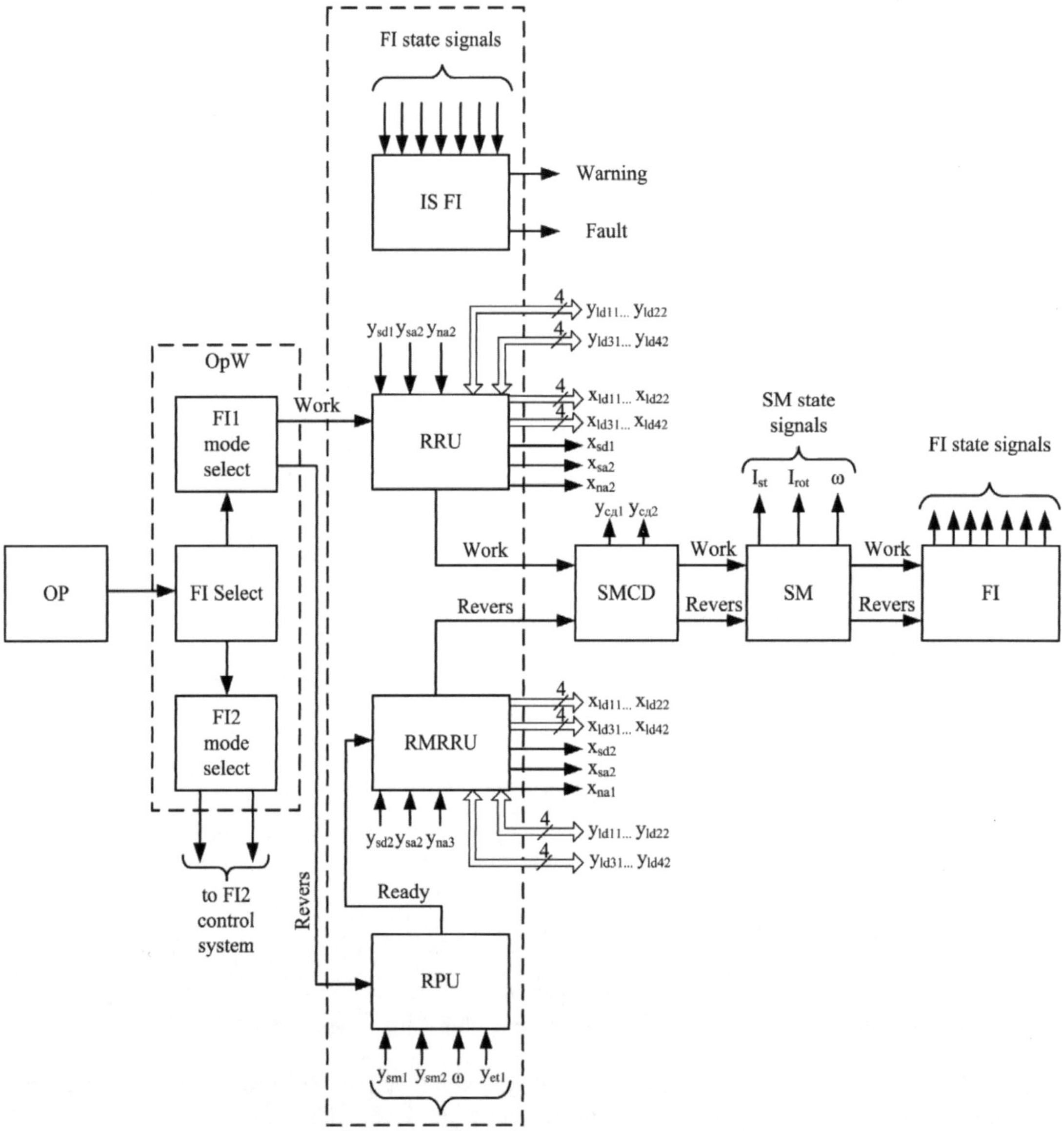

Figure 5. Functional diagram of the fan control device.

Thus, when designing functional structure of the controlling object of the fan unit, the following algorithm can be used:
1. Analysis of functional characteristics of individual elements in the fan installation with the identification of the type of control actions and information about the state of the mechanism;
2. Analysis of the processes associated with the preparation of the mechanism to work and the set up of initial states;

3. Identification of the features of functional structure and system modules that imitates the process of “manual control” (preparation for work – bringing all mechanisms into the initial state, readiness assessment).

4. The structure and functions of the information-control system of the fan installation are determined.

3. Conclusion

The proposed approach to design of a digital control system for a fan installation makes it possible to simplify the automation process considerably, since it allows control modules of a system having the features of standardization and operational independence to be obtained. For example, in case of necessity to control the performance without modifying the remaining modules, an additional module with its own control channel can be added. Each module has the flexibility and the ability to add control functions.

References

[1] Ostrovlyanchik V Yu et al 2018 *Bulletin of KuzSTU* **1** 139–148

[2] Babak G A et al 1982 *Mine Fan Installations of the Main Ventilation. Reference book* (Moscow: Nedra) p 296

[3] *Control Theory. Terminology* 1988 ed B G Volik issue 107 (Moscow: Nauka) p 56

Comparative analysis of options for applying spectral-acoustic method of gas-dynamic phenomena forecasting

A V Shadrin, A A Kontrimas and A S Teleguz

Federal Research Centre of Coal and Coal Chemistry at the Siberian Branch of the Russian Academy of Sciences, 10 Leningradskey ave, Kemerovo city, 650065, Russia

E-mail: avsh-357@mail.ru

Abstract. The options of applying spectral-acoustic method of controlling stress state of a rock massif for preventing dynamic phenomena are analyzed in the paper. Two options are based on the dependence of relationship of high and low frequencies amplitudes of acoustic signal spectrum of operating equipment on a stress state of a massif in the area between the generator and the receiver of the sound. The difference between these two options is in choosing between high-frequency and low-frequency areas of a source sound spectrum. The drawback of both is in failure to use the whole spectrum of a sound that may cause the error in controlling stress state. To correct this drawback a modified option for applying the method based on the dependence of the median of amplitude and frequency characteristics of a sound on actual stresses is justified. The results of testing the method for controlling stress state of a working face space of a stope in the process of a hard roof directional hydraulic fracturing by hardware-software complex based on RIPAS apparatus are introduced.

1. Introduction

Initially spectral acoustic method for controlling stress state of a rock massif was applied as a geophysical method of current outburst danger forecasting [1-3]. That time it was named as a method of forecasting based on amplitude-frequency characteristic of a seam [2]. The given name did not take hold probably because of its length. Another name was taken as an acoustic method [3-4]. The name also didn't take hold as there were many acoustic methods for controlling rock massif and its subject-matter was not expressed in the name. In regulating documents which are in effect at present time this method of forecasting outburst danger is named as a method of forecasting based on artificial acoustic signal parameters [5]. As we see it, the name is not entirely correct as the method is realized through spectral analysis of noises produced by operating mining equipment that went through the controlled rock massif. The sources of the noise are such mining machines as a heading-and-winning machine, a boring bit etc. and their acoustic signals are rather natural. Thus, as the bases of the method is in spectral analysis of acoustic signals the name "spectral-acoustic" is short and expresses its basic idea better.

Physical bases of the spectral-acoustic method is in the idea that the attenuation rate of an acoustic vibration in its first approximation is linearly proportional to the vibration frequency and inversely proportional to average stress that acts in propagation path of the vibrations [6-8]. This implies that with the growth of stresses the attenuation rate of high-frequency harmonics decreases more intensively than the low-frequency ones. That is why the relationship of amplitudes of high-frequency harmonics to low-frequency ones increases with the growth of stresses. This parameter is used in forecasting the outburst danger.

However, the outburst danger depends not only on the forces of rock pressure but also on in-situ pressure of free gas the volume of which does not practically influence on the attenuation rate of a sound signal [9]. Thus, spectral acoustic method controls only rock pressure. To forecast outburst danger the spectral acoustic control is to be enhanced by controlling methods of other basic factors of outburst danger, in particular, methods of controlling in-situ pressure and coal strength of the mostly crushed coal bench. Therewith spectral acoustic method is the basic one and the outburst danger criterion for it is corrected depending on gas pressure and coal strength [7-8, 10-11].

As the necessity to control stress state of a face working space appears not only during the mining of outburst danger seams the spectral-acoustic method is applied or justified for application for forecasting other dynamic phenomena and for monitoring acting stresses while performing specific technologies of mining [12-14].

2. The options for applying spectral-acoustic method

At present time there are two options of applying spectral-acoustic methods. They differ by the choice of frequencies used for determining the coefficient of relative stresses.

The first is characterized by splitting of operating frequency range of a noise source (for example, 20-150 Hz (cps)) into subranges of low and high frequencies [15]. The splitting is done with the help of analogue filters of low and high frequencies. The signals from the filters are detected by amplitude detectors and the relation of high-frequency A_h and low-frequency A_l detected signals is called the coefficient of relative stresses K and equals to [6, 7]:

$$K = \frac{A_\mathrm{h}}{A_\mathrm{l}} = \exp\left(-C\frac{\sigma_\mathrm{lm}}{\sigma_\mathrm{c}}r\right), \qquad (1)$$

where σ_{lm} and σ_c are limit and current values of average stresses, MPa, variable;
r is the distance from the source to the receiver of the signal, m, variable ;
h – notation “high”;
l – notation “low”;
lm – notation “limit”;
c – notation “current”.
Parameter C equals to:

$$C = \frac{\alpha_0\beta(f_\mathrm{h}-f_\mathrm{l})}{f_\mathrm{m}},$$

where α_0 is the attenuation at a middle frequency of working range f_m in stressless position (under stressless condition), m^{-1};
m – notation “middle”;
index 0 means stress absence;

β is nondimensional coefficient of proportionality, characterized by acoustic properties of the massif, variable;

f_h and f_l – are average frequencies chosen for controlling the areas correspondingly of high and low operating subranges of spectrum that mining equipment radiates, Hz.

The second option of choosing the operating frequencies is done as follows: the received (by geophone) acoustic signal is digitalized and by applying fast Fourier transformation the amplitude-frequency characteristic (AFC) of a signal is determined. Harmonic A_{max} with maximal amplitude is determined on it. To the left and to the right of this harmonic harmonics with amplitudes equal to $0{,}5A_{max}$ and $0{,}75A_{max}$ are determined. Subrange of frequencies between the harmonics with amplitudes $0{,}5A_{max}$ and $0{,}75A_{max}$ to the left of the frequency with A_{max} is referred to low-frequency area and similar subrange of frequencies to the right of the harmonics with A_{max} is referred to high-frequency area. The coefficient of relative stresses in this case is determined according to the formula:

$$K = \frac{\sum_j A_j}{\sum_i A_i}, \tag{2}$$

where A_j and A_i are the harmonic amplitudes in high-frequency and low-frequency areas; i .and j - numbers of low and high frequency harmonics respectively.

The peculiarities of both options are considered in details in [16]. One of their significant drawback is the non - use of the entire spectrum of a useful signal. Multiple experimental data show that due to the diversity of mining and geological conditions of coal seam excavation the rise of amplitude-frequency characteristic before a sudden outburst was registered in different frequencies (for example, see [15]). It can bring about the effect when the change of a stress state ahead of the working causes the significant changes of a signal spectrum in the area of the spectrum which is situated out of previously set low and high frequency subband.

To eliminate the drawback we justified the modification of spectral-acoustic method using the whole continuous spectrum of an acoustic signal of an operating equipment within the range between 20-1000 Hz. The usage of higher frequencies can decrease the sensitivity depth of the method ahead of a development working which is determined by minimal length of a wave in the spectrum of a sounding signal where the wave front is determined in accordance with Huygen's principle [17, 18].

The justification is in the following: suppose that after the digitalization of an analogue sounding signal and fast Fourier transformation we get a spectrum of *N* harmonics. The amplitude of n-th harmonic in the distance of *r* from the source can be written as:

$$A_n(r) = A_{n,s} F(r) e^{-\alpha_n r}, \tag{3}$$

where: $A_{n,s}$ –is an amplitude of n-th harmonic at the signal source (noise of an operating equipment); index s – means that the distance between the sound source and the receiver is equal to zero; $F(r)$ – is a function that considers the diagram of signal source directivity (for example, $F(r) = 1$ – for a plane wave; $F(r) = 1/r$ – for spherical wave); α_n – the attenuation rate of n-th harmonic signal; *r*- the distance between the source and the receiver of the signal.

Let's assume that the interval between the neighboring harmonics Δf equals, for example, 20 Hz, and the maximum frequency of chosen by us working range is equal to 1000 Hz.

Then the frequency of n-th harmonic will be equal to

$$f_n = 20\text{n, Hz, n} \in [1; 50]. \tag{4}$$

In first approximation, the attenuation rate αn depends on its frequency f_n and current average stresses σ_c as follows [6]:

$$\alpha_n = \alpha_0 \beta \frac{f_n}{f_0} \frac{\sigma_{lm}}{\sigma_c} = \varepsilon f_n \frac{\sigma_{lm}}{\sigma_c}, \text{ m}^{-1,} \tag{5}$$

where: σ_{lm} – is an average limit stress (maximal possible for a controlled section of a seam that precedes the shattering of a massif caused by dynamic phenomenon) on the propagation path of an acoustic signal, Pa; $\varepsilon = \frac{\alpha_0 \beta}{f_0}$, $m^{-1}Hz^{-1}$.

Then for the selected frequency range the sum of discrete series of amplitudes of an acoustic signal spectral component $A(f, r)$ considering Eqs. (3-5) will be equal to:

$$A(f,r) = \sum_{n=1}^{50} A_{n,0} F(r) e^{-\varepsilon 20 n \frac{\sigma_{lm}}{\sigma_c} r}. \tag{6}$$

IOP Conf. Series: Earth and Environmental Science **206** (2018) 012042 doi:10.1088/1755-1315/206/1/012042

Assuming that amplitude-frequency characteristic of an acoustic signal at the source has a maximum value then its components near the signal transmitter tool can be described by two exponential functions (increasing and decreasing) as follows:

$$A_{n,s} = \begin{Bmatrix} A_{0,s}e^{+\xi 20n}, & n \in [1,14] \\ A_{0,s}e^{-\eta 20n}, & n \in [15,50] \end{Bmatrix}, \quad (7)$$

where $A_{0,s}$ – is an amplitude of hypothetic "zero" harmonic with n = 0, V; the parameters ξ and η – determine the exponents change rate.

Determine parameters ξ and η out of the condition that A_0 = 1,75 V and amplitudes of the following harmonics satisfy the condition $A_{14,0} = A_{15,0} \approx 1{,}0$ V with r=10m and d=1,0 (the condition of joining of two exponents). Then equation (7) is:

$$A_{n,0} = \begin{Bmatrix} 0{,}333e^{+0{,}073n}, & n \in [1,14] \\ 2{,}86e^{-0{,}07n}, & n \in [15,50] \end{Bmatrix}, \quad (8)$$

Substitute Eq.(8) into Eq.(6). The results of calculating the amplitudes of separate harmonics are shown in figure 1 with the following values of input parameters: α_0 = 1,3 m^{-1}; β = 0,07; f_0 = 500 Hz; $F(r)$ = 1; r = 10 m; d= $\frac{\sigma_c}{\sigma_{lm}}$ = 0,2 и 1,0.

Figure 1 demonstrates that together with the stresses growth the amplitudes of high-frequency harmonics increase more significantly than the low-frequency ones and the median of the amplitude-frequency characteristic moves to the right.

For introduced parameters of an acoustic signal we will determine the dependence of a current value of discrete series of spectral component amplitudes median on the relation $\frac{\sigma_c}{\sigma_{lm}}$. Thus according to the definition median will be understood as the root of equation $A(f,r)$ = 0,5, in other words M_c – it is a value of harmonic frequency f_{md} ($M_c \equiv f_{md}$ = 20p), where the following condition is satisfied:

$$\begin{cases} \sum_{n=1}^{p} A_{0,s}e^{-\varepsilon 20n\frac{\sigma_{lm}}{\sigma_c}r} \lesssim 0.5\sum_{n=p+1}^{50} A_{0,s}e^{-\varepsilon 20n\frac{\sigma_{lm}}{\sigma_c}r}, \\ \sum_{n=1}^{p+1} A_{0,s}e^{-\varepsilon 20n\frac{\sigma_{lm}}{\sigma_c}r} > 0.5\sum_{n=p+2}^{50} A_{0,s}e^{-\varepsilon 20n\frac{\sigma_{lm}}{\sigma_c}r}, \end{cases} \quad (9)$$

where p – the number of harmonic;

md – notation "median".

Figure 1 shows that with the growth of average current stresses from 0,2 σ_{lm} to 1,0 σ_{lm} the value of the median grows from 100 to 260 Hz.

Introduce the concept of relative stresses coefficient K_{md} for the given modification of spectral-acoustic method that equals to the ratio of the current value of the median M_c to the reference value M_{rf}:

$$K_{md} = M_c/M_{rf}, \quad (10)$$

where rf – notation "reference".

To simplify this, assume that M_{rf} corresponds to the ratio of current and limit average stresses values d=0.2. The table demonstrates the calculation of K_{md} value from the ratio d.

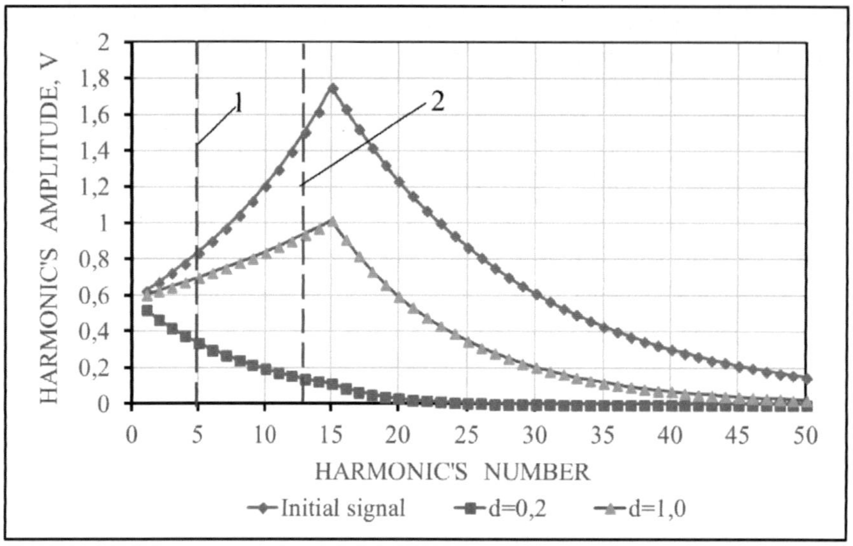

Figure 1. Amplitude-frequency characteristics of the initial signal that went through the area of a 10 m length rock massif with the relation of current and limit average stresses d=0.2 and d=1.0. 1 and 2 are the current values of a median M_c, equal to 100 Hz (5-th harmonic) and 260 Hz (13-th harmonic) for corresponding d=0.2 and d=1.0.

Table 1 shows that in the considered model of acoustic signal that imitates the noise of the operating mining equipment (for example, mining combined-machine) the increase of the current stress ratio to the limit one in 5 times caused the growth of relative stresses coefficient K_{md} in 2.6 times. Therein, to determine the outburst danger indicator the whole spectrum of an acoustic signal in the limits of f = 20-1000 Hz was used and that excluded the possible error during the controlling procedure as the change of the stress state of a face working space brought about the change of amplitudes of spectral components in uncontrolled area of the transmitter tool operating frequencies.

Table 1. The value of the relative stresses coefficient K_{md} for different values d of average current and limit stresses ratio.

$d=\sigma_c/\sigma_{lm}$	$K_{md} = M_c/M_{rf}$
0.2	1.0
0.4	1.8
0.6	2.2
0.8	2.4
1.0	2.6

Functional testing of this modified spectral-acoustic method took place in "Yubileinaya" Mine "TopProm" Ltd. with the help of software-hardware complex based on intrinsically safe portable recorder of an acoustic signal RIPAS, produced by "MNTL RIVAS" company [13]. The task of the researches was in registering the changes of a stress state ahead of the long face caused by directional hydraulic fracturing of the roof performed through the boreholes drilled into the airway. The estimation of the face working space stress state was performed before and after the directional hydraulic fracturing (DHF) process.

Following the measuring technique of RIPAS apparatus geophone was installed on the gallery edge and both sides of it were hammered at a distance of about 2 meters. Then geophone was taken to another place far from the stope and the sounding procedure, similar to the previously described one, was performed.

Signal processing with the purpose of determining the coefficients of relative stresses K according to the algorithm recommended by the Technique [13] and according to the algorithm based on the analysis of amplitude-frequency characteristic (AFC) median shifting K_{md} was performed after the exiting the mine, under laboratory condition. The data are presented in figure 2.

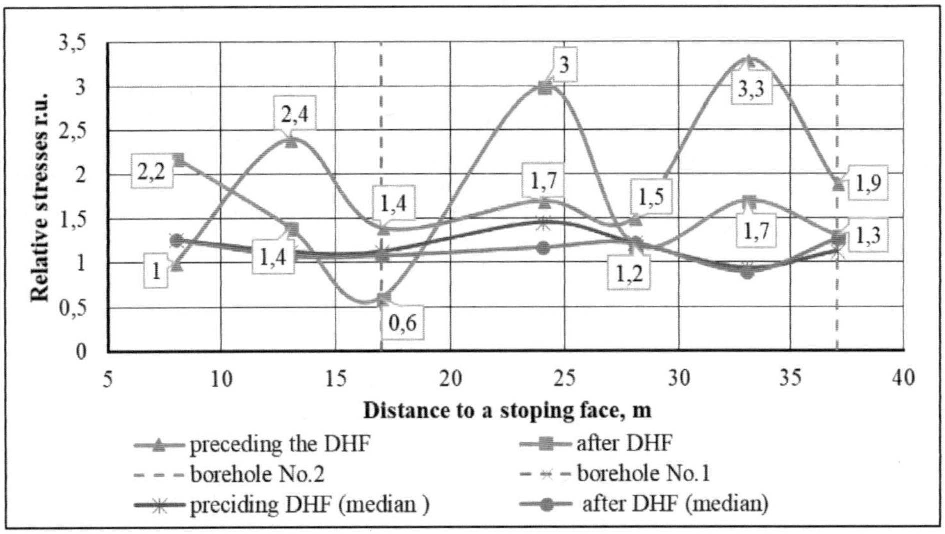

Figure 2. The dependence of the relative stresses coefficient K determined by RIPAS apparatus and K_{md} coefficient calculated on the AFC median shift before and after DHF on a distance to a stoping face.

As it is seen from figure 2 firstly, coefficient K changes drastically along the gallery. Secondly, the coefficient K_{md} is characterized by lesser changeability than K coefficient therewith the direction of changes do not always coincide with K changes. Analyzing these experimental data we can indicated the following:

Noncoincidence of the coefficient K and K_{md} directions as you move away from the stoping face can be explained only by the difference of algorithms applied for their determinations. Additional experiments should be performed for defining the most trustworthy algorithm out of these two.

It is unlikely that such significant changes of K coefficient along the gallery can be caused by changes of a massif stress state. Significant changes of K coefficient can be caused by both different rock jointing in the places where the geophone is installed or by different quality of the geophone contacting with the massif in various points of installation. To avoid the influence of near the face rock jointing on determining the coefficient of relative stresses it is planned to use geophone the design of which allows us to install it into the borehole with the length of 1.0-1.5 m behind the zone of intensive rock jointing.

Changing of K coefficient can be conditioned by the fact that geophone is placed into the closest zone of the transmitter tool where exponential dependence of a signal amplitude on the distance between the transmitter tool and the receiver is not fulfilled [19]. In the indicated literature source, the transmitter tool is point-like whereas the cutting unit of a combined machine cannot be referred to as a point-like due to its significant size. To organize the research on the dependence of acoustic signal amplitude generated by the mining combined machine into the massif in the closest zone while mining is extremely difficult. That is why the research on determining the peculiarities of the acoustic signal behavior in the closest zone was performed on the model.

Electron accelerator "Impulse" was used as an acoustic signal source with the transmitting surface in the form of a circle. Energetic beam of electrons (EBE) generated by the electron accelerator was directed on the metallic target with the piezoelectric element pressed to its backside and with the acoustic trap in form of cone for absorption of a transmitted signal attached to the opposite side. The duration of electron momentum was 5 nanoseconds, its diameter equaled to 10.2 mm and average EBE energy made 0.36 J. Under these EBE parameters and with its interaction with the target an acoustic impulse on thermoelastic mechanism was generated at the surface of the target. The form of the acoustic signal was registered by piezoelectric element and the signal was sent to a storage oscilloscope. Having found the length of the impulse and the velocity of signal propagation the length of the wave of a basic acoustic signal harmonic was determined in the target material. It depended on the target material and for the examined samples it reached from several decimals to one mm. Thus, the closest zone of the transmitter tool reached about 3-5 mm.

Figure 3 represents the results of the experiment. Here we can see that the signal amplitude in the closest zone of a transmitter tool is characterized by its changeability, the value of which depends on the target material.

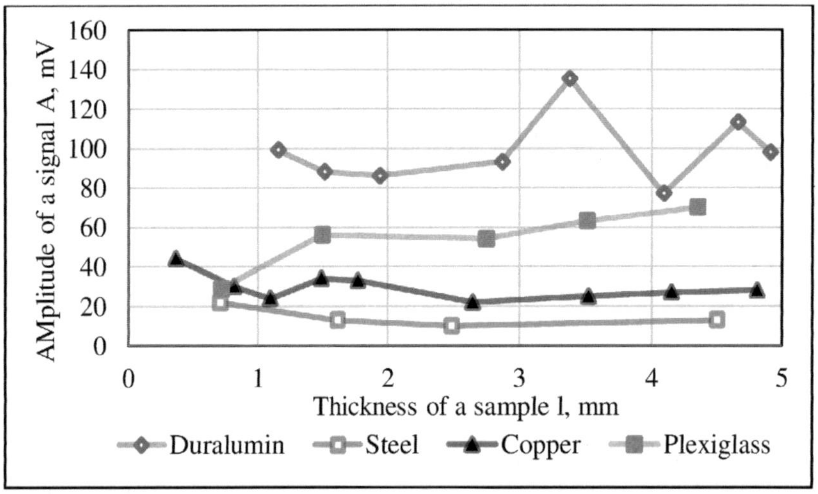

Figure 3. The dependence of acoustic impulse amplitude that went through the sample, on its thickness.

3. Conclusion

The option of applying spectral-acoustic method on the bases of the dependence of AFC median on the stresses uses the whole signal spectrum, that is why it eliminates the forecasting error arisen due to the loss of data in uncontrolled part of a spectrum. However, the sensitivity of it is less than other applied options.

High changeability of the relative stress coefficient can be referred to joint influence of the following factors: various degrees of a massif rock fracturing along the gallery, unequal conditions of the geophone contacting with the massif, peculiarities of the signal propagation in the closest zone of a transmitter tool and measuring errors.

References

[1] Hanson D R 1985 *Mining Engineering* **l** 37 **1** 54–60
[2] Mirrer S V, Khmara O I and Maslennikov Ye V 1987 To the issue of Preventing Sudden Outbursts. Scientific report. *A. A. Skochinsky Institute for Mining* 52–61
[3] Zorin A N, Kovtun Ye D and Kachan I I 1987 *Ugol' Ukrainy* **2** 29–30
[4] Korol V I and Skobenko A V 2013 *Acoustic Method of Forecasting Gas-dynamic Phenomena*

in Coal Mines (Dnepropetrovsk: National mining university) p 181
[5] *Federal norms and regulations in the field of industrial safety "Instruction for the forecast of dynamic phenomena and monitoring of the rock massif during the coal deposits mining"* (approved by the order of the Federal Service for Environmental, Technological and Nuclear Supervision of August 15, 2016 No. 339) p 129
[6] Shadrin A V and Degtyaryeva M V *The Bulletin of the Scientific Center for the Safety of Work in the Coal Industry (Industrial Safety)* **1.2** 55–9
[7] Shadrin A V 2016 *Phisiko-Ttechnicheskiye Problemy Razrabotky Poleznykh Iskopayemikh* **4** 48–62
[8] Shadrin A V 2016 *Journal of Mining Science* vol 52 issue 4 670–82
[9] Shadrin A V 2009 *GIAB* **1** 200–10
[10] Shadrin A V and Bireva Yu A 2016 *Naukoyemkiye technologii razrabotki I ispolzovaniya mineralnykh resursov: scientific journal* **2** 410–16
[11] Shadrin A V and Bireva Yu A 2016 *IOP Conference Series: Earth and Environmental Science* **45** 012008
[12] Shadrin A V and Teleguz A S 2017 *Vestnik Nauchnogo tsentra po bezopasnosti rabot v ugolnoy promyshlennosti (Industrial Safety)* **3** 10–4
[13] The Procedure of Coal Massif Acoustic Sounding Applying Hardware-software Solution 2016 (Moscow: MNTL RIVAS) p 36
[14] Voznesenskey Ye 2012 *A Justification and development of spectral-acoustic method for controlling roof bolting providing opening strength: extended abstract of Cand. Sci. Dissertation* (Moscow: MSMU) p 24
[15] Mirrer S V, Khmara O I and Shadrin A V 1999 *Spectral – Acoustic Forecasting of Coal Seam Outburst Danger (*Kemerovo: Kuzbassvuzizdat) p 92
[16] Shadrin A V 2017 *50 Years of Russian Scientific School of Complex Exploitation of Interior Part of the Earth, International Scientific Practical Conference Materials* (Moscow: IPKON RAN) 85–9
[17] Shadrin A V and Kontrimas A A 2017 *Naukoyemkiye Technologii Razrabotki I Ispolzovaniya Mineralnykh Resursov: Scientific Journal* **3** 408–13
[18] Shadrin A V and Kontrimas A A 2017 *IOP Conference Series: Earth and Environmental Science* **84** 012040
[19] Krautkramer J and Krautkramer G 1991 *Ultrasonic Testing of Materials* (Moscow: Metallurgy) p 752

Walking support control system algorithms testing with brain-computer interface (BCI) and augmented reality (AR) technology integration

M S Nikitenko[1], S S Zhuravlev[2], S V Rudometov[2], B Neogi[3] and A M Belyi[4]

[1]Federal Research Center of Coal and Coal Chemistry, Siberian Branch of the Russian Academy of Sciences, 18 Sovetskiy ave, Kemerovo, 650000, Russia
[2]Institute of Computational Technologies of the Siberian Branch of the Russian Academy of Sciences (ICT SB RAS), 6 Ak. Lavrentieva ave, Novosibirsk, 630090, Russia
[3]CoE & CII Centre for Innovation, JIS College of Engineering (JIS Group India), Block A, Phase III, Dist. Nadia, Kalyani, West Bengal, 741235, India
[4] Siberian State Industrial University, 42 Kirova street, Novokuznetsk, 654006, Russia

E-mail: schum24@rambler.ru

Abstract. The article presents a perspective platform for the creation of robotic complexes for the effective and safe development of hard-to-extract minerals based on the walking support module. It describes the concept of automation control based on combining BCI and AR technology, and a method for testing control algorithms based on the application of the MTSS (Manufacturing and Transportation Simulation System) as part of the hardware and software complex for debugging and testing the applied software of the automated process control system for coal and ore mines. The reported study was partially funded by RFBR according to the research project No. 18-37-00356.

1. Introduction

As a result of the mining complex in Russia and abroad easily accessible and rich deposits exhaustion, the reserves that are characterized as difficult for development mining and geological conditions are being brought into operation - the occurrence is noted in steeply inclined and thick seams, the depth of bedding increases, and the coal seams are characterized by high gas content.

The need to increase the development efficiency of hard-to-extract minerals in difficult mining and geological conditions and in mines dangerous for gas and dust creates the need of transition to the minerals "uninhabited" excavation development technology with automated and automatic equipment developed to the level of robotic complexes [1].

In addition, as part of the long-term program for the Russian coal industry development till 2030, approved by the Russian Federation Government Decree of 24.01.2012 No. 14-p., the main direction of technological development with respect to the underground mining method is also the implementation of measures to develop and implement minerals "uninhabited" extraction systems on the basis of complex mechanization and automation. In accordance with Russia's fuel and energy sector scientific and technological development forecast for the period up to 2035, coal mining

IOP Conf. Series: Earth and Environmental Science **206** (2018) 012043 doi:10.1088/1755-1315/206/1/012043

technology without permanent presence of people in the working space, the introduction of robotic complexes for mineral development, were included in the priority critical technologies.

To date, the mining equipment and technologies which is used to extract minerals from thick steep and steeply inclined seams cannot operate without the constant presence of people at the mine face and preparatory faces due to the mining technology complexity. In this regard, not only new technical development solutions for thick steep seam deposits [2] extraction, but also modern automated equipment control tools are topical. In addition, the new development technological flow sheet [3] and the technical facilities for extracting minerals [4-7], for the equipment control and its diagnostics [8-11], allow them on their basis to create robotic equipment and machines for solving the tasks set.

2. Research objective

A promising platform for the robotic complexes creation for the effective and safe development of hard-to-extract minerals, with the underground method of development is a robotic module of a walking powered support [12, 13].

A hydroficated walking support module (figure 1) is designed to support the roof. Consists of two frames interconnected with each other through the hydraulic travel cylinders. Each frame is equipped with longitudinal top beams and cross beams rigidly fixed to each other, and is based through hydraulic props and supports on the ground.

In the working position, the travelling hydroficated support is strutted between the roof and the ground (thrill). To move the support, strutting is removed from one of the frame's support props, the frame released of strutting, sliding with the tops along the cross beam of the strutted frame, moves one travelling step. At the end of the travelling, pressure is applied to the piston cavities of the supporting hydraulic props and the moved frame is strutted [3].

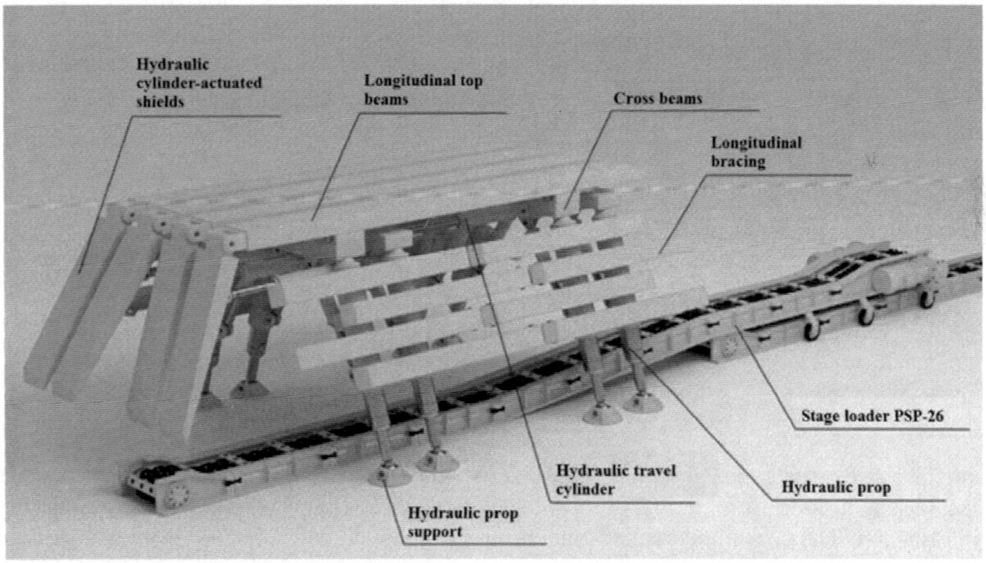

Figure 1. A module of walking powered shielding type support with a conveyor reloader.

The second frame is moved in a similar sequence, completing the travelling cycle of walking. With hydroficated shields we achieve dosing of top coal caving from the roof and controlled fencing of the face zone. Coal transportation along the gallery to the conveyor is carried out by loading machine such as PSP-26.

To build the automated control system for the walking support module, the authors consider it promising to integrate the Brain-Computer Interface (BCI) based technology together with the Augmented Reality (AR) equipment into the system [14].

IOP Conf. Series: Earth and Environmental Science **206** (2018) 012043 doi:10.1088/1755-1315/206/1/012043

Currently, the authors, together with foreign colleagues from JIS Collage of Engineering, where specialists are already working on the control system basic modules study based on BCI [10, 11], are working on the operator's mobile working place organization concept and on integration of BCI equipment and augmented reality in automated control system (ACS) [15, 16].

The main aim is to organize the work of the ACS manager of technological processes (ACS TP) SCADA of mineral deposits underground mining in a new quality, without permanent presence at the working place, reducing the likelihood of error, loss of visual contact with the control object parameters

The main idea is to organize the work of the dispatcher of the process control application for underground development of mineral deposits (ACS of technological processes - ACS TP) in a new quality, without constant binding to the workplace with a reduction in the probability of making an error, loss of visual contact with the parameters of the control object, based on virtualization of control and monitoring instruments.

Within the conceptual framework, the possibility of displaying information about the state of the complex directly in front of the operator with the complex interactive elements control rendering is provided by the augmented reality systems, the access to the main emergency control functions is provided through the adaptive automated control system based on the BCI control interface technology. The general diagram of the walking support control devices is shown in figure 2.

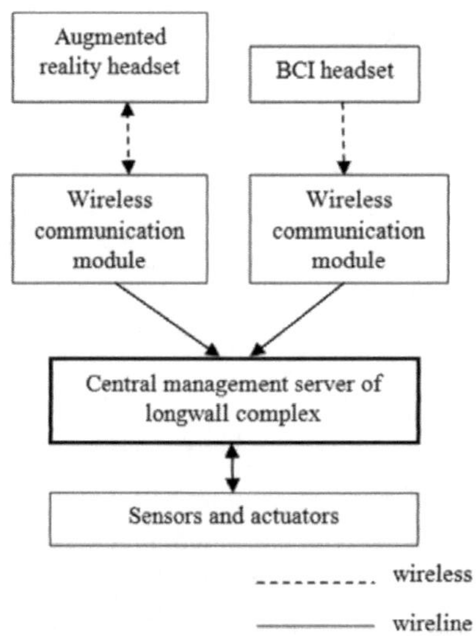

Figure 2. General diagram of the walking support control devices.

Brain-computer interface and augmented reality technology application is designed to solve problems such as a long time period eye contact loss with indicators displaying working complex status information. The opportunity to enter additionally a backup control channel from the dispatcher or the operator of the equipment, to equip better the ergonomics of the operator's workplace and reduce his fatigue.

Hardware based on the BCI has entered the market recently – from 2003. Existing brain-machine interfaces critically investigate in order to assess the maturity of this technology and estimate the improvements that it could yield [17-22]. However, existing brain-machine interfaces is insufficiently researched and tested in real and virtual conditions in the direction of application at industrial enterprises, especially for dispatching tasks. Therefore, the robotic walking support modules creation and integration into the production environment question consideration in the NeuroNet orientation plane and the introduction of digital production elements at mining enterprises and at coal mining machine building enterprises certainly involves the use of simulation modeling.

Integrated debugging and testing of ACS TP applied software at the early stages can significantly improve the reliability and safety of ACS TP, it can reduce the time and cost of starting–up and debugging work and test operation, facilitate the maintenance, upgrading and optimization of the applied software [23].

3. Methodology

Integrated debugging and testing of ACS TP applied software is a complicated task. Main problem is the difficulty to generate the complete set of coordinated signals like from the real technological equipment and the impossibility of artificial situation creation that go beyond normal operating modes at a real object. To solve this problem, the software and hardware complex of debugging and testing of applied software was developed at the ICT SB RAS. It includes simulation models of both technological processes and control programs for the technological process automated control system (ACS TP). They are used for simulating of the technological equipment sensor signals and for the technological process automated control system (ACS TP) components emulation.

There are a number of physical and software signal simulators that can be used in development and testing of ACS TP. Most effective is to have specialized software and hardware complexes that contains in their structure a problem-oriented, mathematical models of automated technological processes that, in turn, ensure the consistent generation of signals from sensors and control signals.

The specialized software and hardware complexes that implement the generation of test signals based on the technological process model or the whole enterprise include:

- WinMOD [24];
- MiMiC [25];
- xPC Target [26];
- Simulation model of the hydraulic unit for testing the algorithms of the ACS TP [27];
- Testing and debugging area for automation means and systems [28];
- Training and control software and hardware complex for chemical technology facilities [29];
- A test bench for aircraft onboard equipment software debugging [30];
- Test bench automated control system for submersible electrical equipment testing [31];
- Complex for automated control system technological process projects development and debugging [32];
- A test bench of semi-realistic simulation for development and for the embedded real-time computer systems [33]
- Development and validation of flight control system for small unmanned aerial vehicles [34];
- Application of the model-based design approach for robot arm software development [35];
- Technology of design and debugging of the control system software of bench nuclear power station with use of tool and software complexes of modeling [36];
- Management of the built-in model in semi-natural simulation for industrial wireless networks of enterprises using WirelessHART standard [37];
- some others.

Similar software was developed in ICT SB RAS. It is called "Software and hardware complex for testing of the applied software for ACS TP for coal and ore mines" [23]. This complex contains simulation modeling system MTSS (Manufacturing and Transportation Simulation System), also developed in ICT SB RAS and already used for technological process simulation models creation [38, 39].

A complex performs testing of the applied software on the principle of replacing the real signals from sensors of the processing equipment with virtual signals, generated on the basis of the simulated parameters of the processing equipment [40].

To solve the tasks of mining technological processes control within the MTSS, a specialized library of technological equipment models (TEM) of coal and ore mines has been developed. The TEM library is implemented using the Eclipse platform and the Java programming language.

Each TEM in the MTSS system consists of the following parts: a conditional graphic image as its visual part; list of parameters; a list of input and output signals; an operation algorithm that in particular involves the relationships between parameters; the list of states in which the TEM can while the modeling evolves; a list of control commands that switches the TEM from one state to another; additional blocks that extend the functions of the TEM (for example, the communication interface with external systems).

4. The expected results discussion and conclusion

Thus, for debugging and testing of BCI technology integration in conjunction with the use of augmented reality tools in the walking support control system with the Hardware and software complex for coal and ore mines ACS TP applied software testing, it is required to:

1. Develop an algorithm for controlling the automation system of walking support using the equipment of the Brain-Computer Interface;
2. Develop the TEM of the supporting-protecting type walking powered support module, in accordance with the technological equipment model description;
3. Develop the technological process simulation model of mineral deposits unmanned mining with an underground method when integrating the concept of the operator's mobile working place in conjunction with the use of augmented reality;
4. Adopt the algorithms and technical support of specialized simulation software and hardware complex for debugging and testing the applied software of the walking support system ACS TP for mineral deposits underground mining.

When carrying out the tasks, a specialized software and hardware complex will be obtained with an integrated simulation model of the mine industrial environment to solve control algorithms testing problems of travelling support complex automation.

Acknowledgements
The reported study was partially funded by Russian Foundation for Basic Research (RFBR) according to the research project No. 18-37-00356 "Development of scientific foundations for the creation of mobile operator position on the basis of combining technologies of the brain computer interface and augmented reality".

References

[1] Yermekov T Ye, Nesipbaev Zh S and Sarsembaev T U 2001 *Proceedings of the University* **1** 46–47
[2] Malakhov Yu V 2017 *Proc. of Int. Sci. and Prac. Conf. on Innovations in the fuel and energy complex and in engineering* pp 62–68
[3] Klishin V I 2013 *Mining Informational and Analytical Bulletin (Scientific and Technical Journal)* **S6** 36–47
[4] Klishin V I and Kokoulin D I *Support for Development of Thick Steep Coal Seams with Sub-level Extraction* patent for utility model RU 1604472 U1 24.11.2015
[5] Klishin V I, Nikitenko S M and Pfarger Ye S 2017 *Proc. of the All-Russian Scientific and Practical Conference of Young Scientists, Postgraduates and Students: Ecology and Safety in the Technosphere: Modern problems and solutions* (Tomsk: Publishing house of Tomsk Polytechnic University) pp 217–220
[6] Klishin V I and Nikitenko S M 2017 *Proc. of the 11th All-Russian Scientific and Practical Conference on Automation systems in education, science and production* (Novokuznetsk: SibeSIU) pp 226–259
[7] Klishin V I, Fokin Yu S, Kokoulin D I and Kubanychbek B 2006 *Development of Thick Seams With Power Supports Using Controllable Coal Release* (Novosibirsk: Science, Siberian Publishing Company RAS)
[8] Sadovets V Yu and Kizilov S A 2016 *Technologies and Materials* **3** 4–7

[9] Nikitenko M S 2016 *IOP Conf. Series: Earth and Environmental Science* **45** 012007 doi:10.1088/1755-1315/45/1/012007
[10] Neogi B et al 2016 *Cognitively Velocity Controlled Vehicle* Patent IND File no. 201631017186 Indian Patent Journal no. 26/2016 http://www.ipindia.nic.in/writereaddata /Portal/IPO Journal/1_350_1/part1.pdf
[11] Neogi B et al 2016 *Thought Concentration Controlled Dexterous Prosthetic Arm for Handicapped* Patent IND File no. 201631017174. Indian Patent Journal no. 26/2016. http://www.ipindia.nic.in/writereaddata/Portal/IPOJournal/1_350_1/part1.pdf
[12] Nikitenko M S, Malakhov Yu V and Nikitenko S M 2017 *High Technology of Development And Use of Mineral Resources: A Scientific Journal* **3** pp 249–251
[13] Nikitenko M S, Malakhov Yu V, Neogi B, Chakraborty P and Banerjee D 2017 *IOP Conf. Series: Earth and Environmental Science* **84** 012002
[14] Kizilov S A, Neogi B, Nkitenko M S, Nikolaev P I and Kuznetsov I S 2017 *Mining Industry Journal* **6 (136)** pp 96–99
[15] Kizilov S A, Nikitenko M S and Neogi B 2018 *IOP Conf. Series: Materials Science and Engineering* **354** 012016
[16] Kizilov S A, Nikitenko M S, Neogi B 2017 *Proc. of the 11th All-Russian Scientific and Practical Conference on Automation Systems in Education, Science and Production* (Novokuznetsk: Publishing Center of SibSIU) pp 44–47
[17] Carlo Menon et al 2009 *Acta Astronautica* **64** 448–456
[18] Joseph DelPreto et al 2018 *Robotics: Science and Systems* (Pittsburgh, PA, USA, June 26-30)
[19] Filipp Gundelakh et al 2018 *MATEC Web of Conferences* **161** 03003
[20] Borisov V, Syskov A, Tetervak V and Kublanov V 2017 *IEEE SIBIRCON* pp 550–555
[21] Fabien Lotte and Cuntai Guan. 2011 *IEEE Transactions on Biomedical Engineering* **58(2)** pp 355–362
[22] Anna Buttfield, Pierre W Ferrez, and Jd R Millan 2006 *IEEE Transactions on Neural Systems and Rehabilitation Engineering* **14(2)** pp 164–168
[23] Zhuravlev S S 2013 *Computer Technologies* **SE 18** pp 150–155
[24] *WinMOD system platform. Hennigsdorf: Mewes & Partner GmbH* 1995 http://winmod.com
[25] *MiMiC simulation software. Chesterfield, MYNAH Technologies, LLC* 2011 http://www.mynah.com/products/mimic
[26] *xPC Target. Natick, MathWorks* 2013 http://www.mathworks.com/products/xpctarget/
[27] Zakharchenko V Ye 2007 *Automation in Industry* **7** 37–40
[28] Ivushkin A A 2007 *Doctor's Thesis: Basics of Coal Industry Objects Automation Control Systems Creation and Introduction* (Novokuznetsk, SibSIU) p 312
[29] Sukharev A V, Golovushkin B A, Labutin A N and Yerofeev Ye V 2011 *Automation in Industry* **7** 25–28
[30] Babaian R R and Osipov A B 2013 *Automation in Industry* **5** 55–56
[31] Komelin A 2004 *Modern Automation Technologies ("STA" magazine)* **3** 16–23
[32] Maslov A and Viskov A 2001 *Modern Automation Technologies ("STA" magazine)* **3** 68–76
[33] Balashov V, Bakhmurov A, Volkanov D, Smeliansky R, Chistolinov M and Yushchenko N 2009 *Proc. of the Fourth All-Russian Scientific and Practical Conference on Simulation Modeling. Theory and Practice (IMMOD-2009)* (St. Petersburg) pp 215–19
[34] Okolnishnikov V V and Rudometov S V 2009 *Proc. of the IVM and MG SB RAS, Informatics series: Materials of the Fifth Asian International Seminar School "Complex systems optimizing problems"* (Novosibirsk) **9** pp 253–57
[35] Meola D, Iannelli L and Glielmo L 2013 *Control & Automation (MED) 21st Mediterranean Conference* pp 357–62
[36] Mossaad Ben Ayed, Lilia Zouari and Mohamed Abid 2017 *Engineering, Technology & Applied Science Research* vol 7 **5** 2017–21

[37] Batrakov S V, Derbukov E I, Efimov V A, Romanova T V, Chernego V A, Chernykh V P and Shaleninov A A 2017 *Technologies provision a life cycle of nuclear power plants* **1(7)** pp 26–38
[38] Chung Duc Tran, Rosdiazli Ibrahim, Vijanth Sagayan Asirvadam, Nordin Saad and Hassan Sabo Miya 2018 *ISA Transactions* **75** pp 236–46
[39] Rudometov S V 2011 *Bulletin of SibSUTIS* **3** pp 14–26
[40] Ryssel U, Ploennigs J, Kabitzsch K and Folie M 2007 *APGES'07* (Salzburg, Austria) pp 4–11

Technology of augmented reality applications in dispatching control of industry processes and mining

A M Belyi[1] and M S Nikitenko[2]

[1]Siberian State Industrial University, 42 Kirova street, Novokuznetsk, 654006, Russia
[2]Federal Research Center of Coal and Coal Chemistry, Siberian Branch of the Russian Academy of Sciences, 18 Sovetskiy ave, Kemerovo, 650000, Russia

E-mail: lefleure92@yandex.ru

Abstract. The article describes approaches to creating products based on Augment Reality technology, examples of their practical application in production, including for dispatching, prospects for development in industry, including in the mining sector. The given examples demonstrate the production and economic efficiency of the introduction of modern computer technologies, including the technology of augmented reality. The article was prepared with the financial support of the Russian Fund for Fundamental Research within the framework of the scientific project No. 18-37-00356.

1. Introduction

The modern stage of civilization technological development is characterized by a significant integration of mobile devices into many spheres of human life. Using mobile device applications, users get fast access to an impressive amount of information and functions, for which before it was necessary to spend more time and use several devices instead of one. The use of Augmented Reality (AR) technologies in these applications allowed not only to improve the interface control, but also to obtain a tangible economic effect.

It is generally believed that the development of virtual reality began in the 50s of the past century. The augmented reality technology development went hand in hand with the virtual one, until 1990, when scientist Tom Codell first proposed the term "augmented reality". In 1992, Lewis Rosenberg developed one of the earliest functioning augmented reality systems for the US Air Force. The Rosenberg exoskeleton allowed the military to control the machines virtually while in the remote control center [1].

2. Research objective

Augmented reality technologies are at the peak of popularity the last few years. Virtual elements or clues: geolocation tags, addresses, recommendations, pointers are superimposed on the real environment with the help of special glasses or mobile devices. Similarly, augmented reality has practical application in industry, only in production we have mechanisms and schemes instead of buildings and streets which are supplemented with data that help the operators, technicians, maintenance personnel to make decisions [2].

Scientists from all over the world are engaged in research into the augmented reality and its implementation in the industry. For example, the Chinese scientist Shou-xiang Zhang in his article "Augmented Reality on Long-Wall Face for Unmanned Mining" examines aspects of introducing augmented reality into technology of unmanned mining [3, 4]. A team of scientists from N. Soete, A. Claeys, S. Hoedt, B. Mahy, J. Cottyn from Gent University in their research paper "Towards Mixed Reality in SCADA Applications" gives an example of SCADA system interaction and applications using technology augmented reality on the basis of an experimental setup [5]. This setup is an automated logistic process, which is very integrated into the Computer Integrated Manufacturing (CIM) structure.

3. Methodology

Solutions that use the augmented reality technology, consist of hardware and software parts. The first includes the means to ensure the virtual presence of a remote specialist on the spot - such a tool can be a pad, a smartphone or special glasses. The IT infrastructure should also be referred to the equipment involved in data transmission, which include switching equipment, storage servers, etc. At the same time, the software part can vary considerably depending on the specific manufacturer. The software part, as a rule, is unique for each project, currently there is no single or at least several major platforms accepted by the developers as a standard. Glasses, a smartphone or a pad project the data on the actual image that the technician needs at a given moment of time, be it diagrams, assembly order, the location of a possible malfunction or the replacement part serial number.

One example of the augmented reality technology successful introduction into the industrial sphere is Boeing aerospace concern, which began to use smart Google Glass, in order to speed up its employees work. One of the labor-intensive processes in the corporation is the cables connection between the aircraft onboard system components. Workers need to maintain concentration and carefully monitor to ensure that the cuts of the required length have been laid on according to the exact pattern and fastened between them into harnesses. They had to follow the assembly instructions, checking information on the laptop and constantly switching between direct work and information retrieval. To speed up the work, a special application for Google Glass was developed: using smart glasses and QR codes, the employee receives a step-by-step instruction for laying the wires, while his hands are free. As a result, the assembly time was reduced by 25%, the number of errors significantly decreased [6].

4. Discussion

One of the brightest domestic AR projects was implemented by Formika Lab. On the basis of the Moscow oil refinery, one of the pilot sites of "Gazprom Neft" for the use of digital technologies, the company proposes to implement a complex IT solution involving AR technologies. Initially, the company planned to launch a training program for the instruments' maintenance.

For the experiment, complicated equipment was specially selected, requiring special qualification of workers - infra-red gas measuring heads. Instruments are designed to control the explosive gases' concentration in the ambient air, and therefore, are critical in such production. Having studied the site and discussed with the workers, Formika Lab found out that the plant uses more than 1100 infrared gas-measuring heads [6]. At the moment the project is undergoing an industrial testing.

The use of AR-technologies is not limited to the automation of services engaged in the maintenance of various industrial installations "workplaces". For example, using the augmented reality technology, it becomes possible to create dispatching simulators, the use of which is possible on real objects of control systems. An example of such simulators is the start, landing and taxiing air traffic controller trainer described by Yeliseev et al. [7], which allows creating training off-optimum situations on a real airfield, and also mastering in the training classes a combination of existing and new air traffic control equipment presented in the form of virtual objects. Such a simulator is described as follows.

The controller is at the workplace and watches the airport's airfield. Along with real planes, virtual planes or other objects-cars, people, etc., controlled by real operators from special consoles, are moving over the field or in the air over the field. Virtual planes seem to be practically the same as real ones. Virtual airplanes can safely participate in training scenes on the ground and in the air, simulating any situation, including emergency ones. Trained controllers communicate with pilots of virtual aircrafts in the usual ways, as shown in figure 1.

Figure 1. Example of application of a simulator in real conditions.

In the conditions of the mining enterprise technological process specific character, the territorial distribution of individual work sections, the decisive importance is acquired by the operational management of both individual work sections and the whole enterprise activities. At the same time, dispatching ensures uniform loading of all the enterprise parts, continuity, rhythmicity and economic fulfillment of all the production cycle main technological processes, uninterrupted operation of auxiliary and servicing sections.

With large industrial and other associations emergence in the sphere of mining, as well as the rapid improvement of transport, dispatching becomes an essential means of operational management of a complex set of technological processes and transport systems.

One of the most vivid examples of dispatching processes in mining production can be considered a single dispatching-analytic center SDAC-SUEK of Siberian Coal Energy Company. Under the control of SDAC there are nine underground mines, two open-cast mines, six auxiliary production units, seven service enterprises, it includes 24 types of different equipment and analytical programs. The dispatcher receives information in the real-time mode from all working extraction faces of underground and open-cast mines both schematically and through video cameras, he has the opportunity to view each object in more details. SDAC introduction allowed to reduce significantly the equipment downtime, to reduce the risks of incidents and breakdowns. So, in 2009, the average daily downtime of conveyor transport of the Kotinskaia mine due to problems with the drive systems in March (the month of maximum production) was about 3.5 hours (there were five conveyors in operation) [8]. In 2015, this indicator was about 0.4 hours and only due to the problems with automation systems (clipping, oxidation, fiber optic communication line breakdown, etc.). A photo of the video wall fragment installed in the central dispatching company is shown in figure 2.

In this context, an example of the augmented reality technology prospective application is its consideration as part of the control system, within the operator's mobile working place organization concept framework and integration with the neurocomputer interface (Brain computer interface, BCI) equipment for the mining industry [9, 10]. The main idea is to combine the technologies of the BCI and the AR for the control system of mineral deposits underground mining technological processes. The possibility to show the complex state information right before the operator with the detailed representation of interactive controls is designed to solve the problem of eye contact long time loss of indicators that display information about the state of operating mining equipment, thus reducing the likelihood of an emergency situation [10, 11].

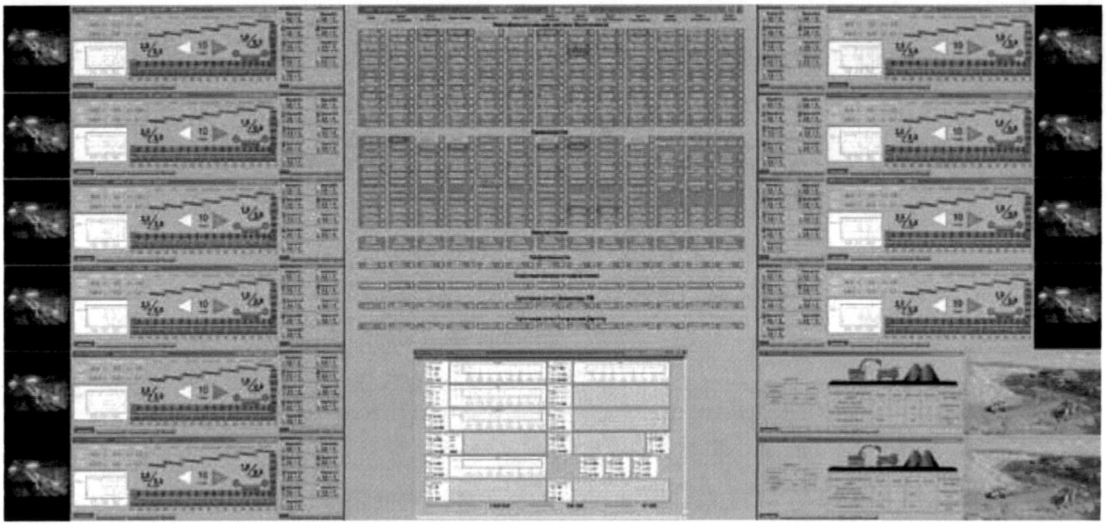

Figure 2. Operative-dispatching control automated systems ODCAS.

However, not all AR-devices can meet the requirements of the production environment. In some cases, the devices must be shock- and explosion-proof. For example, Microsoft Hololens, with which developers often work, in their current configuration cannot be used. The first certified AR-device for production purposes was HMT-1 from the German company RealWear. The reality headset device is ergonomic: it can be attached under the helmet, it does not interfere with the use of protective glasses in the industrial environment. The device allows workers to get access to documents and communicate with other employees, helps with production navigation

A significant limitation for implementing the augmented reality at the enterprises are also the safety requirements - in addition to the fact that glasses and pads must exactly correspond to industrial regulations on explosion protection and shockproofness, in some enterprises it is also inadvisable to use open technology to spread information over the Wi-Fi. Therefore, at the plant territory AR-devices must operate in off-line mode. All the performed operation information is stored on the device. Database is used for data synchronization on devices. You can synchronize data in a specially designated place where the use of Wi-Fi is acceptable.

5. Conclusion

This article demonstrates the diversity of approaches to the creation and application of augmented reality products for industry. Developers are often “carried away” by technological innovations, their characteristics and technical capabilities, while for “industrialists” it is important how much this is useful and promising for the technological process, the employee and the business as a whole. At the same time, it should be noted that “industry”, in turn, does not always have the technological and organizational flexibility to integrate the augmented reality technology into the production and management process.

As a result, we can conclude that the augmented reality technology is firmly embedded in everyday life. At the same time, they can be used not only in the entertainment industry, but also bring significant economic and technological effect in many industries, including the mining sector.

Acknowledgements

The reported study was partially funded by Russian Foundation for Basic Research (RFBR) according to the research project No. 18-37-00356 “Development of scientific foundations for the creation of mobile operator position on the basis of combining technologies of the brain computer interface and augmented reality”.

IOP Conf. Series: Earth and Environmental Science **206** (2018) 012044 doi:10.1088/1755-1315/206/1/012044

References

[1] *Everything You Must Know About VR/AR – Technologies* https://rb.ru/story/vsyo-o-vr-ar/
[2] *Augmented Reality in Russian Industry: Useless or Necessary* https://vc.ru/32831-dopolnennaya-realnost-v-rossiyskoy-promyshlennosti-bespolezna-ili-neobhodima
[3] Shou-xiang Zhang 2011 *Journal of Computers* vol 6 **6** 1213–21
[4] Zhang S X 2011 Applied Mechanics and Materials vol 40-41 388–391
[5] Soete N et al 2015 *Preprints of the 15th IFAC Symposium on Information Control Problems in Manufacturing* (May 11-13, 2015. Ottawa, Canada)
[6] *Augmented Reality in Industry: Case From Russia* (Siberian Federal University) http://icmim.sfu-kras.ru/node/2595
[7] Yeliseev B P et al 2010 *Scientific Bulletin of the Moscow Civil Aviation State Technical University* **161** 5–14
[8] Yutiaieev Yu P 2017 *Coal* **5** 30–36
[9] Kizilov S A et al 2018 *IOP Conference Series: Materials Science and Engineering* **354** 012016
[10] Kizilov S A et al 2017 *Mining Industry Journal* **6(136)** 96–99
[11] Kizilov S A et al 2017 *Proc. of the 11th All-Russian Sci. and Pract. Conf. on Automation Systems in Education, Science and Production* (Novokuznetsk: SibSIU) 44–47

The vector optimization mechanism for resource planning problems

V V Zimin[1], V V Mitkov[1], A V Zimin[1], I G Stepanov[2] and A A Ivushkin[3]

[1]Siberian State Industrial University, 42 Kirova street, Novokuznetsk, 654007, Russia
[2]Novokuznetsk Institute (Branch) of Kemerovo State University, 23 Tsiolkovsky street, Novokuznetsk, 654041, Russia
[3]Consolidated Company "Sibshakhtostroy" LLC, 9 Kuznetskoye shosse, Novokuznetsk, 654034, Russia

E-mail: zimin.1945@mail.ru

Abstract. The mechanism of two-criteria optimization for the resource planning problem is proposed, where the formalization of the problem can be represented by the direct and inverse control problem with a structurally similar description of the objective and cost functions. The set of Pareto-solutions of the two-criteria problem is determined by solution of the direct and inverse problems by the method of dichotomous programming, and the optimum on the set of Pareto-solutions – by the mechanism of complex estimation.

1. Introduction

Many problems of optimal resource planning, solved by industrial companies, are formulated in the form of a scalar direct or inverse control problem [1,2]. Let us consider, for example, the task of planning resources for training personnel operating the ERP system at a mining company [3].

Let $\{\{p_{ji} \mid i=\overline{1,n_j}\} \mid j=\overline{1,m}\}$ be a set of training programs implemented by a consulting company involved in the creation of the ERP-system. Here j – the number of a business process, i – the number of the training program, p_{ji} – the i-th training program for the j-th business process, n_j – the number of training programs for the j-th process, m – the number of business processes.

Let us denote the cost of training of one user within the program by $c_{ji}=c(p_{ji})$, the "expansion" of the user's competence (which we will estimate in points) as a result of training within the program p_{ji} – by $q_{ji}=q(p_{ji})$, the number of users that it is desirable to train within the program p_{ji} – by $k_{ji}=k(p_{ji})$. Let us also assume that the management of each business process imposes a restriction on the minimum number of trained users k_j^*, which allows the initial operation to be started. Let us denote the maximum limit of funds, that the management of a company can spend on the users training, by c^*.

2. The problem formalization and и the solution method

We will introduce a discrete variable x_{ji}, which is equal to 1, if users are to be trained in accordance with the program p_{ji}, and equal to 0 otherwise. Then the direct problem of training costs optimal planning is formulated as follows:

$$q = \sum_{j=1}^{m} \sum_{i=1}^{n_j} k_{ji} q_{ji} x_{ji} \rightarrow \max \qquad (1)$$

$$\sum_{j=1}^{m} \sum_{i=1}^{n_j} k_{ji} c_{ji} x_{ji} \le c^* \qquad (2)$$

$$\sum_{i=1}^{n_j} k_{ji} x_{ji} \ge k_j^*, \quad j = \overline{1, m}. \qquad (3)$$

The solution of problem (1) – (3) is a user training program $\{\{x_{ji} \mid i = \overline{1, n_j}\} \mid j = \overline{1, m}\}$ that maximizes the total "expansion" of users competencies q with a given restraint c^* for the maximum funds limit on allocated for training and that meets the requirements set by the managers of the business processes for the minimum number of users to be trained.

The opposite to the formulated problem (1) – (3) will be the problem:

$$c = \sum_{j=1}^{m} \sum_{i=1}^{n_j} k_{ji} c_{ji} x_{ji} \rightarrow \min \qquad (4)$$

$$\sum_{j=1}^{m} \sum_{i=1}^{n_j} k_{ji} q_{ji} x_{ji} \ge q^* \qquad (5)$$

$$\sum_{i=1}^{n_j} k_{ji} x_{ji} \ge k_j^*, \quad j = \overline{1, m}., \qquad (6)$$

in which it is required to define such training program for users $\{\{x_{ji} \mid i = \overline{1, n_j}\} \mid j = \overline{1, m}\}$, which minimizes the costs c to achieve the given value of the user competence development q^*.

Because of the structural similarity of functions c and q, both problems can be effectively solved by the method of dichotomous programming [4,5,6]. We should note the following important feature of this method. When solving each of the sequence of evaluation problems into which the initial problem is broken (because of this structural similarity), a subset of solutions are formed that are permissible by constraints and are Pareto optimal, that is, they are not dominated by other solutions of this set [7,8]. The set of solutions of the last estimated problems of the direct and inverse problem, due to the noted peculiarity, is a subset of Pareto solutions of the vector optimization problem:

$$(q, c) \rightarrow \max \qquad (7)$$

$$\sum_{i=1}^{n_j} k_{ji} x_{ji} \ge k_j^*, \quad j = \overline{1, m}. \qquad (8)$$

Based on their preferences, the decision maker chooses the best option from this subset. The procedure for determining the optimum on a subset of Pareto solutions can be formalized by applying the mechanism of complex estimation widely used in the theory of management of organizational systems [9, 10].

3. An example of a problem solving and results presentation

Let us consider the case with three business processes (m = 3) and, respectively, with three $(n_1 = 3)$, two $(n_2 = 2)$ and two $(n_3 = 2)$ training programs for these processes. The initial data for the training programs are given in table 1.

Table 1. The initial data of training programs.

	p_{11}	p_{12}	p_{13}	p_{21}	p_{22}	p_{31}	p_{32}
q_{ji}	4	3	5	3	4	5	3
k_{ji}	3	2	3	4	3	4	2
c_{ji}	60	64	90	54	90	90	54
	$k_1 \geq k_1^* = 5$			$k_2 \geq k_2^* = 3$		$k_3 \geq k_3^* = 2$	

We will take for the direct problem $c^* = 1100$, and for the inverse problem $q^* = 55$.

Solving the direct and inverse problem by the method of dichotomous programming, we obtain, respectively, the Pareto solutions presented in tables 2 and 3.

Table 2. Pareto optimal solutions of the direct problem.

q_{123}	45	51	39	36	59	50	53	56
k_{123}	12	14	11	11	14	13	13	15
c_{123}	774	902	722	632	1026	884	974	992
x_1	101	111	011	110	101	110	011	110
x_2	10	10	10	10	10	10	10	10
x_3	01	01	01	01	10	10	10	11

Table 3. Pareto optimal solutions of the inverse problem.

q_{123}	83	77	71	65	59	56
k_{123}	21	19	18	16	14	15
c_{123}	1532	1404	1262	1134	1026	992
x_1	1 1 1	1 0 1	1 1 1	1 0 1	1 0 1	1 1 0
x_2	1 1	1 1	1 0	1 0	1 0	1 0
x_3	1 1	1 1	1 1	1 1	1 0	1 1

The optimal solutions of the direct and inverse problems are underlined.

By uniting the solutions of both problems and excluding those that are dominated by others, we obtain the Pareto solution of the vector optimization problem, table 4 (a and b), figure 1.

Table 4a. Union of Pareto solutions for the direct and inverse problem.

q_{123}	83	77	71	65	59	56

IOP Conf. Series: Earth and Environmental Science **206** (2018) 012045 doi:10.1088/1755-1315/206/1/012045

k_{123}	21	19	18	16	14	15
c_{123}	1532	1404	1262	1134	1026	992
x_1	1 1 1	1 0 1	1 1 1	1 0 1	1 0 1	1 1 0
x_2	1 1	1 1	1 0	1 0	1 0	1 0
x_3	1 1	1 1	1 1	1 1	1 0	1 1

Table 4b. Union of Pareto solutions for the direct and inverse problem.

q_{123}	53	51	50	45	39	36
k_{123}	13	14	13	12	11	11
c_{123}	974	902	884	774	722	632
x_1	0 1 1	1 1 1	1 1 0	1 0 1	0 1 1	1 1 0
x_2	1 0	1 0	1 0	1 0	1 0	1 0
x_3	1 0	0 1	1 0	0 1	0 1	0 1

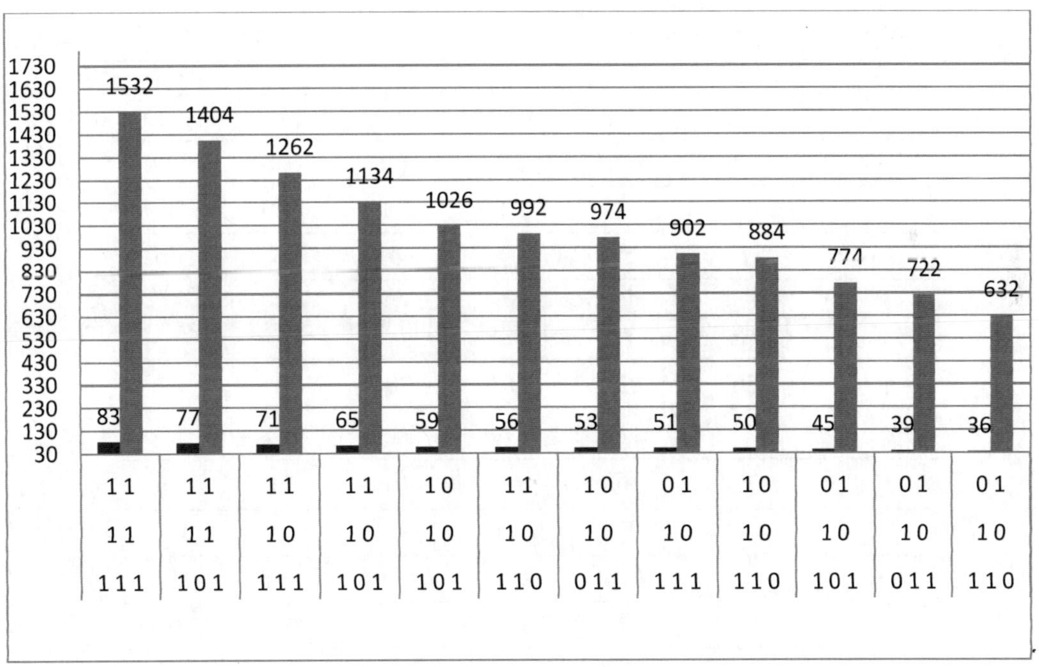

Figure 1. Histogram describing the set of Pareto-solutions to the vector optimization problems.

To formalize the choice of the best Pareto-solution by the decision make, we will apply the mechanism of complex estimation, which, in the annex to the problem under consideration, will include the following procedures:

1. Bringing the scales for measuring local efficiency indicators q and c to a single discrete scale of a given rank R (we assume R = 100):

1.1 Defining the measurement ranges q and c (in our example: q_{min}=36, q_{max}=83, then $q \in [35,85]$; c_{min}=632, c_{max}=1532, we assume $c \in [630,1540]$).

1.2 Defining the lengths of intervals q and c per one point of the measurement scale.

$$\delta_q = \frac{85-35}{100} = 0.5, \quad \delta_c = \frac{1540-630}{100} = 9.1.$$

2. Calculation of values $q^{á}(q)$ è $c^{á}(ñ)$ in the new measurement scale (the direct scale of measurement for $q^{á}(q)$, the inverse scale for $c^{á}(ñ)$).

3. Calculation of the average values of the integral indicator $(q,c)^{á} = \alpha_q q^{á} + \alpha_ñ ñ^{á}$ describing the effectiveness of the solution, obtained on the basis of additive convolution of local indicators and their relative weights given by the person making the decision ($\alpha_q + \alpha_c$=1).

Table 5 shows the results of converting the values of the indicators in table 4 from the natural scales of measurement to a single 100-point scale (the integrated indicator $(q,c)^{á} = \alpha_q q^{á} + \alpha_ñ ñ^{á}$, ($\alpha_q = 0{,}34$ è $\alpha_c = 0{,}66$) is in the place of indicator k).

Table 5a. Evaluation of Pareto solutions in points.

q^b_{123}	96	84	72	60	48	42
$(q,c)^b$	98.64	85.32	71	57	45.36	40.68
c^b_{123}	100	86	70	56	44	40
x_1	1 1 1	1 0 1	1 1 1	1 0 1	1 0 1	1 1 0
x_2	1 1	1 1	1 0	1 0	1 0	1 0
x_3	1 1	1 1	1 1	1 1	1 0	1 1
No.	1	2	3	4	5	6

Table 5a. Evaluation of Pareto solutions in points.

q^b_{123}	36	32	30	20	8	2
$(q,c)^b$	37.32	30.68	28.68	17.36	9.98	1.34
c^b_{123}	38	30	28	16	11	1
x_1	0 1 1	1 1 1	1 1 0	1 0 1	0 1 1	1 1 0
x_2	1 0	1 0	1 0	1 0	1 0	1 0
x_3	1 0	0 1	1 0	0 1	0 1	0 1
No.	7	8	9	10	11	12

Figure 2 shows a histogram describing the Pareto solutions obtained in a single point scale for q and c.

Solutions 3 and 4 are of interest. The third solution (x_1 = 111, x_2 =10, x_3 =11) is the most balanced and is characterized by the fact that spending 70% of the resource maximum we get 72% of the maximum possible result. The fourth solution (x_1 = 101, x_2 =10, x_3 =11) is the best in terms of resource efficiency: 56% of resources provide 60% of the value q (that is, 4% more than resource consumption).

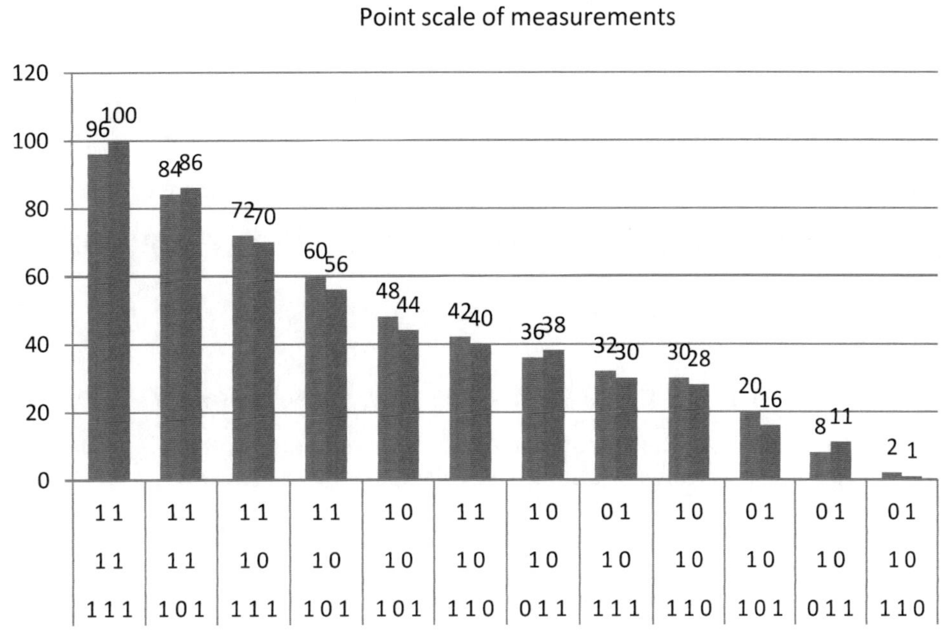

Figure 2. A histogram describing the set of Pareto solutions for the vector optimization problem in a single point scale of measurements.

4. Conclusion

The two-criteria formulation of the resource planning problem is preferable for the decision-maker, as he usually searches for a compromise between costs and the objective indicator. The stated mechanism of two-criteria optimization in many cases allows the decision maker to find such a compromise.

References

[1] Zimin V V, Burkova I V and Zimin A V 2017 *Models and Mechanisms for Managing the Life Cycle of IT Services* (Saarbrucken: LAB LAMBERT Academic Publishing)
[2] Zimin V V and Kiseleva T V 2016 *Proc. of the IV All-Russian Sci. and Pract. Conf. on Modeling and Knowledge-intensive Information Technologies in Technical and Socio-economic Systems* (Novokuznetsk: SibSIU)
[3] Mitkov V V and Zimin V V 2017 *Proc. of the 11th All-Russian Sci. and Pract. Conf. on Automation Systems in Education, Science and Production* (Novokuznetsk: SibSIU) pp 293 – 297
[4] Burkov V N and Novikov D A 1997 *How to Manage Projects: Scientific and Practical Edition* (Moscow: SINTEG-GEO) p 188
[5] Burkova I V 2009 *Automation and Telemechanics* **10** 15–21
[6] Burkova I V, Kosenkov A V and Kharitonova T B 2006 *Control Systems and Information Technologies: Scientific and Technical Journal* **1.1(23)** 123–126
[7] Petrovsky A B 2009 *Theory of Decision-making* (Moscow: Akademiya) p 400
[8] Larichev O I 2003 *Theory and Methods of Decision-making* (Moscow: Logos) p 392
[9] Burkov V N, Kondratiev V D and Shchepkin A V 2012 *Mechanisms for Improving Road Safety* (Moscow: LIBRICOM) p 208
[10] Novikov D A 2007 A Theory of Organizational Systems Management (Moscow: Fizmatgiz) p 584

Determination of geological conditions of gassy coal seams on the basis of seismic acoustic profiling in underground mine workings

D N Zastrelov, A N Kormin, E A Saltymakov, S V Sokolov and O V Taylakov

Federal State Budgetary Institution of Science Coal Institute of the Siberian Branch of the Russian Academy of Science, 10 Leningradskiy ave, Kemerovo, 650065, Russia

E-mail: Zastrelov@uglemetan.ru

Abstract: The article views the experience of mine geophysical prospecting using a seismic screening method of roof rocks from mine workings. The possibility of recording coal seams structural disturbances is estimated based on the analysis of seismic signals velocity fluctuation. The results of coal seam conditions assessment and their influence on the technological process of coal mining are presented.

1. Introduction

The current level of scientific and technical support of a coal mining process allows us to use methods of geophysical measurements for the estimation of coal seams roof parameters, on the basis of artificially generated physical fields of various nature: electric, electromagnetic, and seismoacoustic. Any of these investigations is performed, based on a single principle of certain characteristic changes during waves passing through the roof rocks of the extraction panel.

2. Materials

The main factor limiting the use of the listed types of geophysical equipment in mine workings is the increased safety requirements for equipment operated in coal mines, which are met by very few of the measurement systems available on the market. At the same time, the safe manufacturing of geophysical instruments often significantly limits their technical and operational characteristics. For seismoacoustic screening of a coal-rock massif in mine conditions the authors use an autonomous seismic station "R-1" with the following characteristics: number of registration channels - 1; the maximum capacity of a storage device CompactFlash is 16 GB; amplification of the preamplifier, once - 1, 2, 4, 8, 16, 32, 64; the band of recorded frequencies is 0 ÷ 3200 Hz; the sampling interval is 4, 2, 1, 0.5, 0.25, ms; the timing error is less than ± 1 ms/ 8h. The registration of seismic data is programmed using a computer with a USB-host port and an operating system that supports the FAT file system and storage devices of a USB Mass Storage standard. For the estimation of extraction pillars roof conditions in coal mines one can highlight the sampling step among the mentioned characteristics of the seismic stations "R-1",. The possibility of setting a sampling step to 0.25 ms significantly increases the accuracy of measurements within the extraction panel boundary. For example, consider the station capabilities with sampling intervals: 2, 1, 0.5 and 0.25 ms (table 1).

IOP Conf. Series: Earth and Environmental Science **206** (2018) 012046 doi:10.1088/1755-1315/206/1/012046

Table 1. The possibility of registration, when physico-mechanical parameters of the extraction panel roof fluctuate.

Interval of velocity change in a coal-rock massif	Sampling step 2 ms	Sampling step 1 ms	Sampling step 0.5 ms	Sampling step 0.25 ms
2. 8 km/s– 3.2 km/s	44.8 m	22.4 m	11.2 m	5.6 m
3.2 km/s –3.6 km/s	57.6 m	28.8 m	14.4 m	7.2 m

The Table presents a calculation, showing the minimally recorded length of a rock section with altered physico-mechanical parameters, when performing a seismic screening. Three values of longitudinal seismic waves passing velocity in the roof of the extraction panel are taken into account, resorting to the experience of undertaken mine measurements, characterized as follows: 3.2 km/s - medium-stable condition, average strength parameters, 2.8 km/s - rocks unstable, fissured, 3.6 km/s strong, resistant rocks [1]. Accordingly, in the thickness of the extraction panel medium-stable rocks, when recording data with a sampling step of 0.25 ms, it is possible to fix an unstable, fractured inclusion of a 5.6 m length and a section of strong, stable rocks of a starting 7.2 m length.

3. Methods

The method of research that allows the most use of the autonomous seismic station "P-1" capabilities is a seismic screening [2]. When performing the geophysical works, a certain number of stations with a set interval are linearly distributed across the mine working rim with deepening and installation of a receiving element. Excitation of seismic waves is performed from a parallel mine working with a given step based on the presented schematic diagrams (figure 1).

4. Discussion

Taking into account a minimal interference background possible within the limits of the operating coal mining enterprise, the presented equipment set allows data recording with a maximum travel time length of up to 500 m. Examples of a signal recording in various pickets of oscillation excitation are shown in figure 2. At the same time, two kinds of waves are captured in the roof: longitudinal (a red line) and cross-sectional (a green line) [3].

As the main analyzing parameter for assessing the extraction pillar roof condition, a change of velocity parameters registered between two parallel workings is used [4]:

$$t = \int_{И}^{П} \frac{1}{V} \cdot dl, \tag{1}$$

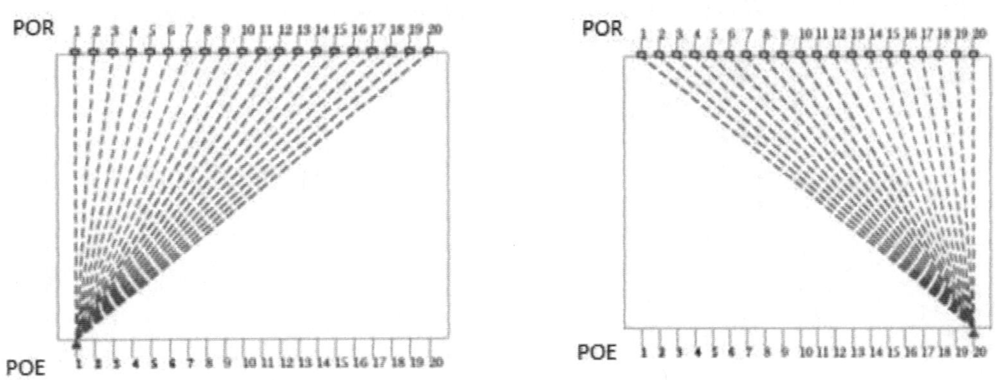

(a) (b)

IOP Conf. Series: Earth and Environmental Science **206** (2018) 012046 doi:10.1088/1755-1315/206/1/012046

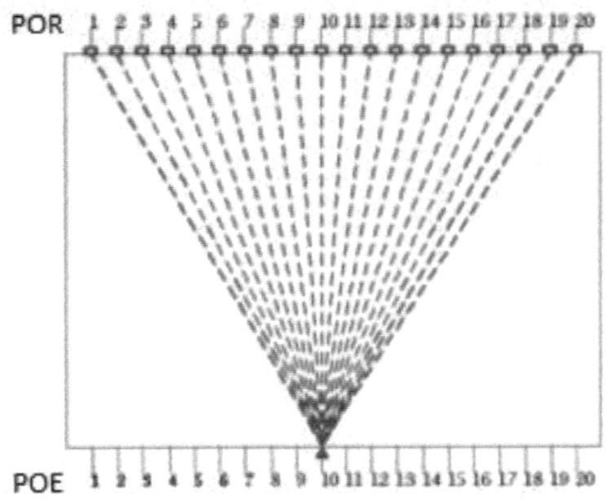

(c)

Figure 1. Schemes of passing waves propagation in the extraction panel body from various pickets of oscillation excitation of (POE) to pickets of oscillation reception of (POR): a) from the first POE; b) from the last POE; c) from the central POE.

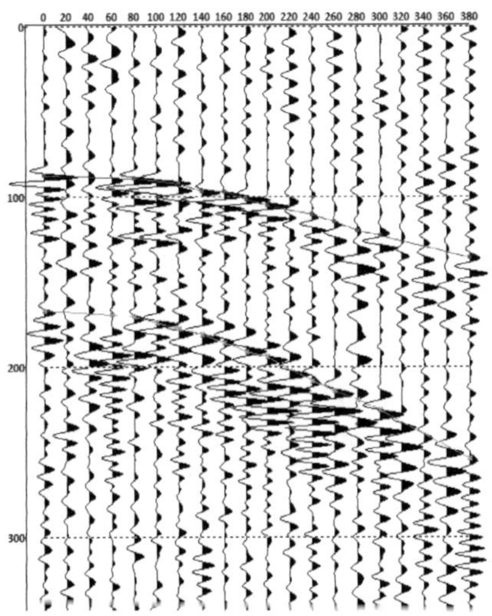

(a)

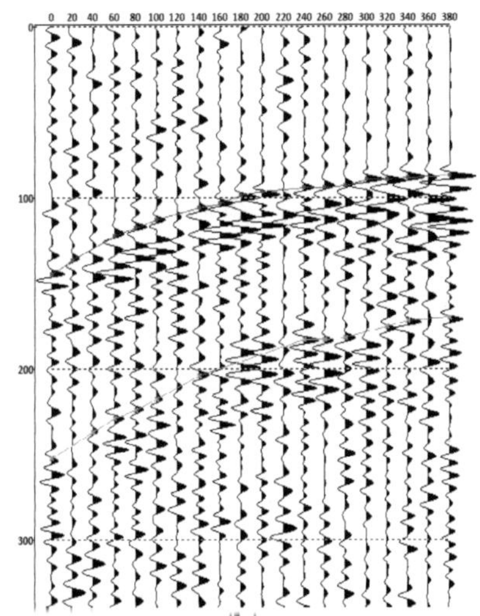

(b)

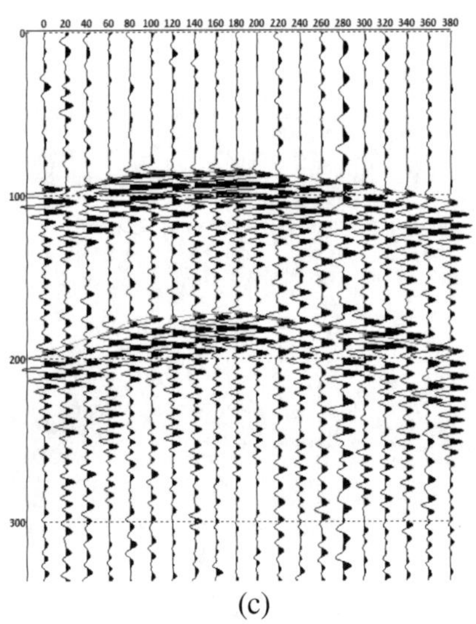

(c)

Figure 2. Initial multichannel seismograms: a) from the first EP; b) from the last EP; c) from the central EP.

t – seismic wave passing time;
V - seismic wave passing velocity;
l – distance from the source (S) to the receiver (R).

In different sections of the extraction panel, the division into which is performed by digitizing and pixelization on N pixels, taking into account the real coordinates, the given dependence is presented like this:

$$t_i = \sum_{j=1}^{N} \frac{1}{V_j} d_{ij} \quad \text{(i=1...n)}, \tag{2}$$

n – number of observations;
d_{ij} – the distance traveled by the ray i in pixel j.

5. Results

The application of the dependence (2) makes it possible to calculate the velocity model of the extraction panel and build on its basis a horizontal geophysical section of velocity characteristics distribution, where a contrast rise or decrease in seismic velocities illustrates the change in the physico-mechanical parameters of the roof rocks. For example, the geophysical section (figure 3) obtained for one of Kuzbass coal mines conditions is characterized by longitudinal seismic waves passing velocities of 3-3.2 km/s, which corresponds to the immediate roof of the coal seam presented by a medium strength argillite.

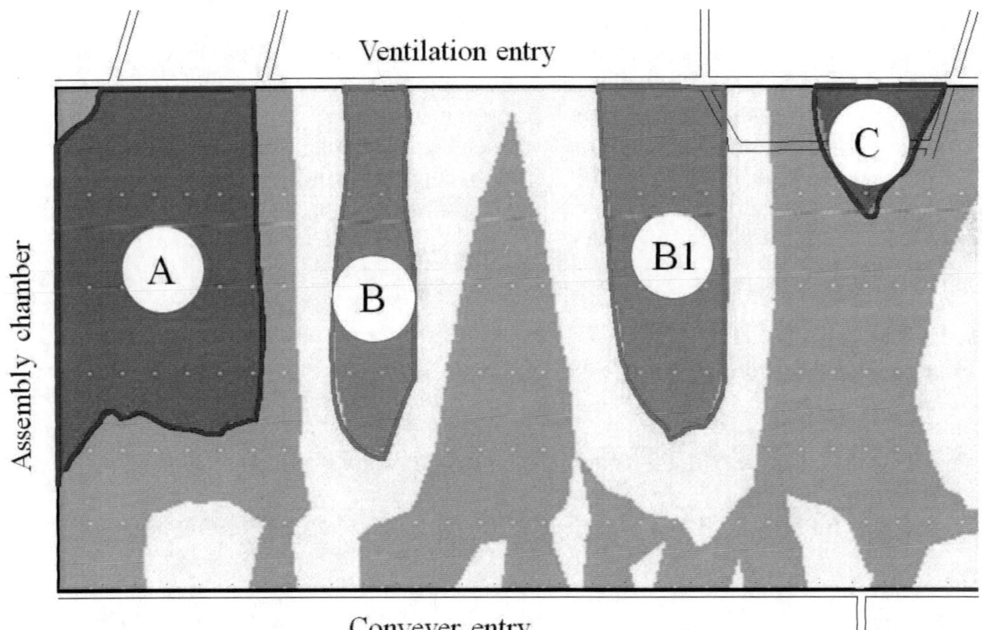

Figure 3. Fragment of horizontal section with velocity parameters distribution in the extraction panel area.

Within the extraction panel studied area the regions, where seismic waves passing velocity decreases in the massif down to 2.8 km / s are identified: A and C. The presence of such regions in the extraction panel area is explained by the increased fracturing due to the assembly chamber location in vicinity of A region and the pump chamber location within the region C boundary.

Also on the geophysical section, regions with the increased seismic velocities are contrasted: B and B1, which characterize manifestations of the main roof, represented by sandstone rocks in the area of the immediate roof thickness lowering down to 2-3 m.

Based on the analysis of the obtained section, the assessment of total conditions was performed in accordance with the VNIMI classification [8, 9]. Taking into account the distribution of velocity parameters in the study area, a medium-stable immediate roof is predicted. Manifestations of unstable roof may correspond to sections which are the subject of influence by regions A and C. Sections of hard-to-break roof are predicted in regions B-B1.

As part of the presented method development, an approbation use of neural network analysis tools was undertaken to improve interpretation operability of unstable and difficult to break roof sections within the studied sections of the extraction panel [10-12]. For this purpose, the horizontal section of velocity parameters distribution within the extraction panel area was divided into intervals: the sections with the most contrasting manifestations of the unstable immediate roof and the hard-to-break main roof were used for training the neural network; for the remaining intervals, a forecast was made.

As the main material for the estimation of active roof parameters of the coal seam the seismic wave propagation velocities determined within the studied section were used. The possibility of employing the amplitude-frequency characteristics of the detected signal was also evaluated, which demonstrated the low expediency of their application under given conditions of seismic vibrations excitation. Within the presented sample of initial data, the neural network analysis of seismic signal velocity parameters confirmed the primary state forecast for the active roof of the extraction panel. The use of a trained neural network to interpret data recorded within the extraction panel of other mines has shown the need for further expansion of a training sampling.

6. Conclusion

The gained practical results of autonomous seismic recorders application for the roof conditions assessment testify to the rationality of using such equipment for addressing issues related to the geophysical survey of the delineated complex mechanized faces (longwalls) in order to identify abnormal tectonic zones and other roof areas with changed geomechanical characteristics. Together with the data of geological prospecting drilling, the results of seismic screening provide an array of data sufficient for the classification of active roofs required for the geophysical support of the coal mining process.

References

[1] Jackson M J and Tweeton D R 2017 *Report of Investigations 9497* Bureau of Mines.

[2] Tailakov O V, Kormin A N, Zastrelov D N, Utkaev E A and Sokolov S V 2016 *The 8th Russian-Chinese Symposium. Coal in 21st Century: Mining, Processing and Safety* pp 324–329.

[3] Tailakov O V, Makeev M P, Kormin A N and Smyslov A I 2015 *Applied Mechanics and Materials* **770** 512–517

[4] Goshco E Y, Salnikov A S and Efimov A S 2014 *6th Saint Petersburg International Conference and Exhibition on Geosciences 2014: Investing in the Future* pp 392–396

[5] Kuznetsov V L, Salnikov A S, Markov VM and Kanareikin B A 2008 *Society of Petroleum Engineers – 70th European Association of Geoscientists and Engineers Conference and Exhibition – Incorporating SPE EUROPEC* pp 3553–57

[6] Salnikov A S, Dunaeva K A, Kanareikin B A, Sagaidachnya O M and Harlamov A S *2012 Proc. of the 5th Saint Petersburg International Conference & Exhibition 2012: Making the Most of the Earth's Resources Proseedings*

[7] Kanareykin B A, Dunaeva K A, Sagaidachnya O M and Salnikov A S 2014 *Proc. of the 6th Saint Petersburg International Conference and Exhibition on Geosciences 2014: Investing in the Future* pp 755–759

[8] Kozlov E A 2004 *Geophysics* **69** 885–897

[9] Kozlov E A, Vinnikovsky V S, Kirseleva O Y and Bovykin A A 2008 *Society of Petroleum Engineers – 70th European Association of Geoscientists and Engineers Conference and Exhibition – Incorporating SPE EUROPEC 2008 70th European Association of Geoscientists and Engineers Conference and Exhibition - Incorporating SPE EUROPEC 2008. Rome* pp 3902–3907

[10] Kozlov E, Liu E R, and Garagash 2007 *Journal of Seismic Exploration* **16** **2-4** 159–176

[11] Tailakov O V, Utkaev E A, Makeev M P, and Smyslov A I 2003 *International Coalbed Methane Symposium University of Alabama College of Continuing Studies; University of Alabama College of Engineering* 298–302

IOP Conf. Series: Earth and Environmental Science **206** (2018) 012047 doi:10.1088/1755-1315/206/1/012047

Hazardous manifestation of gas-dynamic phenomena in the faces of coal mines

V S Zykov and Yu M Filatov

VostNII Scientific Center on Industrial and Ecological Safety in Mining Industry, 3 Institutskaya street, Kemerovo, 650003, Russia

E-mail: v.zykov@nc-vostnii.ru

Abstract. The statistics of coal and gas outbursts in the faces of the Russian coal mines is provided. The mining-geological and mining-engineering conditions for the occurrence of sudden outbursts during the coal mining are considered, and the distinctive features in the factors of their occurrence and behavior are discussed. The criteria for identifying the potentially hazardous zones with the help of geophysical measurements are given. It is shown that it is important to take into account the data on the actual occurrence of outburst hazard in overlying and underlying longwall faces and development workings, as well as the results of the current forecast when preparing the extraction area for mining, as it follows from the results of underground investigations. The features are noted and the concept of the mechanism causing sudden coal and gas outbursts of faces is formulated. According to this concept, the probability of a hazardous situation is minimal at the beginning of the face advancement before the primary roof collapse. After the primary caving, the outburst hazard increases with the face advancement from the place of caving due to increased stresses in the bearing pressure zone, strengthening of the restraint of the nearest roof support, and the formation of a low permeable bridge in the coal seam near the exposed surface of the coal massif, which increases the pressure and free gas volume behind this bridge. The maximum degree of outburst hazard in the cleaning face is reached at the time of secondary roof collapses.

1. Introduction

At the Russian coal mines, the largest number of sudden coal and gas outbursts, about 80%, was registered during the preparation works in the coal seams. Sudden outbursts during coal mining began to occur later and with a lower frequency than when the development of seams was started and development workings were carried in them. The lower outburst hazard of the cleaning faces is due to the peculiarities of the stressed-deformed state of the mountain massif in the vicinity of the mine workings, which will be considered below. However, by the present time, mining operations in the mines are conducted at the depths at which in most faces in the certain geological conditions and mining engineering parameters, an outburst hazard is possible.

The share of outbursts that occurred during the development works from the total number reached approximately 15%. At present, most of the faces work at the depths below the critical ones for sudden outbursts of coal and gas, that is why the problem of early assessment of the gas-dynamic hazard is urgent for them. It concerns mainly the mechanized faces in the flat and inclined seams mined at high speeds with the use of pillar method. In these faces the largest amount of coal is extracted.

2. Statistical data on the sudden coal and gas outbursts in the mines of Russia

The largest number of sudden coal and gas outbursts in the mine workings (47 outbursts) was registered in the mines of the Pechora Basin. Of them 31 sudden outbursts of coal and gas occurred in 1960-1964 at mine No. 1 at a depth 365-460 m from the surface during the development of the seam Dvoynoy. The outbursts intensity at mine No. 1 varied from 1.2 to 37 tonnes of coal, the gas intensity – from 65 to 3800 m^3. The gas outburst reached an extremely high value – 295 m^3 per 1 tonne of the ejected coal.

In the Primorsky Basin in 1970, 13 sudden outbursts of coal with associated gas release were recorded during the development of the seam Modestov II at the mine "Podgorodnenskaya" and the seam Nizhne-Kedrovy at the mine "Tsentralnaya", which, in accordance with the then in force "Instruction on the safe performance of works in the seams hazardous because of their sudden coal (rock) and gas outbursts" during the investigation of gas-dynamic phenomena were attributed to sudden coal and gas outbursts. The outburst intensity reached 35 tonnes of coal and 5000 m^3 of gas.

The total number of sudden outbursts in the faces of the Pechora, Primorsky and Kuznetsk Basins amounted to 69 or 7.4% of the total outbursts in all types of workings, and in the Pechora Basin the share of outbursts in the mining face was 17%.

It should be noted that statistical studies on the mines of Donbass indicate that the greatest number of sudden outbursts in mining faces occurs at depths of 400-500 m [1]. This is confirmed by the manifestation of outburst hazard in the mines of the Pechora basin.

3. Methodological approaches to the establishment of probable zones of sudden coal and gas outbursts in the faces

As it is known, sudden outbursts of coal and gas occur in areas of tectonically disturbed coals in the area of disjunctive or plicative dislocations. Other geological factors that determine the possibility of identifying the outburst hazard zones are the degree of coal and host rocks metamorphism, the petrographic composition and coal seam thickness, the moisture content of the coal, the lithological composition and thickness of the host rocks.

Since coal has reduced strength properties within the range of the hazardous portion, a redistribution of stresses occurs near it. The undisturbed part of the coal seam along the boundaries of the site is in a more stressed state. Therefore, when approaching the boundaries of thc hazardous portion, sharp, abrupt changes in the velocity parameters of seismic waves are observed [2-4].

According to the results of the geophysical survey, areas with anomalously low values of the potential difference, with not less than two times less than the average values for the excavation column, and lowered not less than 20% values of the velocities of the formation waves belong to areas prone to hazardous outbursts [5]. At the boundaries of these sections, anomalous increases in the values of the potential difference are observed – approximately in 2 times compared with the average background and the increase in the seam waves velocities by 20%.

It is important to take into account the data on the actual outbursts manifestation in the process of contouring development workings and the results of determination of the outburst factor at the current forecast in these workings. This follows from the results of the studies [6]. Figure 1 contains the curve of the gradient change in the initial velocity of gas release from control wells g_n in face of longwall No. 3 of the seam Vladimirovsky at the mine "Severnaya" (in Kuzbass) at the approach to the alignment with the position of the face in the underlying longwall No. 2, where a sudden outburst of coal and gas occurred on August 08, 1978.

As it is known, this parameter is a reliable characteristic of the outburst hazard and is used in regulatory forecasting methods. The curve indicates that the maximum gradient was measured at a distance of 2 m from this alignment, that is, it indicated the entrance to the hazardous area.

The sudden coal and gas outburst that occurred at the same mine on April 11, 1970, when the same seam was mined by longwall No. 2, was preceded by seven outbursts in the cross-cut of the II parallel drift, driven for the preparation of longwall No. 2, six – when drilling wells ahead of the longwall. Their intensity for well outbursts was very high – 15-35 tonnes.

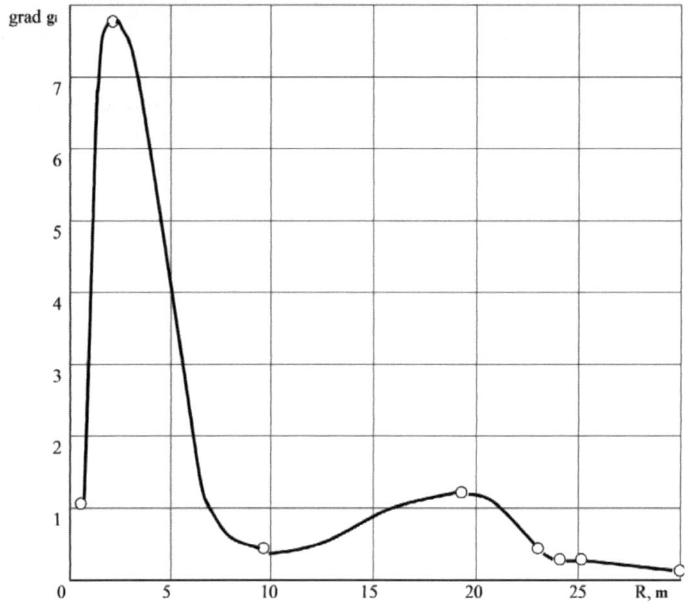

Figure 1. Change in the gradient of initial gas release from the holes in longwall No. 3 of the seam Vladimirovsky at mine "Severnaya", depending on the distance to the alignment with the place of sudden outburst in the underlying longwall No.2.

Similar results were obtained in the Pechora and Primorsky Basins. At the same time, the non-hazardous behavior of longwalls in all these basins corresponded to the non-hazardous values to the outburs hazard indicator in the development workings contouring them. The mining areas, the outburst hazard of which is confirmed by the gas-dynamic phenomena occurring on them near their borders, or by the results of the current outburst hazard forecast in the contouring development workings, are subject to compulsory mining with the help of regional methods to prevent sudden outbursts of coal and gas. The decision on the need to apply regional methods to prevent sudden outbursts in other areas with a disturbed structure is adopted depending on the values of the energy criterion of outburst hazard [6].

4. Consideration of the peculiarities of sudden coal and gas outbursts of the faces and the mechanism of their behavior

The founder of the theory of sudden coal and gas outbursts V.V. Khodot [7] notes the differences in conditions that cause sudden outbursts of coal and gas in the faces of development workings. He supplements the main factors that determine the occurrence of outbursts in the faces with the dynamic load on the coal seams on the side of the host rocks. Having carried out the research in the mines of Donbass, A.F. Shak and S.V. Mirer [8] confirmed that the poor caving of roof, contributes to the occurrence of a hazardous situation. Half of all sudden outbursts in the faces occurs in the lower and upper parts of the longwalls due to a sharp change in the ratio of the elastic and inelastic zones in them [9].

In the process of primary collapse, there is a transition of the type of roof loading from the beam scheme to the cantilevered. As a result, the zone of bearing pressure with a region of increased stresses and a clearly pronounced maximum is formed ahead of the face in the massif. In the coal seam ahead of the face, four zones [10] of the massif condition in the vicinity of the face in terms of outburst hazard are indentified (figure 2).

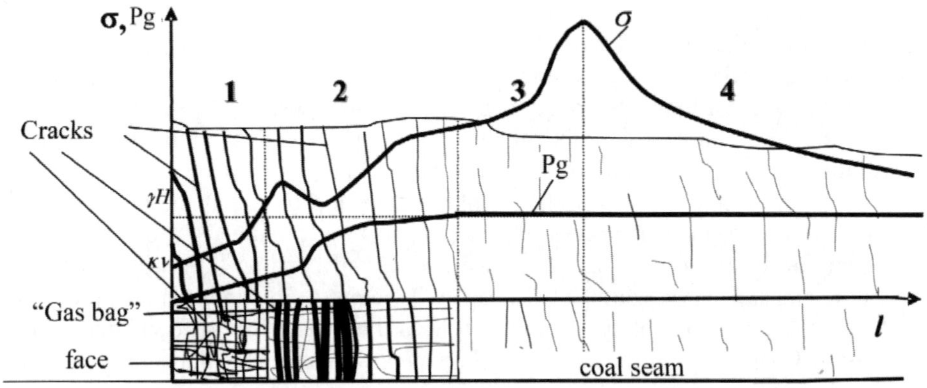

σ –the value of stresses in the massif; Pg – gas pressure in the seam;
l – distance from the face into the depth

Figure 2. The scheme of formation of outburst hazard condition in the massif ahead of the face.

Zone 1, adjacent to the face, is composed of the most destroyed coal due to the destructive effect of the roof rocks. It has very low filtration properties, so the degassing process of the coal massif is greatly extended in time. Zone 2 is a system of roughly parallel to the face elements of the array, consisting of coal blocks, tightly pressed to each other by the host rock. In it, due to the low filtration capacity of zone 1, conditions are created for accumulating large volumes of free gas under pressure. In zone 3, which is the most remote from the face, the preparation of the massif for the transition from the elastic state to the elastic-plastic one takes place. Zone 4 is the undisturbed massif outside the area of influence of the face on the stress-strain state of the coal seam.

The outburst hazard of the mining faces occurs due to the formation of a "gas bag" in zone 2 because of the presence of a bridge pinched by the bent roof between zones 1 and 2 with a very low gas permeability [11].

In addition, in the coal massif in the zone influenced by mining the pressure of free gas in the process of hysteresis of methane sorption by coal increases, caused by low-frequency oscillations of the roof above the seam during its step-by-step caving established by the authors of the work [12, 13].

5. Conclusions

Summarizing the above, we can formulate the following concept of occurrence of sudden coal and gas outbursts in the face. The probability of a hazardous situation is minimal at the beginning of the face advancement. The process of reducing stresses in the seam in the zone, corresponding to the bending zone of the roof, and the formation of a "gas bag" in it is not pronounced due to the stable condition of the roof. Apparently, a sudden outburst in the face before the primary step of caving is possible only when the face encounters a large local tectonic disturbance of a closed type with large gas reserves. However, there is a high probability of outburst at the moment of roof caving because of a sharp decrease in the cohesion forces between the coal seam and the host rocks and coal layers.

After the primary step of caving, the outburst hazard increases as the face advances from the place of caving due to the rise in the stresses in the bearing pressure zone (the intensity of formation of the outburst hazardous cracks and the amount of free gas enclosed in them increases), the formation of a zone of low stresses due to the roof bending (a "gas bag" appears), increased strengthening of the restraint of the nearest roof support (formation of a low permeable bridge in the coal seam near the exposed surface of the coal massif formed, which contributes to an increase in pressure and a free gas volume in the "gas bag").

If the roof collapses at this stage of the massif behavior, then the nearest roof support (coal bridge) will collapse with the sufficient pressure and gas volume in the gas bag, a sudden coal and gas outburst will happen, and at a high gas pressure gradient it will transit to the stage of gas crushing of coal. The process will be strengthened by the fact that instantaneously due to the hysteresis of methane sorption by coal the volume and pressure of the gas in the "gas bag" will increase.

If, however, the advancement from the face will increase and there will be no roof caving, then starting from a certain moment, the process of cracks formation, oriented in a direction close to the perpendicular plane of the layering, begins, but due to the roof bending they will be directed to the gob. Through these cracks the gas from the "gas bag" will begin to flow into the gob, the pressure and the volume of the gas and, consequently, the outburst hazard of the of face in it will begin to decrease. The outburst hazard from this point onward with the face advancement will also decrease. The same will happen after the next cavings.

Thus, from all that has been said, it follows that the maximum degree of outburst hazard in the face is reached during the roof collapses. The degree of hazard is especially great if, by this time, no cracks have already appeared that penetrate into the gob and degasify the zone of intense crack formation 2 (figure 2) with large gas reserves under pressure between the layers of coal massif divided by the cracks quasi-parallel to the face, including the so-called "gas bag", that is, when the formation of zone 2, which is the "driver" of the sudden outburst, reaches the maximum degree of hazard.

According to [10], the face might be hazardous because for its sudden outbursts if $Fa \geq Fo$ and $l \geq \Delta x$, where Fa – is the resultant of the active forces tending to eject coal; Fo – is the resultant of the limiting obstructing forces; l – the depth of penetration; Δx – the value of the degassed coal zone. The values of these parameters are determined depending on geological conditions and mining engineering parameters. However, this model needs to be clarified, since it does not take into account the processes of hysteresis of methane sorption by coal and the formation of a zone of low stresses ahead of the face at the place of roof bending and, consequently, changes in the degree of outburst hazard depending on the amount of face advancement from the caving site.

To take into account the first process, an amendment was made to the model during the development of instructions [2], according to which the value of the gas pressure Pk used in the model for the face advancement intervals in the zone, where the roof collapse is possible, is determined from the expression $Pk = P_0+\Delta P$, where: P_0 – is the seam pressure of gas; ΔP – increase in gas pressure due to the phenomenon of hysteresis of methane sorption by coal. The value of ΔP is determined by the diagram given in the instructions.

In order to take into account the influence of the roof bending on the outburst hazard, a considerable amount of experimental research is required, and the model of sudden outbursts occurrence in the faces should be elaborated on the basis of these research results.

References

[1] Stepanovich G Ya and Nikolin V I 1971 *Safety, Labor Protection and a Mine Rrescue Work* **6** 16

[2] Zykov V S, Egorov P V, Potapov P V, Khveshchuk N M, Sidorchuk V V and Zheltkov I V 2003 *Forecast and Prevention of Sudden Coal and Gas Outbursts in the Mining Faces* (Kemerovo: Kuzbassvuzizdat) p 198

[3] Krey Theodore C 1963 *Geophysics* **28 5** 701–14

[4] Kjartansson E 1979 *Geophysical Research* **84 B9** 4748–4737

[5] Zykov V V 2010 *Sudden Coal and Gas Outbursts and other Gas-dynamic Phenomena in Mines* (Kemerovo: Institute of Coal and Coal Chemistry of SB RAS) p 333

[6] Puzyrev V N and Zykov V S 1983 *Improvement of Mining Operations Safety* (Kemerovo: VostNII) pp 40–34

[7] Khodot V V 1961 *Sudden Outbursts of Coal and Gas* (Moscow: Gosgortekhizdat) p 363

[8] Shak A F and Mirer S V 1974 *Coal of Ukraine* **2** 37–35
[9] Nikolin V I, Voronin V A and Kulakovsky V N 1974 *Coal of Ukraine* **9** 30–29
[10] Murashev V I *1980 Development of Scientific Foundations for Safe Mining Operations in the Mines on the Basis of Scientific Studies of Geomechanical Processes* (Moscow: IGD n.a. A A Skochinsky) p 35
[11] Balashova T A 1988 *Bulletin of KuzSTU* **4** 31–29
[12] Puzyrev V N, Balashova T A and Alekseev D V 1993 *Coal* **6** 48–46
[13] Puzyrev V N, Alekseev D V, Ivanov V V and Balashova T A 1992 *Control of Gas Resease in Mines* (Kemerovo: KuzPI) pp 43–35

IOP Conf. Series: Earth and Environmental Science **206** (2018) 012048 doi:10.1088/1755-1315/206/1/012048

Fully environmentally closed technology of gasification dusted fractions of coal concentration with restoration of metals and alumosilicates from cinders

V P Tsymbal[1], I A Rybenko[1], A A Olennikov[2], A A Rybushkin[1], V I Kozhemyachenko[1] and P A Sechenov[1]

[1]Siberian State Industrial University, 42 Kirova street, Novokuznetsk, 654007, Russia
[2]Tyumen State University, 15a Perekopskaya street, Tyumen, 625003, Russia

E-mail: tsymbal33@mail.ru

Abstract. The project proposes a completely ecologically and energetically closed waste-free technology for gasification of pulverized coal waste in a suspended slag layer, producing a conditioned synthesis gas, simultaneous reduction and recovery of metal from ash and slag, and also obtaining a highly porous aluminosilicate slag (microspheres) suitable as lightweight aggregate, paints and coatings with low thermal conductivity, as well as adsorbent and other applications.

1. Brief description of the project

To solve the problem of a completely ecologically closed separation of coal (waste coal), the jet-emulsion aggregate SER (self-organizing jet-emulsion reactor), previously developed for the direct reduction of pulverized metallurgical wastes and ores, is used [1, 2]. In the first (spherical) reactor of this aggregate, a large interphase chemical interaction surface is created due to the effective sputtering of the incident charge stream (coal, slag-forming) by counter propagating oxygen jets, the gas suspension formed through a gas-dynamically locked channel is fed to the lower part of the vertical cylindrical reactor, where it forms gas-dynamic "failure grating", on which the gas-slag emulsion hangs. Inside this emulsion and incomplete combustion of coal (up to CO) with almost full use of the resulting thermal and chemical energy. Thanks to the organization of the gravity separation mechanism, the products are separated into metal and gas slag, which enters the lattice of the gas reformer via the slag receiver, where steam-coal combustion and gas enrichment is carried out with hydrogen to a volume ratio H_2/CO equal to 1/1, which is already good the result of [3, 4] for the subsequent synthesis of dimethyl ether is at least better than in the Fisher-Tropsch method, where it is still necessary to separate nitrogen [5].

If there are resources of methane or natural gas nearby, then by oxygen conversion of a relatively small amount of methane (15-20% of the gas volume), it is possible to obtain an optimal H_2/CO ratio of 3/2 for the synthesis of dimethyl ether and 2/1 for methanol, corresponding to their chemical formulas. In this case, it is possible to carry out an almost non-waste catalytic synthesis of motor fuel.

2. The device and principle operation of the SER unit

The process is realized in the aggregate shown in figure 1 as follows [6]. The powdered charge consisting of a mixture of coal (waste) and slag-forming (iron oxides) is fed by a metering feeder 2 to the central zone of the reaction chamber 1 where a seal disk is formed at the place where the oxygen flows delivered through the inverted directional lances 3 meet. As a result of the dynamic interaction of jets there is an intensive turbulence of the charge flow and the formation of large surfaces for heterogeneous chemical interaction. At the same time, due to incomplete combustion of a part of coal or other reducing agent in the reaction chamber 1, heating and partial reduction of oxides occur in accordance with the proportion of oxygen supplied.

In this unit, which has a certain universality, as a separate task, it is possible to process, that is, gasification, only coal, without the addition of ore or with a minimum addition of ore, only to reduce the process temperature and produce a foamy slag emulsion. With this version of the technology, simultaneously with the incomplete combustion of coal inside the gas-slag emulsion, metals (iron, manganese, etc.) are recovered from the oxides contained in the coal ash and in slag-forming additives. In this case, the composition of the gas by the sum of CO and H_2 approaches 90%.

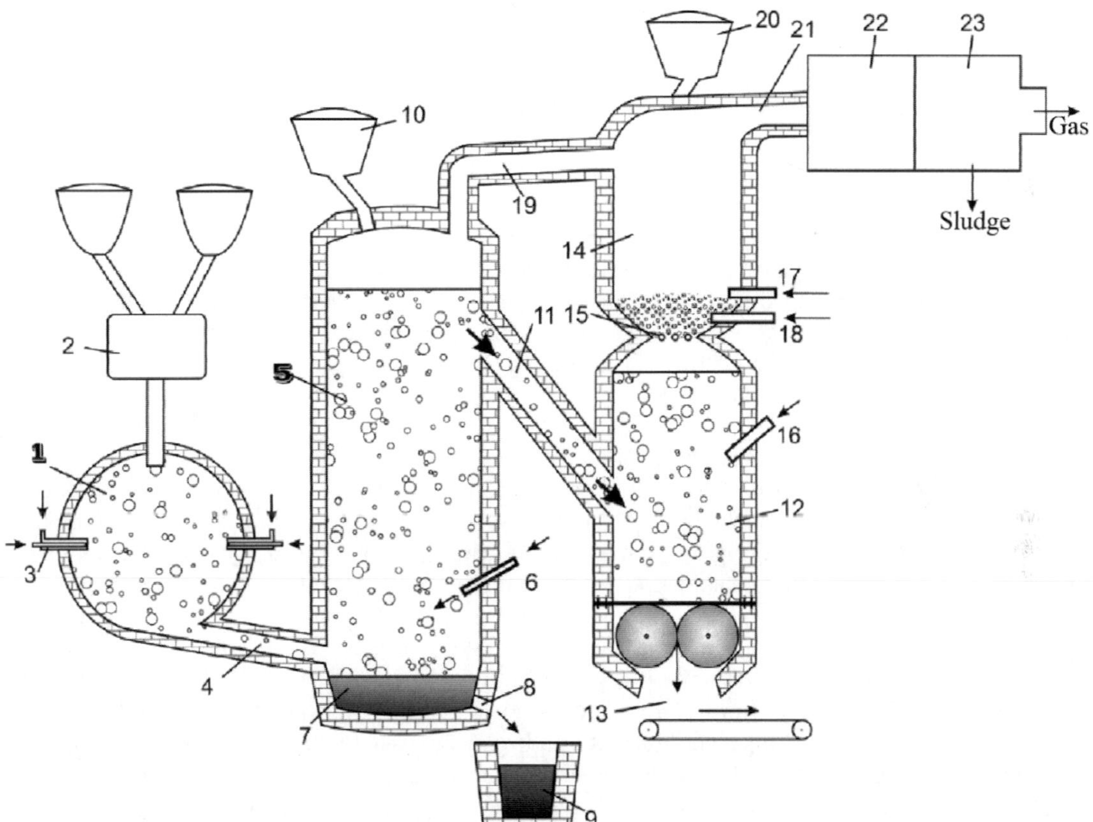

Figure 1. Scheme of wasteless complex gasification.

Technology with the addition of iron oxides (sludge, tailings of enrichment) to coal is very rational, since part of the gaseous oxygen is replaced by oxygen extracted from iron oxides, and the resulting metal or synthesis gas becomes essentially free. At the same time liquid iron and its oxides are simultaneously catalysts, allowing to accelerate the process and increase the yield of some components of synthesis gas.

The use of a vertical column reactor 5 with a lower feed of the reaction gas suspension in combination with a significant deviation of the processes from thermodynamic equilibrium as the refining settler is an important factor that makes it possible to separate the flow of metal deposited in the calculator 7 and the flux of the gas slag emulsion discharged along the inclined channel 11 in a slag receiver 12 where slag and gas are separated. To bring the composition of the associated process gas from the slag receiver 12 through the cooled grate 15 to the gas conversion chamber 14 to the conditioning composition of the synthesis gas suitable for processing vdimethyl ether or methanol under the grate 15, with the continuously expanded material 19 a layer of coke or coal, steam is supplied, and natural gas and oxygen are supplied above the grate 15 to the gas conversion chamber 14, while providing a volume ratio of CO to H_2 in the produced gas 2/3 or 1/2 for production of dimethyl of ether or methanol. The resulting synthesis gas from the chamber 14 through the channel 21 is fed to a cooler (for example, a waste heat boiler) 22, a gas purification 23 and then to the consumer, for example, in a catalytic synthesis unit of dimethyl ether or methanol or for energy purposes. The sludge generated in the gas scrubbing 23 is recycled into the process, which allows the production of smokeless technology.

An advantage of the present invention is that the installation in series with the refiner 5 of the gas composition correction chamber 14 combined with the slag receiver 12 and the slag granulator 13 gives a number of additional effects. First of all, there is the possibility of flexible (practically independent) control of the composition of the gas at the outlet from the refining settler 5 and at the outlet from the aggregate as a whole, more precisely from the chamber 14. Combination of the coal gasifier (chamber 14) with the metallurgical unit (reaction chamber 1 and refining settler 5) and the slag receiver 12 allows us to use the physical heat of associated gas and slag to increase caloric content, and the amount of commodity gas. Thus, on the basis of chamber 14, a chemical regenerator (utilizer) of physical heat is obtained. The increased pressure created in the reaction chamber 1 allows the reaction gas to be pushed through all the elements of the unit, which results in a small specific volume of the unit and eliminates the need for using flow and smoke drivers.

If, however, gas is produced in a special gasifier (gas generator), as in the Lurgi-Fischer-Tropsch method, a considerable amount of thermal energy is spent to maintain the temperature (800-1000^{0}C) in the gasification site and nitrogen heating [5], and the gas is obtained low caloric and requires a large expenditure for the separation of ballast components before catalytic synthesis.

3. Characteristics of the products obtained

Due to the fact that the gas suspension weighs in the column reactor 5 and then thrown out at high speed through the inclined channel 11 to the slag receiver 12, has a very high gas content (more than 99%), it is possible to produce highly porous aluminosilicate microspheres. The parameters of the column reactor, the connecting channel and the blast regime are selected in such a way that the so-called inertial-turbulent regime of motion of the two-phase working mixture created in the reactor-oscillator (the first reactor) is fulfilled. In this case, conditions are created when the slag film is "wound" on gas bubbles and hollow spherical structures (microspheres) are formed. Managing the slag regime, it is possible to obtain films of the required chemical composition, and the installation of a gravity separator after the slag-receiver granulator, allows us to classify the microspheres according to the granulometric composition, that is, to satisfy the most stringent requirements. Quality micro-granules are more expensive than metal. The following slag composition for micro-granules was obtained on the mathematical model (table 1):

Table 1. Chemical composition of microgranules.

Slag	SiO_2	Al_2O_3	Fe_2O_3	CaO	MgO	Na_2O	K_2O	Total
%	62.856	26.727	1.412	2.568	2.568	2.217	2.568	100

After the steam conversion, chamber 14 produces a gas of the following composition (table 2), with an H_2/CO ratio of 1/1, that is already suitable for the catalytic synthesis of motor fuel. If such a task is not put, then this gas is fed to a gas turbine or a gas piston machine and turns into electricity, including that used inside the process cycle for the production of oxygen. This completes the first stage of the project implementation.

Table 2. Composition of gas after steam-oxygen conversion.

Gas	CO	CO_2	N_2	H_2	H_2O	CH_4	Total
% By weight	51.518%	31.355%	0.887%	3.680%	12.559%	0.000%	100.00%
% By volume	35.92%	13.91%	0.62%	35.92%	13.62%	0.00%	100.00%

If the task is to convert the synthesis gas into motor fuel, either the gas of the composition obtained above can be used, or by oxygen conversion of not a sick amount of methane is brought to higher conditions corresponding to the chemical formulas of dimethyl ether (diesel fuel) or methanol. Table 3 shows the calculated composition of such a gas.

Table 3. Composition of gas after additional methane-oxygen correction.

Gas	CO	CO_2	N_2	H_2	H_2O	CH_4	Total
% By weight	58.084%	29.434%	0.880%	5.632%	5.970%	0.000%	100.00%
% By volume	35.03%	11.30%	0.53%	47.55%	5.60%	0.00%	100.00%

The metal (iron-carbon melt) located in the casing 8 is necessary for the stable maintenance of the gas-slag emulsion, inside which carbon and volatile constituents of coal are burned. In addition, iron oxide is a catalyst for the formation of hydrocarbons. Periodically produced from a forehearth metal, in the simplest case is sold as a first-class scrap, the market of which is not limited. But, in principle, all metals contained in the coal sol are concentrated in this metal. This task requires separate consideration.

In the presence of a conditioned synthesis gas, the task of catalytic synthesis does not cause any special difficulties, although, of course, it requires additional, but quickly recouped, capital investments. We do not consider this stage of the possible development of the project here.

In the case of the above project, the resulting synthesis gas can be considered as a by-product, since its cost is completely covered by two other liquid products – aluminosilicate micro-granules and metal, as can be seen from table 4, which presents the main technical and economic indicators from implementation of the pilot sample of the unit with a coal output (waste) of 3.6 tons per hour.

The effectiveness of the project is very high: the payback period is less than a year after launch, and every ton of recycled waste (currently polluting the environment) can generate profit of 17 thousand rubles. It should be emphasized that a completely ecologically and energetically closed technological scheme is created, the input stream in which is only coal (waste), yet other products, including oxygen, are obtained within this technology. This allows you to design the unit in a modular mobile version and place it near waste generation sites or in locations remote from electrical networks.

Table 4. Technical and economic indicators.

Input current	Consump., hour	Consump., year	The price of units, rubles/ton, or thousand m^3	Costs, income for the year, million rubles	Notes
Coal (waste), ton	3.6	25920	400*	10.4	
Oxygen, m^3	1075	322500	-	-	Oxygen station (with compressor up to P=150 atm.) – 8 million rubles

Input current	Consump., hour	Consump., year	The price of units, rubles/ton, or thousand m^3	Costs, income for the year, million rubles	Notes
Products received					
Gas, thousand m^3	7.2	51840	4000	207.9	
Al – Si, ton	0.9	6480	35000	**226.8**	
Metal, ton	0.07	504	18000	**9.1**	
Possible full generation of electricity, MW·h	**10.7**	**77172**	**-**	**-**	**Gas-piston station 6xTCG2032V12 Power 3.4 MW, the cost of 57 million rubles**
Electricity for the production of oxygen, MW·h	1.3	9360	-	-	
Electricity for own needs of aggregates 1-8, MW·h	0.4	2880	-	-	
Remaining exported electricity, MW·h	10.0	64800	4200	**272.2**	
Operating costs				**60.0**	
Profit from sales: el. energy, Al-Si, metal				**448.1**	
Investments				250	**Implementation period, 1.5-2.0 years**

*Note: The deliberately inflated value of waste enrichment is accepted, in fact it is equal only to the cost of loading and delivery.

4. The non-volatile scheme of the SER unit

Figure 2 shows one of the variants of a fully energetically closed circuit. The exhaust gas from unit 1 is supplied to reformer 2, which increases the caloric value of the gas to 12320 kJ/m^3 and changes its chemical composition to the H_2/CO ratio of 1/1. After gas purification 3, the gas enters the gas piston station 4, on the shaft of which there is an electric generator 5 [7]. As can be seen from Table 4, 7200 m^3/h of exhaust gas (after reformer 2) can be fully used in the gas piston station 4 to generate electricity by the generator 5 in the amount of 10.7 MW·h. This makes it possible to completely cover the own needs of aggregates 1-8 (about 0.4 MW·h), produce oxygen in unit 6 (in the amount of 1100 m^3/h) and nitrogen (in the amount of 4138 m^3/h), with an electricity consumption of about 1.3 MW·h. The remaining part of the electricity in the amount of 9.0 MWh and the thermal energy from the cooling systems of gas piston machines, in the amount of 10.7 Gcal/h, can be used by nearby workshops, process units, micro districts, etc. [8].

A part of the gas, in the amount of 200 m^3/h, is pumped by the compressor 7 into the storage gas tank 8 at a pressure of 15 MPa, which acts as a battery. Through a special pressure regulator, the gas enters the gas piston station 4, which makes it possible to ensure its stable operation during short-term idle times of the unit 1.

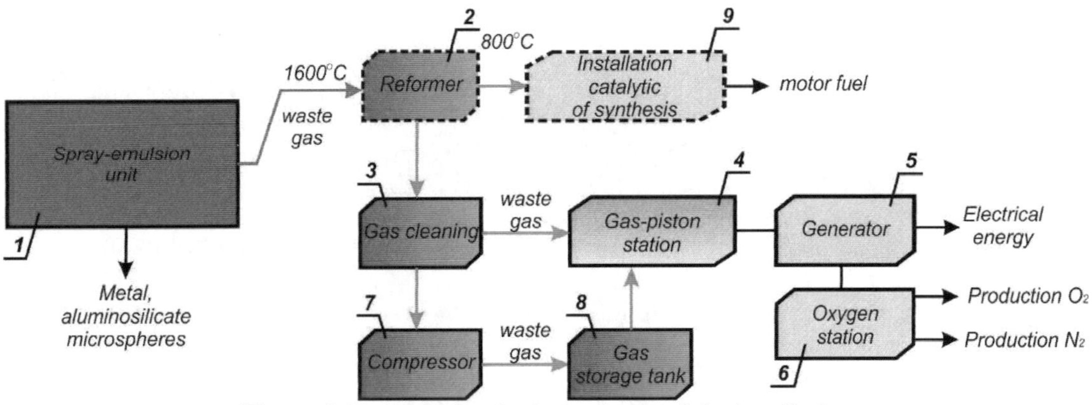

Figure 2. Energotechnological scheme of the installation.

The nitrogen produced at the oxygen station can be used in various production processes, including ammonia synthesis and nitrogen fertilizer production, methane conversion, associated gas processing, as well as for more efficient processing of high-octane components and reduction of sulfur deposits in the production of synthetic motor fuel.

Compared to gas turbines, gas piston engines in the power range up to 4 MW have a higher efficiency and are adapted to work at partial loads [9]. In addition, they are less susceptible to the influence of high ambient temperature, and create less noise and harmful gas emissions [10, 11]. There is already experience in the effective use of reciprocating machines. For example [12], on the territory of the greenhouse complex "UGMK-Agro" an energy complex based on piston machines was built, capable of generating electricity in the volume of 26.44 MW, as well as heat energy, cold and carbon dioxide. The cost of electricity from own generation is 1.9 rubles for kW·h, and from the network 5.4 rubles. This is taking into account the cost of natural gas, on which piston machines operate, and in our case, the gas is obtained from waste coal, for which storage, in principle, you need to pay.

5. Conclusions

Thus, in the presented article it is shown that on the basis of the previously developed spray-emulsion metallurgical SER unit, a completely non-waste technology of coal gasification (processing) and waste enrichment can be created with the production of three liquid products: synthesis gas (or electricity), aluminosilicate micro-granules and iron with rare metals dissolved in it. It should be emphasized that if there are dusty iron-containing waste (dust of gas purification, tailings of enrichment) near the coal enrichment waste, the efficiency of the described technology is increased: the possibility of stable maintenance of the gas-slag emulsion and the flow of the two-phase mixture in the connecting channels is improved, and the consumption of gaseous oxygen due to replacement of oxygen from iron oxides. As can be seen from the table above, the proposed technology has high economic efficiency and quick payback. The implementation of such a technology near the concentrator allows, together with the solution of the problem of processing pulverized waste, to ensure the independence of energy supply.

References

[1] Tsymbal V P, Mochalov S P, Rybenko I A, Aizatulov R C, Sokolov V V, Padalko A G, Kozhemyachenko V I, Krasnoerov S Yu, Shakirov K M, Kalashnikov S N, Ermakova L A,

Olennikov A A, Ognev A M, Schipanov S V, Rybushkin A A and Suzdal'tsev E V 2014 *SER Process – Metallurgical Jet-emulsion Reactor* (Moscow: Metallurgizdat) p 488
[2] Tsymbal V P, Kozhemyachenko V I, Rybenko I A, Padalko A G, Olennikov A A 2015 *Steel in Translation* **45** 550-554
[3] Terent'ev G A, Tyukov V M, Smal' F V 1989 *Motor Fuel From Alternative Raw Materials* (Moscow: Chemistry) p 271
[4] Plate N 2004 *Journal Science and Life* **11** 66-68
[5] Shilling G-D, Bonn B and Kraus U 1986 *Coal Gasification* (Moscow: Nedra) p 175
[6] Tsymbal V P, Mochalov S P, Rybenko I A Tsymbal Yu V *A Method for Direct Reduction of Metals to Produce Synthesis Gas and an Aggregate For Its Implementation* Patent 2371482 Russia C1 priority 03.12.2007 publ. 27.10.2009 Bul **30**
[7] Zaychenko V M and Chernyavsky A A 2010 *Academy of Energy* **4 (36)** 58-63
[8] Olennikov A A and Tsymbal V P 2008 *Steel in Translation* vol 38 **6** 439-443
[9] Gordeev A V and Mikhailov V A 2017 *International Scientific and Practical Conference* pp 155-159
[10] Sosnina E N, Masleeva O V, Pachyurin G V and Filatov D A 2013 *Basic Research* **6-1** 76-80
[11] Rodionova M A, Khrest'yanovskaya M V and Kukolev M I 2017 *Construction of Unique Buildings and Structures* **4(55)** 50-66
[12] *In UGMK a Unique Energy Center Has Been Built* Ore&MetalsWeekly – News of Mining and Metallurgy https://www.ugmk.com/press/news/v-ugmk-postroili-unikalnyy-energotsentr

IOP Conf. Series: Earth and Environmental Science **206** (2018) 012049 doi:10.1088/1755-1315/206/1/012049

Features of tidal deformations before strong distant and closely spaced regional earthquakes

S V Panov[1], M D Parushkin[2], V M Semibalamut[3] and Yu N Fomin[3]

[1]Institute of Laser Physics, Siberian Branch, Russian Academy of Sciences, 13/3 Lavrentyev ave, Novosibirsk, 630090, Russia
[2]Chinakal Institute of Mining, Siberian Branch, Russian Academy of Sciences, 54 Krasny ave, Novosibirsk, Russia
[3]Seismology Division, Federal Research Center, Unified Geophysical Service of the Russian Academy of Sciences, 189 Lenin ave, Obninsk, 249035, Russia

E-mail: mihail.parushkin@yandex.ru

Abstract. The article presents the data of deformation monitoring implemented in the Baikal Rift Zone in the time of initiation of strong earthquakes using the heterodyne laser deformograph. The anomalous behavior of the semidiurnal tide wave amplitude before the region-scale earthquakes is described.

1. Introduction
The Earth crust is an active medium in which mechanical deformation waves continuously originate and propagate within a wide range of vibration periods. Different physical processes on the surface and underground show themselves as seismic, free, tidal, seasonal and induced vibrations creating a complex space–time structure of a nonstationary deformation field. The integrated applied research into deformation vibrations is connected with the problems of prediction of such disastrous events as earthquakes, rock bursts, volcanic activity, etc.

Understanding importance and relevance of high-precision equipment to be used in the studies of deformation processes spurred creation of an original two-coordinate laser deformograph at the Institute of Laser Physics. This equipment allows long-term and continuous detection of small deformation displacements in rocks at a relative sensitivity of 10^{-9}–10^{-10} in a wide frequency range in mine conditions at a spacing up to 100 m [1].

2. Laser deformography system
The equipment and procedure developed at the Institute of Laser Physics for the continuous precision measurement of deformations in the Earth's crust is persistently subjected to upgrading aimed to improve sensitivity and reliability. The basic operating principles of the equipment remain unchanged. The functional configuration of the deformation measurement system is shown in figure 1. The system includes a heterodyne interferometer composed of four independent optical channels three of which are measurement channels and one is a reference channel. The channels are based on an asymmetric Michelson interferometer; the reference beam from the heterodyne laser is offset with respect to frequency by 1 MHz relative to the sample beam to a displacing object. The operating principle of the measurement channels is based on the continuous recording of time history of wave phase due to the

Doppler effect under the beam reflection from the displacing object. Both measurement arms have a length of 25 m.

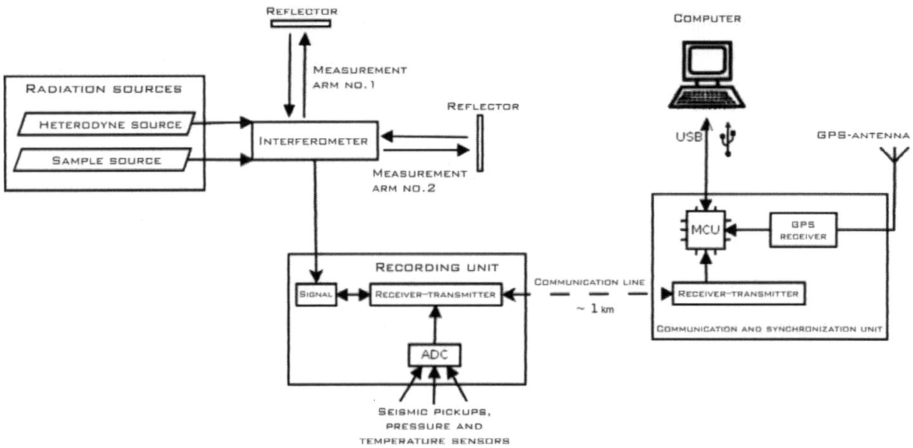

Figure 1. Functional configuration of the heterodyne laser deformography system.

The key feature of the deformograph is the absence of sampling radiation shielding from the action of atmospheric perturbations causing variation in the refraction index of air. To this end, the interferometer has, alongside two recording channels, an optical channel to neutralize atmospheric effects and frequency instability of the lasers. The optical channel includes a rod made of alloy 36N (invar) 1.2 m long with a thermal expansion coefficient $dL/dT = 8 \cdot 10^{-7}$ deg^{-1}. The signal obtained using the neutralization arm is proportional to variation in the laser source wavelength and in the mine air refraction index, and is independent of rock mass deformations. Such approach allows overcoming the major constraint of wide application of laser deformographs in geophysical surveys, namely, resource-hungry continuous-service evacuated beam guides tens meters long.

Computer-aided recording of signals is carried out at a sampling rate of 0.5 Hz. This system ensures reliable recording of small displacements in a wide range of vibration periods from 10^0 to 10^7 s, as well as allows recording of free and tidal vibrations of the Earth, deterministic diurnal variations in micro-deformation noise and features of deformation processes accompanying seismicity.

The site for introduction of the new-modification deformograph and testing of measurement procedures is situated nearby the south-west coast of the Lake Baikal. The Baikal region is selected owing to its seismic activity in the form of numerous earthquakes of different energy classes within a year. The deformograph is installed in the Talaya seismic station tunnel. Continuous recording of deformation process started more than 20 years ago. Over this period of time, the deformation measurement system has proved to be a reliable, high-precision and unfussy instrumentation for deformation monitoring of lithosphere [2]. The long-term observations reveal that the signs of seismic events in the behavior of a deformation signal are greatly governed by two parameters – energy class of an event, K, and distance to the event source, S. During initiation of seismic events with K = 13–14, the signs in the deformation behavior of the crust appear at a distance up to 100 km; for the events with K = 10, such signs can be observed at a distance of 20–30 m.

3. Features of tidal deformations during initiation of seismic events

As it was mentioned above, the high-priority objectives of deformographic monitoring is the analysis of dynamic processes in the crust before a seismic event. A peculiar place among the deformation processes of interest belongs to the lithosphere tides as they can trigger both rock bursts and earthquakes (including induced events). Tides considerably contribute to mass transfer in the Earth's interior [3], which governs evolution of stress state of the crust [4]. Fundamental harmonics of a tide signal are diurnal and semidiurnal tide waves. Their amplitudes depend on the stiffness of the crust rocks, as well as on the mutual location of the Sun and Moon relative to the Earth. Stiffness and density of rocks are governed by rock mass stresses in a general case. The diurnal tidal harmonic is distorted due to daily variations in meteorological parameters, as a rule; for this reason, it is simpler to analyze the semidiurnal tide harmonic.

The differences and the behavioral features of tidal deformations before a seismic event were continuously monitored and recorded starting from the earliest introduction of the deformography system. By way of illustration of a "tidal abnormality" in a deformation process, figure 2 presents the 10 days-long deformographic record. The deformation curve in the figure clearly shows the devastating Kobe Earthquake on 16 Jan 1995 (10:52 UTC) and the abnormal tidal oscillations 1.5 days before the seismic event.

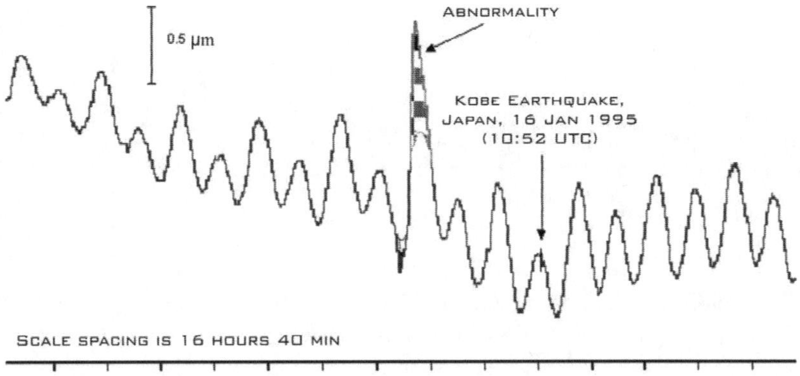

Figure 2. Fragment of deformographic recording of the Kobe Earthquake.

Similar nonstationary perturbations inside a natural deformation process in rocks, including tidal vibrations, were many times observed a few days before large planetary earthquakes, including the Spitak Earthquake in 1988 and the earthquake that hit the eastern coastal line of the Honshu Island in 2011. In view of the large distance between the laser monitoring station and the foci of the strong earthquakes, an attempt was made to explain the mechanisms of the recorded abnormalities based on the assumption that the phenomenon was of the planetary nature and connected with the interaction between the earthquake foci and the solid core of the Earth. However, until reliable comparable data are synchronously recorded by mutually remote geophysical stations, the question on the mechanism of this phenomenon remains open-ended.

Another class of periodically recorded tidal abnormalities are the signs of regional seismic events in the form of the long-term alteration of deformational response of the crust to the Moon/Sun tides by way of the change in the amplitude of an envelope of a semidiurnal tide in the same arm of the deformograph. An example of such behavior is the deformation process dynamics in the period of initiation and release of the Kultuk Earthquake on 27 Aug 2008, which was the largest earthquake in the neighborhood of the Talaya seismic station for the whole monitoring time. The epicenter of that earthquake with the energy class K = 15.5 located in the south of the Baikal lakeside at the distance S = 30 km from the seismic station. Due to the high energy class and relatively short remoteness from the observation station, the earthquake was a rare and very valued event for the analysis. In terms of the earthquake, the features of geodynamic effect of a close-spaced large earthquake on deformation process in the curst were demonstrated in [5]. Amongst deformation signals related with the initiation of a closely spaced earthquake, the anomalous behavior of the amplitude of the semidiurnal deformation tide in the East–West direction was described. That abnormality showed itself as a jump by 20–30 % in the deformation response of the Earth's crust to the joint gravitational tides of the Moon and Sun three days before the Kultuk Earthquake and the subsequent relaxation of the abnormality within a few days after the event.

The described feature is illustrated in figure 3: the curves A and B show the tide wave trains in two measurement arms of the deformograph after filtration in the range of 708–768 min (the lighter shade is the East–West arm, the darker shade is the North–South arm). The amplitude of this signal changes with a period of 14 day approximately depending on the position of the Sun and Moon relative to the Earth. The tide has the maximum amplitude when the Sun, Moon, Earth centers and the monitoring station lie in the same plane. It is seen in the figure that the East–West tide amplitude is smaller than the North–South tide amplitude, which agrees with the theory. However, in the time zone closer to the Baikal earthquake, the effect is opposite. Three days before the seismic event in question, the tide amplitude is noticeably higher in the East–West line than in the orthogonal direction. This fact is also visible in the enveloping curves C and D.

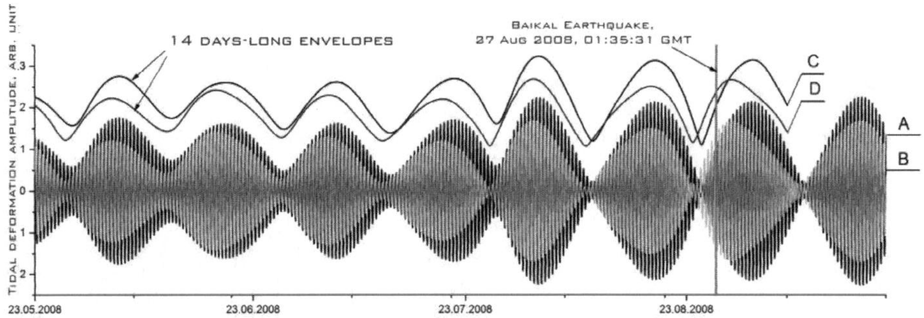

Figure 3. Semidiurnal tidal harmonics in the orthogonal arms of the deformograph and the enveloping curves.

The first half-plot visualizes "natural" train of modulated signals corresponding to the semidiurnal tides. This span contains an anomalous phase delay between the enveloping curves of the semidiurnal tidal deformations, which has been observed in the earlier experiments and is discussed below in this article. The experimental enveloping curves, as against the theory, show the phase delay 1 day long, which is of regular nature [6]. In the meanwhile, it is seen in the graphs that the time delay closer to the earthquake point changes and reaches approximately 4 days.

Figure 4 offers one more example of abnormality in the amplitude of semidiurnal tidal harmonics in the period of initiation of a regional earthquake focus. Such behavior was recorded a weak before the event of energy class K = 12.6 on 27 Sep 2010. The abnormality is traced in the East–West arm, which conforms with the direction from the seismic station to the earthquake epicenter within the accuracy of 10°.

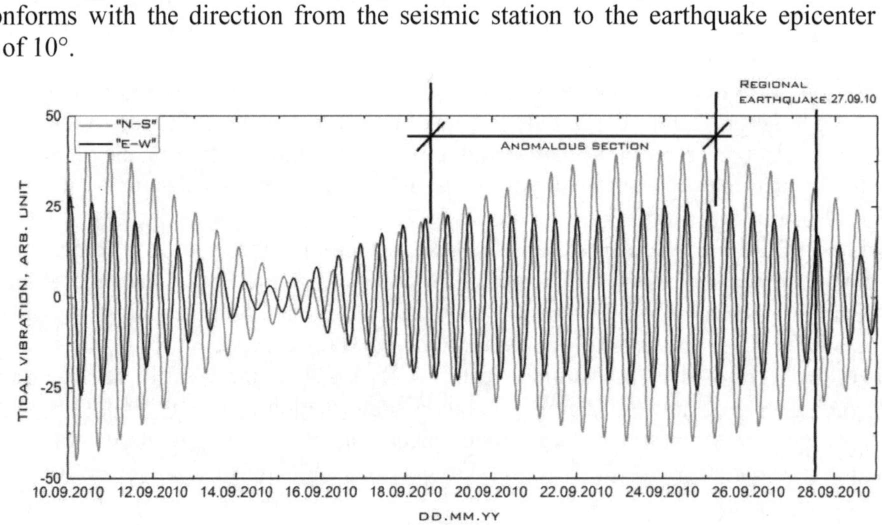

Figure 4. Anomalous behavior of semidiurnal tidal wave amplitude in the East–West arm of the deformograph.

4. Phase differences between enveloping curves of semidiurnal tides for 14 days

Thus, it has been noticed that there is a phase delay of the order of 1 day between envelopes of the semidiurnal tide curves recorded for 14 days. The theoretical curves modeled for various arms of the deformograph coincide, i.e., the theory forecasts none of the phase delays [7]. Figure 5 illustrates the described behavior of epy envelopes. At the top of the figure, there are theoretical enveloping curves of semidiurnal tides, and at the bottom, there are envelopes of the filtered experimental data. Modeling of the crustal tides was carried out using Etgtab 3.0 program from Eterna 3.0. In this program, the parameters of the Talaya seismic station and the deformographic characteristics were set, e.g., coordinates and altitude of the station, displacement measurement direction (azimuth), measurement start time, sampling rate, etc. In this manner, theoretical data can match up with the experimental recording in each measurement arm of the deformograph. Then, the experimental and theoretical results are subjected to digital band-pass filtration in the wanted range of frequencies. It is worthy of mentioning that in figure 5, for the convenience of visualization, the normed enveloping curves are shown, which entails the loss of information on the amplitude differences between the orthogonal tides, observable on the left-hand side of figure 4.

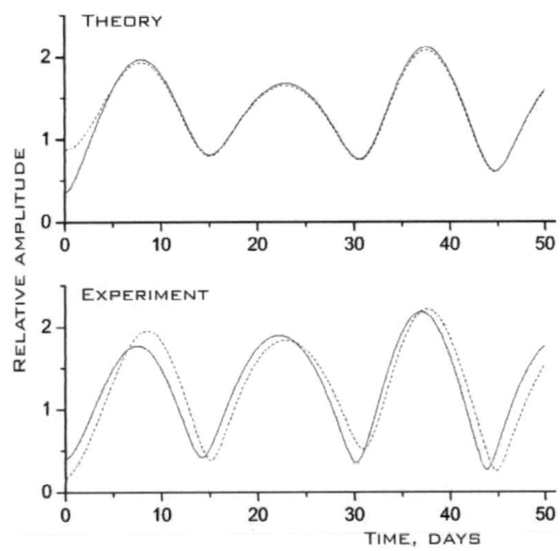

Figure 5. Phase delay between the enveloping semidiurnal tide curves based on the data from the orthogonal arms of the deformagraph (bottom) and no delay in the theoretical data (top).

The detailed study of the delays in the set of data recorded in 14 days proves both presence and value of the delay. Poor accuracy of the estimates is from the noise and perturbations in the recording.

Characteristic delays for the semidiurnal tide are:

- Arm 1 – experiment is delayed by 9–10 h;
- Arm 2 – theory is delayed by 7–9 h.

Characteristic delays for the diurnal tide:

- Arm 1 – experiment is delayed by 10–12 h;
- Arm 2 – theory is delayed by 9–11 h.

The phase delay, unexplained theoretically, should be more investigated as it is a potential sign of large seismic events. The investigations should be based on long-term continuous deformation monitoring using supplementary procedures of observation and data interpretation.

Acknowledgements
The studies in the section *Phase differences between enveloping curves of semidiurnal tides for 14 days* have been supported by the Russian Science Foundation, Project No. 16-17-10121.

References

[1] Panov S V, Parushkin M D, Semibalamut M V and Fomin Yu N 2016 *Heterodyne Laser Deformograph for Geophysical Surveys. Current Problems of Modern Seismology Tashkent* pp 293–299
[2] Orlov V A, Panov S V, Parushkin M D, Semibalamut M V and Fomin Yu N 2013 *Conference Proceedings on Geodynamics and Stress State of the Earth's Interior* (Novosibirsk: IGD SO RAN) pp 25–31
[3] Revuzhenko A F 2013 *Tidal Waves and Directional Mass Transport in the Earth System* (Novosibirsk: Nauka) p 204
[4] Adushkin V V and Spivak A A 2014 *Physical Fields in Subsurface Geophysics* (Moscow: GEROS) p 360
[5] Orlov V A, Panov S V, Parushkin M D and Fomin Yu N 2010 *Conference Proceedings on Geodynamics and Stress State of the Earth's Interior* (Novosibirsk: IGD SO RAN) pp 30–37
[6] Orlov V A and Piatigorets A V 2014 *APEIE_2004 Proceedings* **1** 148–159
[7] Melchior P 1968 *The Tides of the Planet Earth* (Oxford: Pergamon Press)

IOP Conf. Series: Earth and Environmental Science **206** (2018) 012050 doi:10.1088/1755-1315/206/1/012050

Management of current environmental costs contributing to reduce eco-economic risks

T V Kiseleva[1] and V G Mikhailov[2]

[1]Siberian State Industrial University, 42 Kirova street, Novokuznetsk, 654007, Russia
[2]T.F. Gorbachev Kuzbass State Technical University, 28 Vesennyaya street, Kemerovo, 650000, Russia

E-mail: kis@siu.sibsiu.ru

Abstract. The analyzes of the known approaches to the distribution of environmental and economic resources, including the current environmental costs, in connection with their impact on reducing the level of eco-economic risks, is carried put in the article. The analysis of the dynamics and structure of current environmental costs of large industrial enterprises is carried out, and the dependence of the level of excess emission charges on the magnitude of current environmental costs is graphically presented. A mechanism for adjusting the current system of distribution of such costs has been developed to lead to their proportional reduction relative to an increase in the values of risk level indicators. In order to prepare for an effective management decision, an algorithm for managing the system of distribution of current environmental costs is presented. The analysis of the impact of the adjusted system of distribution of current environmental costs on the technical and economic performance of the enterprise was carried out. The work done is of practical importance for large industrial enterprises that have a complex negative impact on the environment, which significantly affects their technical, economic and financial performance.

1. Introduction

The current environmental costs (CEC), including all types of costs of enforcement of environmental protection measures, ensuring the current work of production processes and specific equipment to reduce or eliminate negative anthropogenic impact are very important for the efficient eco-economic performance of a modern enterprise [1-5]

These costs are reported by enterprises in a special form [6] and are differentiated in the following areas:

- air protection and climate change mitigation;
- wastewater collection and treatment;
- waste management;
- protection and rehabilitation of land, surface and ground water;
- protection of the environment from noise, vibration and other types of physical impact;
- preservation of biodiversity and protection of natural areas;
- ensuring radiation safety of the environment;
- research and development activities to reduce negative anthropogenic impacts on the environment;
- other activities in the field of environmental protection.

Under conditions of a rapidly changing environment, limited economic resources and maintaining the required level of competitiveness [7-11], an effective distribution of such costs, which can be based on the "current environmental costs - the level of risk" dependence, is essential. This is due to the fact that the process of eco-economic risk assessment is of particular relevance in modern conditions of tightening national environmental legislation and increasing the sensitivity of the technical and economic performance of an enterprise to its eco-economic characteristics.

2. Materials and methods

The object of study is industrial enterprises engaged in coal conversion and having a complex negative impact on the environment (air emissions, discharges into water sources, production and consumption waste generation). The subject of the research is the system of distribution of CEC. The study is based on an analysis of modern sources on the problems of the effective allocation of environmental and economic resources, one of which is the CEC. Particular attention is paid to the study of the methodology for the allocation of such costs based on indicators related to the assessment of eco-economic risks. For the purpose of practical implementation of the author's methodology, statistical processing of environmental data of industrial enterprises was carried out on the basis of official reporting forms. On the basis of the developed algorithm of the improved system of distribution of CEC, their impact on the technical and economic performance of the enterprise was analyzed. The work also used elements of system analysis and the results obtained by experts in the field of management of environmental and economic systems at various levels.

3. Results and discussion

To build a system of distribution of current costs on the principles of adequate assessment of eco-economic risks, an analysis of the known mechanisms for evaluating and managing specific types of risks was carried out. Fundamental studies on this issue are given in [12–14], where not only risk assessment methods are considered, but also the most efficient risk management mechanisms at enterprises (fines, risk payment, risk reduction financing, recovery of risk reduction costs and others).

The integrated risk assessment procedure based on discrete scales to reduce the risk rate of a risk event to the required level with minimal costs is proposed in the paper [15].

Studies [16] show that the risk magnitude is defined as the product of damage by the probability of its occurrence, obtained on the basis of the expert evaluation method. This approach should also be used for eco-economic risks, which in many cases are transformed into traditional financial and economic risks.

A significant problem of adequate calculation of eco-economic risks is the choice of a method for assessing the probability of an adverse event. In practice, three such basic methods as statistical, analytical and expert evaluation are most often used.

For the most qualitative and accurate assessment of the probability of occurrence of adverse events, it is necessary to use all methods simultaneously with checking the convergence of results.

A number of experts believe that in addition to the above methods, simulation modeling [17], ensuring greater reliability of the result obtained, is of practical importance.

One of the problems of assessing eco-economic risks is an adequate interpretation of the result. Some experts [18] consider eco-economic risk zones as a normal allocation, similar to the image of speculative risks in the financial sphere, when any deviation from the "normal" situation is considered as a risky event (profit, damage, gain, loss, etc.).

The practical application of the theory of eco-economic risks is reflected in the analysis of environmental and economic performance of an enterprise [2, 10], when one of the simple options for assessing this type of risk is to determine the proportion of excessive environmental impact fees in the total amount of fees. Analysis of data of large industrial enterprises showed that this value can reach up to 90 %, characterizing the inefficiency of the environmental policy, and negatively affecting the technical, economic and financial performance of the enterprise, reduces its profits. Another aspect of this problem is inadequate rationing of the potential environment impact when the compliance with regulations is impossible due to the technology applied.

The reduction of eco-economic risks at the enterprise is largely associated with internal CEC (figure 1).

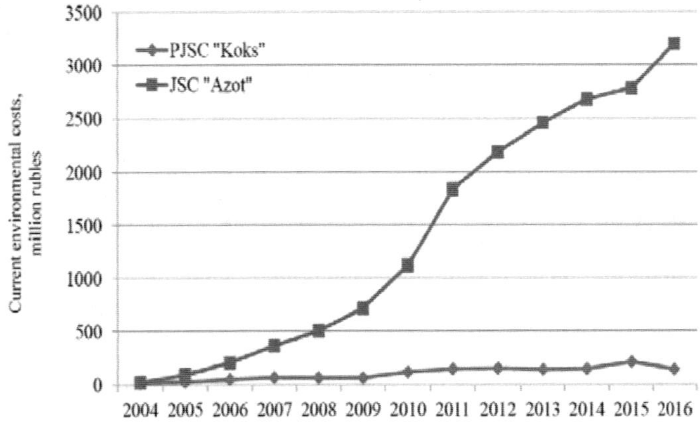

Figure 1. Dynamics of current environmental costs of JSC "Azot" и PJSC "Koks".

Figure 1 shows that the ranges of changes in CEC of these enterprises differ significantly (from 20.2 to 212.4 million rubles for PJSC "Koks" and from 23.8 to 3198.0 million rubles for JSC "Azot"). The increase in CEC of JSC "Azot" is associated with a highly diversified environmental impact and the construction of a closed water cycle system. Another reason is the increase in depreciation charges for the restoration of basic production assets for environmental protection. Figures 2 and 3 show the structure of CEC of the JSC "Azot и PJSC "Koks" enterprises [19].

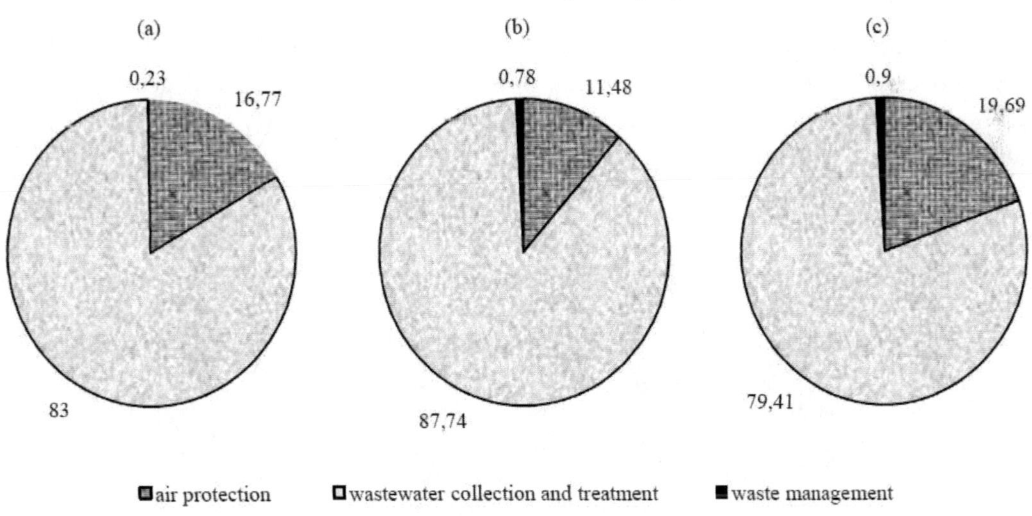

Figure 2. Structure of current environmental costs of PJSC "Koks" enterprise in (a) 2014, (b) 2015, (c) 2016, %.

Figure 2 shows that the cost structure is changing, in particular, there is an increase in the share of current waste management costs from 0.23 % in 2014 to 0.9 % in 2016.

The air protection costs change "nonlinearly" – a decline in 2015 (11.48 %), and then a sharp increase in 2016 to 19.69 %. The reverse trend is observed at current costs associated with the wastewater collection and treatment. The high proportion of this type of cost (up to 87.74 % in 2015) is due to the need for efficient operation of the closed water cycle system.

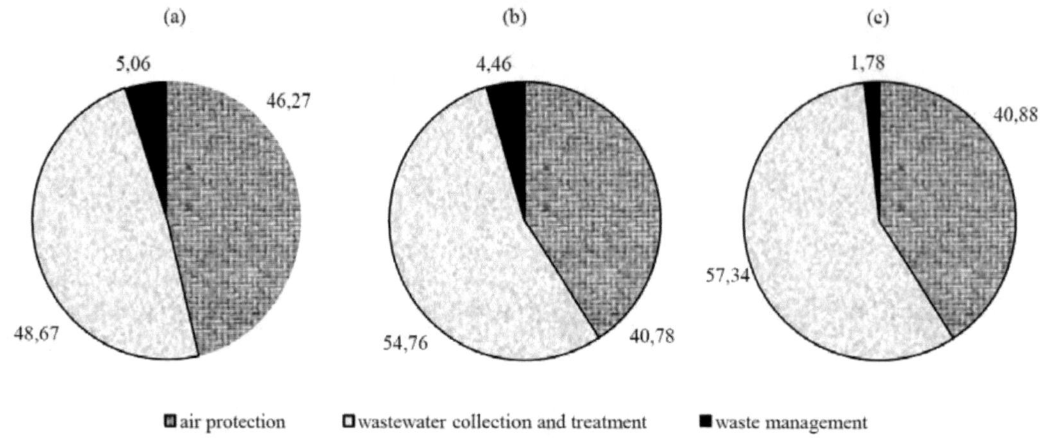

Figure 3. Structure of current environmental costs of JSC "Azot" enterprise in (a) 2014, (b) 2015, (c) 2016, %.

The analysis of CEC of JSC "Azot" shows steady trends in reducing waste management costs (from 5.06 % in 2014 to 1.78 % in 2016) and increasing costs for wastewater collection and treatment (from 48.67 % in 2014 to 57.34 % in 2016). The growth of this type of costs is associated with the installation of a closed water cycle system at the plant. Current air protection costs decreased in 2015 and remained approximately at this level in 2016. This distribution of CEC of JSC "Azot" is associated with the peculiarities of its production program and diversification of the environmental impact types.

Figure 4 shows the correlation field and the empirical regression line of the relationship between CEC and the specific weight of the excessive environmental impact fee in the total fee value ($SW_{j,t}$), which is closely related to the risk level. Points presented on all analyzed charts mean information obtained for a particular enterprise for a calendar year.

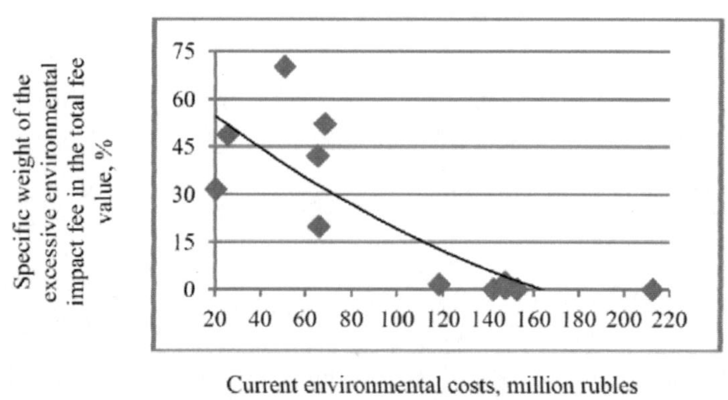

Figure 4. Correlation field and the empirical regression line of the relationship between $SW_{j,t}$ and current environmental costs at PJSC "Koks".

The data on the correlation field have a large scatter, which, apparently, is explained by different environmental policies at the enterprise in different years. The values in the left part of the correlation field show a significant variation of $SW_{j,t}$ with relatively small changes in the CEC. It is noteworthy that there is a large variation in $SW_{j,t}$ values (from 19.8 to 52.06 %) with almost the same values of CEC. The remaining values in the correlation field cover the period from 2010 to 2016, when the CEC significantly increased (from 118.3 to 212.4 million rubles). With such funding, there was a sharp decline of $SW_{j,t}$ which for these years fluctuated around zero.

The considered periods characterize the situation of changes in environmental policy since 2010. This was due to the tightening of environmental legislation. As a result, the company switched to the BAT (best available technologies) standards, ensuring a minimum negative load on the environment. Due to standard, the company has a close water cycle system, and as a result, since 2012, PJSC "Koks" has stopped the wastewater discharge. This trend of tightening environmental legislation is to improve the system of environmental impact charges, which has been in effect since January 1, 2016 and is aimed at additional fines for enterprises that do not implemented the transition to the BAT standards.

Figure 5 shows the correlation field of the relationship between similar variables for JSC "Azot".

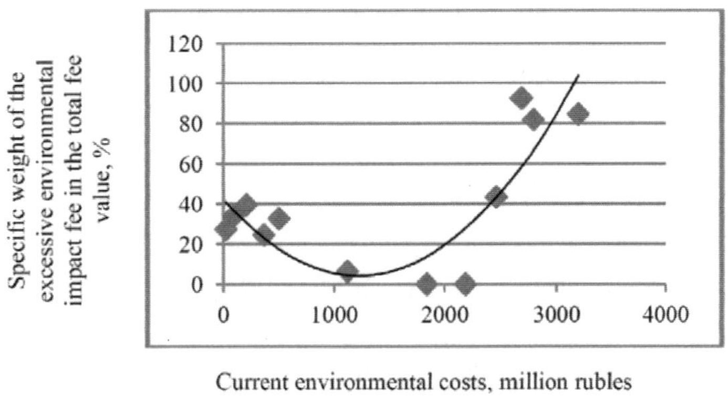

Figure 5. Correlation field and the empirical regression line of the relationship between $SW_{j,t}$ and current environmental costs at JSC "Azot".

Figure 5 shows that the dependence is non-linear in nature, in particular, until 2013 there is an inverse relationship, when an increase in current costs leads to a decrease in $SW_{j,t}$. From 2013, a further increase in current costs is accompanied by a sharp increase in $SW_{j,t}$ to 92.58 %, which is explained by a sharp increase in fees for excessive pollution of water sources and may be caused by large-scale technical re-equipment of sewage treatment plants, which was planned to be completed in 2017.

Due to the fact that the analysis showed a lack of efficiency of the CEC in several enterprises, it is proposed to adjust the current system of distribution of these costs by proportionally decreasing them relative to an increase in indicators related to the level of risk. In this case, for the criteria presented in [20], the adjusted value of the CEC of the planned period for the *j*-th element of the environment ($CEC_{j,1}$) is determined by the formula (1):

$$CEC_{j,1} = CEC_{j,0} \cdot [1-(CIR_{\beta,j,0} - SIR_{\beta,j})] \, , \qquad (1)$$

where $CEC_{j,0}$ – value of current environmental costs in *0*-th time period for *j*-th element of the environment, rubles.; $[1-(CIR_{\beta,j,0} - SIR_{\beta,j})]$ – reduction ratio of CEC in *0*-th time period for *j*-th element of the environment ($RR_{CECj,0}$),the operating time of the current system of distribution of CEC is taken as *0*-th period, and the next year with the adjusted value of the CEC is taken as *1*-st period (for example, if the *0*-th period is 2016, then the *1*-st is 2017); $CIR_{\beta,j,0}$ calculated indicator associated with the level of risk for the *β*-th criterion and the *j*-th element of the environment in the *0*-th time period; $SIR_{\beta,j}$ – the standard indicator associated with the level of risk for the *β*-th criterion and the *j*-th element of the environment, determined on the basis of the expert evaluation method.

The value of the minimum required level of current environmental costs for items of expenditure in the *0*-th time period for the *j*-th element of the environment ($V^{min}_{CECj,0}$) is determined using the expert evaluation method, based on the ecological and economic feasibility of the required amount of CEC.

If the calculated and regulatory indicator associated with the risk level has a dimension of %, then the adjusted value of current costs is determined by the formula (2):

$$CEC_{j,1} = CEC_{j,0} \cdot [100-(CIR^{*}_{\beta,j,0} - SIR^{*}_{\beta,j})] , \quad (2)$$

where $CIR^{*}_{\beta,j,0}$ – calculated indicator и $SIR^{*}_{\beta,j}$ – regulatory indicator, associated with the risk level, having a dimension of %.

For other criteria, the adjusted value of CEC is determined by formulas (3) and (4):

$$CEC_{j,1} = CEC_{j,0} \cdot [1-(SIR_{\beta,j} - CIR_{\beta,j,0})] , \quad (3)$$

where $[1-(SIR_{\beta,j} - CIR_{\beta,j,0})]$ – $RR_{CECj,0}$ indicator;

$$CEC_{j,1} = CEC_{j,0} \cdot [100-(SIR^{*}_{\beta,j} - CIR^{*}_{\beta,j,0})] , \quad (4)$$

where $[100-(SIR^{*}_{\beta,j} - CIR^{*}_{\beta,j,0})]$ – $RR_{CECj,0}$ indicator.

The decrease in the value of current environmental costs (ΔCEC_j) is determined by the formula (5):

$$\Delta CEC_j = CEC_{j,0} - CEC_{j,1} . \quad (5)$$

The suggested limitations for using this algorithm:

$CEC_{j,1} \le CEC_{j,0}$ provided that the inflation rate for the study period does not change.

Consider the formulation of the problem of distribution of current environmental costs, taking into the account the efficiency of their use.

Given:

1. The current system of distribution of CEC by items of expenditure.
2. Indicators of the efficiency of the use of CEC associated with the risk level [20].
3. The technical and economic performance (TEP) of the enterprise.
4. Limitations: 1) $CEC_{j,1} \le CEC_{j,0}$; 2) $RR_{CECj,0} \ge V^{min}_{CECj,0}$.
5. Criterion: the annual eco-economic efficiency of the system of CEC distribution.

Required: to build an algorithm for managing the system of CEC distribution meeting the limitations and maximizing the criterion.

Figure 6 shows a flowchart of the algorithm for distribution of CEC, taking into account the efficiency of their use.

Unit 1 provides for the input of all necessary data, including the CEC, criteria for the efficiency of the distribution of CEC, and other indicators. In Unit 2, the current system of distribution of CEC by item of expenditure is presented. Unit 3 is needed to select the criterion for the efficiency of the adjustment of CEC. In logical Unit 4, on the basis of previous calculations, a conclusion is made about the efficiency (or inefficiency) of using CEC. In the case of efficient use of CEC, the existing distribution system, presented in Unit 2, is preserved. Otherwise, first $V^{min}_{CECj,0}$ (Unit 5) is determined and compared with $RR_{CECj,0}$ (Unit 6). If the result of this comparison a positive, the adjustment of the value of CEC for the individual elements of the environment is made in Unit 7, and if negative - in Unit 8. Unit 9 corrects the entire system of distribution of CEC by item of expenditure. An analysis of the impact of the adjusted system of distribution of CEC on the main technical and economic performance of the enterprise is presented in Unit 10.

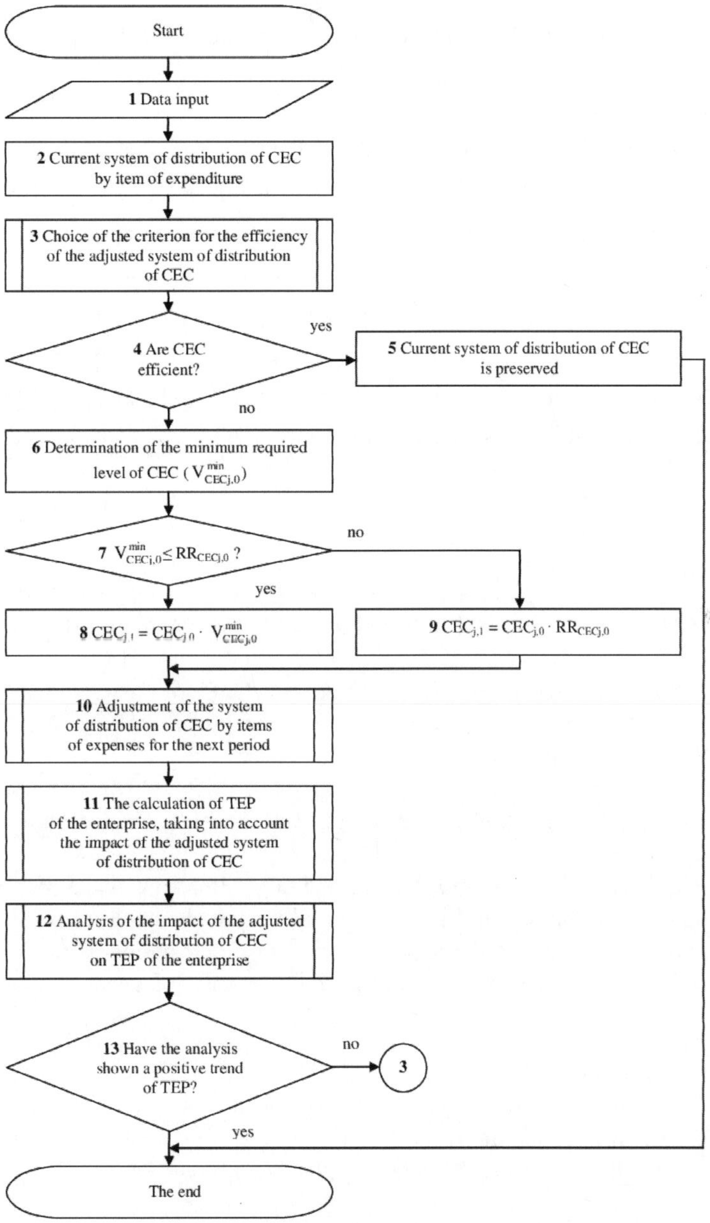

Figure 6. Flowchart of the algorithm for distribution of current environmental costs, taking into account the efficiency of their use.

The calculations carried out for JSC "Azot" show a wide range of changes in the values of performance related to the level of risk determined using criteria - integral indicators of the proportion of excessive environmental impact fees in the total amount of fees (SW^{int}) and growth rate of environmental impact fees (GR_{eif}^{int}). The SW^{int} values vary in the range from 0 to 0.93, and the criterion GR_{eif}^{int} values vary over wider limits from 0.24 to 3.94. The regulatory values of the indicators associated with the risk level are 0.2 for SW^{int} and 1.0 for GR_{eif}^{int}, and the minimum required level of CEC is equal to 0.5, which is determined using the expert evaluation method. These values significantly affect the adjustment (decrease) in $CEC_{j,1}$ value. As a result, the range of change in ΔCEC_j is from 13.1 to 1599.0 million rubles. The ΔCEC_j value leads to a decrease in cost, increase in profits, and increase in product competitiveness. Calculations showed that ΔCEC_j, presented as a percentage of profits, ranges from 3.16 to 24.66.

Calculations made according to the data of PJSC "Koks" also indicate a wide range of changes in indicators that affect the risk level and are determined using SW^{int}, GR_{eif}^{int} criteria and the integral indicator of the growth rate of the economic damage compensation coefficient (GR_{edcc}^{int}).

The scatter of $SW^{инт}$ and GR_{eif}^{int} values has a smaller value than those for the previous enterprise and varies, respectively, in the interval from 0 to 0.70 and from 0.69 to 2.71.

In turn, the values of the GR_{edcc}^{int} criterion vary in a very wide range from 0.43 to 17.21 with $SIR_{\beta,j} \geq 1.0$, also determined using the expert evaluation method. As a result, the ΔCEC_j change interval for the enterprise PJSC "Koks" has a smaller range from 1.03 to 106.2 million rubles, which is a percentage of the profit, varying in the range from 0.02 to 2.33.

4. Conclusion

The study leads to the following conclusions:

- the analysis of known approaches to the allocation of environmental and economic resources, including CEC, in connection with their impact on reducing the level of eco-economic risks, is carried out;
- the analysis of the dynamics and structure of CEC of large industrial enterprises was conducted and the dependence of the excessive environmental impact fees on the value of CEC was graphically presented;
- the adjusted system of distribution of the values of the CEC was developed, leading to their proportional reduction relative to the increase in the performance values related to the risk level;
- the problem statement is presented and the algorithm for managing the system for the distribution of the values of the CEC is built, taking into account the efficiency of their use;
- the analysis of the impact of the adjusted system of distribution of the values of the CEC on the main technical and economic performance of the enterprise was conducted.

References

[1] Trushina G S 2017 *Ugol* **3** 52–55
[2] Mikhailov V G, Kiseleva T V, Karasev V A, Mikhailov G S and Skukin V A 2017 *IOP Conf. Series: Earth and Env. Sci.* **66** 012008
[3] Kiseleva T V, Mikhailov V G and Mikhailov 2017 *IOP Conf. Series: Earth and Env. Sci.* **84** 012044
[4] Prokopenko S A, Ludzish V S, Lesin Y V, Tyulenev M A and Sushko A V 2017 *Journal of Mining Science* vol 53 **1** 92–98
[5] Fridman Y A, Rechko G N and Pimonov A G 2017 *Regional Research of Russia* vol 7 **4**

333-341
[6] Chernikova O P and Baranov P P 2018 *Gornyi Zhurnal* **3** 82–85
[7] Tumin V M and Koryakov A G 2013 *Middle East Journal of Scientific Research* 2013 vol 17 **9** 1350–1355
[8] Zhernov E, Nekhoda E 2018 *E3S Web of Conf.* **41** 04009
[9] Zolotukhin V, Zolotukhina N, Yazevich M, Rodionov A and Kozyreva M 2017 *E3S Web of Conf.* **21** 04008
[10] Zhironkin S, Khloptsov D, Skrylnikova N, Petinenko I and Zhironkina O 2018 *E3S Web of Conf.* **41** 04010
[11] Novichikhin A V, Fryanov V N, Petrova T V, Pavlova L D and Temlyantsev M V 2017 *IOP Conf. Series: Earth and Env. Sci.* **84** 012036
[12] Burkov V N, Novikov D A and Shchepkin A V 2015 *Studies in Systems, Decision and Control* **10** 1-24
[13] Burkov V N, Novikov D A and Shchepkin A V 2015 *Studies in Systems, Decision and Control* **10** 117–154
[14] Titarenko B, Hasnaoui A and Titarenko R 2018 *IOP Conf. Series: Materials Sci. and Eng.* vol 365 **4** 042019
[15] Poryadina V, Burkov V and Barkalov S 2018 *MATEC Web of Conf.* **170** 01122
[16] Makhutov N, Gadenin M 2016 *Ecology and Industry of Russia* vol 20 **3** 47–51
[17] Tikhomirov N, Tikhomirova T, Khamitov E and Ponomarev V 2017 *European Research Studies Journal* vol 20 **2** 321–333
[18] Murzin A D and Anopchenko T Y 2014 *Asian Social Science* 2014 vol 10 **15** 249–254
[19] Mikhailov V, Kiseleva T, Bugrova S, Muromtseva A and Mikhailova Ya 2017 *E3S Web of Conf.* **21** 02004
[20] Kiseleva T V, Mikhailov V G and Stepanov T V 2018 *Control Systems and Information Technology* vol 73 **3** 32–37

Methodical specifics of scenarios development for increasing the level of environmental safety (in the conditions of Novokuznetsk city)

A V Shorokhova, A V Novichikhin and E N Yuryeva

Siberian State Industrial University, 42 Kirova street, Novokuznetsk, 654007, Russia

E-mail: shoroxova_a@mail.ru

Abstract. The procedure for developing scenarios of waste processing, land reclamation and creation of recreational zones is proposed. Three types of scenarios for improving the level of environmental safety in mining areas have been developed. On the basis of the software complex, forecast estimates of the scenario indicators have been obtained, the selection of Pareto-optimal solutions and their ranking according to the levels of environmental safety have been carried out.

1. Introduction

Currently, the level of environmental safety in most areas is decreasing due to the permanent growth of waste production. The overwhelming majority of man-caused resources are located in the Siberian Federal District – 18 billion tonnes or more than 57% of the total volume [1, 2]. At the same time, more than 80% of industrial wastes are generated in the process of extraction and processing of minerals [3, 4]. Accordingly, the unfavorable environmental situation has developed and is progressing in the mining areas, especially in the city of Novokuznetsk because of the high concentration of industries where its old industrial areas (for example, Abashevsky and Kuznetsky districts) are especially dangerous.

2. Methods of research

It conditions the necessity to search for effective solutions to overcome the above mentioned negative tendencies. So, it is proposed to use the scenario approach [5]. This task stipulates the development of a procedure for forming the scenarios of waste processing, land reclamation and the creation of recreational zones. In this connection, the task of scenarios development can be presented as follows [6]:

Task 1. A list of options for waste recycling and land reclamation is determined $j, j \in [1; J]$. Each j-th option is characterized by a certain set of technologies Th(j) and indicators $F(j) = \{F_l(j), l \in [1; L]\}$, with $L = 5$: the economic effect is F_1; the area of reclaimed lands – F_2; the amount of pollution resulting from negative production activities – F_3; the population with statutory indicator – F_4; the prevented pollution per capita – F5.

It is required to form a set of scenarios A_n, $n \in [1; N]$ for waste processing with the creation of recreational areas on the reclaimed lands from j-th option, $j \in [1; J]$, which allow the negative effects of production to be eliminated in accordance with the statutory indicators $\{F^*_l\}$:

$$\sum_{t=1}^{T}\sum_{n=1}^{N}\sum_{i=1}^{I} Kp_{tni} \rightarrow \min \text{ if } A_n = \bigcup_{j=1}^{J} Th(j),$$

provided that:

$$F^*_1(A)<F_1(A);\ F^*_2(A)<F_2(A);\ F^*_3(A)>F_3(A);\ F^*_4(A)<F_4(A);\ F^*_5(A)<F_5(A). \quad (1)$$

To solve this problem, we propose a procedure for creating scenarios for processing waste, land reclamation and creation of recreational zones (figure 1).

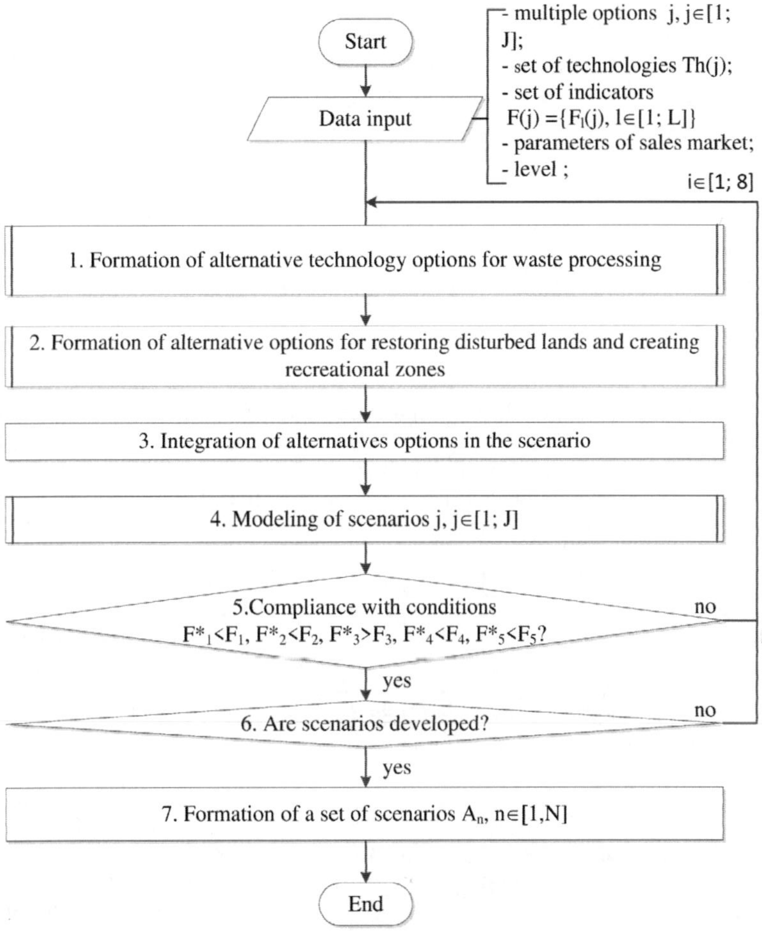

Figure 1. Procedure for the formation of waste treatment scenarios, land reclamation and creation of recreational zones [7].

The number of iterations of the procedure is $i\in[1;\ 8]$. After $i=8$, the procedure stops. If no solution is found, the source data (a set of technologies and their parameters), constraints are changed. The cycle is restarted to find the solution.

Based on the proposed procedure, three types of scenarios have been developed:

- pessimistic scenario is characterized by a low level of investments into the implementation of non-waste technologies and the creation of recreation areas (the total amount of financing is up to 100 million rubles), the total capacity of the waste recycling facility is less than 0.7 million tonnes per year;

IOP Conf. Series: Earth and Environmental Science **206** (2018) 012051 doi:10.1088/1755-1315/206/1/012051

- moderate scenario is characterized by an average level of investments into the implementation of non-waste technologies and the creation of recreation areas (the total amount of financing is 100-195 million rubles), the total capacity of the waste recycling facility is 0.7-1.0 million tonnes per year;
- optimistic scenario is characterized by a high level of investments into the implementation of non-waste technologies and the creation of recreation areas (total financing over 195 million rubles), the total capacity of the waste recycling facility is more than 1.0 million tonnes per year.

3. Results and discussion

During the scenarios development, the level of possible funding was taken into account – in this regard, moderate scenarios for increasing the level of environmental safety in Novokuznetsk in the conditions of coal preparation plants [5] (table 1) are considered, taking into account the best domestic and world practices for processing coal sludges [8-10] and rocks [11-13].

Table 1. Scenarios for increasing the level of environmental safety in Novokuznetsk in the conditions of coal preparation plants.

Script ID	Production	Indicators
C1	Coal concentrate obtained by gravity enrichment methods (separators with a shaken bed), and materials for construction needs.	The productivity of faculties is 0.7 and 0.2 million tonnes per year. The volume of investments is 120 million rubles.
C2	Coal concentrate obtained by the oil granulation method and materials for construction needs. It is possible to use waste industrial and machine oil for the enrichment of coal slurries. The use of oil granulation increases the heat from combustion of the combustible material.	The productivity of faculties is 0.8 and 0.2 million tonnes per year. The volume of investments is 135 million rubles.
C3	Coal concentrate, obtained by gravity methods of enrichment in heavy mediums, and bricks for construction needs.	The productivity of faculties is 0.8 and 0.2 million tonnes per year. The volume of investments is 156 million rubles.
C4	Coal concentrate obtained by gravity enrichment methods (oil granulation), and construction bricks.	The productivity of faculties is 0.8 and 0.2 million tonnes per year. The volume of investments is 135 million rubles.
C5	Coal concentrate obtained by oil granulation and clinker with preliminary crushing and firing.	The productivity of faculties is 0.8 and 0.2 million tonnes per year. The volume of investments is 165 million rubles.
C6	Coal concentrate obtained by the oil granulation method and additives to concrete.	The productivity of faculties is 0.7 and 0.3 million tonnes per year. The volume of investments is 136 million rubles.
C7	Coal concentrate obtained by gravity enrichment methods (in hydrocyclones), and additives to concrete.	The productivity of faculties is 0.7 and 0.3 million tonnes per year. The volume of investments is 178 million rubles.
C8	Fertilizers for agriculture.	The productivity of faculties is 0.8 and 0.2 million tonnes per year. The volume of investments is 100 million rubles.

The implementation of scenarios is considered in the period 2019-2032. Based on the software complex implemented in Scilab environment, the forecast estimates of the scenarios indicators [5] were obtained (table 2).

Table 2. Indicators of scenarios for improving the level of environmental safety in Novokuznetsk in the conditions of coal preparation plants.

Scenario	F_1, mln. rub	F_2, thous. m^2	F_3, t/rub	F_4, person	F_5, t/person
C1	822.3	9.4	0.049	710	19.4
C2	820.8	9.8	0.029	754	20.9
C3	821.4	9.6	0.045	750	21.9
C4	823.8	9.2	0.042	720	20.7
C5	801.7	9.1	0.039	735	20.6
C6	819.3	9.3	0.034	732	21.2
C7	817.6	9.1	0.04	725	20.1
C8	810.4	8.9	0.054	701	18.5

Based on the obtained indicators of scenarios and in accordance with the developed selection procedure [6, 14], Pareto-optimal solutions were chosen aimed at reducing the technogenic load of mining enterprises and a high level of environmental safety (figure 2).

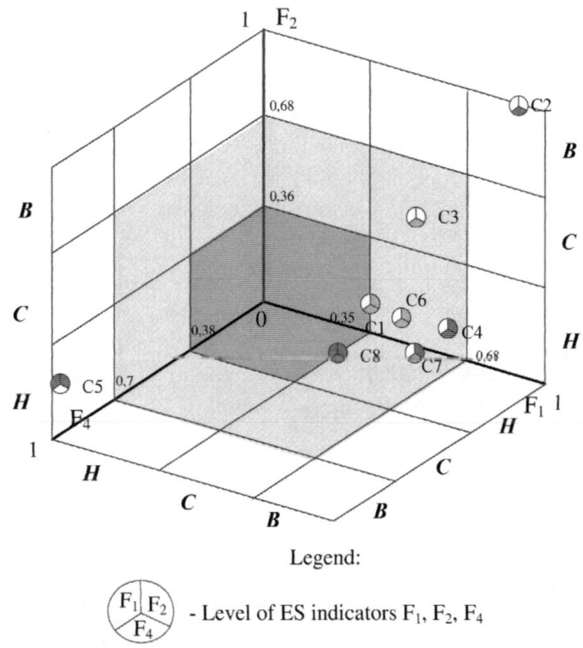

Figure 2. Definition of Pareto-optimal scenarios for increasing the level of ecological safety in Novokuznetsk.

The economic effect (F_1), the area of the reclaimed land (F_2), the population with normative indices (F_4), determined by the specifics of the mining area and the focus on coordinating the interests of the state, investors and production workers, were taken as criteria. Pareto-optimal scenarios are: C1, C2, C3, C4, C5, C6, C7.

Due to the large number of Pareto-optimal scenarios for the coordination of environmental, social, economic interests and ensuring a high level of environmental safety, their ranking is proposed. The problem of *ranking Pareto optimal set of scenarios* can be formulated as follows:

Problem 2. There are many Pareto-optimal scenarios A_m, $m \in [1, M]$. Each m-th scenario is characterized by normalized criteria fx, $x \in [1, X]$, consisting of three levels of environmental safety:

low (L) $fn_x \in [0; Y_{x1}]$; moderate (M) $fy_x \in [Y_{x1};Y_{x2}]$; and high (H) $fv_x \in [Y_{x2};1]$, where Yx_1 and Yx_2 are the values of the boundaries between low and moderate, moderate and high levels of environmental safety, respectively. Moreover, each m-th scenario is characterized by w number of high, g number of moderate, b number of low levels of environmental safety [6].

Scenario ranking *is required*. The scenario $A_1^{rank(1)}$, surpassing the scenario $A_m^{rank(z)}$ by w number of high level of environmental safety, is assigned the first rank if the following is fulfilled:

$$\sum_{t=1}^{T}\sum_{n=1}^{N}\sum_{i=1}^{I}(Kp_{tni} + Kz_{tni} - Gi_{tni}) \rightarrow \min, \tag{2}$$

provided that:

$$w(A_1^{rank(1)}) \geq w(A_m^{rank(z)}) \text{ if } z \in [1, 3]. \tag{3}$$

If necessary, the g-th and b-th number of moderate and low levels of environmental safety are additionally taken into account:

$$g(A_1) \geq g(A_m),\ b(A_1) \geq b(A_m). \tag{4}$$

Scenarios, characterized by the number w=3, are priority, w=2 – prospective, w=1 – stabilizing. Within the framework of this work, stabilizing scenarios are understood that provide adjustment and regulation of the level of environmental safety. Prospective scenarios are aimed at increasing the level of environmental safety. The values Yx_1 and Yx_2 are determined by the expert decision method [15] in accordance with the social, economic and environmental conditions of the mining area. Ranked Pareto-optimal scenarios are presented in table 3.

Table 3. Ranking of Pareto-optimal scenarios for increasing the level of environmental safety in Novokuznetsk.

Scenario	F_1	F_2	F_4	Pareto-optimal scenarios	Level of environmental safety			Place
					F_1	F_2	F_4	
C1	0.932	0.556	0.585	+	H	M	M	3
C2	0.864	1	0	+	H	H	L	2
C3	0.891	0.778	0.453	+	H	H	M	1
C4	1	0.333	0.358	+	H	L	L	4
C5	0	0.222	1	+	L	L	H	4
C6	0.796	0.444	0.641	+	H	M	M	3
C7	0.719	0.277	0.925	+	H	L	H	2
C8	0.394	0	0.17	-	L	L	L	-

From table 3 it follows that Pareto-optimal scenarios of the first rank are C3, of the second one is C2, of the third is C1, C6, of the fourth is C4, C5.

The scenarios of high ranks are aimed at eliminating the negative impact of coal preparation plants, increase in the economic effect, improvement by reducing environmental fees, the quality of life of the population through the creation of new jobs and recreational areas.

4. Conclusion

Thus, the proposed methodological features of the scenarios development take into account the investment potential, the waste characteristics, and are aimed at maintaining the main production, creating additional products from waste that is in demand on the market. Increase in the level of environmental safety in Novokuznetsk in the conditions of coal preparation plants is planned through recycling waste and reclamation of disturbed lands with the receipt of additional products: coal concentrate from substandard fines (sludge), materials for construction needs from the rock.

References

[1] *State report "On the state and on the protection of the environment of the Russian Federation in 2011"* (Ministry of Natural Resources and Ecology of the Russian Federation. State reports) http://www.mnr.gov.ru/regulatory/list.php?part=1392
[2] *Statistical information on environmental protection in the regions of the Siberian Federal District (SFO) for 2015* (Kemerovo) http://ecokem.ru/wp-content/uploads/2016/09/СФО-за-20151.pdf
[3] *Generation, use and neutralization of production and consumption wastes by types of economic activity in the Russian Federation* (Federal Service of State Statistics. Official statistics) http://www.gks.ru/wps/wcm/connect/rosstat_main/rosstat/ru/statistics/environment/
[4] *Regions of Russia. Socio-economic indicators 2015: Statistical Compilation* (Moscow: Goskomstat of Russia) p 1266
[5] Shorokhova A V and Novichikhin A V 2016 *Economics and Management Systems Management* **4** 93–100
[6] Novichikhin A V and Shorokhova A V 2017 *Steel in Translation* 47 **7** 456–462
[7] Shorokhova A V and Novihichin A V 2016 *Economics and Management Systems Management* **4.1(22)** 194–200
[8] Astakhov A S et al 2009 *Ecological Safety and Efficiency of Nature Management* (Moscow: Mining Book) p 323
[9] Alexandrova T N et al 2012 *Mountain Informational and Analytical Bulletin* **9** 284–289
[10] Bagimova N A and Popov S M 2010 *Mountain Informational and Analytical Bulletin* **8** 239–242
[11] Bardovsky A D 2007 *Mountain Informational and Analytical Bulletin* **1** 318–321
[12] Melentiev G B and Malinina E N 2008 *Ecology of Industrial Production* **2** 51–65
[13] Nagin A S 2010 *Mountain Informational and Analytical Bulletin* **1** 55–59
[14] Novichikhin A V and Shorokhova A V 2017 *Steel in Translation* 47 **7** 456–462
[15] Trakhtengerts E A 1998 *Computer Decision* (Moscow: SYNTEG) p 376

IOP Conf. Series: Earth and Environmental Science **206** (2018) 012052 doi:10.1088/1755-1315/206/1/012052

The mechanism for assessing the personnel professional competencies at a loading and transport enterprise

S V Shishkina[1], Yu D Pristupa[1], L D Pavlova[2] and V N Fryanov[2]

[1]LLC "Consolidated PTU of Kuzbass", 13 Kirova ave, Leninsk-Kuznetskiy, Russia, 652518
[2]Siberian State Industrial University, 42 Kirova street, Novokuznetsk, 654007, Russia

E-mail: ld_pavlova@mail.ru

Abstract. The article describes the mechanism for assessing the personal and communicative professional competencies of personnel, corresponding to labor functions and the development strategy of a loading and transport enterprise. According to the results of the questionnaire of 21 respondents 44 meaningful competences were chosen from 257 proposed competencies. It is established that the use of the communicative competencies of personnel due to the emergence of a synergistic effect is more productive. The effectiveness of the work of the industrial control of labor protection department was defined within the range 58 - 98%. However, high production risks of incidents give rise to concerns, since during 23.7% of working time the labor functions are performed with violations of the requirements of professional standards. The results of the professional competences ranking are presented and distribution histograms on the importance of competences for improving labor productivity and reducing the risk of incidents in a loading and transport enterprise are constructed. The algorithm for sampling reduction by means of identification of the most significant competences is developed at loss of the minimum volume of the information. The probability of significant professional competencies for an enterprise exceeds 0.9.

1. Introduction

One of the main parameters that determine the personnel qualification is personal professional competence. These competencies are realized, as a rule, during performance of labor functions within the structure of production infrastructure elements, that is, services, sections, stations, storage facilities and other workplaces. The effectiveness of the interaction of personal competencies of individuals is generally assessed by the quantity and quality of the aggregate product created by the team with a minimal risk of incidents. In these conditions, the need for the development of personnel communicative competencies arises [1-4].

Under the personnel communicative competence the manager and worker should be understood as its certain elements, who are obliged to interact with other people when performing a separate or complex of labor functions within the prescribed modes of operation of the production facilities. The compulsory components of communicative competence are communicative abilities, knowledge, skills, motivation, as well as a synergetic effect when these qualities are jointly realized.

2. Methods of research

To develop a mechanism for assessing the personnel professional competencies at a loading and transport enterprise (LTE), various options for improving the personnel competence were considered, for example, skills development while stimulating the growth of labor productivity or reducing the risk of incidents, hiring or firing when replacing workers with low competencies by employees with competences that increase competence. It is established that for each organization methods and mechanisms for increasing competence have their own peculiarities that are difficult to generalize and apply to other enterprises. Therefore, for LTE conditions, a mechanism has been developed to increase the personnel competence with a combination of personal and communicative competencies, depending on the forms of work organization and work functions.

Based on the analysis of the known methodological approaches and personnel management schemes [5-7], it is established that the use of communicative competences due to the appearance of a synergistic effect is more productive. To identify the links between personal and communicative competencies and assess their compliance with labor functions, complex full-scale studies were conducted according to the following scheme ("Delphi" method) [8]:

- development of a questionnaire form for expert assessment of the competencies importance by managers and line executors;
- development of the main provisions for the scoring system for assessing the importance weight of competencies;
- substantiation of the key indicators determining achievement of labor functions and LTE performance criteria;
- selection of experts;
- conducting an expert survey;
- processing the consistency of the results of expert assessment;
- development of a communicative competencies model based on the synthesis of individual competencies relevant to the corresponding labor functions.

3. Results and discussion

Prior to the survey, the experts were provided with the results of preliminary studies on the compliance of the activities of individual services and of LTE management with the requirements and opinion of the staff as a whole. Within the framework of the production control (PC) and labor protection (LP) services, a preliminary survey on the state of work safety was conducted.

For the expert survey two groups of experts were chosen: managers and line performers. The experts were asked to assess professional personal competencies according to the following indicators of influence: the risk of the incident occurrence, increase in labor productivity with the identification of target indicators, key competencies.

The list and types of competences were specified in the initial version of the questionnaire, but in the process of its discussion with experts new professional competences were added. Thus, only 257 competencies were included into the questionnaire. And only 44 competencies were identified as important by the experts. In total, an expert group of 21 people took part in the survey, who submitted 444 responses. After carrying out the expert evaluation in accordance with the questionnaire, the following results of the expert survey presented in figure 1 were obtained.

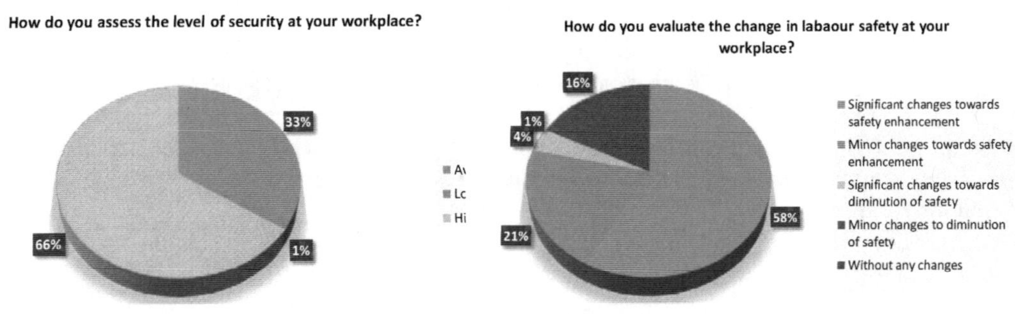

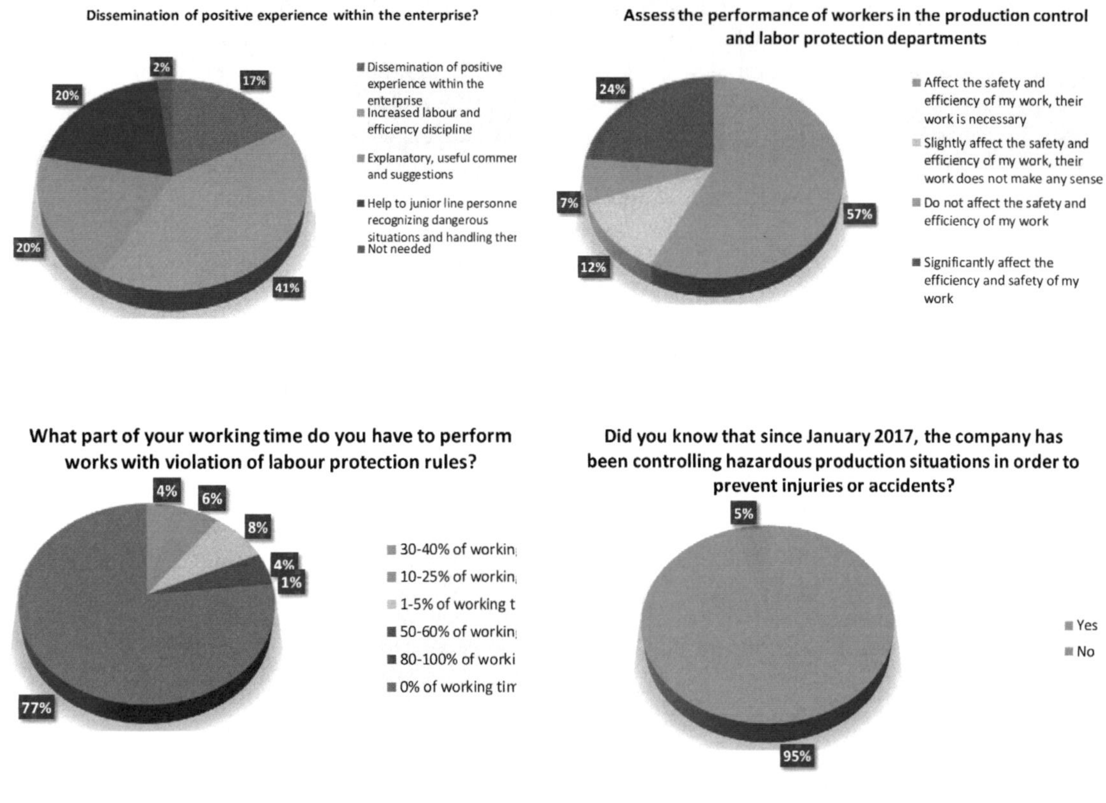

Figure 1. The results of statistical analysis of questionnaires on the effectiveness of industrial control and safety systems.

As follows from the diagrams given in figure 1, the performance of the production control (PC) and labor protection (LT) services is estimated by the staff positively within the range from 58 to 98%. However, high production risks of incidents give rise to concerns, since during 23.7% of working time the labor functions are performed with violations of the requirements of professional standards.

The probability of incidents exceeds 0.2, which is a high indicator. Obviously, the dependence of personal and communicative competencies on the level of individual, collective, social and technical and other risks should be revealed. For this purpose, in the memo for experts it is offered to highlight the positive or negative impact of each competence on the probability of incidents and the level of labor productivity during performance of labor functions.

To assess the competences importance, their ranking was carried out by the number of expert answers. Figure 2a shows the distribution histogram of the number of answers for each personal competence by the criterion of minimum incidents, and in figure 2b – by the criterion of labor productivity.

Given the large number of competencies and the range of their assessments, the task of reducing the sampling arises by highlighting the most significant competencies with the loss of a minimum amount of information. If we consider each competence as a random event, and the number of experts' assessments for each competence as the probability of this event, then we can find the law of distribution for these probabilities.

The scheme of changes in competencies assessments is close to the exponential distribution of the probability density [9-10], that is:

$$f(x) = a\mathrm{e}^{-\lambda x}, \quad (1)$$

where $f(x)$ is the distribution density of competencies assessments; a , λ – empirical coefficients; x – the ranking number of the competency in the graphs of figure 2.

To identify the significant competencies in the graphs of figure 2, the rule $3\sigma_x$ was used. On the graphs the boundary $3\sigma_x$is indicated by a dashed line.

The results of the statistical analysis of competency assessments for the histograms shown in figure 2 are presented in table 1.

Table 1. Results of statistical analysis of competencies assessments.

Name of the indicator					Probability of meaningful competences $P_{3\sigma}$
mathematical expectation m_x	average quadratic deviation σ_x	$3\sigma_x$	empirical coefficients		
			a	λ	
3.00	4.69	14.07	13.00	0.18	0.920
3.00	3.98	11.94	14.67	0.22	0.913

According to the results obtained, it can be argued that the probability of selecting meaningful competencies from the general expert sampling exceeds 0.9. Therefore, the personal competencies are recommended for the development of a model of communicative competencies.

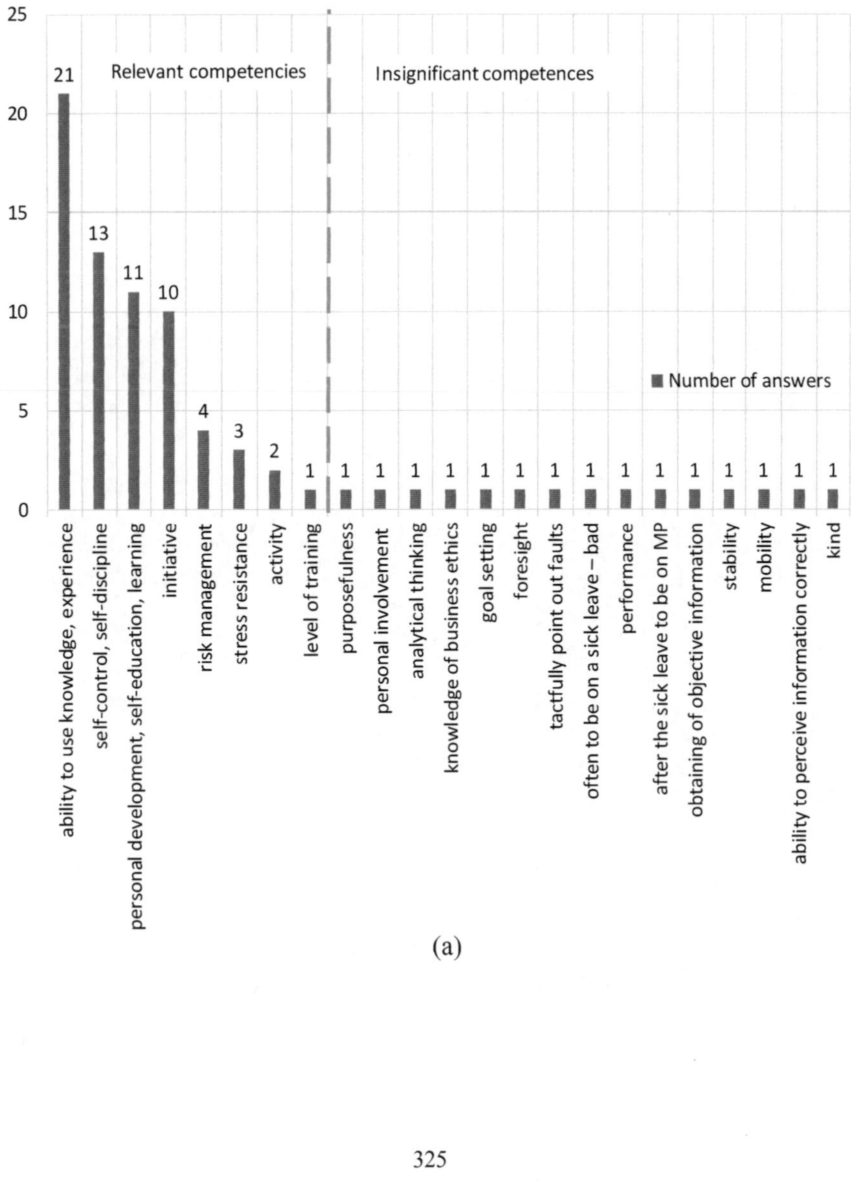

(a)

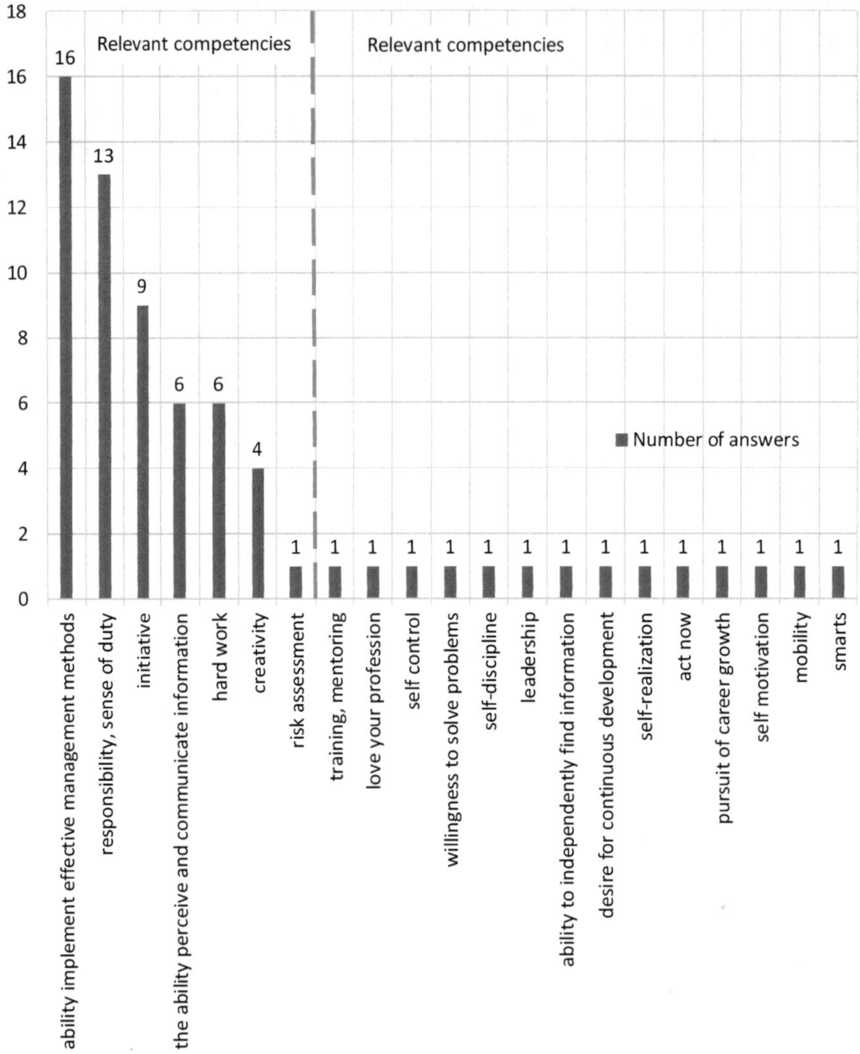

(b)

Figure 2. The scores distribution of personal professional competencies of the personnel: a – by the level of influence on the risk of incidents, $3\sigma_x = 14.07$; b – by level of influence on labor productivity $3\sigma_x = 11.94$.

4. Conclusion

A mechanism for assessing personal professional competencies, corresponding to labor functions and the development strategy at a loading and transport enterprise, has been developed. The recommendations for the use of personal professional competencies identified by ranking results during formation of communicative competences are given.

References

[1] Shishkina S V, Pristupa Yu D, Pavlova L D and Fryanov V N 2016 *Proc. of the IV All-Russian Sci. and Pract. Conf. on Modeling and High-Tech Information Technologies in Technical and Socio-Economic Systems* (Novokuznetsk: SibSIU) pp 8–14

[2] Shishkina S V, Pristupa Yu D, Pavlova L D and Fryanov V N 2017 *Proc. of the XI All-Russian*

Sci. and Pract. Conf. on Automation Systems in Education. Science and Production AS'2017 (Novokuznetsk: SibSIU) pp 284–90
[3] Shishkina S V, Pristupa Yu D, Pavlova L D and Fryanov V N 2017 *IOP Conference Series: Earth and Environmental Science* **84** 012027
[4] Barkalov S A, Kalinina N Yu and Nasonova T V 2017 *Economics and Management Systems Management* **3.1 (25)** 110–6
[5] Guts A and Frolova Yu V 2007 *Mathematical Methods in Sociology. Series: Synergetics: from the Past to the Future* (Moscow: LKI) p 216
[6] Kochiev T B, Novikov D A and Kochieva T B 2000 *Basic Stimulation Systems* (Moscow: Apostrof) 108
[7] Labunsky L V 2003 *Development of the Competence of Personnel of the Mining Enterprise* (Ekaterinburg: UB RAS) p 230
[8] Babicheva N B, Kiseleva T V and Liakhovets M V 2011 *Management Systems and Information Technologies* **3.2(45)** 236–41
[9] Ayvazyan S A, Buchstaber V M, Enyukov I S and Meshalkin L D 1989 *Applied Statistics* (Moscow: Finansy and Statistika) p 607
[10] Ventzel E S 1998 *Theory of Probability* (Moscow: Vyshaya Shkola) p 576

Shift of the world energy sector development paradigm: consequences for Russia

S M Nikitenko[1,3], E V Goosen [1,2] and A E Kontorovich[1,4]

[1]Federal Research Center of Coal and Coal Chemistry, Siberian Branch of the Russian Academy of Sciences, 18 Sovetskiy ave, Kemerovo, 650000, Russia
[2]Kemerovo State University, 6 Krasnaya street, Kemerovo, 650043, Russia
[3]Kemerovo Institute (branch) of Plekhanov Russian University of Economics, 39 Kuznetskiy ave, Kemerovo, Russia, 650992
[4]Trofimuk Institute of Petroleum Geology and Geophysics of Siberian Branch of Russian Academy of Sciences, 3 Koptug ave, Novosibirsk, Russia, 630090

E-mail: nsm.nis@mail.ru

Abstract. The current stage of society development is characterized by rapid changes in the global fuel and energy complex (FEC). Supply chains, added value chains, and the structure of global markets are changing under the influence of the exhaustion of easily available resources, price fluctuations, the emergence of equipment and technologies for extracting previously inaccessible energy resources. Russia is one of the largest countries, and one of the richest in energy resources. The profound changes taking place in the world fuel and energy markets are a challenge for Russia, which can both dramatically accelerate its development (acting as an "open window of opportunity"), and throw it back to the periphery of the world economy. In these circumstances, it is very important to understand the essence of these changes and to find the best way for the country to respond to these changes. The purpose of the article is to identify the essence of the profound changes taking place in the global energy sector and to find the optimal ways of responding to them both for the Russian fuel and energy sector and the Russian economy as a whole..

1. Introduction

The current stage of society development is characterized by rapid changes in the global energy sector. Supply chains, added value chains, and the structure of global markets are changing under the influence of the exhaustion of easily available resources, price fluctuations, the emergence of equipment and technologies for extracting previously inaccessible energy resources (shale oil and gas, bituminous sands, etc.), as well as the technologies for more efficient and environmentally friendly use of natural resources (for example, the use of water-coal suspensions as fuel, clean coal combustion technologies, etc.). Estimations of the available stocks in different countries and in the world at large, and the conceptions of the types of deposits where they are concentrated and from which they can be effectively extracted are changing. All this has a serious impact on countries with significant reserves of energy resources specializing in their production. The transformation of the energy sector as an important source of competitiveness has a serious impact on the development trajectory of entire regions and the entire world economy.

Russia is one of the largest countries, and one of the richest in energy resources. The profound changes taking place in the world fuel and energy markets are a challenge for Russia, which can both dramatically accelerate its development (acting as an "open window of opportunity"), and throw it back to the periphery of the world economy. In these circumstances, it is very important to understand the essence of these changes and to find the best way for the country to respond to these changes.

The purpose of the article is to identify the essence of the profound changes taking place in the global energy sector and to find the optimal ways of responding to them both for the Russian fuel and energy sector and the Russian economy as a whole.

2. Data and methods

To analyze the current and emerging paradigms of the world energy sector development, the data of the world energy statistics posted on the websites of IAE (International Energy Agency) and BP were used. The information on the state of the Russian fuel and energy sector was obtained from the official websites of the Federal State Statistics Service of the Russian Federation and the Ministry of Energy of the Russian Federation.

3. Discussion of the new paradigm of energy sector development

Access to energy and electricity has always had a profound impact on the economy. So the local energy crisis in England in the 18th century, caused by the exhaustion of wood resources, led to the transition to coal and created conditions for industrialization. The global problem of energy shortage in the 1970s forced to turn to intensive energy-saving consumption. In modern conditions, the world is experiencing one of the most acute global energy crises, the reason for which was the need to abandon the extensive production and consumption of mineral energy resources and transition to the model of integrated development of energy resources.

At the turn of the century, the problem of global depletion of conventional fossil fuels became the subject of lively discussions [1]. Initially, both experts and practitioners saw the solution in the traditional extensive expansion of the energy sector – the active search for large deposits and new alternative energy sources, as had happened before, when coal replaced wood, and then oil replaced coal. The role of new global energy sources was predicted for natural gas, biofuels, solar and wind energy [2 5]. However, this did not happen. The unprecedented rise in mineral energy resource prices (which increased by more than three times) from the late 1990s to 2008 allowed to begin active development of hard-to recover reserves of hydrocarbons and to expand the stored reserves. Thus, the oil and gas sector began to expand due to heavy oil and nonconventional tight oil and gas resources, shale oil and gas, pre-salt hydrocarbons, etc. [6].

However, all these new sources of resources only led to the decrease in energy prices in 2014 (figure 1), but none of them could take the place of a new global fuel (figure 2). Unlike previous periods, when the fuel balance was achieved by reducing the production of one resource and the increasing the production of another, the solution was found in reducing energy intensity and creating hybrid resources with predetermined properties. As a result, energy demand fell and the number of consumers working on mixed energy resources (hybrid cars, hybrid power plants, etc.) began to grow rapidly. This allows us to conclude that in the near future the leading energy resource will be a range of hybrid interchangeable resources with predetermined properties. This can explain the fact that the present-day fuel balance is not so much dominated by substitution, but rather by diversification: the equalization of the shares of coal, oil, gas and non-mineral primary energy sources.

The change in the fuel balance has led to major changes in the mechanisms for its maintenance. Firstly, long-term contracts were gradually abandoned and replaced by short-term spot contracts. Secondly, the market of one global energy resource, determining the place and role of other resources in the global fuel balance, was gradually replaced by a global "permeable" market of many interchangeable energy resources – a kind of "mix energy", whose main task is to ensure the possibility of choosing the optimal resource from a set of available resources, and the energy security, stability and balance of the energy sector.

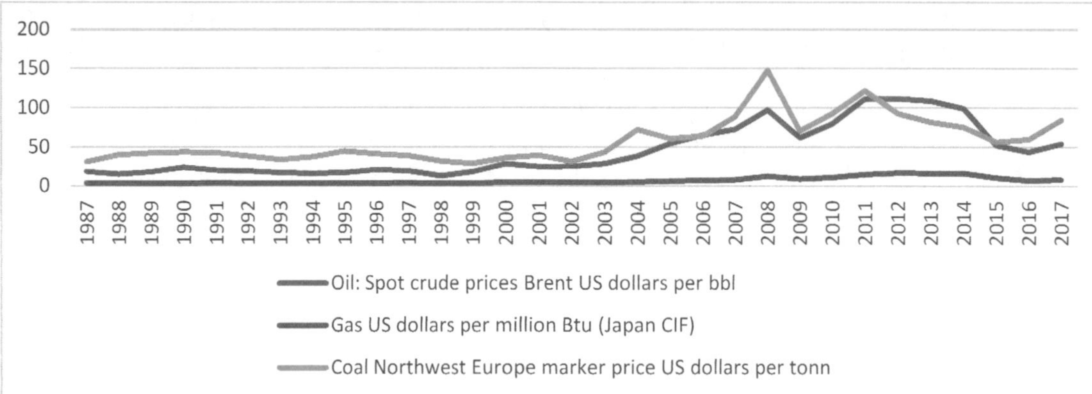

Figure 1. Dynamics of energy resource prices, 1987-2017[7].

In turn, it allowed increasing the number of consumers working on the basis of hybrid technologies [10]. Under these conditions, despite the continued general trend of mineral energy consumption exceeding the supply, the traditional model of the "seller's market" (which is typical for natural resource sectors) was replaced by the "buyer's market" with all its features: price volatility, and increased requirements for flexibility of supply and quality of resources [11]. The influence of midstream and downstream in providing market flexibility is growing. All this leads to the conclusion that a certain mixstream of subsectors has formed for the modern energy sector [10].

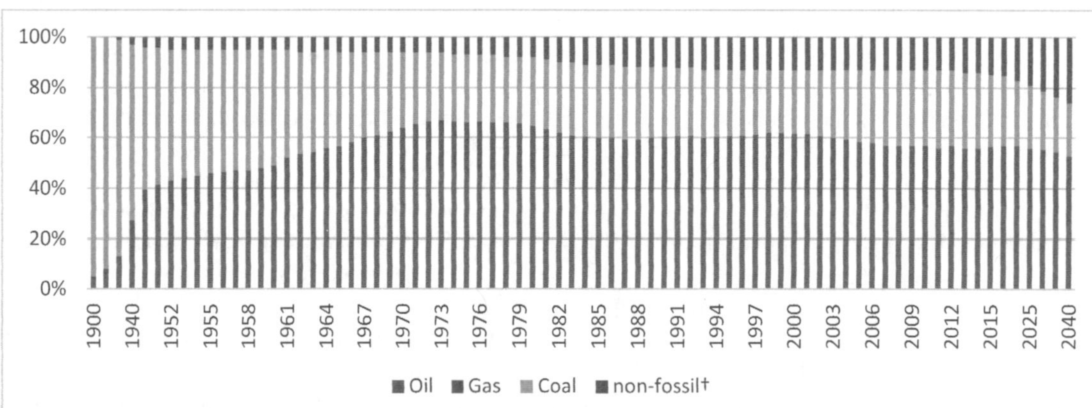

Figure 2. Dynamics of energy resource prices, 1987-2017[7].

The development of the energy sector is seriously influenced by the unfolding new industrial revolution – Industry 4.0. The global energy sector is on the verge of major technological and structural changes. Digitalization, the "Internet of things", advanced robotics, artificial intelligence, machine learning, blockchain and the creation of new materials have fundamentally changed he technological base of the sector. Due to automation, sensors and Internet connectivity, they form a huge potential to improve efficiency and reduce costs in the energy sector. According to the assessment of the World Economic Forum, digitalization of the oil and gas industry alone could generate additional revenue of $1.6 trillion by 2026. New technologies modify supply chains and impose new requirements to the main actors, as well as to the rules of their interaction among themselves and with the actors of related industries [10].

A special term reflecting the new approach to cooperation between organizations –'coopetition' – has even appeared in the studies on the problems of modern management. Scientists and business practices emphasize that today, in order to be successful, the company must rely on both cooperation and competition in its activities, and the former is given the major role. The traditional "win-lose" paradigm of actors' competition is replaced by a more complex strategy, defined by a set of "win–win–win" relations, in which the success of the company is determined not only by the dynamics of profit, but also by public access to rare resources, social responsibility, etc. [12].

All of the above processes are most active in the manufacturing sectors of the economy. However, they also affect the energy sector. Global vertically integrated companies (VICs), whose competitiveness is largely determined by their ability to build up-to-date, high-value supply chains are replaced by national and regional companies included in horizontal networks, whose competitiveness is largely determined by flexibility and the ability to cooperate. In the past decades, there has been a transition of direct control over the world oil and gas resources from "supermajors" (the largest transnational non-state oil and gas companies) in favor of the national and regional oil companies, which influenced the growth of their competitive potential [13].

The wide transformation of the energy sector leads to the fact that large energy companies are increasingly cooperating with companies in knowledge-intensive industries. Thus, only during 2017 BP acquired *Beyond Limits*, a start-up based on artificial intelligence and cognitive computing, which adapts NASA's upstream-technologies, designed for deep space exploration, for use in the fuel and energy sector. *Chevron* is actively developing graphic processors for seismic data visualization and creation of three-dimensional models of fields. The main goal is to determine the most suitable places for drilling. *Shell* develops machine learning algorithms for seismic exploration to automatically detect and classify the geological structures in onshore and offshore oil and gas fields. All this indicates that the process of development exceed a profound transformation of the energy sector: instead, barriers between industries and borders between suppliers, producers and consumers will be destroyed. These global changes suggest a complete change in the paradigm of development of the energy sector and related industries [14]. They almost completely put an end to the old approach to the extraction and use of energy resources, whose essence lay in the consistent development of new mineral reserves, while the main emphasis was on the exploration and development of large and giant deposits in the first place. These global changes also require active involvement of a new conceptual approach in the development process [15-17].

Table 1 lists the major changes in the global energy sector over the recent years and compares the old and new paradigms of its development.

Table 1. Paradigms of energy sector development [15-22].

Criteria for comparison	The old paradigm of energy sector development	The new paradigm of energy sector development
Reserves	Large and giant deposits of conventional fossil fuels	Medium and small deposits partly with nonconventional fossil fuels
Basic resource	Coal – until the 1930s, oil – until the 1980s	Hybrid resources with predetermined properties – mix energy
Fuel balance	Based on one global energy resource (e.g. coal or oil)	Multicomponent balance based on all the major sources of primary energy; both the subsectoral and the regional structures are important
Key subsector	Upstream (exploration and extraction)	All the segments of the energy sector (mixstream)
Key technologies	Technologies of extensive extraction of mineral energy resources. Standardization and limited automation of the extraction of mineral energy resources.	Technologies of comprehensive exploitation of mineral resources Industry 4.0 technologies.: digitalization, Big data and industrial 'Internet of things" Just-in-time approach and lean production in transportation and logistics
Approach to organization of production in the	Mainly extensive development with elements of intensification of basic	Mainly intensive development on the basis of the account of local features

IOP Conf. Series: Earth and Environmental Science **206** (2018) 012053 doi:10.1088/1755-1315/206/1/012053

sector	production Process production and place-neutral approach. Rigid borders between subsectors, spheres and industries	Place-based project approach Flexibility and mobility of resource extraction and supply based on blocks and modules Blurring of borders between the subsectors of the energy sector and related industries
Supply chains	Closed, vertical, high, ascending The leading management role belongs to the extracting company Global	Open, low, multiple, descending The leading management role belongs to service, project and research structures National and localized
Market	Seller's market Global market	Buyer's market Increasing role of national and local markets
Major actors	Global private vertically integrated companies (VICs)	Different companies, including national companies and small companies
Key form of interaction	Global competition for the available reserves	Cooperation based on common access to unique non-material assets and skills
Major source of competitiveness	High quality and availability of the reserves Resource prices Decreasing extraction costs	Flexibility of extraction and supply of energy resources Quality of resources and non-monetary factors Decreasing the costs of all the segments of the energy sector (mixstream)
Basic approach	Natural rent from using mineral energy resources	Decrease in the role of the natural rent and increase in the role of entrepreneurial income
Approaches to state regulation of the energy sector	Minimal interference in the market Indirect administrative measures aimed at imposing severe budgetary constraints. In some cases, the redistribution of rents through national welfare funds.	Growing intervention of the state aimed at support of new technologies, competences, and cooperation, including PPP mechanisms. Active participation of the state in the development of national and local markets. Support for small and medium-sized businesses in the energy sector

4. Conclusions

In many countries and regions that are rich in natural resources, the transition to a new development path has been slow and mixed. These countries include Russia in the first place.

Russia has been and will remain the largest energy power for the foreseeable future. According to BP Statistical Review of World Energy 2018, by the end of 2017, it accounted for 15.5% of proven coal reserves, 6.3% of oil and 18.1% of natural gas [9].

Our country belongs to the so-called resource-dependent countries, which means that the well-being of its economy is largely determined by the income from the production and export of energy resources [23-25]. Over the past 15 years, the share of extractive industries in GDP has fluctuated around 11-12%, and 40-50% of the Federal budget was provided by the sale of energy resources, which accounted for about 70% of the country's exports.

The change of the subsoil use paradigm and the transition of the world economy to integrated subsoil development poses a number of challenges for Russia. The most important of them is whether it will remain a country capable of pursuing independent foreign and domestic economic policies, or will become only a "raw material appendage" of developed countries, constantly catching up with more developed countries in terms of technological and socio-economic development. In the near future, Russia's place in the international division of labor and the prospects for economic development will largely depend on how effectively its "fuel and energy complex" (the term for the energy sector used in Russia) will function. Therefore, it is so important to link the solution of the FEC development problems with the solution of the social, economic, scientific, technical, regional and other problems of Russia's development, and to ensure its gradual transition to a new model of development of the energy sector and of the Russian economy as a whole.

Currently, the direction of the Russian FEC's development is opposite to that of the world's. This is due to the peculiarities of its production, market and institutional structures. Process production on the basis of closed upward vertical technological chains and extraterritoriality of VIC value chains blocks the development of related industries and domestic markets, promotes replication of extensive export-oriented development of fuel and energy industries, and preserves the enclave nature of resource regions' development. Under these conditions, the mere digitalization of production processes and restructuring of the largest VICs are not a solution. It is necessary to find and form the new actors capable of creating the preconditions for the transition to a new model of development. Despite all the negative trends in the Russian FEC, there is still a niche where it is possible to move to a new development paradigm - a segment of small independent resource companies operating mainly on the domestic market. In Russia, there are institutional and legal mechanisms that, after a little refinement, will allow to start implementing investment projects on the basis of PPP mechanisms with the participation of such small independent resource companies [27]. Various forms of partnership between government, business and science should play a crucial role in this.

Acknowledgements
The research was supported by the Russian Scientific Fund (research project No. 16-18-10182 "Formation of the organizational-economic mechanisms of integrated development of mineral resources based on the partnership between science, business and authorities in the resource type regions").

References

[1] Mitchell J, Marcel V and Mitchell B 2012 *What Next for the Oil and Gas Industry?* (London: Chatham House) p 128
[2] 2011 *International Energy Outlook* (US, Washington, DOE/EIA) p 292
[3] *World energy in the context of globalization: challenges for Russia. Scientific Leader Academician N A Simony* 2007 (Moscow: IMEMO RAN) p 153
[4] *World Energy Outlook* http://www.worldenergyoutlook.org/publications/weo-2011/es_russian.pdf
[5] Smeets N 2015 *Where Does the Wind Blow?* N V Toganova and Yu D Kvashnin eds (Moscow: IMEMO RAS) p 123
[6] Shafranik Yu K and Kryukov V A 2016. *Russia's Oil and Gas Sector: a Difficult Path to Diversity* Yu K Shafranik and V A Kryukov eds (Moscow) p 272
[7] *World Energy Balances 2018* https://webstore.iea.org/world-energy-balances-2018-overview
[8] *BP Energy Outlook 2018 edition* https://www.bp.com/content/dam/bp/en/corporate/powerpoint/energy-economics/energy-outlook/bp-energy-outlook-2018.pptx
[9] *BP Statistical Review of World Energy all data 1965-2017* https://www.bp.com/content/dam/bp/en/corporate/excel/energy-economics/statistical-review/bp-stats-review-2018-all-data.xlsx
[10] *Capturing Value From Disruption Technology and Innovation in an Era of Energy Transformation* 2016 (PwC global power & utilities. PwC) p 36
[11] *World of 2035. Global Outlook* 2018 Dynkin A A ed (Moscow: Magistr)
[12] Chernova E G and Romanova S V 2017 *Bulletin of St. Petersburg State University. Economy* **33(4)** 622–640
[13] *Oil and Gas Trends. Adjusting Business Models to a Period of Recovery* 2017 (PwC) p 16
[14] *The industry's future: Destruction of* barriers 2018 (PwC) p 16

[15] Kontorovich A E, Eder L V and Filimonova I V 2017 *IOP Conference Series: Earth and Environmental Science* **53** 012010
[16] Filimonova I V, Eder L V, Mishenin M V and Mamakhatov T M 2017 *IOP Conf. Series: Earth and Environmental Science* **84** 012011
[17] Nikitenko S M, Goosen E V et al 2017 *IOP Conf. Series: Materials Science and Engineering* **253** 012023
[18] *BP Technology Outlook 2018* https://www.bp.com/content/dam/bp/en/corporate/pdf/technology/bp-technology-outlook-2018.pdf
[19] Kontorovich A E and Eder L V 2015 *Mineral Recourses of Russia. Economics and Management* **5** 8–17
[20] *Transformation of World Oil Market* 2016 ed. Zhukov S V (Moscow: IMEMO RAN)
[21] Shafranik Yu K and Kryukov V A 2016. *Russia's Oil and Gas Sector: a Difficult Path to Diversity* Shafranik Yu K and Kryukov V A eds (Moscow) p 272
[22] *Post Hydrocarbon Economy: Issues of Transformation 2017* https://www.fief.ru/img/files/Postuglevodorodnaya_economika_2017.pdf
[23] Polterovich V, Popov V and Tonis A 2007 *Economic Policy, Quality of Institutions, and Mechanisms of Resource Curse* (M) p 98
[24] Chirkova E 2017 *The Iinfluence of Institutions on the Development of Resource Economies* (M) p 66
[25] Kagan E S and Goosen E V 2017 *IOP Conf. Series: Earth and Environmental Science* **84** 012016
[26] *Execution of the federal budget and budgets of the budgetary system of the Russian Federation for 2017 (preliminary results)* 2018 (Moscow: Ministry of Finance) p162
[27] Nikitenko S M and Goosen E V 2017 *IOP Conf. Series: Earth and Environmental Science* **53** 012018
[28] Pakhomova E O, Goosen E V and Nikitenko S M 2017 *IOP Conf. Series: Earth and Environmental Science* **84** 012014

IOP Conf. Series: Earth and Environmental Science **206** (2018) 012054 doi:10.1088/1755-1315/206/1/012054

Fuel and energy complex and methods for assessing the harm from air pollution

A S Tulupov[1] and I V Petrov[2,3]

[1]Market Economy Institute of Russian Academy of Sciences, 47 Nakhimovsky ave, Moscow, 117418, Russia
[2]Institute of Comprehensive Exploitation of Mineral Resources Russian Academy of Sciences, 4 Kryukovsky tupik, Moscow, 111020, Russia
[3]Financial University under the Government of the Russian Federation, 49 Leningradsky ave, Moscow, 125993, Russia

E-mail: tul@bk.ru

Abstract. A significant impact of the fuel and energy complex of Russia on the components of the environment is shown; in particular, the air pollution. The necessity of adopting the methodology for calculating the amount of harm from air pollution is substantiated. The analysis of the draft of Methodology for the Estimation of the Amount of Environmental Damage Caused by Air Pollution developed by the Ministry of Natural Resources and Environment of the Russian Federation is performed. The estimations of the amount of damage caused to the environment by excess emissions from stationary air pollution sources, as well as from agricultural burning and from combustion at waste disposal sites, including landfills, sites of temporary waste accumulation and unauthorized landfills. The directions of improvement have been developed both in terms of structure and content, theoretical and calculated part of the Methodology proposed by the Ministry of Natural Resources and Environment of the Russian Federation.

1. Introduction

The fuel and energy complex has a significant impact on the components of the environment. In figure 1, according to the Federal State Statistics Service, the dynamics of air emissions of pollutants from stationary sources by types of economic activity is provided.

The environmental footprint caused by the economic activity “mining” is clearly seen in the figure. At the same time, more than 90% of pollutants are emitted due to the extraction of fuel and energy minerals.

Any kind of negative impact on the components of the environment requires an appropriate economic estimation. We have collected and analyzed more than 200 methods of assessing damage from environmental pollution or, in modern terms, harm from violation of environmental legislation for the last fifty-year period 1967-2018 [1-5]. It should be noted that most of the methods are not approved at the federal level, they have the status of unofficial scientific and research developments. During the past decade, the Ministry of Natural Resources and Environment of the Russian Federation set the goal to close up a gap. So, in 2007 two Methodologies were approved – calculation of the amount of damage caused by water pollution [6], and violations of the forest legislation [7].

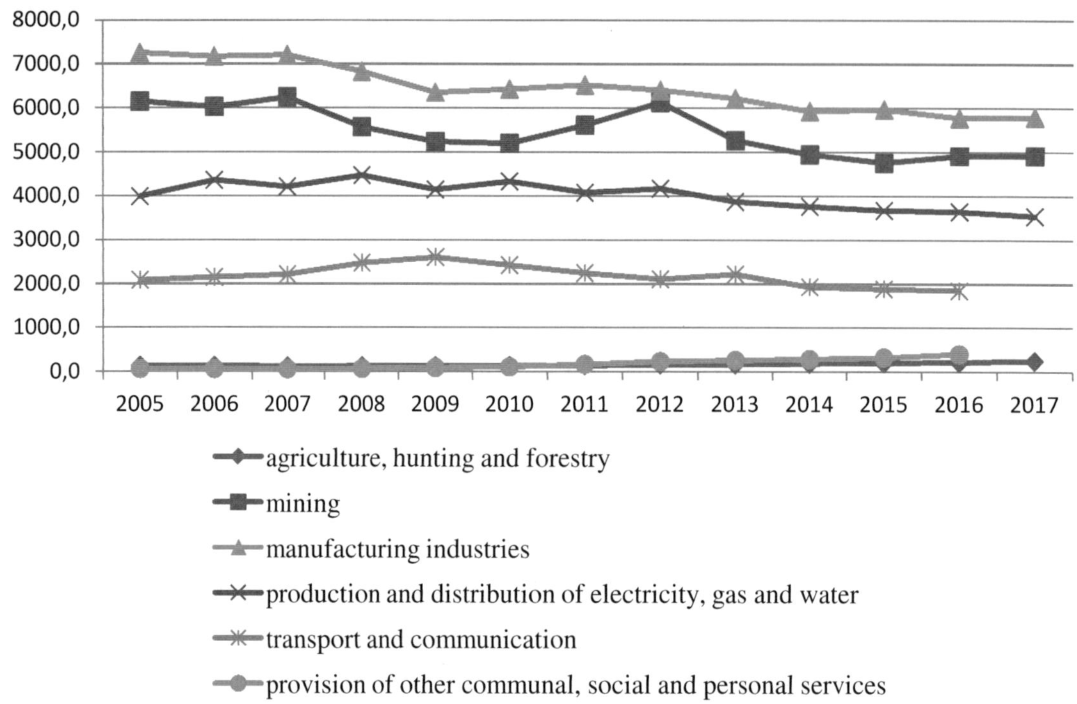

Figure 1. Emissions of pollutants into the atmosphere from stationary sources by types of economic activity, thousand tonnes.

In 2008, the methodology was approved [8] that allows damage caused to fauna objects listed in the Red Book of the Russian Federation to be estimated, as well as to other objects of the animal world that do not belong to hunting and fishing objects and their habitat. In 2009, the methodology [6] was adjusted taking into account the experience of its two-year application in practice, the complaints of business entities; and it was reapproved [9]. In 2010, a methodology for soils was adopted [10], and in 2011 for aquatic biological [11] and hunting resources [12]. Since 2012 there have been no changes. Everyone was waiting for the adoption of the necessary methodology for assessing the amount of damage caused by air pollution. This methodology was needed not only to assess damage and resolve litigation caused by incidents and accidents related to air pollution, but also to develop project documentation taking into account the environmental factor, as well as in the areas of environmental auditing and insurance.

The opponents of the adoption of this necessary document represented by business and large industrial enterprises that have significant environmental impacts justified their position by the fact that harmful substances emitted into the atmosphere are dissipated and damaged forests, soils, water, wildlife and other components of the environment, the methods for that have already been developed and are working. At the same time, in our opinion, even information about air pollution, for example, in large cities, was ignored. And before the harmful substances get dissipated, a great number of people would be exposed to harmful substances that are hazardous to health.

Until now, the damage from air pollution, unlike a number of other environmental components, had to be calculated at the rates of pollution charges, which even with a multiple increase did not show the real extent of the harm done. The methodological approach, in contrast to the calculation of the necessary costs for the restoration of the violated right, takes into account broader aspects of the harm caused, including long-term consequences.

In 2018, the Ministry of Natural Resources and Environment of Russia announced the Methodology for the Estimation of the Amount of Environmental Damage Caused by Air Pollution

[13]. This methodology is still at the final stage, in the so-called draft state. Acquaintance with this methodology, including the performance of calculations, made it possible to form an opinion about the document proposed by the Ministry of Natural Resources and Environment of Russia, as well as to formulate recommendations for the completion and improvement of this methodology.

2. Methods of research

The system analysis of the Methodology for the Estimation of the Amount of Environmental Damage Caused by Air Pollution proposed by the Ministry of Natural Resources and Environment of Russia, taking into account the existing methodological support for assessing harm and damage from both air pollution (see, for example, [14-17]) and other components of the environment, as well as approved and unofficial documents was performed.

The necessity to take into account the features of specific developing territories of the fuel and energy complex, the Arctic Region in particular, was identified [18]. It is also necessary to take into account the approaches applied in the countries of the Eurasian Economic Union [19].

The environment and economic analysis of theoretical provisions, the estimation and methodological tools of the Methodology proposed by the Ministry of Natural Resources and Environment of Russia was carried out. In particular, the estimation of the amount of environmental damage caused by air pollutants (Section 3 of the Methodology) of 1, 2, 3, 4 classes of hazard, NO_x, CO, benz (a) pyrene, as well as hydrocarbons and other substances with the determined reference impact safe level (RSIL) in the air of populated areas was performed.

Also, calculations of the amount of damage caused to the environment by air pollution as a result of combustion at waste disposal sites, including landfills, temporary waste storage sites and unauthorized landfills (Section 4 of the Methodology), as well as as a result of agricultural burning (Section 5 of the Methodology) were performed.

The special attention is paid to the verification of examples of calculations on the amount of damage, given in Appendix 2 of the Methodology.

3. Results and discussion

The methodology harmoniously fits into the accepted and applied to date list of methodological support for assessing damage from pollution of water [9], soil [10], forests [7], aquatic biological resources [11], wildlife [8], hunting resources [12]], etc. Paragraph 4 of the Methodology takes into account the biological cycle of substances (pollutants) in nature.

Section 1 states that the Methodology "is designed to estimate the damaged caused to the environment by air pollution in monetary terms ... including natural disasters, ..., accident and emergency situations of natural and industry-related nature ...". In our opinion, for such cases a separate section with appropriate calculation support is needed.

Section 2 contains the provisions for calculating the weights of pollutant emissions that are taken into account when calculating the amount of damage caused to the environment by air pollution. The formulas given in this section, in our opinion, duplicate each other. If the subject of economic activity does not have a special permit and the established norms of MPE and/or TDP, the value indicator of the standard in formula (1) is not automatically taken into account and equated to zero. The distinction between formulas (1) and (2), in our opinion, is not necessary, because formula (1) and the indices written under the formulas are repeated.

At the same time, in this section the procedure for calculating the weight of pollutant emissions by their concentrations, which is often used for these purposes, is not described here. It is necessary to supplement this section of the Methodology with formulas, or to make exact references to regulatory sources.

The Methodologies also should take into account the need to stimulate the transition of enterprises of the fuel and energy complex to the best available technologies [20].

IOP Conf. Series: Earth and Environmental Science **206** (2018) 012054 doi:10.1088/1755-1315/206/1/012054

Figure 2 shows the results of calculations on the amount of specific damage caused to the environment by air pollution, performed in accordance with the formula from Section 3 of the Methodology.

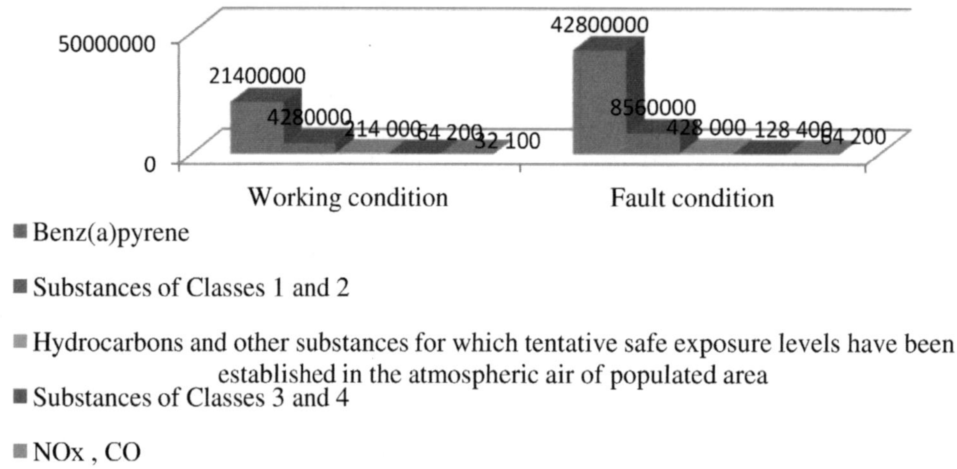

Figure 2. The calculated amount of damage per 1 tonne of pollutants (for cases of serviceable operation, or for gas cleaning installations not included into the design documentation, or for failure or non-use (disconnection)) or for gas purification installations not included into the design documentation), rubles.

Section 4 contains the rules for calculating the amount of damage caused to the environment by air pollution from combustion at waste disposal sites, including landfills, temporary waste storage sites and unauthorized landfills.

As shows the calculation according to formula (4) of the Methodology, the specific damage for one tonne of burnt waste in prices of 2018 will be 81 000 rubles.

Section 5 regulates the calculation of the amount of damage caused to the environment by air pollution as a result of agricultural burning.

According to formula (6), in the prices of 2018, the damage from 1 square meter of agricultural burning will be 2 rubles. This value, in our opinion, even taking into account the real areas of agricultural burnings, is understated.

Appendix 1 of the Methodology, containing fees for calculating the amount of damage, in our opinion, generalizes the force of exposure to pollutants, that results in the average damage values.

Appendix 2 of the Methodology provides examples of calculating the amount of environmental damage caused by air pollution. At the same time, in the calculation examples for the faulty gas cleaning equipment a cleaning factor, which does not correspond to the coefficient from paragraph 14 of the Methodology, is applied. In addition, in this Appendix, in the calculation examples the fee for calculating the amount of damage caused by the excess emission for nitrogen dioxide NO_2 and sulfur dioxide SO_2 does not correspond to the values given in Appendix 1 of the Methodology.

4. Conclusions

The study of the theoretical part of the Methodology proposed by the Ministry of Natural Resources and Environment of Russia, as well as the implementation of calculation procedures according to the basic formulas from the Methodology, made it possible to conclude that the methodology is relevant and necessary. But, at the same time, it needs to be corrected and refined.

First, it is necessary to devote a separate section to the calculation of the amount of damage caused to the environment by air pollution due to incident and emergency situations.

Second, Section 2 of the Methodology should be corrected: to eliminate duplications in formulas (1) and (2), and also to supplement this section with the provisions for calculating the weight of pollutant emissions by their concentrations.

Third, in Section 5, it is necessary to increase the fee per square meter of agricultural burning.

Fourth, in Appendix 1 it is necessary to structure the fee rate in more detail depending on the harmful substances, which would allow for more differentiated calculations.

Fifth, in Appendix 2, it is necessary to adjust the applied coefficients for compliance with the main text of the Methodology.

In general, the adoption of the Methodology proposed by the Ministry of Natural Resources and Environment of the Russian Federation for calculating the amount of damage to the environment caused by air pollution will help to make significant progress towards the construction of an integral system of compensation for damage caused by violations of legislation in the field of environmental protection.

Acknowledgements
The work was supported by the Department of Humanitarian and Social Sciences of the Russian Foundation for Basic Research, project No. 17-02-00245-OGN "Formalization and assessment of probability factors and damage caused by environmental pollution".

References
[1] Vitukhin A D and Tulupov A S 2017 *Proc. of All-Russian Conf. on Innovative Development of Territories* (Cherepovets: CSU) pp 101–5
[2] Tulupov A S 2017 *Mining Journal* **8** 61–5
[3] Porfiryev B and Tulupov A 2017 *Studies on Russian Economic Development* **6** 600–7
[4] Goncharenko S, Le B, Stoyanova I and Petrov I 2014 *Mountain Journal* **9** 143–6
[5] Petrov I and Stoyanova and 2014 *Monitoring. Science and Technology* **2** 54–8
[6] *2007 Methodology of calculating the amount of damage caused to water due to violations of water legislation* (Moscow: Order of the The Ministry of Natural Resources and Environment of the Russian Federation No. **71**)
[7] *2007 Methodology of calculating the amount of damage caused to forests, including forest plantations, or trees, bushes and lianas not attributed to forest plantations, due to violation of forest legislation* (Moscow: Resolution of the Government of the Russian Federation No. **273**)
[8] *2008 Methodology of calculating the amount of damage caused to fauna objects listed in the Red Book of the Russian Federation, as well as to other fauna objects that do not belong to hunting and fishing object and their habitat* (Moscow: Order of the The Ministry of Natural Resources and Environment of the Russian Federation No. 107)
[9] *2009 Method for calculating the amount of damage caused to water due to violations of water legislation* (Moscow: Order of The Ministry of Natural Resources and Environment of the Russian Federation No. 87)
[10] 2010 *Method of calculating the amount of damage caused to soils as an object of environmental protection* (Moscow: Order of The Ministry of Natural Resources and Environment of the Russian Federation No. 238)
[11] *2011 Methodology for calculating the amount of damage caused to aquatic biological resources* (Moscow: Order of The Ministry of Natural Resources and Environment of the Russian Federation No. 948)
[12] *2011 Methodology for calculating the amount of damage caused to hunting resources* (Moscow: Order of The Ministry of Natural Resources and Environment of the Russian Federation No. 948)

[13] *2018 Methodology of calculating the amount of damage to the environment caused by air pollution (draft)* (Moscow: Order of The Ministry of Natural Resources and Environment of the Russian Federation)
[14] *2005 Methodology for determining the amount of environmental damage caused by air pollution from stationary sources of pollution on the territory of Moscow* (Moscow: Resolution of the Government of the Russian Federation No. 94)
[15] *2005 Methodology for determining the amount of environmental damage caused by air pollution from burnings on the territory of Moscow* (Moscow: Resolution of the Government of the Russian Federation No. 689)
[16] *2003 Temporary methodology for determining the amount of environmental damage caused by exceeding the specified standards of pollutants emission during the operation of vehicles* (Tatarstan: Order of the Ministry of Environment and Natural Resources of the Republic No. 660)
[17] *2008 Method for calculating the amount of damage to the environment caused by air pollution* (Altai Territory: Order of the Department of Natural Resources and Environmental Protection No. 116)
[18] Vartanov A Z, Galchenko Yu P, Kalabin G V, Petrov I V and Fedash A V 2018 *Vestnik MGSU* **2(113)** 240–8
[19] Vartanov A Z, Petrov I V and Fedash A V 2017 *Mining Journal* **11** 14–7
[20] Kazakov V B, Kalacheva L V, Petrov I V and Surat I L 2017 *Coal* **6(1095)** 48–50

IOP Conf. Series: Earth and Environmental Science **206** (2018) 012055 doi:10.1088/1755-1315/206/1/012055

Search for "transforming innovations" for Russian regions and barriers to their implementation

E V Goosen[1,2], O N Kavkaeva[1], E O Pakhomova[1] and A P Protsenko[1]

[1]Federal Research Center of Coal and Coal Chemistry, Siberian Branch of the Russian Academy of Sciences, 18 Sovetskiy ave, Kemerovo, 650000, Russia
[2]Kemerovo State University, 6 Krasnaya street, Kemerovo, 650043, Russia

E-mail: egoosen@yandex.ru

Abstract. The paper reviews the approaches to defining "Resource regions" and substantiates the authors' own definition of "transforming innovations". Patent work activities of Kemerovo region's mining companies were analyzed to identify the transforming innovations.

1. Introduction
In the present-day economic studies resource regions are increasingly seen not as the "curse" of the Russian economy, but rather as its potential growth points. This trend was particularly evident in the stages of development of the approaches to defining resource regions in the economic literature. Traditionally, resource regions have been studied as areas with significant reserves of mineral resources, land, forest and other natural resources. A region was classified as a "resource region" based on the evaluation of the features, quality and amount of reserves, while the efficiency and comprehensive use of the available resources were not taken into account. This approach was most widespread in the Soviet era. Further development of approaches to understanding resource regions is associated with the beginning of the formation of the market approach to the use of natural resources in Russia and an attempt to assess the resource potential (resource reserves and the extent of their involvement in public production). Within the framework of this approach, the key characteristics are the structure of the GRP of the region and the share of raw materials in it, and the level and features of the development of material infrastructure that supports extraction, transportation and processing of natural resources. The strength of this approach is the quantitative assessment of the involvement of natural resources in public production while the weak point is its inability to explain the sources of formation and mechanisms of development of resource regions.

2. Data and methods
The key point in the search for tools of resource regions' development is to determine the "resource-type" of Russian regions. Let us turn to the method proposed by E.S. Kagan and E.V. Goosen, based on the approach of V.P. Orlov [1], who determined the "resource-type" of Russian regions basing on the share of the value added in the "Resource extraction" section in the GRP [2]. E.S. Kagan and E.V. Goosen calculated the degree of the regions' resource dependence using the K-rate (the ratio of the value of the share of extractive industries in the region's GRP (K_{reg}) to the value of the share of extractive industries in the Russian Federation's GDP (K_{Rus}), which allowed comparing the regions' "resource-type". If K>1, it means that the region is a resource region. Table 1 presents 35 Russian

resource regions with varying degrees of resource dependence. Thus, in Russia more than a quarter of the regions are classified as resource regions.

Significant reserves of mineral resources are concentrated in resource regions, and their export is a source of growth of the regions' welfare, but at the same time it can cause the formation of the "enclave economy". Therefore, it is necessary to find the sources and mechanisms of transition to a new balanced development path. "Transforming innovations" play a critical role to play in this process. The authors of this paper suggest their own approach to the concept of "transforming innovations" basing on the microeconomic model of the replication of "dual enclave economy" (based on the theory of value chains) proposed by S.M. Nikitenko, E.V. Goosen, and K.S. Sablin [4,5].

By "transforming innovations" we mean the innovations that contribute to the change of the region's development path towards its balance. "Transformative innovations" possess the following distinguishing features: they are available to medium-sized businesses (innovation is a competitive advantage of medium-sized businesses); they contribute to the formation of branched horizontal descending value chains in resource extraction and related industries; they create conditions for the localization of manufacturing industries in the region, which are the basis for the development of the regional market on the basis of inter-firm interaction and diversification of the economy; they never emerge on by themselves and always have an organizational or institutional component; they also form another market environment and other market relations.

The aim of this study is to identify the "transforming innovations" in resource regions based on the analysis of patent activity of large mining companies localized in their territory. The study focuses on Kemerovo Region (also known as Kuzbass – after 'Kuznetsk Coal Basin'), Russia's largest coal mining region. Kemerovo Region is a resource region, and its economy is based on the extraction of minerals, which amounted 29.7% of the GRP in 2016, according to Kemerovo Region's Office of Federal State Statistic Service. We selected the two largest coal mining companies: Public company "EVRAZ" and Joint-stock company "Siberian coal energy company" (JSC "SUEK"). The data sources were the patent database "Contour focus" and the companies' websites.

In 2017, Russia produced 408.9 million tons of coal (+23 million tons (+5.6%) compared to 2016), of which 22% were coking grades: 88.6 million tons (-4 million tons (-4%) compared to 2016) [3]. In Kuzbass, coal production amounted to 241.51 million tons (+13.5 million tons (+5.5%) compared to 2016) [3], which is 59% of the total Russian production, including 75% of the Russian production of coking grades. Kuzbass accounts for 70% of the total investment in the coal industry [6,7]. The largest coal companies are SUEK (whose market share is 26%), UMMC (13%), SDS-Coal (7%), Mechel (5%) and EVRAZ (5%) [8].

Table 1. Dynamics of the structure of the regions' resource dependence in 2005-2014 [3].

		2005	2010	2011	2012	2013	2014
1.	Nenets Autonomous Okrug	5.80	7.56	6.60	6.73	7.03	7.01
2.	Khanty-Mansiysk Autonomous Okrug – Yugra	5.85	6.06	5.98	6.06	6.07	6.31
3.	Sakhalin Region	1.73	5.70	5.41	5.49	5.66	6.20
4.	Tyumen Region	4.68	4.80	4.63	4.87	4.86	5.11
5.	Yamalo-Nenets Autonomous Okrug	4.80	4.61	4.29	4.64	4.87	4.74

6.	Republic of Sakha (Yakutia)	3.09	3.86	3.88	3.82	3.98	4.20
7.	Chukotka Autonomous Okrug	0.59	3.67	3.71	3.38	2.90	4.05
8.	Orenburg Region	2.89	3.45	3.16	3.30	3.80	3.40
9.	Komi Republic	2.68	3.22	2.98	2.88	3.00	3.17
10.	Tomsk Region	2.77	2.30	2.69	2.79	2.69	2.69
11.	Republic of Udmurtia	2.09	2.25	2.30	2.29	2.33	2.28
12.	Kemerovo Region	2.12	3.02	3.09	2.39	2.06	2.04
13.	Astrakhan Region	0.21	0.34	1.56	1.74	1.97	2.03
14.	Irkutsk Region	0.30	0.72	1.16	1.40	1.56	1.88
15.	Tatarstan Republic	2.41	2.08	1.98	1.95	1.90	1.87
16.	Magadan Region	2.12	1.98	2.24	1.63	1.61	1.62
17.	Krasnoyarsk Kray	0.30	1.74	1.48	1.38	1.59	1.59
18.	Perm Kray	1.23	1.30	1.39	1.62	1.58	1.46
19.	Samara Region	0.84	1.14	1.25	1.31	1.24	1.34
20.	Belgorod Region	1.70	1.62	1.87	1.43	1.40	1.17
21.	Murmansk Region	0.84	1.46	1.66	1.44	1.67	1.15
22.	Republic of Karelia	1.51	1.23	1.27	1.21	1.15	1.13
23.	Amur Region	0.41	0.98	1.40	1.33	1.07	1.06
24.	Republic of Khakassia	0.57	1.48	1.40	1.06	1.10	1.03
25.	Kursk Region	1.41	1.16	1.33	1.13	1.09	0.88
26.	Transbaikal Kray	0.59	1.23	0.71	0.81	0.93	0.73
27.	Republic of Tyva	0.40	0.66	0.55	0.54	0.43	0.61
28.	Volgograd Region	0.53	0.54	0.55	0.53	0.55	0.43
29.	Khabarovsk Kray	0.37	0.43	0.49	0.58	0.53	0.42
30.	Kaliningrad Region	1.15	0.57	0.51	0.43	0.40	0.33
31.	Republic of Buryatia	0.35	0.47	0.54	0.45	0.38	0.33
32.	Bashkortostan Republic	0.99	0.85	0.43	0.26	0.26	0.30
33.	Republic of Ingushetia	0.80	0.18	0.17	0.17	0.19	0.16
34.	Chechen Republic	0.79	0.26	0.21	0.17	0.18	0.16
35.	Republic of Kalmykia	0.51	0.33	0.26	0.27	0.27	0.13

At the first stage of our study, based on the analysis of the information contained in the database "Contour focus", detailed supply chains of both companies (SUEK and EVRAZ) were built, the most innovative and active organizations in their structures were identified, their role and place were shown.

Public company "EVRAZ" [9] is a vertically integrated steel and mining company with assets in Russia, the USA, Canada, the Czech Republic, Italy and Kazakhstan. Its own base of iron ore and coking coal almost completely meets the internal needs of Eurasia. In total, the PC "EVRAZ" has 23 units, three of which are engaged in coal mining: JSC "Yuzhkuzbassugol" (Kemerovo region), PJSC "Raspadskaya" (Kemerovo region), and Mezhegeyugol (the Republic of Tyva.) In total, the company owns 9 mines and open pits, and 3 processing plants. Coal production volumes in 2016 amounted to 22.1 million – 6 % of the Russian total (ranking 5th in Russia), of which 22.5 million tons was

IOP Conf. Series: Earth and Environmental Science **206** (2018) 012055 doi:10.1088/1755-1315/206/1/012055

produced in Kemerovo Region. In 2017, EVRAZ produced a quarter of the Russian coking coal (22.4 million tons) became its largest producer [9].

EVRAZ is a vertically integrated company which produces iron ore and coal further used for producing metal. A large part of the extracted coal is used for the own needs of EVRAZ. A smaller part of the coal goes to the housing and utilities sector (see figure 1).

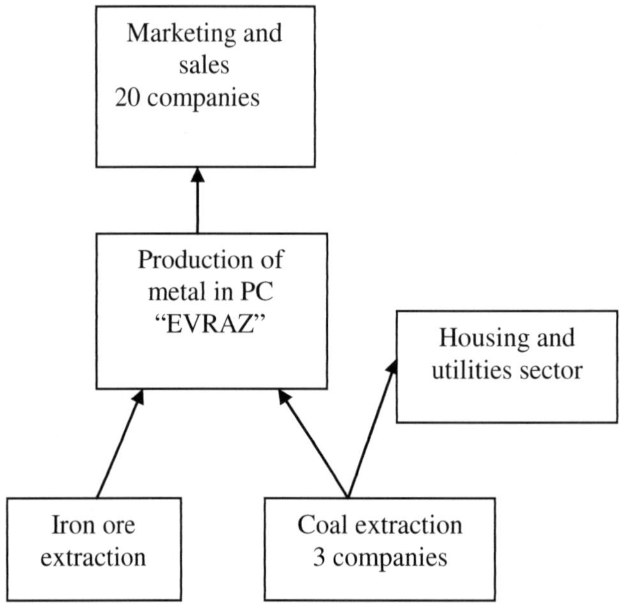

Figure 1. Supply chain of the Public company "EVRAZ".

Joint stock company "Siberian coal energy company" (JSC "SUEK") [10] is one of the largest coal mining companies in the world (ranking 5th in terms of coal exports in 2016 – 51.9 million tons) and a leading coal producer in Russia (ranking 1st in terms of coal production in 2016 – 105.4 million tons, which exceeds the 2015 amount by 7.77%). Of these, 37.7 million tons were produced in 2016 in Kemerovo Region (ranking 1st place among the other regions of presence). The company owns 27 mines and open pits, 9 enrichment plants and systems, and 3 ports. JSC "SUEK" has 55 units, 36 of which are involved in coal production.

SUEK is the main supplier of thermal coal to the Russian market, its share is 40%. The delivery of coal to the leading energy companies which supply electricity and heat to industrial enterprises and homes across the country is facilitated by proximity to major consumers and the company's large car fleet.

SUEK is a vertically integrated company that controls the entire operational cycle – from coal mining, processing and transportation to sales of products to consumers around the world through a network of sales offices. The company produces coal in large open-cuts and mines of Siberia and the Far East, and its enrichment plants and systems allow improving the quality of products. SUEK has ports and railway infrastructure, which helps deliver products effectively to consumers in Russia and to foreign countries in the Atlantic and the Asia-Pacific regions (see figure 2).

At the second stage of the study, the search and analysis of patents owned by the organizations included in the supply chain, taking into account affiliated companies, was carried out.

Only 2 of 23 structural units of PC EVRAZ (8.7 %) are involved patent activity: they are JSC "United Coal Company "Yuzhkuzbassugol" and PJSC "Raspadskaya". As of 01.01.2018, EVRAZ has received four patents, none of which are justified in force (see table 2).

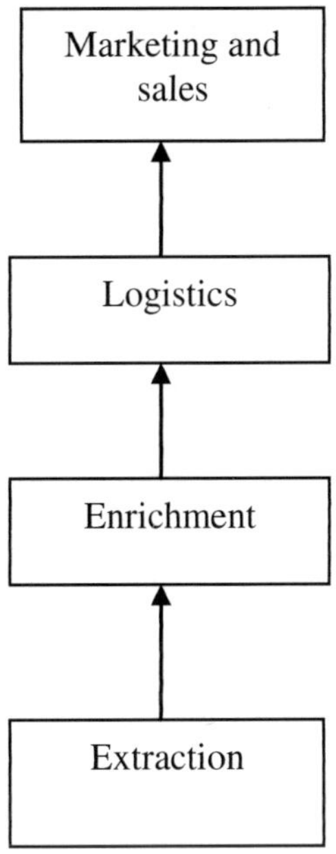

Figure 2. Supply chain of JSC "SUEK".

Table 2. Patents of PC "EVRAZ".

Name of company	The company's position in the added value chain	Number of useful model patents
JSC "United coal company "Yuzhkuzbassugol" (Novokuznetsk, Kemerovo Region)	R&D, exploration, extraction, enrichment, transportation, maintenance and technological support	2
PJSC "Raspadskaya", LTD "Montazhnik Raspadskoy" (Mezhdurechensk District, Kemerovo Region)	Development and production of mining equipment, maintenance and technological support	2
Total number of patents		4

Seven out of the 55 structural units of JSC "SUEK" (12.7%) show patent activity. In JSC "SUEK" the R&D strategy is concentrated within one specialized scientific institute – "Sibniiugleobogashchenie", which owns 70% of the patents in the coal industry, and 30% of the mining divisions' patents (see table 3). As of 01.01.2018, only 10 out of 13 patents (76.9%) are justified in force.

3. Results and discussion

The analysis of the companies' selected patents showed that the main direction of their research and technological activity is the development of technologies and equipment related to the enrichment of coal.

At the same time, these patents were registered over the past three years, and the lifetime of the technology is 8 years, therefore, the companies are constantly improving their own developments. Moreover, the patents are valid, i.e. justified in force, which proves that they are used in coal production.

At the third stage of our study, the existing patents were evaluated in terms of their "breakthrough character". The studied patents cannot be attributed to "transforming innovations". According to experts, the patents refer to the obsolete technology, and match the level of the 1970s – so, they cannot be attributed to "transforming innovations" nor can they shape the value chain.

Table 3. Patents of JSC "SUEK".

Name of company	The company's position in the added value chain	Number of patents for invention	Number of useful model patents
LTD "Siberian research institution for coal enrichment" (LTD "Sibniiugleobogashchenie") (Moscow)	R&D	1	0
Prokopyevsk branch of LTD "Sibniiugleobogashchenie" (Prokopyevsk, Kemerovo Region)	R&D	7	0
JSC "Tugnuisky Open Pit" (Sagan-Nur, Republic of Buryatia)	Extraction	1	0
JSC "Berezovsky Open Pit" (Rodniki, Sharypovo District, Krasnoyarsk Kray)	Extraction	1	0
JSC "SUEK-Kuzbass" (Leninsk-Kuznetsky, Kemerovo Region)	Transportation	0	2
JSC "United Production and Transport Department of Kuzbass" (Leninsk-Kuznetsky, Kemerovo Region)	Transportation	0	0
Murmansk Commercial Seaport (Murmansk, Murmansk Region)	Transportation	0	1
Total number of patents	13		

To create and implement "transforming innovations" in the coal industry, firstly, it is necessary to cooperate with coal companies from related industries, which will increase the added value of the final product. The closest to the coal industry in Kemerovo Region are the enterprises of mechanical engineering, small chemical production, energy industry, and developers of automated control algorithms. Secondly, new forms of cooperation are needed, for example, innovative enterprises in related industries to form value chains with the region's largest coal companies which will develop new technologies taking into account the identified strategies for conducting research activities. Thirdly, measures are needed to encourage companies to form and use "transforming innovations".

4. Conclusion

Thus, the barriers to the formation of "transforming innovations" in resource regions are as follows. The first one is the focus on the place based approach, as it makes no sense to introduce the latest technologies where the level of technological development is low. To overcome it, it is necessary to modernize the entire complex of equipment of coal mining enterprises, the qualifications of their staff, etc. Secondly, the strategy of patent activity showed that large vertical companies are not interested in the development of R&D. a possible solution will be to use smart specialization in the companies' strategies [11]. The third barrier is the lack of a clear policy for the resource regions' development strategy. On the one hand, in the regions where the rent allows solving social issues, the federal authorities do not see a threat to the socio-economic development of the regions; on the other hand, the example of Kemerovo Region clearly demonstrates the need to change the resource region's development path.

Acknowledgements

The research received support from the Russian Science Foundation project №17-78-20218 "Space specialization and integrated development of resource-type regions".

References

[1] Orlov V P 2007 *Mineral Recourses of Russia. Economics and Management* **2** 3–5
[2] Kagan E S and Goosen E V 2017 *Instutional transformation of the economy: space and time. Proceedings of the 5th International conference* **1** 57–64

[3] *Passport of Kemerovo Region. Approved by the Decree of Kemerovo Region Administration Board No.1472-p dated 29 December 2005* (amended edition approved by the Decree of Kemerovo Region Administration Board No.587-p dated 13 December 2017)
[4] Kagan E S and Goosen E V 2017 *IOP Conference Series: Earth and Environmental Science* **84** 012016
[5] Nikitenko S M, Goosen E V and Sablin K S 2016 *IOP Conference Series: Earth and Environmental Science* **45** 012001
[6] *Website of the Central Dispatching Department of the Fuel and Energy Complex* http://www.cdu.ru/tek_russia/articles/5/464/.
[7] *2035 Forecast of the Research and Technological Development of the Russian Fuel and Energy Complex* pp 33 -34 https://minenergo.gov.ru/node/6366
[8] *JSC "SUEK": Integrated Annual Report* 2017 http://www.suek.ru/upload/iblock/bfb/SUEK_AnnualReport2017_RUS.pdf
[9] *Website of Public company "EVRAZ"* http://www.evraz.com/ru/
[10] *Website of JSC "SUEK"* http://www.suek.ru/
[11] Eder L V, Filimonova I V et al 2017 *International Multidisciplinary Scientific GeoConference Surveying Geology and Mining Ecology Management, SGEM* **17(53)** 727–34

Sustainable development of resource regions on the declining stage of production: innovative technologies and economic efficiency

L V Eder[1,2,3], A V Komarova[1,2], I V Filimonova[1,2,3], I V Provornaya[1,2] and J C Benitez Garcia[2]

[1]Trofimuk Institute of Petroleum Geology and Geophysics, Siberian Branch of the Russian Academy of Sciences, 3 Koptyug ave, Novosibirsk, 630090, Russia
[2]Novosibirsk State University, 2 Pirogova street, Novosibirsk, 630090, Russia
[3]Federal Research Center of Coal and Coal Chemistry, Siberian Branch of the Russian Academy of Sciences, 18 Sovetskiy ave, Kemerovo, 650000, Russia

E-mail: EderLV@yandex.ru

Abstract. The article analyzes the problems of sustainable development of Russia's resource regions, where the extraction of natural resources is at a declining stage. The economic indicators of the development projects with application of advanced extraction technologies to increase production efficiency were estimated. It was revealed that high costs of installation and application of advanced technologies do not allow for the full application, especially under the conditions of highly volatile oil prices and limitations on usage of foreign technologies. Measures to optimize the structure of costs and increase the overall efficiency of implementation of innovations in resource regions were proposed.

1. Introduction

The drastic drop in oil prices has affected different sectors of the petroleum industry in Russia but mainly the upstream sector [1]. In this scenario, investments in new or ongoing petroleum projects faces different challenges due to high development cost for some areas and limited budget structure of companies. As a response, each project has to be adjusted taking into consideration the market expectation of low oil price outlook to avoid being postponed or canceled [2].

The research aim is efficiency evaluation of application of oil production technologies for oilfield during declining stage of production. To achieve this goal, the following tasks were accomplished: (1) to analyze current condition of the oil industry in Russia and define main problems of developed regions, (2) to develop scenarios of technology implementation to increase future oil production of the oilfield in the Western Siberia, (3) carry out calculation and analyze the results for evaluation of efficiency for the chosen scenarios.

Nowadays, advanced production technologies are considered a vital part of the oil industry in Russia [3]. Under sanctions, Russia is facing a complicated situation since it is necessary to develop import substitution in the oil sector and create a Russian technological base for a new project in term of development of new and traditional fields (Bobylev, 2016) [4]. Such techniques as horizontal drilling and hydraulic fracturing have played an important role in enhancement of reservoir

IOP Conf. Series: Earth and Environmental Science **206** (2018) 012056 doi:10.1088/1755-1315/206/1/012056

productivity in fields which have reached a peak in production, as well as in development of unconventional resources (Akangbou et al. 2016) [5].

Some issues during the oil production are results of water invasion into the reservoir. For most significant deposits, the average of water cut of developed deposits may be more than 80% (Bailey et al. 2000) [6]. Considering that water flooding is a technique for maintain reservoir pressure and obtain additional oil recovery (Hourfar et al. 2017), it may present a progressive increase in the ratio of produced water to the oil extraction in oilfields in Russia [7].

Most of the companies in the oil and gas sector use two methods associated with the investment valuation; (a) allocation of capital in a systematic manner and (b) implementation of specific methods of investment analysis. Therefore, the investment process in the development of oil projects must consider essential aspects such as; valuation, strategic thinking and risk exposure (Lima & Suslick 2005; Lyons et al. 2016) [8]. Evaluation's indicators are usually calculating according to particular formulas and equations, classified into two categories: the first category considers the value of money over time such as Net Present Value (NPV), Internal Rate of Return (IRR), Discounted Payback and Profitability Index. The second category covers non-discounted methods such as Break-even Analysis, Cost/Benefit Ratio, Exposure Point, and Sensitivity Analysis (Hisham, 2010) [9].

2. Methods of research

The evaluation of economic efficiency of project development was carried out via complex model of geological and economic evaluation. The model includes evaluation of production profile, based on physical properties of rocks and liquids; assessment of main items of cash flow such as revenues, capital and operational expenditures, taxes based on analysis of current and forecasted market situation; calculation of resulting indicators such as cash flow, profits, NPV, IRR [10].

The estimation of the annual production allows creating the production profile of an oilfield which has a pi-shaped pattern and consists of three stages (build up, plateau, and decline). Moreover, the graphical representation of the oil production profile shows can be estimated through a function (formula 1).

$$Qt = \begin{cases} r * t + s, t_1 < t < t_2 \\ u,\ t_2 > t < t_3 \\ -w * t + x,\ t_3 < t < t_4 \end{cases} \quad (1)$$

Where, Q_t – annual oil production; r, s, u, w, x – constants; t_i – start and end of oil production for each stage (i = *1, 2, 3, 4… n*).

According to the line of the research, forecasted production for basic strategy was varied according to the probable implications of technologies.

3. Results and discussion

During the first stage of research, extensive analysis of Russian oil and gas industry was carried out to define main problems connected with technological development of extraction industries of resource regions.

The traditional macroregions of the Russian oil industry are Western Siberia and European part of Russia. Western Siberia remains the leader in oil production with a production level of 57.3% of the total Russian output [11]. However, the share of the region tend to decrease during the past decade. Share of European part has remained stable for the recent years and had the level of is 29.9% in 2017. Regions with the most dynamically developing oil production are East Siberia and the Far East. The share of these regions was less than 3% in the total oil production in 2008, and has risen up to 12.8% in 2017 (figure 1).

The currents problems in Russian oil industry include growing number of mature oilfields in declining stage of production; high percentage of water produced from the reservoir in mature and new fields due to the applied technologies of oil production; low development level of hard-to-access areas, mainly in Arctic Region; low production of unconventional resources.

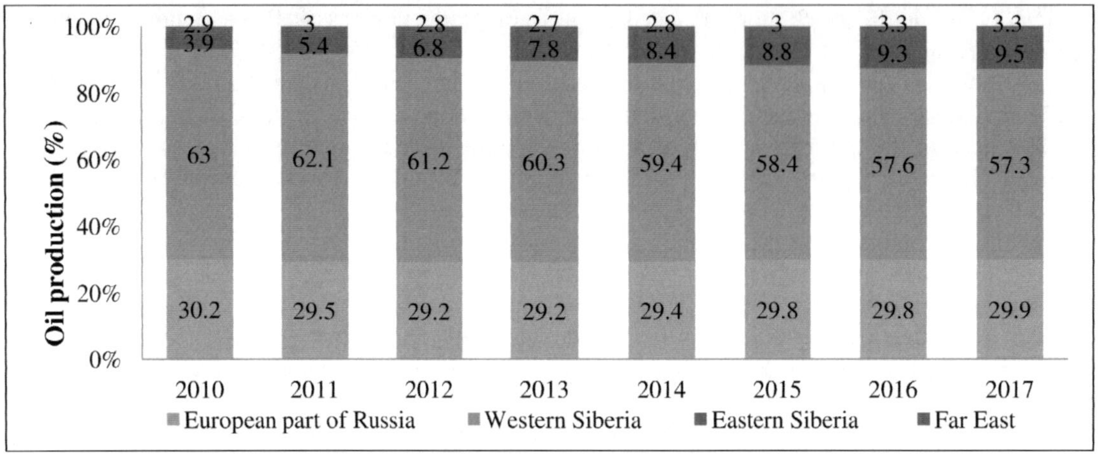

Figure 1. Shares of macroregions in Oil Production of Russia in 2017, %.

The increase in production in traditional areas of oil production can became possible due to the active introduction of new technologies in fields with a high degree of depletion, water cut and lower quality of crudes characterized by high viscosity and sulfur content.

During the next stage, the typical development project of Koltogor oilfield in Western Siberia was chosen for evaluation of economic efficiency from point of view of implementation of advanced technologies of oil production [12].

Seven strategies, reflecting different levels of technologies implementation and consequently different extraction profiles and cumulative oil production, were justified and modeled on the basis of the characteristics of reservoir and liquids [13]. The cumulative production in the scenario based on free flow extraction is 14.6 million tons of oil, which is accounted for extraction of 42% of recoverable oil. The maximum implementation of artificial list system allows for extraction of up to 62% of recoverable reserves in the planning horizon of the project (figure 2).

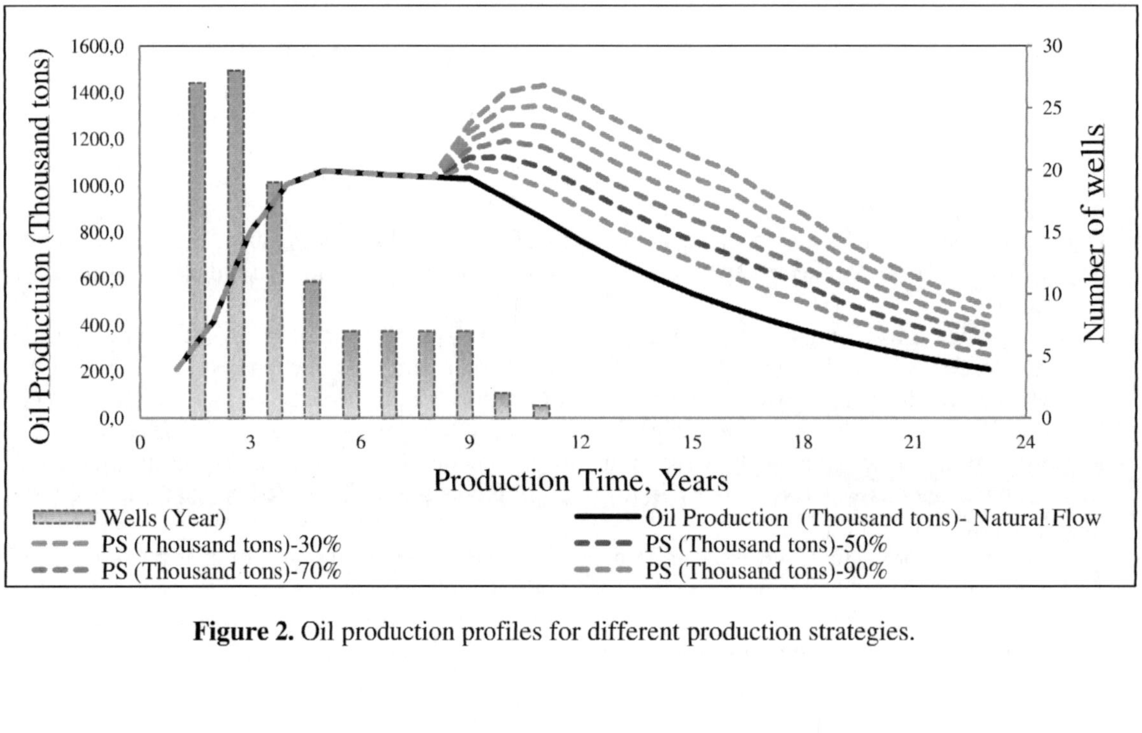

Figure 2. Oil production profiles for different production strategies.

Four essential elements (capital expenditure, cost of production, product price, product schedule) of the economic evaluation were considered and developed into a cash flow model spread over 30 years. Implementation of advanced technologies in oil production (such as ALS) influences the efficiency of the project in different directions: positive influence on revenues, negative influence on CAPEX and OPEX, which results in the existence of a point of inflection, reflecting the optimal level of technology implementation under current conditions (table 1).

Table 1. Main economic indicators of implementation of production strategies, billion rubles.

Indicator	PS-(30%)	PS-(50%)	PS-(70%)	PS-(90%)	PS-(110%)	PS-(130%)
Oil production (Million Tons)	16.26	17.34	18.42	19.50	20.58	21.66
Revenues	352. 2	375.4	398.8	422.1	445.5	468.8
CAPEX	90.0	90.6	91.2	93.1	93.1	93.8
OPEX	214.0	218.4	224.4	250.7	261.8	295.3
NPV	7291.4	9392.3	11243.0	9958.8	9672.1	9357.6
IRR, %	17%	18%	19%	18%	18%	17%

For the current economic and financial situation it was revealed that high costs of installation and application of advanced technologies do not allow for its full application. Shown results were calculated under positive scenario for oil prices at level of 85 dollars per barrel. However, application of the model under lower prices 45-65 dollars per barrel showed different results with negative NPV for most productive strategies.

In the current conditions with unstable oil prices and restrictions on the use of foreign technologies and equipment, optimization and cost reduction, as well as the development and use of import substitution technologies at all stages of oil and gas production, are of particular importance. At the same time, the development, testing and introduction of new processes and equipment can take a considerable time, which is the reason for the need to reduce total costs in the short-term period due to other cost categories [14]. One of these categories are tax payments, the regulation of which can be implemented quickly and selectively for individual projects and regions.

Another constraining factor is overall low level of complex development of most of the resource regions of Russia [15]. In absence of developed intraregional economy and innovative type of specialization, this can lead to decreasing production efficiency and lowering economic growth. Directions of innovative development should include processing, related and associated industries, such as chemistry, engineering, construction, services, financial sector, IT industries.

4. Conclusion

Russia faces significant challenges in issues related to increasing the oil production in the country, since a part of the currently developing oilfields are in the decline phase, and new fields can face an early depletion due to the rates and methods of oil production. This situation affects in the short and long run different regions of the country, especially traditionally developed region of Western Siberia which produces almost 58 % of the total oil in Russia.

The oil production in the country is expected to decline after 2020 due to a peak in production of Brownfields and Greenfields and existing problems in developed producing regions such as declining rate of production, high water cut and deterioration of reserves in mature fields.

Implementation of advanced technologies in oil production influences the efficiency of the project in different directions: positive influence on revenues, negative influence on CAPEX and OPEX, which results in the existence of a point of inflection, reflecting the optimal level of technology implementation under current conditions. Evaluation result for Western Siberia shows that high costs of installation and application of advanced technologies do not allow for the full implementation, especially under the conditions of highly volatile oil prices and limitations on usage of foreign technologies.

Acknowledgements
The research was supported by the grant No. 17-78-20218 of the Russian Science Foundation.

References

[1] Eder L V, Filimonova I V, Provornaya I V and Nemov V Y 2017 *IOP Conference Series: Earth and Environmental Science* **84(1)** 012012
[2] Kontorovich A E, Eder L V and Filimonova I V 2017 *IOP Conference Series: Earth and Environmental Science* **84(1)** 012010
[3] Eder L V, Kontorovich A E, Filimonova I V, Provornaya I V and Goosen E V 2017 *International Multidisciplinary Scientific GeoConference Surveying Geology and Mining Ecology Management, SGEM* **17 (53)** 719–26
[4] Bobylev Y 2016 *Problems of Economic Transition* **58 (11-12)** 965–87
[5] Akangbou H, Burby M and Nasr G 2017 *Journal of Petroleum Science and Engineering* **150** 128–36
[6] Bailey B, Crabtree M, Elphick J, Kuchuk F, Roodhart L, Romano C and Tyrie J 2000 *Oilfield Review* **Autumn** 20–35
[7] Hourfar F, Moshori B, Salahshoor K and Elkamel A 2017 *Computers and Chemical Engineering* **106** 339–54
[8] Lima G and Suslick S 2005 *SPE* 94665
[9] Yas H 2010 *Journal of Petroleum Researches & Studies* **1**
[10] Santos S, Gaspar A and Schoizer D 2017 *Journal of Petroleum Science and Engineering* **151** 116–27
[11] Henderson J, Grushevenko E 2017 *The Oxford Institute for Energy Studies (OIES)* **February** 1–19
[12] Hook M, Soderbergh B, Jakobsson K and Aleklett K 2009 *Natural Resources Rearch* **18(1)** 39–56
[13] Guyaguler B, Horne R 2004 *Reservoir evaluation & Engineering* **February** 24–32
[14] Eder L V, Filimonova I V, Provornaya I V, Nemov V U and Nikitenko S M 2017 *International Multidisciplinary Scientific GeoConference Surveying Geology and Mining Ecology Management, SGEM* **17 (53)** 727–34
[15] Eder L V, Filimonova I V, Provornaya I V, Komarova A V and Nikitenko S M 2017 *17th International Multidisciplinary Scientific GeoConference SGEM* **17(15)** 365–72